# CAMBRIDGE LIBRARY COLLECTION

*Books of enduring scholarly value*

## Darwin

Two hundred years after his birth and 150 years after the publication of 'On the Origin of Species', Charles Darwin and his theories are still the focus of worldwide attention. This series offers not only works by Darwin, but also the writings of his mentors in Cambridge and elsewhere, and a survey of the impassioned scientific, philosophical and theological debates sparked by his 'dangerous idea'.

## Zoonomia

Erasmus Darwin (1731–1802) is remembered not only as the grandfather of Charles but as a pioneering scientist in his own right. A friend and correspondent of Josiah Wedgwood, Joseph Priestley and Matthew Boulton, he practised medicine in Lichfield, but also wrote prolifically on scientific subjects. He organised the translation of Linnaeus from Latin into English prose, coining many plant names in the process, and also wrote a version in verse, The Loves of Plants. The aim of his Zoonomia, published in two volumes (1794–6), is to 'reduce the facts belonging to animal life into classes, orders, genera, and species; and by comparing them with each other, to unravel the theory of diseases'. The first volume describes human physiology, especially importance of motion, both voluntary and involuntary; the second is a detailed description of the symptoms of, the and the cures for, diseases, categorised according to his physiological classes. Many of his proposed treatments are not for the squeamish, but his attempt to classify symptoms and link them to human anatomy and physiology marked a new stage in the development of the science of medicine; and his theory that all living things may perhaps have evolved from 'a single living filament' prefigures in a remarkable way the subsequent work of his more famous grandson.

Cambridge University Press has long been a pioneer in the reissuing of out-of-print titles from its own backlist, producing digital reprints of books that are still sought after by scholars and students but could not be reprinted economically using traditional technology. The Cambridge Library Collection extends this activity to a wider range of books which are still of importance to researchers and professionals, either for the source material they contain, or as landmarks in the history of their academic discipline.

Drawing from the world-renowned collections in the Cambridge University Library, and guided by the advice of experts in each subject area, Cambridge University Press is using state-of-the-art scanning machines in its own Printing House to capture the content of each book selected for inclusion. The files are processed to give a consistently clear, crisp image, and the books finished to the high quality standard for which the Press is recognised around the world. The latest print-on-demand technology ensures that the books will remain available indefinitely, and that orders for single or multiple copies can quickly be supplied.

The Cambridge Library Collection will bring back to life books of enduring scholarly value (including out-of-copyright works originally issued by other publishers) across a wide range of disciplines in the humanities and social sciences and in science and technology.

# Zoonomia

*Or, the Laws of Organic Life*

VOLUME 2

ERASMUS DARWIN

CAMBRIDGE UNIVERSITY PRESS

Cambridge, New York, Melbourne, Madrid, Cape Town, Singapore,
São Paolo, Delhi, Dubai, Tokyo

Published in the United States of America by Cambridge University Press, New York

www.cambridge.org
Information on this title: www.cambridge.org/9781108005500



This edition first published 1794-6
This digitally printed version 2009

ISBN 978-1-108-00550-0 Paperback

# ZOONOMIA;

OR,

# THE LAWS

OF

# ORGANIC LIFE.

## VOL. II.

*By ERASMUS DARWIN, M.D. F.R.S.*
AUTHOR OF THE BOTANIC GARDEN.

---

Principiò cœlum, ac terras, campoſque liquentes,
Lucentemque globum lunæ, titaniaque aſtra,
Spiritus intùs alit, totamque infuſa per artus
Mens agitat molem, et magno ſe corpore miſcet.

VIRG. ÆN. vi.

Earth, on whoſe lap a thouſand nations tread,
And Ocean, brooding his prolific bed,
Night's changeful orb, blue pole, and ſilvery zones,
Where other worlds encircle other ſuns,
One Mind inhabits, one diffuſive Soul
Wields the large limbs, and mingles with the whole.

---

LONDON:
PRINTED FOR J. JOHNSON, IN ST. PAUL'S CHURCH-YARD.

1796.

# ZOONOMIA;

OR,

# THE LAWS OF ORGANIC LIFE.

## PART II.

CONTAINING

A CATALOGUE OF DISEASES

DISTRIBUTED INTO

NATURAL CLASSES ACCORDING TO THEIR PROXIMATE CAUSES,

WITH THEIR

SUBSEQUENT ORDERS, GENERA, AND SPECIES,

AND WITH

THEIR METHODS OF CURE.

---

Hæc, ut potero, explicabo; nec tamen, quasi Pythius Apollo, certa ut sint et fixa, quæ dixero; sed ut Homunculus unus e multis probabiliora conjecturâ sequens.

Cic. Tusc. Disp. l. 1. 9.

---

# PREFACE.

ALL diſeaſes originate in the exuberance, deficiency, or retrograde action, of the faculties of the ſenſorium, as their proximate cauſe; and conſiſt in the diſordered motions of the fibres of the body, as the proximate effect of the exertions of thoſe diſordered faculties.

The ſenſorium poſſeſſes four diſtinct powers, or faculties, which are occaſionally exerted, and produce all the motions of the fibrous parts of the body; theſe are the faculties of producing fibrous motions in conſequence of irritation which is excited by external bodies; in conſequence of ſenſation which is excited by pleaſure or pain; in conſequence of volition which is excited by deſire or averſion; and in conſequence of aſſociation which is excited by other fibrous motions. We are hence ſupplied with four natural claſſes of diſeaſes derived from their proximate cauſes; which we ſhall term thoſe of irritation, thoſe of ſenſation, thoſe of volition, and thoſe of aſſociation.

In the ſubſequent claſſification of diſeaſes I have not adhered to the methods of any of thoſe, who have preceded me; the principal of whom are the great names of Sauvages and Cullen; but have neverthelefs availed myfelf, as much as I could, of their definitions and diftinctions.

The effential characteriftic of a difeafe confifts in its proximate caufe, as is well obferved by Doctor Cullen, in his Nofologia Methodica, T. ii. Prolegom. p. xxix. Similitudo quidem morborum in fimilitudine caufæ eorum proximæ, qualifcunque fit, reverâ confiftit. I have taken the proximate caufe for the claffic character. The characters of the orders are taken from the excefs, or deficiency, or retrograde action, or other properties of the proximate caufe. The genus is generally derived from the proximate effect. And the fpecies generally from the locality of the difeafe in the fyftem.

Many fpecies in this fyftem are termed genera in the fyftems of other writers; and the fpecies of thofe writers are in confequence here termed varieties. Thus in Dr. Cullen's Nofologia the variola or fmall-pox is termed a genus, and the diftinct and confluent kinds are termed fpecies. But as the infection from the diftinct kind frequently produces the confluent kind, and that of the confluent kind frequently produces the diftinct; it would feem more analogous to botanical arrangement, which thefe nofologifts profefs to imitate, to call the diftinct and confluent fmall-pox varieties than fpecies. Becaufe the fpecies of plants in botanical fyftems propagate others fimilar to themfelves; which does not uniformly occur in fuch vegetable productions as are termed varieties.

In

In ſome other genera of noſologiſts the ſpecies have no analogy to each other, either in reſpect to their proximate cauſe, or to their proximate effect, though they may be ſomewhat ſimilar in leſs eſſential properties; thus the thin and ſaline diſcharge from the noſtrils on going into the cold air of a froſty morning, which is owing to the deficient action of the abſorbent veſſels of the noſtrils, is one ſpecies; and the viſcid mucus diſcharged from the ſecerning veſſels of the ſame membrane, when inflamed, is another ſpecies of the ſame genus, Catarrhus. Which bear no analogy either in reſpect to their immediate cauſe or to their immediate effect.

The uſes of the method here offered to the public of claſſing diſeaſes according to their proximate cauſes are, firſt, more diſtinctly to underſtand their nature by comparing their eſſential properties. Secondly, to facilitate the knowledge of the methods of cure; ſince in natural claſſification of diſeaſes the ſpecies of each genus, and indeed the genera of each order, a few perhaps excepted, require the ſame general medical treatment. And laſtly, to diſcover the nature and the name of any diſeaſe previouſly unknown to the phyſician; which I am perſuaded will be more readily and more certainly done by this natural ſyſtem, than by the artificial claſſifications already publiſhed.

The common names of diſeaſes are not well adapted to any kind of claſſification, and leaſt of all to this from their proximate cauſes. Some of their names in common language are taken from the remote cauſe, as worms, ſtone of the bladder; others from the remote effect, as diarrhœa, ſalivation,

 hydro-

hydrocephalus; others from ſome accidental ſymptom of the diſeaſe, as tooth-ach, head-ach, heart-burn; in which the pain is only a concomitant circumſtance of the exceſs or deficiency of fibrous actions, and not the cauſe of them Others again are taken from the deformity occaſioned in conſequence of the unnatural fibrous motions, which conſtitute diſeaſes, as tumours, eruptions, extenuations; all theſe therefore improperly give names to diſeaſes; and ſome difficulty is thus occaſioned to the reader in endeavouring to diſcover to what claſs ſuch diſorders belong.

Another difficulty attending the names of diſeaſes is, that one name frequently includes more than one diſeaſe, either exiſting at the ſame time or in ſucceſſion. Thus the pain of the bowels from worms is cauſed by the increaſed action of the membrane from the ſtimulus of thoſe animals; but the convulſions, which ſometimes ſucceed theſe pains in children, are cauſed by the conſequent volition, and belong to another claſs.

To diſcover under what claſs any diſeaſe ſhould be arranged, we muſt firſt inveſtigate the proximate cauſe; thus the pain of the tooth-ach is not the cauſe of any diſeaſed motions, but the effect; the tooth-ach therefore does not belong to the claſs of Senſation. As the pain is cauſed by increaſed or decreaſed action of the membranes of the tooth, and theſe actions are owing to the increaſe or decreaſe of irritation, the diſeaſe is to be placed in the claſs of irritation.

To diſcover the order it muſt be inquired, whether the pain be owing to increaſed or defective motion of the pained membrane; which is known by the concomitant heat or coldneſs of the part. In tooth-ach without inflammation there is generally a coldneſs attends the cheek in its vicinity; as may be perceived by the hand of the patient himſelf, compared with the oppoſite cheek. Hence odontalgia is found to belong to the order of decreaſed irritation. The genus and ſpecies muſt be found by inſpecting the ſynopſis of the ſecond order of the claſs of Irritation. See Claſs I. 2. 4. 12.

This may be further elucidated by conſidering the natural operation of parturition; the pain is occaſioned by the increaſed action or diſtention of the veſſels of the uterus, in conſequence of the ſtimulus of the fetus; and is therefore cauſed by increaſed irritation; but the action of the abdominal muſcles in its excluſion are cauſed by the pain, and belong to the claſs of increaſed ſenſation. See Claſs II. 1. 1. 12. Hence the difficulty of determining, under what claſs of diſeaſes parturition ſhould be arranged; conſiſts in there being two kinds of diſeaſed actions comprehended under one word; which have each their different proximate cauſe.

In Sect. XXXIX. 8. 4. and in Claſs II. 1. 1. 1. we have endeavoured to give names to four links of animal cauſation, which conveniently apply to the claſſification of diſeaſes; thus in common nictitation, or winking with the eyes without our attention to it, the increaſed irritation is the proximate cauſe; the ſtimulus of the air on the dry cornea is the remote cauſe; the cloſing of the eyelid is the proximate effect; and the diffuſion of tears over the eye-ball is the remote effect. In some cases two more links of causation may be introduced; one of them may be termed the pre-remote cause; as the warmth or motion of the atmosphere, which causes greater exhalation from the cornea. And the other the post-remote effect; as the renewed pellucidity of the cornea; and thus six links of causation may be expressed in words.

But if amid these remote links of animal causation any of the four powers or faculties of the sensorium be introduced, the reasoning is not just according to the method here proposed; for these powers of the sensorium are always the proximate causes of the contractions of animal fibres; and therefore in true language cannot be termed their remote causes. From this criterion it may always be determined, whether more diseases than one are comprehended under one name; a circumstance which has much impeded the investigation of the causes, and cures of diseases.

Thus the term fever, is generally given to a collection of morbid symptoms; which are indeed so many distinct diseases, that sometimes appear together, and sometimes separately; hence it has no determinate meaning, except it signifies simply a quick pulse, which continues for some hours; in which sense it is here used.

In naming diseases I have endeavoured to avoid the affectation of making new compound Greek words, where others equally expressive could be procured: as a short periphrasis is easier to be understood, and less burthensome to the memory.

In

In the Methodus Medendi, which is marked by M. M. at the end of many of the ſpecies of diſeaſes, the words incitantia, ſorbentia, torpentia, &c. refer to the ſubſequent articles of the Materia Medica, explaining the operations of medicines.

The remote cauſes of many diſeaſes, their periods, and many circumſtances concerning them, are treated of in the preceding volume; the deſcriptions of many of them, which I have omitted for the ſake of brevity, may be ſeen in the Noſologia Methodica of Sauvages, and in the Synopſis Noſologiæ of Dr. Cullen, and in the authors to which they refer.

In this arduous undertaking the author ſolicits the candour of the critical reader; as he cannot but foreſee, that many errors will be diſcovered, many additional ſpecies will require to be inſerted; and others to be tranſplanted, or eraſed. If he could expend another forty years in the practice of medicine, he makes no doubt, but that he could bring this work nearer perfection, and thence render it more worthy the attention of philoſophers.——As it is, he is induced to hope, that ſome advantages will be derived from it to the ſcience of medicine, and conſequent utility to the public, and leaves the completion of his plan to the induſtry of future generations.

Derby, *Jan.* 1, 1796.

# ZOONOMIA.

## PART II.

## CLASSES OF DISEASES.

I. DISEASES OF IRRITATION.

II. DISEASES OF SENSATION.

III. DISEASES OF VOLITION.

IV. DISEASES OF ASSOCIATION.

## The Orders and Genera of the First Class of Diseases.

# CLASS I.

DISEASES OF IRRITATION.

## ORDO I.

*Increased Irritation.*

### GENERA.

1. With increased actions of the sanguiferous system.
2. With increased actions of the secerning system.
3. With increased actions of the absorbent system.
4. With increased actions of other cavities and membranes.
5. With increased actions of the organs of sense.

## ORDO II.

*Decreased Irritation.*

### GENERA.

1. With decreased actions of the sanguiferous system.
2. With decreased actions of the secerning system.
3. With decreased actions of the absorbent system.
4. With decreased actions of other cavities and membranes.
5. With decreased actions of the organs of sense.

## ORDO III.

*Retrograde Irritative Motions.*

### GENERA.

1. Of the alimentary canal.
2. Of the absorbent system.
3. Of the sanguiferous system.

The

*The Orders, Genera, and Species, of the First Class of Diseases.*

---

# CLASS I.

## DISEASES OF IRRITATION.

## ORDO I.

*Increased Irritation.*

### GENUS I.

*With increased Actions of the Sanguiferous System.*

#### SPECIES.

| | |
|---|---|
| 1. *Febris irritativa.* | Irritative fever. |
| 2. *Ebrietas.* | Drunkenness. |
| 3. *Hæmorrhagia arteriosa.* | Arterial hæmorrhage. |
| 4. *Hæmoptoe arteriosa.* | Spitting of arterial blood. |
| 5. *Hæmorrhagia narium.* | Bleeding from the nose. |

### GENUS II.

*With increased Actions of the Secerning System.*

#### SPECIES.

| | |
|---|---|
| 1. *Calor febrilis.* | Febrile heat. |
| 2. *Rubor febrilis.* | Febrile redness. |
| 3. *Sudor calidus.* | Warm sweat. |

| | |
|---|---|
| *Sudor febrilis.* | Sweat in fevers. |
| —— *a labore.* | ——- from exercise. |
| —— *ab igne.* | ——- from fire. |
| —— *a medicamentis.* | ——- from medicines. |
| 4. *Urina uberior colorata.* | Copious coloured urine. |
| 5. *Diarrhœa calida.* | Warm diarrhœa. |
| ——— *febrilis.* | ——- from fever. |
| ——— *crapulosa.* | ——- from indigestion. |
| ——— *infantum.* | ——- of infants. |
| 6. *Salivatio calida.* | ——- salivation. |
| 7. *Catarrhus calidus.* | ——- catarrh. |
| 8. *Expectoratio calida.* | ——- expectoration. |
| 9. *Exsudatio pone aures.* | Discharge behind the ears. |
| 10. *Gonorrhœa calida.* | Warm gonorrhœa. |
| 11. *Fluor albus calidus.* | ——- fluor albus. |
| 12. *Hæmorrhois alba.* | White piles. |
| 13. *Serum e visicatorio.* | Discharge from a blister. |
| 14. *Perspiratio fœtida.* | Fetid perspiration. |
| 15. *Crines novi.* | New hairs. |

## GENUS III.

*With increased Actions of the Absorbent System.*

### SPECIES.

| | |
|---|---|
| 1. *Lingua arida.* | Dry tongue. |
| 2. *Fauces aridæ.* | Dry throat. |
| 3. *Nares aridi.* | Dry nostrils. |
| 4. *Expectoratio solida.* | Solid expectoration. |
| 5. *Constipatio alvi.* | Costiveness. |
| 6. *Cutis arida.* | Dry skin. |
| 7. *Urina parcior colorata.* | Diminished coloured urine. |

8. *Calculus*

| | | |
|---|---|---|
| 8. | *Calculus felleus et icterus.* | Gall-ſtone and jaundice. |
| 9. | ——— *renis.* | Stone of the kidney. |
| 10. | ——— *veſicæ.* | Stone of the bladder. |
| 11. | ——— *arthriticus.* | Gout-ſtone. |
| 12. | *Rheumatiſmus chronicus.* | Chronic rheumatiſm. |
| 13. | *Cicatrix vulnerum.* | Healing of ulcers. |
| 14. | *Corneæ obfuſcatio.* | Scar on the cornea. |

## GENUS IV.

*With increaſed Actions of other Cavities and Membranes.*

### SPECIES.

| | | |
|---|---|---|
| 1. | *Nictitatio irritativa.* | Irritative nictitation. |
| 2. | *Deglutitio irritativa.* | Irritative deglutition. |
| 3. | *Reſpiratio et tuſſis.* | Reſpiration and cough. |
| 4. | *Excluſio bilis.* | Excluſion of the bile. |
| 5. | *Dentitio.* | Toothing. |
| 6. | *Priapiſmus.* | Priapiſm. |
| 7. | *Diſtenſio mamularum.* | Diſtention of the nipples. |
| 8. | *Deſcenſus uteri.* | Deſcent of the uterus. |
| 9. | *Prolapſus ani.* | Deſcent of the rectum. |
| 10. | *Lumbricus.* | Round worm. |
| 11. | *Tænia.* | Tape-worm. |
| 12. | *Aſcarides.* | Thread-worms. |
| 13. | *Dracunculus.* | Guinea-worm. |
| 14. | *Morpiones.* | Crab-lice. |
| 15. | *Pediculi.* | Lice. |

## GENUS V.

*With increased Actions of the Organs of Sense.*

### SPECIES.

1. *Visus acrior.* Acuter sight.
2. *Auditus acrior.* ——- hearing.
3. *Olfactus acrior.* ——- smell
4. *Gustus acrior.* ——- taste.
5. *Tactus acrior.* ——- touch.
6. *Sensus caloris acrior.* ——- sense of heat.
7. —— *extensionis acrior.* ——- sense of extension.
8. *Titillatio.* Tickling.
9. *Pruritus.* Itching.
10. *Dolor urens.* Smarting.
11. *Consternatio.* Surprise.

## ORDO. II.

*Decreased Irritation.*

## GENUS I.

*With decreased Actions of the Sanguiferous System.*

### SPECIES.

1. *Febris inirritativa.* Inirritative fever.
2. *Paresis inirritativa.* ———— debility.
3. *Somnus interruptus.* Interrupted sleep.
4. *Syncope.* Fainting.
5. *Hæmorrhagia venosa.* Venous hæmorrhage.

| | |
|---|---|
| 6. *Hæmorrhois cruenta.* | Bleeding piles. |
| 7. *Hæmorrhagia renum.* | ——— from the kidneys. |
| 8. ——— *hepatis.* | ——— from the liver. |
| 9. *Hæmoptoe venoſa.* | Spitting of venous blood. |
| 10. *Palpitatio cordis.* | Palpitation of the heart. |
| 11. *Menorrhagia.* | Exuberant menſtruation. |
| 12. *Dyſmenorrhagia.* | Deficient menſtruation. |
| 13. *Lochia nimia.* | Too great lochia. |
| 14. *Abortio ſpontanea.* | Spontaneous abortion. |
| 15. *Scorbutus.* | Scurvy. |
| 16. *Vibices.* | Extravaſations of blood. |
| 17. *Petechiæ.* | Purple ſpots. |

## GENUS II.

*With decreaſed Actions of the Secerning Syſtem.*

### SPECIES.

| | |
|---|---|
| 1. *Frigus febrile.* | Coldneſs in fevers. |
| ——— *chronicum.* | ——— permanent. |
| 2. *Palior fugitivus.* | Paleneſs fugitive. |
| ——— *permanens.* | ——— permanent. |
| 3. *Pus parcius.* | Diminiſhed pus. |
| 4. *Mucus parcior.* | Diminiſhed mucus. |
| 5. *Urina parcior pallida.* | Pale diminiſhed urine. |
| 6. *Torpor hepaticus.* | Torpor of the liver. |
| 7. *Torpor pancreatis.* | Torpor of the pancreas. |
| 8. *Torpor renis.* | To por of the kidney. |
| 9. *Punctæ mucoſæ vultus.* | Mucous ſpots on the face. |
| 10. *Maculæ cutis fulvæ.* | Tawny blots on the ſkin. |
| 11. *Canities.* | Grey hairs. |
| 12. *Callus.* | Callus. |
| 13. *Cataracta.* | Cataract. |

| | |
|---|---|
| 14. *Innutritio oſſium.* | Innutrition of the bones. |
| 15. *Rachitis.* | Rickets. |
| 16. *Spina diſtortio.* | Diſtortion of the ſpine. |
| 17. *Claudicatio coxaria.* | Lameneſs of the hip. |
| 18. *Spina protuberans.* | Protuberant ſpine. |
| 19. *Spina bifida.* | Divided ſpine. |
| 20. *Defectus palati.* | Defect of the palate. |

## GENUS III.

*With decreaſed Actions of the Abſorbent Syſtem.*

### SPECIES.

| | |
|---|---|
| 1. *Mucus faucium frigidus.* | Cold mucus from the throat. |
| 2. *Sudor frigidus.* | —— ſweat. |
| 3. *Catarrhus frigidus.* | —— catarrh. |
| 4. *Expectoratio frigida.* | —— expectoration. |
| 5. *Urina uberior pallida.* | Copious pale urine. |
| 6. *Diarrhœa frigida.* | Cold diarrhœa. |
| 7. *Fluor albus frigidus.* | —— fluor albus. |
| 8. *Gonorrhœa frigida.* | —— gonorrhœa. |
| 9. *Hepatis tumor.* | Swelling of the liver. |
| 10. *Chloroſis.* | Green ſickneſs. |
| 11. *Hydrocele.* | Dropſy of the vagina teſtis. |
| 12. *Hydrocephalus internus.* | ——— of the brain. |
| 13. *Aſcites.* | ——— of the belly. |
| 14. *Hydrothorax.* | ——— of the cheſt. |
| 15. *Hydrops ovarii.* | ——— of the ovary. |
| 16. *Anaſarca pulmonum.* | ——— of the lungs. |
| 17. *Obeſitas.* | Corpulency. |
| 18. *Splenis tumor.* | Swelling of the ſpleen. |

| | |
|---|---|
| 19. *Genu tumor albus.* | White ſwelling of the knee. |
| 20. *Bronchocele.* | Swelled throat. |
| 21. *Scrophula.* | King's evil. |
| 22. *Schirrus.* | Schirrus. |
| 23. ——— *recti.* | ——— of the rectum. |
| 24. ——— *urethræ.* | ——— of the urethra. |
| 25. ——— *æſophagi.* | ——— of the throat. |
| 26. *Lacteorum inirritabilitas.* | Inirritability of the lacteals. |
| 27. *Lymphaticorum inirritabilitas.* | Inirritability of the lymphatics. |

## GENUS IV.

*With decreaſed Actions of other Cavities and Membranes.*

### SPECIES.

| | |
|---|---|
| 1. *Sitis calida.* | Thirſt warm. |
| —— *frigida.* | —— cold. |
| 2. *Eſuries.* | Hunger. |
| 3. *Nauſea ſicca.* | Dry nauſea. |
| 4. *Ægritudo ventriculi.* | Sickneſs of ſtomach. |
| 5. *Cardialgia.* | Heart-burn. |
| 6. *Arthritis ventriculi.* | Gout of the ſtomach. |
| 7. *Colica flatulenta.* | Flatulent colic. |
| 8. *Colica ſaturnina.* | Colic from lead. |
| 9. *Tympanitis.* | Tympany. |
| 10. *Hypochondriaſis.* | Hypochondriaciſm. |
| 11. *Cephalæa frigida.* | Cold head-ach. |
| 12. *Odontalgia.* | Tooth-ach. |
| 13. *Otalgia.* | Ear-ach. |
| 14. *Pleurodyne chronica.* | Chronical pain of the ſide. |

| | |
|---|---|
| 15. *Sciatica frigida.* | Cold ſciatica. |
| 16. *Lumbago frigida.* | ——lumbago. |
| 17. *Hyſteralgia frigida.* | —— pain of the uterus. |
| 18. *Proctalgia frigida.* | —— pain of the rectum. |
| 19. *Veſicæ felleæ inirritibilitas et icterus.* | Inirritability of the gall-bladder and jaundice. |

## GENUS V.

*With decreaſed Actions of the Organs of Senſe.*

### SPECIES.

| | |
|---|---|
| 1. *Stultitia inirritabilis.* | Folly from inirritability. |
| 2. *Viſus imminutus.* | Impaired viſion. |
| 3. *Muſcæ volitantes.* | Dark moving ſpecks. |
| 4. *Strabiſmus.* | Squinting. |
| 5. *Amauroſis.* | Palſy of the optic nerve. |
| 6. *Auditus imminutus.* | Impaired hearing. |
| 7. *Olfactus imminutus.* | ——— ſmell. |
| 8. *Guſtus imminutus.* | ——— taſte. |
| 9. *Tactu, imminutus.* | ——— touch. |
| 10. *Stupor.* | Stupor. |

## ORDO III.

### *Retrograde Irritative Motions.*

### GENUS I.

#### *Of the Alimentary Canal.*

#### SPECIES.

| | |
|---|---|
| 1. *Ruminatio.* | Chewing the cud. |
| 2. *Ructus.* | Eructation. |
| 3. *Apepsia.* | Indigestion, water-qualm. |
| 4. *Vomitus.* | Vomiting. |
| 5. *Cholera.* | Cholera. |
| 6. *Ileus.* | Iliac passion. |
| 7. *Globus hystericus.* | Hysteric strangulation. |
| 8. *Vomendi conamen inane.* | Vain efforts to vomit. |
| 9. *Borborigmus.* | Gurgling of the bowels. |
| 10. *Hysteria.* | Hysteric disease. |
| 11. *Hydrophobia.* | Dread of water. |

### GENUS II.

#### *Of the Absorbent System.*

#### SPECIES.

| | |
|---|---|
| 1. *Catarrhus lymphaticus.* | Lymphatic catarrh. |
| 2. *Salivatio lymphatica.* | Lymphatic salivation. |
| 3. *Nausea humida.* | Moist nausea. |
| 4. *Diarrhœa lymphatica.* | Lymphatic flux. |
| 5. *Diarrhœa chylifera.* | Flux of chyle. |

| | |
|---|---|
| 6. *Diabætes.* | Diabetes. |
| 7. *Sudor lymphaticus.* | Lymphatic ſweat. |
| 8. *Sudor aſthmaticus.* | Aſthmatic ſweat. |
| 9. *Tranſlatio puris.* | Tranſlation of matter. |
| 10. ——— *lactis.* | ——— of milk. |
| 11. ——— *urinæ.* | ——— of urine. |

## GENUS III.

*Of the Sanguiferous Syſtem.*

### SPECIES.

| | |
|---|---|
| 1. *Capillarium motus retrogreſſus.* | Retrograde motion of the capillaries. |
| 2. *Palpitatio cordis.* | Palpitation of the heart. |
| 3. *Anhelatio ſpaſmodica.* | Spaſmodic panting. |

# CLASS I.

## DISEASES OF IRRITATION.

## ORDO I.

*Increafed Irritation.*

## GENUS I.

*With increafed Actions of the Sanguiferous Syftem.*

THE irritability of the whole, or of part, of our fyftem is perpetually changing; thefe viciffitudes of irritability and of inirritability are believed to depend on the accumulation or exhauftion of the fenforial power, as their proximate caufe; and on the difference of the prefent ftimulus, and of that which we had previoufly been accuftomed to, as their remote caufe. Thus a fmaller degree of heat produces pain and inflammation in our hands, after they have been for a time immerfed in fnow; which is owing to the accumulation of fenforial power in the moving fibres of the cutaneous veffels during their previous quiefcence, when they were benumbed with cold. And we feel ourfelves cold in the ufual temperature of the atmofphere on coming out of a warm room; which is owing to the exhauftion of fenforial power in the moving fibres of the veffels of the fkin by their previous increafed activity, into which they were excited by unufual heat.

Hence the cold fits of fever are the occafion of the fucceeding hot ones; and the hot fits contribute to occafion in their turn the fucceeding cold ones. And though the increafe of ftimulus, as of heat, exercife, or diftention, will produce an increafed action of the ftimulated fibres; in the fame manner as it is produced by the increafed irritability which

which was occaſioned by a previous defect of ſtimulus; yet as the exceſſes of irritation from the ſtimulus of external things are more eaſily avoided than the deficiencies of it; the diſeaſes of this country, except thoſe which are the conſequences of drunkenneſs, or of immoderate exerciſe, more frequently begin with torpor than with orgaſm; that is, with inactivity of ſome parts, or of the whole of the ſyſtem, and conſequent coldneſs, than with increaſed activity, and conſequent heat.

If the hot fit be the conſequence of the cold one, it may be aſked if they are proportionate to each other: it is probable that they are, where no part is deſtroyed by the cold fit, as in mortification or death. But we have no meaſure to diſtinguiſh this, except the time of their duration; whereas the extent of the torpor over a greater or leſs part of the ſyſtem, which occaſions the cold fit; or of the exertion which occaſions the hot one; as well as the degree of ſuch torpor or exertion, are perhaps more material than the time of their duration. Beſides this ſome muſcles are leſs liable to accumulate ſenſorial power during their torpor, than others, as the locomotive muſcles compared with the capillary arteries; on all which accounts a long cold fit may often be followed by a ſhort hot one.

## SPECIES.

1. *Febris irritativa.* Irritative fever. This is the ſynocha of ſome writers, it is attended with ſtrong pulſe without inflammation; and in this circumſtance differs from the febris inirritativa of Claſs I. 2. 1. 1. which is attended with weak pulſe without inflammation. The increaſed frequency of the pulſation of the heart and arteries conſtitutes fever; during the cold fit theſe pulſations are always weak, as the energy of action is then decreaſed throughout the whole ſyſtem; and therefore the general arterial ſtrength cannot be determined by the

touch, till the cold part of the paroxyſm ceaſes. This determination is ſometimes attended with difficulty; as ſtrong and weak are only comparative degrees of the greater or leſs reſiſtance of the pulſation of the artery to the compreſſion of the finger. But the greater or leſs frequency of the pulſations affords a collateral evidence in thoſe caſes, where the degree of ſtrength is not very diſtinguiſhable, which may aſſiſt our judgment concerning it. Since a moderately ſtrong pulſe, when the patient is in a recumbent poſture, and not hurried in mind, ſeldom exceeds 120 ſtrokes in a minute; whereas a weak one often exceeds 130 in a recumbent poſture, and 150 in an erect one, in thoſe fevers, which are termed nervous or putrid. See Sect. XII. 1. 4.

The increaſed frequency of the pulſation of the heart and arteries, as it is occaſioned either by exceſs or defect of ſtimulus, or of ſenſorial power, exiſts both in the cold and hot fits of fever; but when the cold fit ceaſes, and the pulſe becomes ſtrong and full as well as quick, in conſequence of the increaſed irritability of the heart and arteries, it conſtitutes the irritative fever, or ſynocha. It is attended with conſiderable heat during the paroxyſm, and generally terminates in a quarter of a lunation, without any diſturbance of the faculties of the mind. See Claſs IV. 1. 1. 8.

M. M. Veneſection. Emetics. Cathartics. Cool the patient in the hot fit, and warm him in the cold one. Reſt. Torpentia.

2. *Ebrietas.* Drunkenneſs. By the ſtimulus of wine or opium the whole arterial ſyſtem, as well as every other part of the moving ſyſtem, is excited into increaſed action. All the ſecretions, and with them the production of ſenſorial power itſelf in the brain, ſeem to be for a time increaſed, with an additional quantity of heat, and of pleaſureable ſenſation. See Sect. XXI. on this ſubject. This explains, why at the commencement of the warm paroxyſm of ſome fevers the patient is in greater ſpirits, or vivacity; becauſe, as in drunkenneſs, the irritative motions are all increaſed, and a greater production of ſenſation,

ſation is the conſequence, which when in a certain degree, is pleaſureable, as in the diurnal fever of weak people. Sect. XXXVI. 3. 1.

3. *Hæmorrhagia arterioſa.* Arterial hæmorrhage. Bleeding with a quick, ſtrong, and full pulſe. The hæmorrhages from the lungs, and from the noſe, are the moſt frequent of theſe; but it ſometimes happens, that a ſmall artery but half divided, or the puncture of a leech, will continue to bleed pertinaciouſly.

M. M. Veneſection. Cathartic with calomel. Divide the wounded artery. Bind ſponge on the puncture. If coffee or charcoal internally? If air with leſs oxygen?

4. *Hæmoptoe arterioſa.* Spitting of arterial blood. Blood ſpit up from the lungs is florid, becauſe it has juſt been expoſed to the influence of the air in its paſſage through the extremities of the pulmonary artery; it is frothy, from the admixture of air with it in the bronchia. The patients frequently vomit at the ſame time from the diſagreeable titillation of blood about the fauces; and are thence liable to believe, that the blood is rejected from the ſtomach.

Sometimes an hæmoptoe for ſeveral ſucceſſive days returns in gouty perſons without danger, and ſeems to ſupply the place of the gouty paroxyſms. Is not the liver always diſeaſed previous to the hæmoptoe, as in ſeveral other hæmorrhages? See Claſs I. 2. 1. 9.

M. M. Veneſection, a purge, a bliſter, diluents, torpentia; and afterwards ſorbentia, as the bark, the acid of vitriol, and opium. An emetic is ſaid to ſtop a pulmonary hæmorrhage, which it may effect, as ſickneſs decreaſes the circulation, as is very evident in the great ſickneſs ſometimes produced by too large a doſe of digitalis purpurea.

Dr. Ruſh ſays, a table ſpoonful or two of common ſalt is ſucceſsful in hæmoptoe; this may be owing to its ſtimulating the abſorbent ſyſtems, both the lymphatic, and the venous. Should the patient re-

ſpire

ſpire air with leſs oxygen ? or be made ſick by whirling round in a chair ſuſpended by a rope ? One immerſion in cold water, or a ſudden ſprinkling all over with cold water, would probably ſtop a pulmonary hæmorrhage. See Sect. XXVII. 1.

5. *Hæmorrhagia narium. Epiſtaxis.* Bleeding at the noſe in elderly ſubjects moſt frequently attends thoſe, whoſe livers are enlarged or inflamed by the too frequent uſe of fermented liquors.

In boys it occurs perhaps ſimply from redundancy of blood; and in young girls ſometimes precedes the approach of the catamenia; and then it ſhews a diſpoſition contrary to chloroſis; which ariſes from a deficiency of red blood.

M. M. It is ſtopped by plunging the head into cold water, with powdered ſalt haſtily diſſolved in it ; or ſometimes by lint ſtrewed over with wheat flour put up the noſtrils; or by a ſolution of ſteel in brandy applied to the veſſel by means of lint. The cure in other reſpects as in hæmoptoe; when the bleeding recurs at certain periods, after veneſection, and evacuation by calomel, and a bliſter, the bark and ſteel muſt be given, as in intermittent fevers. See Section XXVII. 1.

## ORDO I.

*Increaſed Irritation.*

## GENUS II.

*With increaſed Actions of the Secerning Syſtem.*

THESE are always attended with increaſe of partial or of general heat; for the ſecreted fluids are not ſimply ſeparated from the blood, but are new combinations; as they did not previouſly exiſt as ſuch in the blood veſſels. But all new combinations give out heat chemically; hence the origin of animal heat, which is always increaſed in proportion to the ſecretion of the part affected, or to the general quantity of the ſecretions. Nevertheleſs there is reaſon to believe, that as we have a ſenſe purpoſely to diſtinguiſh the preſence of greater or leſs quantities of heat, as mentioned in Sect. XIV. 6. ſo we may have certain minute glands for the ſecretion of this fluid, as the brain is believed to ſecrete the ſenſorial power, which would more eaſily account for the inſtantaneous production of the bluſh of ſhame, and of anger. This ſubject deſerves further inveſtigation.

### SPECIES.

1. *Calor febrilis.* The heat in fevers ariſes from the increaſe of ſome ſecretion, either of the natural fluids, as in irritative fevers; or of new fluids, as in infectious fevers; or of new veſſels, as in inflammatory fevers. The pain of heat is a conſequence of the increaſed extenſion or contraction of the fibres expoſed to ſo great a ſtimulus. See CLASS I. 1. 5. 6.

2. *Rubor febrilis.* Febrile rednefs. When the cold fit of fever terminates, and the pulfations of the heart and arteries become ftrong as well as quick from the increafe of their irritability after their late quiefcence, the blood is impelled forwards into the fine extremities of the arteries, and the anaftomozing capillaries, quicker than the extremities of the veins can abforb and return it to the heart. Hence the pulfe at the wrift becomes full, as well as quick and ftrong, and the fkin glows with arterial blood, and the veins become empty and lefs vifible.

In elderly people the force of the heart and arteries becomes lefs, while the abforbent power of the veins remains the fame; whence the capillary veffels part with the blood, as foon as it is received, and the fkin in confequence becomes paler; it is alfo probable, that in more advanced life fome of the finer branches of the arteries coalefce, and become impervious, and thus add to the opacity of the fkin.

3. *Sudor calidus.* Warm fweat may be divided into four varieties, according to their remote caufes. *Firft*, the perfpirable matter is fecreted in as great quantity during the hot fit of fever, as towards the end of it, when the fweat is feen upon the fkin. But during the hot fit the cutaneous abforbents act alfo with increafed energy, and the exhalation is likewife increafed by the greater heat of the fkin; and hence it does not appear in drops on the furface, but is in part reabforbed, and in part diffipated in the atmofphere. But as the mouths of the cutaneous abforbents are expofed to the cool air or bedclothes; whilft thofe of the capillary glands, which fecrete the perfpirable matter, are expofed to the warmth of the circulating blood; the former, as foon as the fever-fit begins to decline, lofe their increafed action firft; and hence the abforption of the fweat is diminifhed, whilft the increafed fecretion of it continues for fome hours afterwards, which occafions it to ftand in drops upon the fkin.

As the fkin becomes cooler, the evaporation of the perfpirable matter becomes leſs, as well as the abſorption of it. And hence the diſſipation of aqueous fluid from the body, and the conſequent thirſt, are perhaps greater during the hot fit, than during the ſubſequent ſweat. For the ſweats do not occur, according to Dr. Alexander's experiments, till the ſkin is cooled from 112 to 108 degrees of heat; that is, till the paroxyſm begins to decline. From this it appears, that the ſweats are not critical to the hot fit, any more than the hot fit can be called critical to the cold one; but ſimply, that they are the natural conſequence of the decline of the hot fit, commencing with the decreaſed action of the abſorbent ſyſtem, and the decreaſed evaporation from the ſkin. And from hence it may be concluded, that a fever-fit is not in general an effort of nature to reſtore health, as Sydenham conſidered it, but a neceſſary conſequence of the previous torpor; and that the cauſes of fevers would be leſs detrimental, if the fever itſelf could be prevented from exiſting; as appears in the cool treatment of the ſmall-pox.

It muſt be noted that the profuſe ſweats on the ſkin are more frequent at the decline of fever-fits than the copious urine, or looſe ſtools, which are mentioned below; as the cutaneous abſorbents, being expoſed to the cool air, loſe their increaſed action ſooner than the urinary or inteſtinal abſorbents; which open into the warm cavities of the bladder and inteſtines; but which are nevertheleſs often affected by their ſympathy with the cutaneous abſorbents. Hence few fevers terminate without a moiſture of the ſkin; whence aroſe the fatal practice of forcing ſweats by the external warmth of air or bedclothes in fevers; for external warmth increaſes the action of the cutaneous capillaries more than that of the other ſecerning veſſels; becauſe the latter are habituated to 98 degrees of heat, the internal warmth of the body; whereas the cutaneous capillaries being nearer the ſurface are habitually kept cooler by the contact of the external air. Sweats thus produced by heat in confined rooms are ſtill more detrimental; as the air becomes then not only deprived of a part of

its

its oxygene by frequent reſpiration, but is loaded with animal effluvia as well as with moiſture, till it can receive no more; and in conſequence, while the cutaneous ſecretion ſtands upon the ſkin in drops for want of exhalation, the lungs are expoſed to an inſalubrious atmoſphere.

I do not deny, that ſweating may be ſo managed as to be ſerviceable in preventing the return of the cold paroxyſm of fevers; like the warm bath, or any other permanent ſtimulus, as wine, or opium, or the bark. For this purpoſe it ſhould be continued till paſt the time of the expected cold fit, ſupported by moderate doſes of wine-whey, with ſpirit of hartſhorn, and moderate degrees of warmth. Its ſalutary effect, when thus managed, was probably one cauſe of its having been ſo much attended to; and the fetid ſmell, which when profuſe is liable to accompany it, gave occaſion to the belief, that the ſuppoſed material cauſe of the diſeaſe was thus eliminated from the circulation.

When too great external heat is applied, the ſyſtem is weakened by exceſs of action, and the torpor which cauſes the cold paroxyſm recurs ſooner and more violently. For though ſome ſtimuli, as of opium and alcohol, at the ſame time that they exhauſt the ſenſorial power by promoting increaſe of fibrous action, may alſo increaſe the production or ſecretion of it in the brain, yet experience teaches us, that the exhauſtion far out-balances the increaſed production, as is evinced by the general debility, which ſucceeds intoxication.

In reſpect to the fetor attending copious continued ſweats, it is owing to the animalized part of this fluid being kept in that degree of warmth, which moſt favours putrefaction, and not ſuffered to exhale into the atmoſphere. Broth, or other animal mucus, kept in ſimilar circumſtances, would in the ſame time acquire a putrid ſmell; yet has this error frequently produced miliary eruptions, and increaſed every kind of inflammatory or ſenſitive fever.

The eaſe, which the patient experiences during ſweating, if it be not

not produced by much external heat, is ſimilar to that of the warm bath; which by its ſtimulus applied to the cutaneous veſſels, which are generally cooler than the internal parts of the ſyſtem, excites them into greater action; and pleaſureable ſenſation is the conſequence of theſe increaſed actions of the veſſels of the ſkin. From conſidering all theſe circumſtances, it appears that it is not the evacuation by ſweats, but the continued ſtimulus, which cauſes and ſupports thoſe ſweats, which is ſerviceable in preventing the returns of fever-fits. And that ſweats too long continued, or induced by too great ſtimulus of warmth, clothes, or medicines, greatly injure the patient by increaſing inflammation, or by exhauſting the ſenſorial power. See Claſs I. 1. 2. 14.

*Secondly*, The ſweats produced by exerciſe or labour are of the warm kind; as they originate from the increaſed action of the capillaries of the ſkin, owing to their being more powerfully ſtimulated by the greater velocity of the blood, and by a greater quantity of it paſſing through them in a given time. For the blood during violent exerciſe is carried forwards by the action of the muſcles faſter in the arteries, than it can be taken up by the veins; as appears by the redneſs of the ſkin. And from the conſequent ſweats, it is evinced, that the ſecretory veſſels' of the ſkin during exerciſe pour out the perſpirable matter faſter, than the mouths of the abſorbent veſſels can drink it up. Which mouths are not expoſed to the increaſed muſcular action, or to the ſtimulus of the increaſed velocity and quantity of the blood, but to the cool air.

*Thirdly*, the increaſed ſecretion of perſpirable matter occaſioned by the ſtimulus of external heat belongs likewiſe to this place; as it is cauſed by the increaſed motions of the capillary veſſels; which thus ſeparate from the blood more perſpirable matter, than the mouths of their correſpondent abſorbent veſſels can take up; though theſe alſo are ſtimulated by external heat into more energetic action. If the air be ſtationary, as in a ſmall room, or bed with cloſed curtains, the ſweat

ſweat ſtands in drops on the ſkin for want of a quicker exhalation proportioned to the quicker ſecretion.

A *fourth* variety of warm perſpiration is that occaſioned by ſtimulating drugs, of which opium and alcohol are the moſt powerful; and next to theſe the ſpices, volatile alkali, and neutral ſalts, eſpecially ſea ſalt; that much of the aqueous part of the blood is diſſipated by the uſe of theſe drugs, is evinced by the great thirſt, which occurs a few hours after the uſe of them. See Art. III. 2. 12. and Art. III. 2. 1.

We may from hence underſtand, that the increaſe of this ſecretion of perſpirable matter by artificial means, muſt be followed by debility and emaciation. When this is done by taking much ſalt, or ſalted meat, the ſea-ſcurvy is produced; which conſiſts in the inirritability of the bibulous terminations of the veins ariſing from the capillaries; ſee Claſs I. 2. 1. 14. The ſcrophula, or inirritability of the lymphatic glands, ſeems alſo to be occaſionally induced by an exceſs in eating ſalt added to food of bad nouriſhment. See Claſs I. 2. 3. 21. If an exceſs of perſpiration is induced by warm or ſtimulant clothing, as by wearing flannel in contact with the ſkin in the ſummer months, a perpetual febricula is excited, both by the preventing the acceſs of cool air to the ſkin, and by perpetually goading it by the numerous and hard points of the ends of the wool; which when applied to the tender ſkins of young children, frequently produce the red gum, as it is called; and in grown people, either an eryſipelas, or a miliary eruption, attended with fever. See Claſs II. 1. 3. 12.

Shirts made of cotton or calico ſtimulate the ſkin too much by the points of the fibres, though leſs than flannel; whence cotton handkerchiefs make the noſe ſore by frequent uſe. The fibres of cotton are, I ſuppoſe, ten times ſhorter than thoſe of flax, and the number of points in conſequence twenty times the number; and though the manufacturers ſinge their calicoes on a red-hot iron cylinder, yet I have

have more than once ſeen an eryſipelas induced or increaſed by the ſtimulus of calico, as well as of flannel.

The increaſe of perſpiration by heat either of clothes, or of fire, contributes much to emaciate the body; as is well known to jockeys, who, when they are a ſtone or two too heavy for riding, find the quickeſt way to leſſen their weight is by ſweating themſelves between blankets in a warm room; but this likewiſe is a practice by no means to be recommended, as it weakens the ſyſtem by the exceſs of ſo general a ſtimulus, brings on a premature old age, and ſhortens the ſpan of life; as may be further deduced from the quick maturity, and ſhortneſs of the lives, of the inhabitants of Hindoſtan, and other tropical climates.

M. Buffon made a curious experiment to ſhew this circumſtance. He took a numerous brood of the butterflies of ſilkworms, ſome hundreds of which left their eggs on the ſame day and hour; theſe he divided into two parcels; and placing one parcel in the ſouth window, and the other in the north window of his houſe, he obſerved, that thoſe in the colder ſituation lived many days longer than thoſe in the warmer one. From theſe obſervations it appears, that the wearing of flannel clothing next the ſkin, which is now ſo much in faſhion, however uſeful it may be in the winter to thoſe, who have cold extremities, bad digeſtions, or habitual coughs, muſt greatly debilitate them, if worn in the warm months, producing fevers, eruptions, and premature old age. See Sect. XXXVII. 5. Claſs I. 1. 2. 14. Art. III. 2. 1.

4. *Urina uberior colorata.* Copious coloured urine. Towards the end of fever-fits a large quantity of high coloured urine is voided, the kidneys continuing to act ſtrongly, after the increaſed action of the abſorbents of the bladder is ſomewhat diminiſhed. If the abſorbents continue alſo to act ſtrongly, the urine is higher coloured, and ſo loaded

loaded as to deposit, when cool, an earthy sediment, erroneously thought to be the material cause of the disease; but is simply owing to the secretion of the kidnies being great from their increased action; and the thinner parts of it being absorbed by the increased action of the lymphatics, which are spread very thick on the neck of the bladder; for the urine, as well as perhaps all the other secreted fluids, is produced from the kidnies in a very dilute state; as appears in those, who from the stimulus of a stone, or other cause, evacuate their urine too frequently; which is then pale from its not having remained in the bladder long enough for the more aqueous part to have been reabsorbed. The general use of this urinary absorption to the animal œconomy is evinced from the urinary bladders of fish, which would otherwise be unnecessary. High coloured urine in large quantity shews only, that the secreting vessels of the kidnies, and the absorbents of the bladder, have acted with greater energy. When there is much earthy sediment, it shews, that the absorbents have acted proportionally stronger, and have consequently left the urine in a less dilute state. In this urine the transparent sediment or cloud is mucous; the opake sediment is probably coagulable lymph from the blood changed by an animal or chemical process. The floating scum is oil. The angular concretions to the sides of the pot, formed as the urine cools, is microcosmic salt. Does the adhesive blue matter on the sides of the glass, or the blue circle on it at the edge of the upper surface of the urine, consist of Prussian blue?

5. *Diarrhœa calida.* Warm diarrhœa. This species may be divided into three varieties deduced from their remote causes, under the names of diarrhœa febrilis, diarrhœa crapulosa, and diarrhœa infantum. The febrile diarrhœa appears at the end of fever-fits, and is erroneously called critical, like the copious urine, and the sweats; whereas it arises from the increased action of those secerning organs, which pour

 their

their fluids into the inteſtinal canal (as the liver, pancreas, and mucous glands), continuing longer than the increaſed action of the inteſtinal abſorbents. In this diarrhœa there is no appearance of curdled chyle in the ſtools, as occurs in cholera. I. 3. 1. 5.

The *diarrhœa crapuloſa*, or diarrhœa from indigeſtion, occurs when too great a quantity of food or liquid has been taken; which not being compleatly digeſted, ſtimulates the inteſtines like any other extraneous acrid material; and thus produces an increaſe of the ſecretions into them of mucus, pancreatic juice, and bile. When the contents of the bowels are ſtill more ſtimulant, as when draſtic purges, or very putreſcent diet, have been taken, a cholera is induced. See Sect. XXIX. 4.

The *diarrhœa infantum*, or diarrhœa of infants, is generally owing to too great acidity in their bowels. Milk is found curdled in the ſtomachs of all animals, old as well as young, and even of carnivorous ones, as of hawks. (Spallanzani.) And it is the gaſtric juice of the calf, which is employed to curdle milk in the proceſs of making cheeſe. Milk is the natural food for children, and muſt curdle in their ſtomachs previous to digeſtion; and as this curdling of the milk deſtroys a part of the acid juices of the ſtomach, there is no reaſon for diſcontinuing the uſe of it, though it is occaſionally ejected in a curdled ſtate. A child of a week old, which had been taken from the breaſt of its dying mother, and had by ſome uncommon error been ſuffered to take no food but water-gruel, became ſick and griped in twenty-four hours, and was convulſed on the ſecond day, and died on the third! When all young quadrupeds, as well as children, have this natural food of milk prepared for them, the analogy is ſo ſtrong in favour of its ſalubrity, that a perſon ſhould have powerful teſtimony indeed of its diſagreeing, before he adviſes the diſcontinuance of the uſe of it to young children in health, and much more ſo in ſickneſs. The farmers loſe many of their calves, which are brought up by gruel, or gruel and old milk; and among the poor children of Derby, who are thus fed, hundreds are ſtarved into the ſcrophula, and either periſh, or live in a ſtate of wretched debility.

When young children are brought up without a breaſt, they ſhould for the firſt two months have no food but new milk; ſince the addition of any kind of bread or flour is liable to ferment, and produce too much acidity; as appears by the conſequent diarrhœa with green dejections and gripes; the colour is owing to a mixture of acid with the natural quantity of bile, and the pain to its ſtimulus. And they ſhould never be fed as they lie upon their backs, as in that poſture they are neceſſitated to ſwallow all that is put into their mouths; but when they are fed, as they are ſitting up, or raiſed up, when they have had enough, they can permit the reſt to run out of their mouths. This circumſtance is of great importance to the health of thoſe children, who are reared by the ſpoon, ſince if too much food is given them, indigeſtion, and gripes, and diarrhœa, is the conſequence; and if too little, they become emaciated; and of this exact quantity their own palates judge the beſt.

M. M. In this laſt caſe of the diarrhœa of children, the food ſhould be new milk, which by curdling deſtroys part of the acid, which coagulates it. Chalk about four grains every ſix hours, with one drop of ſpirit of hartſhorn, and half a drop of laudanum: But a bliſter about the ſize of a ſhilling is of the greateſt ſervice by reſtoring the power of digeſtion. See Article III. 2. 1. in the ſubſequent Materia Medica.

6. *Salivatio calida.* Warm ſalivation. Increaſed ſecretion of ſaliva. This may be effected either by ſtimulating the mouth of the gland by mercury taken internally; or by ſtimulating the excretory duct of the gland by pyrethrum, or tobacco; or ſimply by the movement of the muſcles, which lie over the gland, as in maſticating any taſteleſs ſubſtance, as a lock of wool, or maſtic.

In about the middle of nervous fevers a great ſpitting of ſaliva ſometimes occurs, which has been thought critical; but as it continues ſometimes two or even three weeks without the relief of the patient, it may be concluded to ariſe from ſome accidental circumſtance, perhaps not unſimilar to the hyſteric ptyaliſms mentioned in Claſs I. 3. 2. 2. See Sect. XXIV.

M. M. Cool air, diluents, warm bath, evacuations.

7. *Catharrhus calidus.* Warm catarrh. Conſiſts in an increaſed ſecretion of mucus from the noſtrils without inflammation. This diſeaſe, which is called a cold in the head, is frequently produced by cold air acting for ſome time on the membranes, which line the noſtrils, as it paſſes to the lungs in reſpiration. Whence a torpor of the action of the mucous glands is firſt introduced, as in I. 2. 3. 3. and an orgaſm or increaſed action ſucceeds in conſequence. Afterwards this orgaſm and torpor are liable to alternate with each other for ſome time like the cold and hot fits of ague, attended with deficient or exuberant ſecretion of mucus in the noſtrils.

At other times it ariſes from reverſe ſympathy with ſome extenſive parts of the ſkin, which have been expoſed too long to cold, as of the head, or feet. In conſequence of the torpor of theſe cutaneous capillaries thoſe of the mucous membrane of the noſtrils act with greater energy by reverſe ſympathy; and thence ſecrete more mucus from the blood. At the ſame time the abſorbents, acting alſo with greater energy by their reverſe ſympathy with thoſe of ſome diſtant part of the ſkin, abſorb the thinner parts of the mucus more haſtily; whence the mucus is both thicker and in greater quantity. Other curious circumſtances attend this diſeaſe; the membrane becomes at times ſo thickened by its increaſed action in ſecreting the mucus, that the patient cannot breathe through his noſtrils. In this ſituation if he warms his whole ſkin ſuddenly by fire or bed-clothes, or by drinking warm tea, the increaſed action of the membrane ceaſes by its reverſe

ſympathy

ſympathy with the ſkin; or by the retraction of the ſenſorial power to other parts of the ſyſtem; and the patient can breathe again through the noſtrils. The ſame ſometimes occurs for a time on going into the cold air by the deduction of heat from the mucous membrane, and its conſequent inactivity or torpor. Similar to this when the face and breaſt have been very hot and red, previous to the eruption of the ſmall-pox by inoculation, and that even when expoſed to cool air, I have obſerved the feet have been cold; till on covering them with warm flannel, as the feet have become warm, the face has cooled. See Sect. XXXV. 1. 3. Claſs II. 1. 3. 5. IV. 2. 2. 10. IV. 1. 1. 5.

M. M. Evacuations, abſtinence, oil externally on the noſe, warm diluent fluids, warm ſhoes, warm night-cap.

8. *Expectoratio calida.* Warm expectoration conſiſts of the increaſed ſecretion of mucus from the membrane, which lines the bronchiæ, or air-cells of the lungs, without inflammation. This increaſed mucus is ejected by the action of coughing, and is called a cold, and reſembles the catarrh of the preceding article; with which it is frequently combined.

M. M. Inhale the ſteam of warm water, evacuations, warm bath, afterwards opium, ſorbentia.

9. *Exſudatio pone aures.* A diſcharge behind the ears. This chiefly affects children, and is a morbid ſecretion; as appears from its fetor; for if it was owing to defect of abſorption, it would be ſaline, and not fetid; if a morbid action has continued a conſiderable time, it ſhould not be ſtopped too ſuddenly; ſince in that caſe ſome other morbid action is liable to ſucceed in its ſtead. Thus children are believed to have had cholics, or even convulſions, conſequent to the

too ſudden healing of theſe morbid effuſions behind their ears. The rationale of this is to be explained from a medical fact, which I have frequently obſerved; and that is, that a bliſter on the back greatly ſtrengthens the power of digeſtion, and removes the heart-burn in adults, and green ſtools in children. The ſtimulus of the bliſter produces ſenſation in the veſſels of the ſkin; with this additional ſenſorial power theſe veſſels act more ſtrongly; and with theſe the veſſels of the internal membranes of the ſtomach and bowels act with greater energy from their direct ſympathy with them. Now the acrid diſcharge behind the ears of children produces ſenſation on that part of the ſkin, and ſo far acts as a ſmall bliſter. When this is ſuddenly ſtopped, a debility of the digeſtive power of the ſtomach ſucceeds from the want of this accuſtomed ſtimulus, with flatulency, green ſtools, gripes, and ſometimes conſequent convulſions. See Claſs II. 1. 5. 6. and II. 1. 4. 6.

M. M. If the matter be abſorbed, and produces ſwelling of the lymphatics of the neck, it ſhould be cured as ſoon as poſſible by duſting the part with white lead, ceruſſa, in very fine powder; and to prevent any ill conſequence an iſſue ſhould be kept for about a month in the arm; or a purgative medicine ſhould be taken every other day for three or four times, which ſhould conſiſt of a grain of calomel, and three or four grains of rhubarb, and as much chalk. If there be no appearance of abſorption, it is better only to keep the parts clean by waſhing them with warm water morning and evening; or putting fuller's earth on them; eſpecially till the time of toothing is paſt. The tinea, or ſcald head, and a leprous eruption, which often appears behind the ears, are different diſeaſes.

10. *Gonorrhœa calida.* Warm gleet. Increaſed diſcharge of mucus from the urethra or proſtrate gland without venereal deſire, or venereal infection. See Claſs I. 2. 3. 8.

M. M. Cantharides,

M. M. Cantharides, balſams, rhubarb, bliſter in perinæum, cold bath, injections of metallic ſalts, flannel ſhirt, change of the form of the accuſtomed chair or ſaddle of the patient.

11. *Fluor albus calidus.* Warm fluor albus. Increaſed ſecretion of mucus in the vagina or uterus without venereal deſire or venereal infection. It is diſtinguiſhed from the fluor albus frigidus by the increaſed ſenſe of warmth in the part, and by the greater opacity or ſpiſſitude of the material diſcharged; as the thinner parts are reabſorbed by the increaſed action of the abſorbents, along with the ſaline part, whence no ſmarting or excoriation attends it.

M. M. Mucilage, as iſinglaſs, hartſhorn jelly, gum arabic. Ten grains of rhubarb every night. Callico or flannel ſhift, opium, balſams. See Claſs I. 2. 3. 7.

12. *Hæmorrhois alba.* White piles. An increaſed diſcharge of mucus from the rectum frequently miſtaken for matter; is ſaid to continue a few weeks, and recur like the bleeding piles; and to obey lunar influence. See Claſs I. 2. 1. 6.

M. M. Abſtinence from vinous ſpirit. Balſam of copaiva. Spice ſwallowed in large fragments, as ten or fifteen black pepper-corns cut in half, and taken after dinner and ſupper. Ward's paſte, conſiſting of black pepper and the powdered root of Helenium Enula.

13. *Serum e veſicatorio.* Diſcharge from a bliſter. The excretory ducts of glands terminate in membranes, and are endued with great irritability, and many of them with ſenſibility; the latter perhaps in conſequence of their facility of being excitable into great action; inſtances of this are the terminations of the gall-duct in the duodenum, and of the ſalivary and lachrymal glands in the mouth and eye; which produce a greater ſecretion of their adapted fluids, when the ends of their excretory ducts are ſtimulated.

The

The external ſkin conſiſts of the excretory ducts of the capillaries, with the mouths of the abſorbents; when theſe are ſtimulated by the application of cantharides, or by a ſlice of the freſh root of bryonia alba bound on it, the capillary glands pour an increaſed quantity of fluid upon the ſkin by their increaſed action; and the abſorbent veſſels imbibe a greater quantity of the more fluid and ſaline part of it; whence a thick mucous or ſerous fluid is depoſited between the ſkin and cuticle.

14. *Perſpiratio fœtida.* Fetid perſpiration. The uſes of the perſpirable matter are to keep the ſkin ſoft and pliant, for the purpoſes of its eaſier flexibility during the activity of our limbs in locomotion, and for the preſervation of the accuracy of the ſenſe of touch, which is diffuſed under the whole ſurface of it to guard us againſt the injuries of external bodies; in the ſame manner as the ſecretion of tears is deſigned to preſerve the cornea of the eye moiſt, and in conſequence tranſparent; yet has this cutaneous mucus been believed by many to be an excrement; and I know not how many fanciful theories have been built on its ſuppoſed obſtruction. Such as the origin of catarrhs, coughs, inflammations, eryſypelas, and herpes.

To all theſe it may be ſufficient to anſwer, that the antient Grecians oiled themſelves all over; that ſome nations have painted themſelves all over, as the Picts of this iſland; that the Hottentots ſmear themſelves all over with greaſe. And laſtly, that many of our own heads at this day are covered with the flour of wheat and the fat of hogs, according to the tyranny of a filthy and waſteful faſhion, and all this without inconvenience. To this muſt be added the ſtrict analogy between the uſe of the perſpirable matter and the mucous fluids, which are poured for ſimilar purpoſes upon all the internal membranes of the body; and beſides its being in its natural ſtate inodorous; which is not ſo with the other excretions of feces, or of urine.

In

In ſome conſtitutions the perſpirable matter of the lungs acquires a diſagreeable odour; in others the axillà, and in others the feet, emit diſguſtful effluvia; like the ſecretions of thoſe glands, which have been called odoriferæ; as thoſe, which contain the caſtor in the beaver, and thoſe within the rectum of dogs, the mucus of which has been ſuppoſed to guard them againſt the great coſtiveneſs, which they are liable to in hot ſummers; and which has been thought to occaſion canine madneſs, but which, like their white excrement, is more probably owing to the deficient ſecretion of bile. Whether theſe odoriferous particles attend the perſpirable matter in conſequence of the increaſed action of the capillary glands, and can properly be called excrementitous; that is, whether any thing is eliminated, which could be hurtful if retained; or whether they may only contain ſome of the eſſential oil of the animal; like the ſmell, which adheres to one's hand on ſtroking the hides of ſome dogs; or like the effluvia, which is left upon the ground, from the feet of men and other creatures; and is perceptible by the nicer organs of the dogs, which hunt them, may admit of doubt.

M. M. Waſh the parts twice a day with ſoap and water; with lime water; cover the feet with oiled ſilk ſocks, which muſt be waſhed night and morning. Cover them with charcoal recently made red hot, and beaten into fine powder and ſifted, as ſoon as cold, and kept well corked in a bottle, to be waſhed off and renewed twice a day. Internally rhubarb grains vi. or viii. every night, ſo as to procure a ſtool or two extraordinary every day, and thus by increaſing one evacuation to decreaſe another. Cool dreſs, diluting liquids?

15. *Crines novi.* New hairs. The black points on the faces of ſome people conſiſt of mucus, which is become viſcid, and which adheres in the excretory ducts of the glands of the ſkin; as deſcribed in Claſs I. 2. 2. 9. and which may be preſſed out by the fingers, and reſembles little worms. Similar to this would ſeem the fabri-

 cation

cation of ſilk, and of cobweb by the ſilk worm and ſpider; which is a ſecreted matter preſſed through holes, which are the excretory ducts of glands. And it is probable, that the production of hair on many parts of the body, and at different periods of life, may be effected by a ſimilar proceſs; and more eſpecially as every hair may be conſidered as a ſlender flexible horn, and is an appendage of the ſkin. See Sect. XXXIX. 3. 2. Now as there is a ſenſitive ſympathy between the glands, which ſecrete the ſemen, and the throat, as appears in the mumps; ſee Hydrophobia, Claſs IV. 1. 2. 7. and Parotitis, Claſs IV. 1. 2. 19. the growth of the beard at puberty ſeems to be cauſed by the greater action of the cutaneous glands about the chin and pubes in conſequence of their ſympathy with thoſe of the teſtes. But this does not occur to the female ſex at their time of puberty, becauſe the ſenſitive ſympathy in them ſeems to exiſt between the ſubmaxillary glands, and the pectoral ones; which ſecrete the milk, and afford pleaſure both by that ſecretion, and by the erection of the mamulæ, or nipples; and by delivering the milk into the mouth of the child; this ſenſitive ſympathy of the pectoral and ſubmaxillary glands in women is alſo obſervable in the Parotitis, or mumps, as above referred to.

When hairs grow on the face or arms ſo as to be diſagreeable, they may be thus readily removed without pain or any ill conſequence. Warm the ends of a pair of nippers or forceps, and ſtick on them a little roſin, or burgundy pitch; by theſe means each ſingle hair may be taken faſt hold of; and if it be then plucked off ſlowly, it gives pain; but if plucked off ſuddenly, it gives no pain at all; becauſe the vis inertiæ of the part of the ſkin, to which it adheres, is not overcome; and it is not in conſequence ſeparated from the cellular membrane under it. Some of the hairs may return, which are thus plucked off, or others may be induced to grow near them; but in a little time they may be thus ſafely deſtroyed; which is much to be preferred to the methods ſaid to be uſed in Turkey to eradicate hair;

ſuch

ſuch as a mixture of orpiment and quick lime; or of liver of ſulphur in ſolution; which injure the ſkin, if they are not very nicely managed; and the hair is liable to grow again as after ſhaving; or to become white, if the roots of it have been much inflamed by the cauſticity of the application. See Claſs I. 2. 2. 11. on grey hairs.

## ORDO I.

*Increaſed Irritation.*

## GENUS III.

*With increaſed Actions of the Abſorbent Syſtem.*

These are not attended with ſo great increaſe of heat as in the former genus, becauſe the fluids probably undergo leſs chemical change in the glands of the abſorbent ſyſtem; nor are the glands of the abſorbent veſſels ſo numerous or ſo extenſive as thoſe of the ſecerning ones. Yet that ſome heat is produced by the increaſed action of the abſorbents appears from the greater general warmth of the ſkin and extremities of feeble patients after the exhibition of the peruvian bark, and other medicines of the article Sorbentia.

## SPECIES.

1. *Lingua arida.* Dry tongue occurs in thoſe fevers, where the expired air is warmer than natural; and happens to all thoſe, who ſleep with their mouths open; the currents of air in reſpiration increaſing the evaporation. There is alſo a dryneſs in the mouth from the increaſed action of the abſorbent veſſels, when a ſloe or a crab-apple are maſticated; and after the perſpiration has been much increaſed by eating ſalt or ſpice, or after other copious ſecretions; as after drunkenneſs, cathartics, or fever fits, the mucus of the mouth becomes viſcid, and in ſmall quantity, from the increaſed abſorption, adhering to the tongue like a white ſlough. In the diabætes, where the thirſt is very great, this ſlough adheres more pertinaciouſly, and

becomes black or brown, being coloured after a few days by our aliment or drink. The inſpiſſated mucus on the tongue of thoſe, who ſleep with their mouths open, is ſometimes reddened as if mixed with blood, and ſometimes a little blood follows the expuition of it from the fauces owing to its great adheſion. When this mucus adheres long to the papillæ of the tongue, the ſaliva, which it contains in its interſtices, like a ſponge, is liable to become putrid, and to acquire a bitter taſte, like other putrid animal ſubſtances; which is generally miſtaken for an indication of the preſence of bile.

M. M. Warm ſubacid liquids. See Claſs I. 2. 5. 8.

2. *Fauces aridæ*. Dry throat. The expuition of a frothy mucus with great and perpetual hawking occurs in hydrophobia, and is very diſtreſſing to the patient; which may be owing to the increaſed irritability or ſenſibility of the upper part of the œſophagus, which will not permit any fluid to reſt on it.

It affects ſome people after intoxication, when the lungs remain ſlightly inflamed, and by the greater heat of the air in expiration the mucus becomes too haſtily evaporated, and is expectorated with difficulty in the ſtate of white froth.

I knew a perſon, who for twenty years always waked with his tongue and throat quite dry; ſo that he was neceſſitated to take a ſpoonful of water, as ſoon as he awoke; otherwiſe a little blood always followed the forcible expuition of the indurated mucus from his fauces. See Claſs II. 1. 3. 17.

M. M. Steel-ſprings fixed to the night-cap ſo as to ſuſpend the lower jaw and keep it cloſed; or ſprings of elaſtic gum. Or a pot of water ſuſpended over the bed, with a piece of liſt, or woollen-cloth, depending from it, and held in the mouth; which will act like a ſyphon, and ſlowly ſupply moiſture, or barley water ſhould be frequently ſyringed into the mouth of the patient.

3. *Nares*

3. *Nares aridi.* Dry noſtrils with the mucus hardening upon their internal ſurface, ſo as to cover them with a kind of ſkin or ſcale, owing to the increaſed action of the abſorbents of this membrane; or to the too great dryneſs of the air, which paſſes into the lungs; or too great heat of it in its expiration.

When air is ſo dry as to loſe its tranſparency; as when a tremulous motion of it can be ſeen over corn fields in a hot ſummer's day; or when a dry miſt, or want of tranſparency of the air, is viſible in very hot weather; the ſenſe of ſmell is at the ſame time imperfect from the dryneſs of the membrane, beneath which it is ſpread.

4. *Expectoratio ſolida.* Solid expectoration. The mucus of the lungs becomes hardened by the increaſed abſorption, ſo that it adheres and forms a kind of lining in the air-cells, and is ſometimes ſpit up in the form of branching veſſels, which are called polypi of the lungs. See Tranſact. of the College, London. There is a rattling or weezing of the breath, but it is not at firſt attended with inflammation.

The Cynanche trachealis, or Croup, of Dr. Cullen, or Angina polypoſa of Michaelis, if they differ from the peripneumony of infants, ſeem to belong to this genus. When the difficulty of reſpiration is great, veneſection is immediately neceſſary, and then an emetic, and a bliſter. And the child ſhould be kept nearly upright in bed as much as may be. See Tonſillitis, Claſs II. 1. 3. 3.

M. M. Diluents, emetics, eſſence of antimony, fœtid gums, onions, warm bath for half an hour every day for a month. Inhaling the ſteam of water, with or without volatile alcali. Soap.

5. *Conſtipatio alvi.* Coſtiveneſs from increaſed action of the inteſtinal abſorbents. The feces are hardened in lumps called ſcybala; which are ſometimes obliged to be extracted from the rectum with a kind of marrow ſpoon. This is ſaid to have happened from the patient having taken much ruſt of iron. The mucus is alſo harden-

ed

ed ſo as to line the inteſtines, and to come away in ſkins, rolled up as they paſs along, ſo as to reſemble worms, for which they are frequently miſtaken; and ſometimes it is evacuated in ſtill larger pieces, ſo as to counterfeit the form of the inteſtines, and has been miſtaken for a portion of them. Balls of this kind, nearly as heavy as marble, and conſiderably hard, from two inches to five in diameter, are frequently found in the bowels of horſes. Similar balls found in goats have been called Bezoar.

M. M. Cathartics, Diluents, fruit, oil, ſoap, ſulphur, warm bath. Sprinkling with cold water, cool clothing. See Claſs I. 2. 4. 18.

6. *Cutis arida.* Dry ſkin. This dry ſkin is not attended with coldneſs as in the beginning of fever-fits. Where this cutaneous abſorption is great, and the ſecreted material upon it viſcid, as on the hairy ſcalp, the ſkin becomes covered with hardened mucus; which adheres ſo as not to be eaſily removed, as the ſcurf on the head; but is not attended with inflammation like the Tinea, or Lepra. The moiſture, which appears on the ſkin beneath reſinous or oily plaſters, or which is ſeen to adhere to ſuch plaſters, is owing to their preventing the exhalation of the perſpirable matter, and not to their increaſing the production of it, as ſome have idly imagined.

M. M. Warm bathing, oil externally, oil-ſkin gloves, reſinous plaſters. Wax.

7. *Urina parca colorata.* Diminiſhed urine, which is high coloured, and depoſits an earthy ſediment, when cold, is owing to the great action of the urinary abſorbents. See Claſs I. 1. 2. 4. In ſome dropſies the cutaneous abſorbents are paralytic, as well as thoſe opening into the cellular membrane; and hence, no moiſture being acquired from the atmoſphere, or from the cellular membrane, great thirſt is excited; and great abſorption from all parts, where the abſorbents are ſtill capable of action. Hence the urine is in very ſmall quantity,

 and

and of deep colour, with copious ſediment; and the kidneys are erroneouſly blamed for not doing their office; ſtimulant diuretic medicines are given in vain; and very frequently the unhappy patient is reſtrained from quenching his thirſt, and dies a martyr to falſe theory.

M. M. Diluent liquids, and warm bathing, are the natural cure of this ſymptom; but it generally attends thoſe dropſies, which are ſeldom curable; as they are owing to a paralyſis both of the cutaneous and cellular lymphatics.

8 *Calculus felleus*. Gall-ſtone. From the too haſty abſorption of the thinner parts of the bile, the remainder is left too viſcid, and cryſtallizes into lumps; which, if too large to paſs, obſtruct the ductus choledochus, producing pain at the pit of the ſtomach, and jaundice. When the indurated bile is not harder than a boiled pea, it may paſs through the bile-duct with difficulty by changing its form; and thus gives thoſe pains, which have been called ſpaſms of the ſtomach; and yet theſe viſcid lumps of bile may afterwards diſſolve, and not be viſible among the feces.

In two inſtances I have ſeen from thirty to fifty gall-ſtones voided after taking an oil vomit as below. They were about the ſize of peas, and diſtinguiſhable when dry by their being inflammable like bad wax, when put into the flame of a candle. For other cauſes of jaundice, ſee Claſs I. 2. 4. 19.

M. M. Diluents, daily warm bathing. Ether mixed with yolk of egg and water. Unboiled acrid vegetables, as lettice, cabbage, muſtard, and creſſes. When in violent pain, four ounces of oil of olives, or of almonds, ſhould be ſwallowed; and as much more in a quarter of an hour, whether it ſtays or not. The patient ſhould lie on the circumference of a large barrel, firſt on one ſide, and then on the other. Electric ſhocks through the gall-duct. Factitious Selter's water made by diſſolving one dram of Sal Soda in a pint of water; to half a pint of which made luke-warm add ten drops of marine acid;

to

to be drank as ſoon as mixed, twice a day for ſome months. Opium muſt be uſed to quiet the pain, if the oil does not ſucceed, as two grains, and another grain in half an hour if neceſſary. See Claſs IV. 2. 2. 4.

9. *Calculus renis.* Stone of the kidney. The pain in the loins and along the courſe of the ureter from a ſtone is attended with retraction of the teſticle in men, and numbneſs on the inſide of the thigh in women. It is diſtinguiſhed from the lumbago or ſciatica, as theſe latter are ſeldom attended with vomiting, and have pain on the outſide of the thigh, ſometimes quite down to the ankle or heel. See Herpes and Nephritis.

Where the abſorption of the thinner parts of the ſecretion takes place too haſtily in the kidnies, the hardened mucus, and conſequent calculous concretions, ſometimes totally ſtop up the tubuli uriniferi; and no urine is ſecreted. Of this many die, who have drank much vinous ſpirit, and ſome of them recover by voiding a quantity of white mucus, like chalk and water; and others by voiding a great quantity of ſand, or ſmall calculi. This hardened mucus frequently becomes the nucleus of a ſtone in the bladder. The ſalts of the urine, called microcoſmic ſalt, are often miſtaken for gravel, but are diſtinguiſhable both by their angles of cryſtallization, their adheſion to the ſides or bottom of the pot, and by their not being formed till the urine cools. Whereas the particles of gravel are generally without angles, and always drop to the bottom of the veſſel, immediately as the water is voided.

Though the proximate cauſe of the formation of the calculous concretions of the kidneys, and of chalk-ſtones in the gout, and of the inſoluble concretions of coagulable lymph, which are found on membranes, which have been inflamed in peripneumony, or rheumatiſm, conſiſts in the too great action of the abſorbent veſſels of thoſe parts; yet the remote cauſe in theſe caſes is probably owing to the inflammation of the membranes; which at that time are believed to ſecrete a material more liable to coagulate or concrete, than they would otherwiſe produce by increaſed action alone without the production of new veſſels, which conſtitutes inflammation. As defined in Claſs II. 1. 2.

The fluids ſecreted from the mucous membranes of animals are of various kinds and conſiſtencies. Hair, ſilk, ſcales, horns, finger-nails, are owing to natural proceſſes. Gall-ſtones, ſtones found in the inteſtines of horſes, ſcurf of the ſkin in leproſy, ſtones of the kidnies and bladder, the callus from the inflamed perioſteum, which unites broken bones, the calcareous cement, which repairs the injured ſhells of ſnails, the calcareous cruſt on the eggs of birds, the annually renewed ſhells of crabs, are all inſtances of productions from mucous membranes, afterwards indurated by abſorption of their thinner parts.

All theſe concretions contain phoſphoric acid, mucus, and calcareous earth in different proportions; and are probably ſo far analogous in reſpect to their component parts as well as their mode of formation. Some calcareous earth has been diſcovered after putrefaction in the coagulable lymph of animals. Fordyce's Elements of Practice: A little calcareous earth was detected by Scheel or Bergman in the calculus of the bladder with much phoſphoric acid, and a great quantity of phoſphoric acid is ſhewn to exiſt in oyſter-ſhells by their becoming luminous on expoſing them a while to the ſun's light after calcination; as in the experiments of Wilſon. Botanic Garden, P. 1. Canto 1. l. 182, note. The exchange of which phoſphoric acid for carbonic acid, or fixed air, converts ſhells into limeſtone, producing mountains of marble, or calcareous ſtrata..

Now as the hard lumps of calcareous matter, termed crabs' eyes, which are found in the ſtomachs of thoſe animals previous to the annual renewal of their ſhells, are rediſſolved, probably by their gaſtric acid, and again depoſited for that purpoſe; may it not be concluded, that the ſtone of the bladder might be diſſolved by the gaſtric juice of fiſh of prey, as of crabs, or pike; or of voracious young birds, as

young

young rooks or hawks, or even of calves? Could not these experiments be tried by collecting the gastric juice by putting bits of sponge down the throats of young crows, and retracting them by a string in the manner of Spallanzani? or putting pieces of calculus down the throat of a living crow, or pike, and observing if they become digested? and lastly could not gastric juice, if it should appear to be a solvent, be injected and born in the bladder without injury by means of catheters of elastic resin, or caoutchouc?

M. M. Diluents. Cool dress. Frequent change of posture. Frequent horizontal rest in the day. Bathe the loins every morning with a sponge and cold water. Aerated alcaline water internally. Abstinence from all fermented or spirituous liquors. Whatever increases perspiration injures these patients, as it dissipates the aqueous particles, which ought to dilute the urine. When the constitution begins to produce gravel, it may I believe be certainly prevented by a total abstinence from fermented or spirituous liquors; by drinking much aqueous fluids; as toast and water, tea, milk and water, lemonade; and lastly by thin clothing, and sleeping on a hardish bed, that the patient may not lie too long on one side. See Class IV. 2. 2. 2. There is reason to believe, that the daily use of opium contributes to produce gravel in the kidnies by increasing absorption, when they are inflamed; in the same manner as is done by fermented or spirituous liquor. See Class I. 3. 2. 11.

When the kidnies are so obstructed with gravel, that no urine passes into the bladder; which is known by the external appearance of the lower part of the abdomen, which, when the bladder is full, seems as if contracted by a cord between the navel and the bladder; and by the tension on the region of the bladder distinguishable by the touch; or by the introduction of the catheter; the following methods of cure are frequently successful. Venesection to six or eight ounces, ten grains of calomel, and an infusion of senna with salts and oil, every three hours, till stools are procured. Then an emetic. After the patient has been thus evacuated, a bliſter on the loins ſhould be uſed ; and from ten to twenty electric ſhocks ſhould be paſſed through the kidnies, as large as can be eaſily borne, once or twice a day. Along with this method the warm bath ſhould be uſed for an hour once or twice a day. After repeated evacuations a clyſter, conſiſting of two drams of turpentine diſſolved by yolk of egg, and ſixty drops of tincture of opium, ſhould be uſed at night, and repeated, with cathartic medicines interpoſed, every night, or alternate nights. Aerated ſolution of alcali ſhould be taken internally, and balſam of copaiva, three or four times a day. Some of theſe patients recover after having made no water for nine or ten days.

If a ſtone ſticks in the ureter with inceſſant vomiting, ten grains of calomel muſt be given in ſmall pills as above ; and ſome hours afterwards infuſion of ſenna and ſalts and oil, if it can be made to ſtay on the ſtomach. And after the purge has operated four or five times, an opiate is to be given, if the pain continues, conſiſting of two grains of opium. If this does not ſucceed, ten or twenty electric ſhocks through the kidney ſhould be tried, and the purgative repeated, and afterwards the opiate. The patient ſhould be frequently put into the warm bath for an hour at a time. Eighty or an hundred drops of laudanum given in a glyſter, with two drams of turpentine, is to be preferred to the two grains given by the ſtomach as above, when the pain and vomiting are very urgent.

10. *Calculus veſicæ.* Stone of the bladder. The nucleus, or kernel, of theſe concretions is always formed in the kidney, as above deſcribed; and paſſing down the ureter into the bladder, is there perpetually increaſed by the mucus and ſalts ſecreted from the arterial ſyſtem, or by the mucus of the bladder, diſpoſed in concentric ſtrata. The ſtones found in the bowels of horſes are alſo formed on a nucleus, and conſiſt of concentric ſpheres; as appears in ſawing them through the middle. But as theſe are formed by the indurated mucus of the



inteſtines

inteſtines alone without the urinary ſalts, it is probable a difference would be found on their analyſis.

As the ſtones of the bladder are of various degrees of hardneſs, and probably differ from each other in the proportions at leaſt of their component parts; when a patient, who labours under this afflicting diſeaſe, voids any ſmall bits of gravel; theſe ſhould be kept in warm ſolutions of cauſtic alcali, or of mild alcali well aerated; and if they diſſolve in theſe ſolutions, it would afford greater hopes, that that which remains in the bladder, might be affected by theſe medicines taken by the ſtomach, or injected into the bladder.

To prevent the increaſe of a ſtone in the bladder much diluent drink ſhould be taken; as half a pint of water warmed to about eighty degrees, three or four times a day: which will not only prevent the growth of it, by preventing any microcoſmic ſalts from being precipitated from the urine, and by keeping the mucus ſuſpended in it; but will alſo diminiſh the ſtone already formed, by ſoftening, and waſhing away its ſurface. To this muſt be added cool dreſs, and cool bed-clothes, as directed above in the calculus renis.

When the ſtone is puſhed againſt or into the neck of the bladder, great pain is produced; this may ſometimes be relieved by the introduction of a bougie to puſh the ſtone back into the fundus of the bladder. Sometimes by change of poſture, or by an opiate either taken into the ſtomach, or by a clyſter.

A dram of ſal ſoda, or of ſalt of tartar, diſſolved in a pint of water, and well ſaturated with carbonic acid (fixed air), by means of Dr. Nooth's glaſs-apparatus, and drank every day, or twice a day, is the moſt efficacious internal medicine yet diſcovered, which can be eaſily taken without any general injury to the conſtitution. An aerated alcaline water of this kind is ſold under the name of factitious Seltzer water, by J. Schweppe, at N° 8, King's-ſtreet, Holborn, London; which I am told is better prepared than can be eaſily done in the uſual glaſs-veſſels, probably by employing a greater preſſure in wooden ones.

Lythotomy

Lythotomy is the laſt recourſe. Will the gaſtric juice of animals diſſolve calculi? Will fermenting vegetable juices, as ſweet-wort, or ſugar and water in the act of fermentation with yeſt, diſſolve any kind of animal concretions?

11. *Calculus arthriticus.* Gout-ſtones are formed on inflamed membranes, like thoſe of the kidnies above deſcribed, by the too haſty abſorption of the thinner and ſaline parts of the mucus. Similar concretions have been produced in the lungs, and even in the pericardium; and it is probable, that the oſſification, as it is called, of the minute arteries, which is ſaid to attend old age, and to precede ſome mortifications of the extremities, may be a proceſs of this kind.

As gout-ſtones lie near the ſurface, it is probable, that ether, frequently applied in their early ſtate, might render them ſo liquid as to permit their reabſorption; which the ſtimulus of the ether might at the ſame time encourage.

12. *Rheumatiſmus chronicus.* Chronic rheumatiſm. After the acute rheumatiſm ſome inſpiſſated mucus, or material ſimilar to chalk-ſtones of the gout, which was ſecreted on the inflamed membrane, is probably left, owing to the too haſty abſorption of the thinner and ſaline part of it; and by lying on the faſcia, which covers ſome of the muſcles, pains them, when they move and rub againſt it, like any extraneous material.

The pain of the ſhoulder, which attends inflammations of the upper membrane of the liver, and the pains of the arms, which attend aſthma dolorificum, or dropſy of the pericardium, are diſtinguiſhed from the chronic rheumatiſm, as in the latter the pain only occurs on moving the affected muſcles.

M. M. Warm bath, cold bath, bandage of emplaſtrum de minio put on tight, ſo as to compreſs the part. Cover the part with flannel. With oiled ſilk. Rub it with common oil frequently. With ether.

ether. A bliſter. A warmer climate. Veneſection. A grain of calomel and a grain of opium for ten ſucceſſive nights. The Peruvian bark.

13. *Cicatrix vulnerum.* The ſcar after wounds. In the healing of ulcers the matter is firſt thickened by increaſing the abſorption in them; and then leſſened, till all the matter is abſorbed, which is brought by the arteries, inſtead of being depoſed in the ulcer.

M. M. This is promoted by bandage, by the ſorbentia externally, as powder of bark, white lead; ſolution of ſugar of lead. And by the ſorbentia internally after evacuations. See Sect. XXXIII. 3. 2.

In thoſe ulcers, which are made by the contact of external fire, the violent action of the fibres, which occaſions the pain, is liable to continue, after the external heat is withdrawn. This ſhould be relieved by external cold, as of ſnow, ſalt and water recently mixed, ether, or ſpirits of wine ſuffered to evaporate on the part.

The cicatrix of an ulcer generally proceeds from the edges of it; but in large ones frequently from the middle, or commences in ſeveral places at the ſame time; which probably contributes to the unevenneſs of large ſcars.

14. *Corneæ obfuſcatio.* Opacity of the cornea. There are few people, who have paſſed the middle of life, who have not at ſome time ſuffered ſome ſlight ſcratches or injuries of the cornea, which by not healing with a perfectly ſmooth ſurface, occaſion ſome refractions of light, which may be conveniently ſeen in the following manner: fill a tea-ſaucer with cream and tea, or with milk, and holding it to your lips, as if going to drink it, the imperfections of the cornea will appear like lines or blotches on the ſurface of the fluid, with a leſs white appearance than that ſurface. Thoſe blemiſhes of the eye are diſtinguiſhed from the muſcæ volitantes deſcribed in Claſs I. 2. 5. 3. by their being invariably ſeen at any time, when you look for them.

Ulcers may frequently be ſeen on the cornea after ophthalmy, like little pits or indentations beneath the ſurface of it: in this caſe no external application ſhould be uſed, leſt the ſcar ſhould be left uneven; but the cure ſhould be confined to the internal uſe of thirty grains of bark twice a day, and from five to ten drops of laudanum at night, with five grains of rhubarb, if neceſſary.

After ulcers of the cornea, which have been large, the inequalities and opacity of the cicatrix obſcures the ſight; in this caſe could not a ſmall piece of the cornea be cut out by a kind of trephine about the ſize of a thick briſtle, or a ſmall crow-quill, and would it not heal with a tranſparent ſcar? This experiment is worth trying, and might be done by a piece of hollow ſteel wire with a ſharp edge, through which might be introduced a pointed ſteel ſcrew; the ſcrew to be introduced through the opake cornea to hold it up, and preſs it againſt the cutting edge of the hollow wire or cylinder; if the ſcar ſhould heal without loſing its tranſparency, many blind people might be made to ſee tolerably well by this ſlight and not painful operation. An experiment I wiſh ſtrongly to recommend to ſome ingenious ſurgeon or oculiſt.

ORDO

## ORDO I.

*Increaſed Irritation.*

## GENUS IV.

*With increaſed Actions of other Cavities and Membranes.*

## SPECIES.

1. *Nectitatio irritativa.* Winking of the eyes is performed every minute without our attention, for the purpoſe of cleaning and moiſtening the eye-ball; as further ſpoken of in Claſs II. 1. 1. 8. When the cornea becomes too dry, it becomes at the ſame time leſs tranſparent; which is owing to the pores of it being then too large, ſo that the particles of light are refracted by the edges of each pore, inſtead of paſſing through it; in the ſame manner as light is refracted by paſſing near the edge of a knife. When theſe pores are filled with water, the cornea becomes again tranſparent. This want of tranſparency of the cornea is viſible ſometimes in dying people, owing to their inirritability, and conſequent neglect of nictitation.

The increaſe of tranſparency by filling the pores with fluid is ſeen by ſoaking white paper in oil; which from an opake body becomes very tranſparent, and accounts for a curious atmoſpheric phenomenon; when there exiſts a dry miſt in a morning ſo as to render diſtant objects leſs diſtinct, it is a ſign of a dry day; when diſtant objects are ſeen very diſtinct, it is a ſign of rain. See Botan. Garden, Part I. add. note xxv. The particles of air are probably larger than thoſe of water, as water will paſs through leather and paper, which will confine air; hence when the atmoſphere is much deprived of moiſture, the pores of the dry air are ſo large, that the rays of light are refracted by their edges inſtead of paſſing through them. But when as much moiſture is added as can be perfectly diſſolved, the air becomes tranſparent; and opake again, when a part of this moiſture collects into ſmall ſpherules previous to its precipitation. This alſo accounts for the want of tranſparency of the air, which is ſeen in tremulous motions over corn-fields on hot ſummer-days, or over brick-kilns, after the flame is extinguiſhed, while the furnace ſtill remains hot.

2. *Deglutitio irritativa.* The deglutition of our ſaliva is performed frequently without our attention, and is then an irritative action in conſequence of the ſtimulus of it in the mouth. Or perhaps ſometimes for the purpoſe of diffuſing a part of it over the dry membranes of the fauces and pharinx; in the ſame manner as tears are diffuſed over the cornea of the eye by the act of nictitation to clean or moiſten it.

3. *Reſpiratio et Tuſſis irritativæ.* In the acts of reſpiration and of coughing there is an increaſed motion of the air-cells of the lungs owing to ſome ſtimulating cauſe, as deſcribed above in Claſs I. 1. 2. 8. and I. 1. 3. 4. and which are frequently performed without our attention or conſciouſneſs, and are then irritative actions; and thus differ from thoſe deſcribed in Claſs II. 1. 1. 2. and 5. To theſe increaſed actions of the air-cells are ſuperadded thoſe of the intercoſtal muſcles and diaphragm by irritative aſſociation. When any unnatural ſtimulus acts ſo violently on the organs of reſpiration as to induce pain, the ſenſorial power of ſenſation becomes added to that of irritation, and inflammation of the membranes of them is a general conſequence.

4. *Excluſio bilis.* The excluſion of the bile from the gall-bladder, and its derivation into the duodenum, is an irritative action in conſequence of the ſtimulus of the aliment on the extremity of the biliary duct, which terminates in the inteſtine. The increaſed ſecretion of tears

tears is occasioned in a similar manner by any stimulating material in the eyes; which affects the excretory ducts of the lacrymal glands. A pain of the external membrane of the eye sometimes attends any unusual stimulus of it, then the sensorial power of sensation becomes added to that of irritation, and a superficial inflammation is induced.

5. *Dentitio*. Toothing. The pain of toothing often begins much earlier than is suspected; and is liable to produce convulsions; which are sometimes relieved, when the gum swells, and becomes inflamed; at other times a diarrhœa supervenes, which is generally esteemed a savourable circumstance, and seems to prevent the convulsions by supplying another means of relieving the pain of dentition by irritative exertion; and a consequent temporary exhaustion of sensorial power. See Class I. 1. 2. 5. Sect. XXXV. 2. 1.

The convulsions from toothing generally commence long before the appearance of the teeth; but as the two middle incisors of the lower jaw generally appear first, and then those of the upper, it is adviseable to lance the gums over these longitudinally in respect to the jaw-bones, and quite down to the periosteum, and through it.

As the convulsions attending the commencement of toothing are not only dangerous to life in their greatest degree, but are liable to induce stupor or insensibility by their continuance even in a less degree, the most efficacious means should be used to cure them.

M. M. Lance the gum of the expected teeth quite through the periosteum longitudinally. Venesection by the lancet or by two or three leeches. One grain of calomel as a purge. Tincture of jalap, five or six drops in water every three hours till it purges, to be repeated daily. After evacuations a small blister on the back or behind the ears. And lastly, two or three drops of laudanum according to the age of the child. Warm bath. See Class III. 1. 1. 5. and 6.

6. *Priapismus chronicus.* I have seen two cases, where an erection of the penis, as hard as horn, continued two or three weeks without any venereal desires, but not without some pain; the easiest attitude of the patients was lying upon their backs with their knees up. At length the corpus cavernosum urethræ became soft, and in another day or two the whole subsided. In one of them a bougie was introduced, hoping to remove some bit of gravel from the caput gallinaginis, camphor, warm bathing, opium, lime-water, cold aspersion, bleeding in the veins of the penis, were tried in vain. One of them had been a free drinker, had much gutta rosacea on his face, and died suddenly a few months after his recovery from this complaint. Was it a paralysis of the terminations of the veins, which absorb the blood from the tumid penis? or from the stimulus of indurated semen in the seminal vessels? In the latter case some venereal desires should have attended. Class III. 1. 2. 16.

The priapismus, which occurs to vigorous people in a morning before they awake, has been called the signum salutis, or banner of health, and is occasioned by the increase of our irritability or sensibility during sleep, as explained in Sect. XVIII. 15.

7. *Distentio mamularum.* The distention of the nipples of lactescent women is at first owing to the stimulus of the milk. See Sect. XIV. 8. and Sect. XVI. 5. See Class II. 1. 7. 10.

8. *Descensus uteri.* This is a very frequent complaint after bad labours, the fundus uteri becomes inverted and descends like the prolapsus ani.

M. M. All the usual pessaries are very inconvenient and ineffectual. A piece of soft sponge about two inches diameter introduced into the vagina gives great ease to these patients, and supports the uterus; it should have a string put through it to retract it by.

There are alſo peſſaries now made of elaſtic gum, which are ſaid to be eaſily worn, and to be convenient, from their having a perforation in their centre.

9. *Prolapſus ani.* The lower part of the rectum becomes inverted, and deſcends after every ſtool chiefly in children; and thus ſtimulates the ſphincter ani like any other extraneous body.

M. M. It ſhould be duſted over with very fine powder of gum ſandarach, and then replaced. Aſtringent fomentations; as an infuſion of oak-bark, or a ſlight ſolution of alum. Horizontal reſt frequently in the day.

10. *Lumbricus.* Round worm. The round worm is ſuſpected in children when the belly is tumid, and the countenance bloated and pale, with ſwelling of the upper lip. The generation of theſe worms is promoted by the too dilute ſtate of the bile, as is evident in the fleuke-worm found in the biliary ducts and ſubſtance of the liver in ſheep; and in water-rats, in the livers of which laſt animals they were lately detected in large numbers by Dr. Capelle. Tranſactions of the college at Philadelphia, v. i.

Now as the dilute ſtate of the bile depends on the deficiency of the abſorption of its thinner parts, it appears, that the tumid belly, and bloated countenance, and ſwelled upper lip, are a concomitant circumſtance attending the general inactivity of the abſorbent ſyſtem; which is therefore to be eſteemed the remote cauſe of the generation of worms.

The ſimplicity of the ſtructure of worms probably enables them to exiſt in more various temperatures of heat; and their being endued with life prevents them from being deſtroyed by digeſtion in the ſtomach, probably in the ſame manner as the powers of life prevent the fermentation and putrefaction of the ſtomach itſelf. Hence I conclude,

clude, that worms are originally taken into our alimentary canal from without; as I believe ſimilar worms of all kinds are to be found out of the body.

M. M. The round worm is deſtroyed by a cathartic with four or ſix grains of calomel; and afterwards by giving ſix or eight grains of filings of iron twice a day for a fortnight. See Hepatis tumor, Claſs I. 2. 3. 9. As worms are liable to come away in fevers, whether of the hectic or putrid kind, could they be removed by purulent matter, or rotten egg, or putrid fleſh, ſince in thoſe fevers from the enfeebled action of the inteſtines the fæces become highly putrid?

11. *Tænia.* Tape-worm conſiſts of a chain of animals extending from the ſtomach to the anus. See Sect. XXXIX. 2. 3. It frequently exiſts in cats, rats, and geeſe, and probably in many other animals.

The worms of this genus poſſeſs a wonderful power of retaining life. Two of them, which were voided by a pointer dog in conſequence of violent purgatives, each of which were ſeveral feet in length, had boiling water poured on them in a baſon; which ſeemed not much to inconvenience them. When the water was cool, they were taken out and put into gin or whiſkey of the ſtrongeſt kind, in which their life and activity continued unimpaired; and they were at length killed by adding to the ſpirit a quantity of corroſive ſublimate. Medic. Comment. for 1791, p. 370.

The tape-worm is cured by an amalgama of tin and quickſilver, ſuch as is uſed on the back of looking-glaſſes; an ounce ſhould be taken every two hours, till a pound is taken; and then a briſk cathartic of Glauber's ſalt two ounces, and common ſalts one ounce, diſſolved in two wine pints of water, half a pint to be taken every hour till it purges. The worm extends from the ſtomach to the anus, and the amalgama tears it from the inteſtine by mechanical preſſure, acting upon it the whole way. Electric ſhocks through the duodenum greatly

greatly aſſiſts the operation. Large doſes of tin in powder. Iron filings in large doſes. The powder of fern-root ſeems to be of no uſe, as recommended by M. Noufflier.

12. *Aſcarides.* Thread-worms. Theſe worms are ſaid to be more frequent in ſome parts of this kingdom than in others, as near the fens of Lincolnſhire. Do they eſcape from the body and become flies, like the bott-worm in horſes? Do they crawl from one child to another in the ſame bed? Are they acquired from flies or worms, which are ſeen in putrid neceſſary houſes, as theſe worms as well as the tape-worms, are probably acquired from without? this may account for their re-appearance a few weeks or months after they have been deſtroyed; or can this happen from the eggs or parts of them remaining?

Aſcarides appear to be of two kinds, the common ſmall ones like a thread; which has a very ſharp head, as appears in the microſcope; and which is ſo tender, that the cold air ſoon renders it motionleſs; and a larger kind above an inch long, and nearly as thick as a very ſmall crow-quill, and which is very hard in reſpect to its texture, and very tenacious of life. One of theſe laſt was brought to me, and was immediately immerſed in a ſtrong ſolution of ſugar of lead, and lived in it a very long time without apparent inconvenience.

M. M. Aſcarides are ſaid to be weakened by twenty grains of cinnabar and five of rhubarb taken every night, but not to be cured by this proceſs. As theſe worms are found only in the rectum, variety of clyſters have been recommended. I was informed of a caſe, where ſolutions of mercurial ointment were uſed as a clyſter every night for a month without ſucceſs. Clyſters of Harrowgate water are recomended, either of the natural, or of the factitious, as deſcribed below, which might have a greater proportion of liver of ſulphur in it. As the cold air ſoon deſtroys them, after they are voided, could clyſters of iced water be uſed with advantage? or of ſpirit of wine and water? or of ether and water? Might not a piece of candle, about an inch long,

long, or two ſuch pieces, ſmeared with mercurial ointment, and introduced into the anus at night, or twice a day, be effectual by compreſſing their nidus, as well as by the poiſon of the mercury.

The clyſters ſhould be large in quantity, that they may paſs high in the rectum, as two drams of tobacco boiled a minute in a pint of water. Or perhaps what might be ſtill more efficacious and leſs inconvenient, the ſmoke of tobacco injected by a proper apparatus every night, or alternate nights, for ſix or eight weeks. This was long ſince recommended, I think by Mr. Turner of Liverpool; and the reaſon it has not ſucceeded, I believe to have been owing to the imperfections of the joints of the common apparatus for injecting the ſmoke of tobacco, ſo that it did not paſs into the inteſtine, though it was ſuppoſed to do ſo, as I once obſerved. The ſmoke ſhould be received from the apparatus into a large bladder; and it may then be certainly injected like the common clyſter with ſufficient force; otherwiſe oiled leathers ſhould be nicely put round the joints of the machine; and a wet cloth round the injecting pipe to prevent the return of the ſmoke by the ſides of it. Clyſters of carbonated hydrogen gas, or of other factitious airs, might be tried.

Harrowgate water taken into the ſtomach, ſo as to induce ſix or ſeven ſtools every morning, for four or ſix weeks, is perhaps the moſt efficacious method in common uſe. A factitious Harrowgate water may be made probably of greater efficacy than the natural, by diſſolving one ounce of marine ſalt, (called bay ſalt) and half an ounce of magneſia Glauber's ſalt, (called Epſom ſalt, or bitter purging ſalt) in twenty-eight ounces of water. A quarter or half a pint of this is to be taken every hour; or two hours in the morning, till it operates, with a tea-ſpoonful of a ſolution of liver of ſulphur, which is to be made by putting an ounce of hepar ſulphuris into half a pint of water. See Claſs IV. 1. 2. 9.

13. *Dracunculus.* A thin worm brought from the coaſt of Guinea. It is found in the interſtices of the muſcles, and is many yards long; it

it makes a ſmall ulcer; which is cured by extracting an inch of the worm a day, and wrapping the extracted part ſlowly round a bit of tobacco pipe till next day, ſo as not to break it. I have twice ſeen long worms, like a thick horſe-hair, in water in July in this country, which appeared hard and jointed.

14. *Morpiones.* Crab-lice. The excrement of this animal ſtains the linen, and appears like diluted blood.

M. M. Spirit of wine. Mercurial ointment, ſhaving the part. Oil deſtroys other inſects, if they be quite covered with it, as the ticks on dogs, and would probably therefore deſtroy theſe. Its manner of operation is by ſtopping up or filling their ſpiracula, or breathing pores; a few drops of oil poured on a waſp, ſo as to cover it, deſtroys it in a few ſeconds.

15. *Pediculi.* Lice. There is ſaid to be a diſeaſe, in which theſe animals are propagated in indeſtructible numbers, ſo as to deſtroy the patient.

M. M. Cleanlineſs, mercurial ointment, ſtavis acria in powder, or the tincture of it in ſpirit of wine. Spirit of wine alone? Bath of oil?

## ORDO I.

*Increaſed Irritation.*

## GENUS V.

*With Increaſed Actions of the Organs of Senſe.*

### SPECIES.

1. *Viſus acrior.* Acuter ſight. There have been inſtances of people, who could ſee better in the gloom of the evening, than in the ſtronger light of the day; like owls, and bats, and many quadrupeds, and flying inſects. When the eye is inflamed, great light becomes eminently painful, owing to the increaſed irritative motions of the retina, and the conſequent increaſed ſenſation. Thus when the eye is dazzled with ſudden light, the pain is not owing to the motion of the iris; for it is the contraction of the iris, which relieves the pain from ſudden light; but to the too violent contractions of the moving fibres, which conſtitute the extremities of the optic nerve.

2. *Auditus acrior.* The irritative ideas of hearing are ſo increaſed in energy as to excite our attention. This happens in ſome diſeaſes of the epileptic kind, and in ſome fevers. Hence the whiſpering of the currents of air in a room, the reſpiration of the company, and noiſes before unperceived, become troubleſome; and ſounds louder than uſual, or unexpected, produce ſtarting, and convulſions.

M. M. Put oil of almonds into the ears. Stop the meatus auditorius with cotton wool. Set the feet of the patient's bed on cuſhions, or ſuſpend it by cords from the ceiling.

3. *Olfactus*

3. *Olfactus acrior.* The irritative ideas of ſmell from the increaſed action of the olfactive nerve excite our attention. Hence common odours are diſagreeable; and are perceived from variety of objects, which were before thought inodorous. Theſe are commonly believed to be hallucinations of the ſenſe.

M. M. Snuff ſtarch up the noſtrils.

4. *Guſtus acrior.* The irritative ideas of taſte, as of our own ſaliva, and even of the atmoſpheric air, excite our attention; and common taſtes are diſagreeably ſtrong.

M. M. Water. Mucilage. Vegetable acids. Scrape the tongue clean. Rub it with a ſage-leaf and vinegar.

5. *Tactus acrior.* The irritative ideas of the nerves of touch excite our attention: hence our own preſſure on the parts, we reſt upon, becomes uneaſy with univerſal ſoreneſs.

M. M. Soft feather-bed. Combed wool put under the patients, which rolls under them, as they turn, and thus prevents their friction againſt the ſheets. Drawers of ſoft leather. Plaſters of cerate with calamy.

6. *Senſus caloris acrior.* Acuter ſenſe of heat occurs in ſome diſeaſes, and that even when the perceptible heat does not appear greater than natural to the hand of another perſon. See Claſs I. 1. 2. See Sect. XIV. 8. All the above increaſed actions of our organs of ſenſe ſeparately or jointly accompany ſome fevers, and ſome epileptic diſeaſes; the patients complaining of the perception of the leaſt light, noiſes in their ears, bad ſmells in the room, and bad taſtes in their mouths, with ſoreneſs, numbneſs, and other uneaſy feels, and with diſagreeable ſenſations of general or partial heat.

7. *Senſus extenſionis acrior.* Acuter ſenſe of extenſion. The ſenſe

 of

of extenſion was ſpoken of in Sect. XIV. 7. and XXXII. 4. The defect of diſtention in the arterial ſyſtem is accompanied with faintneſs; and its exceſs with ſenſations of fulneſs, or weight, or preſſure. This however refers only to the vaſcular muſcles, which are diſtended by their appropriated fluids; but the longitudinal muſcles are alſo affected by different quantities of extenſion, and become violently painful by the exceſs of it.

Theſe pains of muſcles and of membranes are generally divided into acute and dull pains. The former are generally owing to increaſe of extenſion, as in pricking the ſkin with a needle; and the latter generally to defect of extenſion, as in cold head-aches; but if the edge of a knife, or point of a pin, be gradually preſſed againſt the fibres of muſcles or membranes, there would ſeem to be three ſtates or ſtages of this extenſion of the fibres; which have acquired names according to the degree or kind of ſenſation produced by the extenſion of them; theſe are 1. titillation or tickling. 2. itching, and the 3. ſmarting; as deſcribed below. See Sect. XIV. 9.

8. *Titillatio.* Tickling is a pleaſureable pain of the ſenſe of extenſion above mentioned, and therefore excites laughter; as deſcribed in Sect. XXXIV. 1. 4. The tickling of the noſtrils, which precedes the efforts of ſneezing, is owing to the increaſed irritation occaſioned by external ſtimulus; and is attended with a pleaſureable ſenſation in conſequence of the increaſed action of the part. When this action is exerted in a greater degree, the ſenſation becomes painful, and the convulſion of ſneezing enſues; as the pain in tickling the ſoles of the feet of children is relieved by laughter.

A lady after a bruiſe on her noſe by a fall was affected with inceſſant ſneezing, and relieved by ſnuffing ſtarch up her noſtrils. Perpetual ſneeezings in the meaſles, and in catarrhs from cold, are owing to the ſtimulus of the ſaline part of the mucous effuſion on the membrane of the noſtrils. See Claſs II. 1. 1. 2.

9. *Pruritus.*

9. *Pruritus.* Itching ſeems to be a greater degree of titillation, and to be owing to the ſtimulus of ſome acrid material, as the matter of the itch; or of the herpes on the ſcrotum, and about the anus; or from thoſe univerſal eruptions, which attend ſome elderly people, who have drank much vinous ſpirit. It occurs alſo, when inflammations are declining, as in the healing of bliſters, or in the cure of ophthalmia, as the action of the veſſels is yet ſo great as to produce ſenſation; which, like the titillations that occaſion laughter, is perpetually changing from pleaſure to pain.

When the natural efforts of ſcratching do not relieve the pain of itching, it ſometimes increaſes ſo as to induce convulſions and madneſs. As in the furor uterinus, and ſatyriaſis, and in the ſphincter ani and ſcrotum. See Claſs II. 1. 4. 14. IV. 2. 2. 6.

M. M. Warm bath. Fomentation. Alcohol externally. Poultice. Oiled ſilk. Mercurial ointments on ſmall ſurfaces at once. See Claſs II. 1. 4. 12. Solutions of lead on ſmall ſurfaces at once.

10. *Dolor urens.* Smarting follows the edge of a knife in making a wound, and ſeems to be owing to the diſtention of a part of a fibre, till it breaks. A ſmarting of the ſkin is liable to affect the ſcars left by herpes or ſhingles; and the callous parts of the bottoms of the feet; and around the baſes of corns on the toes; and frequently extends after ſciatica along the outſide of the thigh, and of the leg, and part of the foot. All theſe may be owing to the ſtimulus of extenſion, by blood or ſerum being forced into veſſels nearly coaleſced.

M. M. Emplaſtrum de minio put like a bandage on the part. Warm fomentation. Oil and camphor rubbed on the part. Oil-ſilk covering. A bliſter on the part. Ether, or alcohol, ſuffered to evaporate on the part.

11. *Conſternatio.* Surpriſe. As our eyes acquaint us at the ſame time with leſs than half of the objects, which ſurround us, we have learned

learned to confide much in the organ of hearing to warn us of approaching dangers. Hence it happens, that if any ſound ſtrikes us, which we cannot immediately account for, our fears are inſtantly alarmed. Thus in great debility of body, the loud clapping of a door, or the fall of a fire-ſhovel, produces alarm, and ſometimes even convulſions; the ſame occurs from unexpected ſights, and in the dark from unexpected objects of touch.

In theſe caſes the irritability is leſs than natural, though it is erroneouſly ſuppoſed to be greater; and the mind is buſied in exciting a train of ideas inattentive to external objects; when this train of ideas is diſſevered by any unexpected ſtimulus, ſurpriſe is excited; as explained in Sect. XVII. 3. 7: and XVIII. 17: then as the ſenſibility in theſe caſes is greater, fear becomes ſuperadded to the ſurpriſe; and convulſions in conſequence of the pain of fear. See Sect. XIX. 2.

The proximate cauſe of ſurpriſe is the increaſed irritation induced by ſome violent ſtimulus, which diſſevers our uſual trains of ideas; but in diſeaſes of inirritability the frequent ſtarting or ſurpriſe from ſounds not uncommon, but rather louder than uſual, as the clapping of a door, ſhews, that the attention of the patient to a train of ſenſitive ideas was previouſly ſtronger than natural, and indicates an incipient delirium; which is therefore worth attending to in febrile diſeaſes.

ORDO

## ORDO II.

*Decreaſed Irritation.*

## GENUS I.

*With decreaſed Action of the Sanguiferous Syſtem.*

THE reader ſhould be here apprized, that the words ſtrength and debility, when applied to animal motions, may properly expreſs the quantity of reſiſtance ſuch motions may overcome; but that, when they are applied to mean the ſuſceptibility or inſuſceptibility of animal fibres to motion, they become metaphorical terms; as in Sect. XII. 2. 1. and would be better expreſſed by the words activity and inactivity.

There are three ſources of animal inactivity; firſt, the defect of the natural quantity of ſtimulus on thoſe fibres, which have been accuſtomed to perpetual ſtimulus; as the arterial and ſecerning ſyſtems. When their accuſtomed ſtimulus is for a while intermitted, as when ſnow is applied to the ſkin of the hands, an accumulation of ſenſorial power is produced; and then a degree of ſtimulus, as of heat, ſomewhat greater than that at preſent applied, though much leſs than the natural quantity, excites the veſſels of the ſkin into violent action. We muſt obſerve, that a deficiency of ſtimulus in thoſe fibres, which are not ſubject to perpetual ſtimulus, as the locomotive muſcles, is not ſucceeded by accumulation of ſenſorial power; theſe therefore are more liable to become permanently inactive after a diminution of ſtimulus; as in ſtrokes of the palſy, this may be called inactivity from defect of ſtimulus.

2. A ſecond ſource of animal inactivity exiſts, when the ſenſorial power in any part of the ſyſtem has been previouſly exhauſted by violent

lent ſtimuli; as the eyes after long expoſure to great light; or the ſtomach, to repeated ſpirituous potation; this may be termed inactivity from exhauſtion of ſenſorial power. See Sect. XII. 2. 1.

3. But there is a third ſource of inactivity owing to the deficient production of ſenſorial power in the brain; and hence ſtimuli ſtronger than natural are required to produce the accuſtomed motions of the arterial ſyſtem; in this caſe there is no accumulation of ſenſorial power produced; as in the inactivity owing to defect of ſtimulus; nor any previous exhauſtion of it, as in the inactivity owing to exceſs of ſtimulus.

This third kind of inactivity cauſes many of the diſeaſes of this genus; which are therefore in general to be remedied by ſuch medicines as promote a greater production of ſenſorial power in the brain; as the incitantia, conſiſting of wine, beer, and opium, in ſmall repeated quantities; and ſecondly of ſuch as ſimply ſtimulate the arterial and glandular ſyſtem into their natural actions; as ſmall repeated bliſters, ſpices, and eſſential oils. And laſtly the ſorbentia, which contribute to ſupply the more permanent ſtrength of the ſyſtem, by promoting the abſorption of nouriſhment from the ſtomach, and inteſtines; and of the ſuperfluous fluid, which attends the ſecretions.

## SPECIES.

1. *Febris inirritativa.* Inirritative fever. This is the typhus mitior, or nervous fever of ſome writers; it is attended with weak pulſe without inflammation, or ſymptoms of putridity, as they have been called. When the production of ſenſorial power in the brain is leſs than uſual, the pulſe becomes quick as well as weak; and the heart ſometimes trembles like the limbs of old age, or of enfeebled drunkards; and when this force of the contractions of the heart and arteries is diminiſhed, the blood is puſhed on with leſs energy, as well

as

as in leſs quantity, and thence its ſtimulus on their ſides is diminiſhed in a duplicate ratio. In compreſſions of the brain, as in apoplexy, the pulſe becomes ſlower and fuller; for in that diſeaſe, as in natural ſleep, the irritative motions of the heart and arteries are not diminiſhed, volition alone is ſuſpended or deſtroyed.

If the abſorption of the terminations of the veins is not equally impaired with the force of the heart and arteries, the blood is taken up by the veins the inſtant it arrives at their extremities; the capillary veſſels are left empty, and there is leſs reſiſtance to the current of the blood from the arteries; hence the pulſe becomes empty, as well as weak and quick; the veins of the ſkin are fuller than the arteries of it; and its appearance becomes pale, bluiſh, and ſhrunk. See Claſs II. 1. 3. 1.

When this pulſe perſiſts many hours, it conſtitutes the febris inirritativa, or typhus, or nervous fever, of ſome writers; it is attended with little heat, the urine is generally of a natural colour, though in leſs quantity; with great proſtration of ſtrength, and much diſturbance of the faculties of the mind. Its immediate cauſe ſeems to be a deficient ſecretion of the ſenſorial power from the inaction of the brain; hence almoſt the whole of the ſenſorial power is expended in the performance of the motions neceſſary to life, and little of it can be ſpared for the voluntary actions of the locomotive muſcles, or organs of ſenſe, ſee Claſs I. 2. 5. 3. Its more remote cauſe may be from a paralyſis or death of ſome other part of the body; as of the ſpleen, when a tumour is felt on the left ſide, as in ſome intermittents; or of the kidnies, when the urine continues pale and in ſmall quantity. Does the reviveſcence of theſe affected parts, or their torpor, recurring at intervals, form the paroxyſms of theſe fevers? and their permanent reviveſcence eſtabliſh the cure? See Claſs IV. 2. 1. 19.

M. M. Wine and opium in ſmall quantities repeated every three hours alternately; ſmall repeated bliſters; warm but freſh air; forbentia; nutrientia; transfusion of blood. Small electric shocks passed through the brain in all directions. Oxygene air?

2. *Paresis inirritativa.* Inirritative debility. A defective action of the irritative motions without increase of the frequency of the pulse. It continues three or four weeks like a fever, and then either terminates in health, or the patient sinks into one kind of apoplexy, and perishes. Many symptoms, which attend inirritative fevers, accompany this disease, as cold hands and feet at periodic times, scurf on the tongue, want of appetite, muddy urine, with pains of the head, and sometimes vertigo, and vomiting.

This disease differs from the inirritative fever by the pulse not being more frequent than in health. The want of appetite and of digestion is a principal symptom, and probably is the cause of the universal debility, which may be occasioned by the want of nourishment. The vertigo is a symptom of inirritability, as shewn in Class IV. 1. 2. 6. the muddy urine is owing to increased absorption from the bladder in consequence of the diminished cutaneous and cellular absorption, as in anasarca, explained in Sect. XXIX. 5. 1. and is therefore a consequence of the inirritability of that part of the system; the foul tongue is owing to an increased absorption of the thinner part of the mucus in consequence of the general deficiency of fluid, which should be absorbed by the skin and stomach. The sickness is owing to decreased action of the stomach, which is probably the primary disease, and is connected with the vertigo.

M. M. An emetic. Calomel, grains iv. once or twice. Then a blister. Peruvian bark. Valerian. Columbo. Steel. Opium and wine in small quantities, repeated alternately every three hours. Small electric percussions through the stomach.

3. *Somnus interruptus.* Interrupted sleep. In some fevers, where the inirritability is very great, when the patient falls asleep, the pulse in

in a few minutes becomes irregular; and the patient awakes in great diſorder, and fear of dying, refuſing to ſleep again from the terror of this uneaſy ſenſation. In this extreme debility there is reaſon to believe, that ſome voluntary power during our waking hours is employed to aid the irritative ſtimuli in carrying on the circulation of the blood through the lungs; in the ſame manner as we uſe voluntary exertions, when we liſten to weak ſounds, or wiſh to view an object by a ſmall light; in ſleep volition is ſuſpended, and the deficient irritation alone is not ſufficient to carry on the pulmonary circulation. This explanation ſeems the moſt probable one, becauſe in caſes of apoplexy the irritative motions of the arterial ſyſtem do not ſeem to be impaired, nor in common ſleep. See Incubus III. 2. 1. 13.

M. M. Opium in very ſmall doſes, as three drops of laudanum. A perſon ſhould watch the patient, and awaken him frequently; or he ſhould meaſure the time between ſlumber and ſlumber by a ſtop-watch, and awaken the patient a little before he would otherwiſe awake; or he ſhould keep his finger on the pulſe, and ſhould forcibly awaken him, as ſoon as it becomes irregular, before the diſorder of the circulation becomes ſo great as to diſturb him. See Claſs I. 2. 1. 9. and Sect. XXVII. 2.

4. *Syncope.* Fainting conſiſts in the decreaſed action of the arterial ſyſtem; which is ſometimes occaſioned by defect of the ſtimulus of diſtention, as after veneſection, or tapping for the dropſy. At other times it ariſes from great emotions of the mind, as in ſudden joy or grief. In theſe caſes the whole ſenſorial power is exerted on theſe intereſting ideas, and becomes exhauſted. Thus during great ſurpriſe or fear the heart ſtops for a time, and then proceeds with throbbing and agitation; and ſometimes the vital motions become ſo deranged, as never to recover their natural ſucceſſive action; as when children have been frightened into convulſions. See Sect. XII. 7. 1.

Miſs ——, a young lady of Stafford, in travelling in a chaiſe was ſo affected by ſeeing the fall of a horſe and poſtillion, in going down a hill, though the carriage was not overturned, that ſhe fainted away, and then became convulſed, and never ſpoke afterwards; though ſhe lived about three days in ſucceſſive convulſions and ſtupor.

5. *Hæmorrhagia venoſa.* A bleeding from the capillaries ariſing from defect of venous abſorption, as in ſome of thoſe fevers commonly termed putrid. When the blood ſtagnates in the cellular membrane, it produces petechiæ from this torpor or paralyſis of the abſorbent mouths of the veins. It muſt be obſerved, that thoſe people who have diſeaſed livers, are more liable to this kind of hæmorrhages, as well as to the hæmorrhagia arterioſa; the former, becauſe patients with diſeaſed livers are more ſubject to paralytic complaints in general, as to hemiplegia, and to dropſy, which is a paralyſis of the lymphatics; and the latter is probably owing to the delay of the circulation in the vena porta by the torpor of this hepatic veſſel, when the liver is not much enlarged; and to its preſſure on the vena cava, when it is much enlarged.

M. M. Vitriolic acid, opium, ſteel, bark. Sponge bound on the part. Steel diſſolved in ſpirit of wine externally. Flour.

6. *Hæmorrhois cruenta.* In the bleeding piles the capillary veſſels of the rectum become diſtended and painful from the defect of the venous abſorption of the part, and at length burſt; or the mucous glands are ſo dilated as to give a paſſage to the blood; it is ſaid to obſerve lunar periods.

M. M. Veneſection, poultices, cathartics, ſpice, cold bath, and ſorbentia. External compreſſion by applying lint, ſponge, or cotton. Internal compreſſion by applying a bit of candle ſmeared with mercurial ointment. Strangulate the tumid piles with a ſilk ſtring. Cut them off. See Claſs I. 2. 3. 22.

Mrs.

Mrs. —— had for twelve or fifteen years, at intervals of a year or lefs, a bleeding from the rectum without pain; which however ftopped fpontaneoufly after fhe became weakened, or by the ufe of injections of brandy and water. Lately the bleeding continued above two months, in the quantity of many ounces a day, till fhe became pale and feeble to an alarming degree. Injections of folutions of lead, of bark and falt of fteel, and of turpentine, with fome internal aftringents, and opiates, were ufed in vain. An injection of the fmoke of tobacco, with ten grains of opium mixed with the tobacco, was ufed, but without effect the two firft times on account of the imperfection of the machine; on the third time it produced great ficknefs, and vertigo, and nearly a fainting fit; from which time the blood entirely ftopped. Was this owing to a fungous excrefcence in the rectum; or to a blood-veffel being burft from the difficulty of the blood paffing through the vena porta from fome hepatic obftruction, and which had continued to bleed fo long? Was it ftopped at laft by the fainting fit? or by the ftimulus of the tobacco?

7. *Hæmorrhagia renum.* Hæmorrhage from the kidnies, when attended with no pain, is owing to defect of venous abforption in the kidney. When attended with pain on motion, it is owing to a bit of gravel in the ureter or pelvis of the kidney; which is a much more frequent difeafe than the former. See Sect. XXVII. 1.

M. M. 1. Venefection in fmall quantity, calomel, bark, fteel, an opiate; cold immerfion up to the navel, the upper part of the body being kept cloathed. Neville-Holt water. 2. Alcalized water aerated. Much diluent liquids. Cool drefs. Cool bed-room.

Cows are much fubject to bloody urine, called foul water by the farmers; in this difeafe about fixty grains of opium with or without as much ruft of iron, given twice a day, in a ball mixed with flour and water, or diffolved in warm water, or warm ale, is, I believe, an efficacious remedy, to which however ſhould be added about two quarts of barley or oats twice a day, and a cover at night, if the weather be cold.

8. *Hæmorrhagia Hepatis.* Hæmorrhage from the liver. It ſometimes happens in thoſe, who have the gutta roſea, or paralytic affections owing to diſeaſed livers induced by the potation of fermented liquors, that a great diſcharge of black viſcid blood occaſionally comes away by ſtool, and ſometimes by vomiting: this the ancients called Melancholia, black bile. If it was bile, a ſmall quantity of it would become yellow or green on dilution with warm water, which was not the caſe in one experiment which I tried; it muſt remain ſome time in the inteſtines from its black colour, when it paſſes downwards, and probably comes from the bile-ducts, and is often a fatal ſymptom. When it is evacuated by vomiting it is leſs dangerous, becauſe it ſhews greater remaining irritability of the inteſtinal canal, and is ſometimes ſalutary to thoſe who have diſeaſed livers.

M. M. An emetic. Rhubarb, ſteel, wine, bark.

9. *Hæmoptoe venoſa.* Venous hæmoptoe frequently attends the beginning of the hereditary conſumptions of dark-eyed people; and in others, whoſe lungs have too little irritability. Theſe ſpittings of blood are generally in very ſmall quantity, as a tea-ſpoonful; and return at firſt periodically, as about once a month; and are leſs dangerous in the female than in the male ſex; as in the former they are often relieved by the natural periods of the menſes. Many of theſe patients are attacked with this pulmonary hæmorrhage in their firſt ſleep; becauſe in feeble people the power of volition is neceſſary, beſides that of irritation, to carry on reſpiration perfectly; but, as volition is ſuſpended during ſleep, a part of the blood is delayed in the veſſels of the lungs, and in conſequence effuſed, and the patient awakes from the diſagreeable ſenſation. See Claſs I. 2. 1. 3. II. 1. 6. 6. III. 2. 1. 10.

M. M. Wake

M. M. Wake the patient every two or three hours by an alarum clock. Give half a grain of opium at going to bed, or twice a day. Onions, garlic, ſlight chalybeates. Iſſues. Leeches applied once a fortnight or month to the hemorrhoidal veins to produce a new habit. Emetics after each period of hæmoptoe, to promote expectoration, and diſlodge any effuſed blood, which might by remaining in the lungs produce ulcers by its putridity. A hard bed, to prevent too ſound ſleep. A periodical emetic or cathartic once a fortnight.

10. *Palpitatio cordis.* The palpitation of the heart frequently attends the hæmoptoe above mentioned; and conſiſts in an ineffectual exertion of the heart to puſh forwards its contents in due time, and with due force. The remote cauſe is frequently ſome impediment to the general circulation; as the torpor of the capillaries in cold paroxyſms of fever, or great adheſions of the lungs. At other times it ariſes from the debility of the action of the heart owing to the deficient ſenſorial power of irritation or of aſſociation, as at the approach of death.

In both theſe caſes of weak exertion the heart feels large to the touch, as it does not completely empty itſelf at each contraction; and on that account contracts more frequently, as deſcribed in Sect. XXXII. 2. 2. Another kind of palpitation may ſometimes ariſe from the retrograde motions of the heart, as in fear. See Claſs I. 3. 1. 2. and IV. 3. 1. 6.

11. *Menorrhagia.* Continued flow of the catamenia. The monthly effuſion of blood from the uterus or vagina is owing to a torpor of the veins of thoſe membranes in conſequence of the defect of venereal ſtimulus; and in this reſpect reſembles the mucus diſcharged in the periodical venereal orgaſm of the female quadupeds, which are ſecluded from the males. The menorrhagia, or continued flow of

 this

this discharge, is owing to a continued defect of the venous absorption of the membranes of the uterus or vagina. See Class IV. 2. 4. 7.

M. M. Venesection in small quantity. A cathartic. Then opium, a grain every night. Steel. Bark. A blister. Topical aspersion with cold water, or cold vinegar.

12. *Dysmenorrhagia.* A difficulty of menstruation attended with pain. In this complaint the torpor of the uterine vessels, which precedes menstruation, is by sympathy accompanied with a torpor of the lumbar membranes, and consequent pain; and frequently with cold extremities, and general debility. The small quantity and difficulty of the discharge is owing to arterial inactivity, as in chlorosis. Whence it happens, that chalybeate medicines are of efficacy both to stop or prevent too great menstruation, and to promote or increase deficient menstruation; as the former is owing to inirritability of the veins, and the latter of the arteries of the uterus. See Article IV. 2. 6. in the Materia Medica.

M. M. Opium, steel, pediluvium. Warm bath.

13. *Lochia nimia.* Too great discharge after delivery. In that unnatural practice of some hasty accoucheurs of introducing the hand into the uterus immediately after the delivery of the child, and forcibly bringing away the placenta, it frequently happens, that a part of it is left behind; and the uterus, not having power to exclude so small a portion of it, is prevented from complete contraction, and a great hæmorrhage ensues. In this circumstance a bandage with a thick compress on the lower part of the belly, by appressing the sides of the uterus on the remaining part of the placenta, is likely to check the hæmorrhage, like the application of a pledget of any soft substance on a bleeding vessel.

In other cases the lochia continues too long, or in too great quantity, owing to the deficiency of venous absorption.

M. M. An

M. M. An enema. An opiate. A bliſter. Slight chalybeates. Peruvian bark. Clothes dipped in cold vinegar and applied externally. Bandages on the limbs to keep more blood in them for a time have been recommended.

14. *Abortio ſpontanea.* Some delicate ladies are perpetually liable to ſpontaneous abortion, before the third, or after the ſeventh, month of geſtation. From ſome of theſe patients I have learnt, that they have awakened with a ſlight degree of difficult reſpiration, ſo as to induce them to riſe haſtily up in bed; and have hence ſuſpected, that this was a tendency to a kind of aſthma, owing to a deficient abſorption of blood in the extremities of the pulmonary or bronchial veins; and have concluded from thence, that there was generally a deficiency of venous abſorption; and that this was the occaſion of their frequent abortion. Which is further countenanced, where a great ſanguinary diſcharge precedes or follows the excluſion of the fetus.

M. M. Opium, bark, chalybeates in ſmall quantity. Change to a warmer climate. I have directed with ſucceſs in four caſes half a grain of opium twice a day for a fortnight, and then a whole grain twice a day during the whole geſtation. One of theſe patients took beſides twenty grains of Peruvian bark for ſeveral weeks. By theſe means being exactly and regularly perſiſted in, a new habit became eſtabliſhed, and the uſual miſcarriages were prevented.

Miſcarriages more frequently happen from eruptive fevers, and from rheumatic ones, than from other inflammatory diſeaſes. I ſaw a moſt violent pleuriſy and hepatitis cured by repeated veneſection about a week or ten days before parturition; yet another lady whom I attended, miſcarried at the end of the chicken pox, with which her children were at the ſame time affected. Miſcarriages towards the termination of the ſmall pox are very frequent, yet there have been a few inſtances of children, who have been born with the eruption on them. The blood in the ſmall pox will not inoculate that diſeaſe, if taken before the commencement of the ſecondary fever; as ſhewn in Sect. XXXIII. 2. 10. becauſe the contagious matter is not yet formed, but after it has been oxygenated through the cuticle in the puſtules, it becomes contagious; and if it be then abſorbed, as in the ſecondary fever, the blood of the mother may become contagious, and infect the child. The ſame mode of reaſoning is applicable to the chicken pox. See Claſs IV. 3. 1. 7.

15. *Scorbutus.* Sea-ſcurvy is cauſed by ſalt diet, the perpetual ſtimulus of which debilitates the venous and abſorbent ſyſtems. Hence the blood is imperfectly taken up by the veins from the capillaries, whence brown and black ſpots appear upon the ſkin without fever. The limbs become livid and edematous, and laſtly ulcers are produced from deficient abſorption. See Sect. XXXIII. 3. 2. and Claſs II. 1. 4. 13. For an account of the ſcurvy of the lungs, ſee Sect. XXVII. 2.

M. M. Freſh animal and vegetable food. Infuſion of malt. New beer. Sugar. Wine. Steel. Bark. Sorbentia. Opium?

16. *Vibices.* Extravaſations of blood become black from their being ſecluded from the air. The extravaſation of blood in bruiſes, or in ſome fevers, or after death in ſome patients, eſpecially in the parts which were expoſed to preſſure, is owing to the fine terminations of the veins having been mechanically compreſſed ſo as to prevent their abſorbing the blood from the capillaries, or to their inactivity from diſeaſe. The blood when extravaſated undergoes a chemical change before it is ſufficiently fluid to be taken up by the lymphatic abſorbents, and in that proceſs changes its colour to green and then yellow.

17. *Petechiæ.* Purple ſpots. Theſe attend fevers with great venous inirritability, and are probably formed by the inability of a ſingle termination of a vein, whence the correſponding capillary becomes

comes ruptured, and effuſes the blood into the cellular membrane round the inert termination of the vein. This is generally eſteemed a ſign of the putrid ſtate of the blood, or that ſtate contrary to the inflammatory one. As it attends ſome inflammatory diſeaſes which are attended with great inirritability, as in the confluent ſmall pox. But it alſo attends the ſcurvy, where no fever exiſts, and it therefore ſimply announces the inactivity of the terminations of ſome veins; and is thence indeed a bad ſymptom in fevers, as a mark of approaching inactivity of the whole ſanguiferous ſyſtem, or death. The blue colour of ſome children's arms or faces in very cold weather is owing in like manner to the torpor of the abſorbent terminations of the veins, whence the blood is accumulated in them, and ſometimes burſts them.

## ORDO II.

*Decreaſed Irritation.*

## GENUS II.

*Decreaſed Action of the Secerning Syſtem.*

THESE are always attended with decreaſe of partial, or of general heat; for as the heat of animal bodies is the conſequence of their various ſecretions, and is perpetually paſſing away into the ambient air, or other bodies in contact with them; when theſe ſecretions become diminiſhed, or ceaſe, the heat of the part or of the whole is ſoon diminiſhed, or ceaſes along with them.

### SPECIES.

1. *Frigus febrile.* Febrile coldneſs. There is reaſon to believe, that the beginning of many fever-fits originates in the quieſcence of ſome part of the abſorbent ſyſtem, eſpecially where they have been owing to external cold; but that, where the coldneſs of the body is not owing to a diminution of external heat, it ariſes from the inaction of ſome part of the ſecerning ſyſtem. Hence ſome parts of the body are hot whilſt other parts are cold; which I ſuppoſe gave occaſion to error in Martyn's Experiments; where he ſays, that the body is as hot in the cold paroxyſms of fevers as at other times.

After the ſenſorial power has been much diminiſhed by great preceding activity of the ſyſtem, as by long continued external heat, or violent exerciſe, a ſudden expoſure to much cold produces a torpor both greater in degree and over a greater portion of the ſyſtem, by ſubtracting

ſubtracting their accuſtomed ſtimulus from parts already much deprived of their irritability. Dr. Franklin in a letter to M. Duberge, the French tranſlator of his works, mentions an inſtance of four young men, who bathed in a cold ſpring after a day's harveſt work; of whom two died on the ſpot, a third on the next morning, and the other ſurvived with difficulty. Hence it would appear, that thoſe, who have to travel in intenſely cold weather, will ſooner periſh, who have previouſly heated themſelves much with drams, than thoſe who have only the ſtimulus of natural food; of which I have heard one well atteſted inſtance. See Article VII. 2. 3. Claſs III. 2. 1. 17.

*Frigus chronicum.* Permanent coldneſs. Coldneſs of the extremities, without fever, with dry pale ſkin, is a ſymptom of general debility, owing to the decreaſed action of the arterial ſyſtem, and of the capillary veſſels; whence the perſpirable matter is ſecreted in leſs quantity, and in conſequence the ſkin is leſs warm. This coldneſs is obſervable at the extremities of the limbs, ears, and noſe, more than in any other parts: as a larger ſurface is here expoſed to the contact of the air, or clothes, and thence the heat is more haſtily carried away.

The pain, which accompanies the coldneſs of the ſkin, is owing to the deficient exertion of the ſubcutaneous veſſels, and probably to the accumulation of ſenſorial power in the extremities of their nerves. See Sect. XII. 5. 3. XIV. 6. XXXII. 3. and Claſs I. 2. 4. 1.

M. M. A bliſter. Incitantia, nutrientia, ſorbentia. Exerciſe. Clothes. Fire. Joy. Anger.

2. *Pallor fugitivus.* The fugitive paleneſs, which accompanies the coldneſs of the extremities, is owing to a leſs quantity of blood paſſing through the capillaries of the ſkin in a given time; where the abſorbent power of the veins is at the ſame time much diminiſhed, a part of the blood lingers at their junction with the capillary arteries, and a bluiſh tinge is mixed with the paleneſs; as is ſeen in the looſe

ſkin

ſkin under the eye-lids, and is always a mark of temporary debility. See Claſs II. 1. 4. 4. Where the paleneſs of the ſkin is owing to the deficiency of red globules in the blood, it is joined with a yellowiſh tinge; which is the colour of the ſerum, with which the blood then abounds, as in chloroſis, and in torpor or paralyſis of the liver, and is often miſtaken for a ſuperabundance of bile.

A permanent paleneſs of the ſkin is owing to the coaleſcence of the minute arteries, as in old age. See Claſs I. 2. 2. 9. There is another ſource of paleneſs from the increaſed abſorption of the terminations of the veins, as when vinegar is applied to the lips. See Sect. XXVII. 1. and another from the retrograde motions of the capillaries and fine extremities of the arteries. See Claſs II. 3. 1. 1.

M. M. A bliſter, nutrientia, incitantia, exerciſe, oxygene gas.

3. *Pus parcius.* Diminiſhed pus. Dryneſs of ulcers. In the cold fits of fever all the ſecretions are diminiſhed, whether natural or artificial, as their quantity depends on the actions of the glands or capillaries, which then ſhare in the univerſal inaction of the ſyſtem. Hence the dryneſs of iſſues and bliſters in great debility, and before the approach of death, is owing to deficient ſecretion, and not to increaſed abſorption.

M. M. Opium, wine in very ſmall quantities, Peruvian bark.

4. *Mucus parcior.* Diminiſhed mucus. Dryneſs of the mouth and noſtrils. This alſo occurs in the cold fits of intermittents. In theſe caſes I have alſo found the tongue cold to the touch of the finger, and the breath to the back of one's hand, when oppoſed to it, which are very inauſpicious ſymptoms, and generally fatal. In fevers with inirritability it is generally eſteemed a good ſymptom, when the noſtrils and tongue become moiſt after having been previouſly dry; as it ſhews an increaſed action of the mucous glands of thoſe membranes, which were before torpid. And the contrary to this is the facies

I Hippocratica,

Hippocratica, or countenance ſo well deſcribed by Hippocrates, which is pale, cold, and ſhrunk; all which are owing to the inactivity of the ſecerning veſſels, the paleneſs from there being leſs red blood paſſing through the capillaries, the coldneſs of the ſkin from there being leſs ſecretion of perſpirable matter, and the ſhrunk appearance from there being leſs mucus ſecreted into the cells of the cellular membrane. See Claſs IV. 2. 4. 11.

M. M. Bliſters. Incitantia.

5. *Urina parcior pallida.* Paucity of pale urine, as in the cold fits of intermittents; it appears in ſome nervous fevers throughout the whole diſeaſe, and ſeems to proceed from a palſy of the kidnies; which probably was the cauſe of the fever, as the fever ſometimes ceaſes, when that ſymptom is removed: hence the ſtraw-coloured urine in this fever is ſo far ſalutary, as it ſhews the unimpaired action of the kidnies.

M. M. Balſams, eſſential oil, aſparagus, rhubarb, a bliſter. Cantharides internally.

6. *Torpor hepaticus.* Paucity of bile from a partial inaction of the liver; hence the bombycinous colour of the ſkin, grey ſtools, urine not yellow, indigeſtion, debility, followed by tympany, dropſy, and death.

This paralyſis or inirritability of the liver often deſtroys thoſe who have been long habituated to much fermented liquor, and have ſuddenly omitted the uſe of it. It alſo deſtroys plumbers, and houſe-painters, and in them ſeems a ſubſtitute for the colica ſaturnina. See Sect. XXX.

M. M. Aloe and calomel, then the bark, and chalybeates. Mercurial ointment rubbed on the region of the liver. Rhubarb, three or four grains, with opium half a grain to a grain twice a day. Equitation, warm bath for half an hour every day.

7. *Torpor*

7. *Torpor Páncreatis.* Torpor of the pancreas. I ſaw what I conjectured to be a tumour of the pancreas with indigeſtion, and which terminated in the death of the patient. He had been for many years a great conſumer of tobacco, inſomuch that he chewed that noxious drug all the morning, and ſmoaked it all the afternoon. As the ſecretion from the pancreas reſembles ſaliva in its general appearance, and probably in its office of aſſiſting digeſtion, by preventing the fermentation of the aliment; as would appear by the experiments of Pringle and Macbride; there is reaſon to ſuſpect, that a ſympathy may exiſt between the ſalivary and pancreatic glands; and that the perpetual ſtimulus of the former by tobacco might in proceſs of time injure the latter. See Tobacco, Article III. 2. 2.

8. *Torpor renis.* Inirritability or paralyſis of the kidnies is probably frequently miſtaken for gravel in them. Several, who have lived rather intemperately in reſpect to fermented or ſpirituous liquors, become ſuddenly ſeized about the age of ſixty, or later, with a total ſtoppage of urine; though they have previouſly had no ſymptoms of gravel. In theſe caſes there is no water in the bladder; as is known by the introduction of the catheter, of which thoſe made of elaſtic gum are ſaid to be preferable to metallic ones; or it may generally be known by the ſhape of the abdomen, either by the eye or hand. Bougies and catheters of elaſtic gum are ſold at N° 37, Red Lion-ſtreet, Holborn, London.

M. M. Electric ſhocks, warm bath. Emetics. See calculus renis, Claſs I. 1. 3. 9. When no gravel has been previouſly obſerved, and the patient has been a wine-drinker rather than an ale-drinker, the caſe is generally owing to inirritability of the tubuli uriniferi, and is frequently fatal. See Claſs I. 2. 4. 20.

9. ***Punctæ mucoſæ vultûs.*** Mucous ſpots on the face. Theſe are

owing

owing to the inactivity of the excretory ducts of the mucous glands; the thinner part of this ſecretion exhales, and the remainder becomes inſpiſſated, and lodges in the duct; the extremity of which becomes black by expoſure to the air.

M. M. They may be preſſed out by the finger-nails. Warm water. Ether frequently applied. Bliſter on the part?

10. *Maculæ cutis fulvæ.* Morphew or freckles. Tawny blotches on the ſkin of the face and arms of elderly people, and frequently on their legs after ſlight eryſipelas. The freckles on the face of younger people, who have red hair, ſeem to be a ſimilar production, and ſeem all to be cauſed by the coaleſcence of the minute arteries or capillaries of the part. In a ſcar after a wound the integument is only opake; but in theſe blotches, which are called morphew and freckles, the ſmall veſſels ſeem to have become inactive with ſome of the ſerum of the blood ſtagnating in them, from whence their colour. See Claſs III. 1. 2. 12.

M. M. Warm bathing. A bliſter on the part?

11. *Canities.* Grey hair. In the injection of the veſſels of animals for the purpoſes of anatomical preparations, the colour of the injected fluid will not paſs into many very minute veſſels; which nevertheleſs uncoloured water, or ſpirits, or quickſilver will permeate. The ſame occurs in the filtration of ſome coloured fluids through paper, or very fine ſand, where the colouring matter is not perfectly diſſolved, but only diffuſed through the liquid. This has led ſome to imagine, that the cauſe of the whiteneſs of the hair in elderly people may ariſe from the diminution, or greater tenuity, of the glandular veſſels, which ſecrete the mucus, which hardens into hair; and that the ſame difference of the tenuity of the ſecerning veſſels may poſſibly make the difference of colour of the ſilk from different ſilk-worms, which is of all ſhades from yellow to white.

But as the fecreted fluids are not the confequence of mechanical filtration, but of animal felection; we muft look out for another caufe, which muft be found in the decreafing activity of the glands, as we advance in life; and which affects many of our other fecretions as well as that of the mucus, which forms the hair. Hence grey hairs are produced on the faces of horfes by whatever injures the glands at their roots, as by corrofive blifters; and frequently on the human fubject by external injuries on the head; and fometimes by fevers. And as the grey colour of hair confifts in its want of tranfparency, like water converted into fnow; there is reafon to fuppofe, that a defect of fecreted moifture fimply may be the caufe of this kind of opacity, as explained in Cataracta, Clafs I. 2. 2. 13.

M. M. Whatever prevents the inirritability and infenfibility of the fyftem, that is, whatever prevents the approach of old age, will fo far counteract the production of grey hairs, which is a fymptom of it. For this purpofe in people, who are not corpulent, and perhaps in thofe who are fo, the warm bath twice or thrice a week is particularly ferviceable. See Sect. XXXIX. 5. 1. on the colours of animals, and Clafs I. 1. 2. 15.

12. *Callus.* The callous fkin on the hands and feet of laborious people is owing to the extreme veffels coalefcing from the perpetual preffure they are expofed to.

As we advance in life, the finer arteries lofe their power of action, and their fides grow together; hence the palenefs of the fkins of elderly people, and the lofs of that bloom, which is owing to the numerous fine arteries, and the tranfparency of the fkin, that enclofes them.

M. M. Warm bath. Paring the thick fkin with a knife. Smoothing it with a pumice ftone. Cover the part with oiled filk to prevent the evaporation of the perfpirable matter, and thus to keep it moift.

13. *Cataracta.*

13. *Cataracta* is an opacity of the cryſtalline lens of the eye. It is a diſeaſe of light-coloured eyes, as the gutta ſerena is of dark ones. On cutting off with ſciſſars the cornea of a calf's eye, and holding it in the palm of one's hand, ſo as to gain a proper light, the artery, which ſupplies nutriment to the cryſtalline humour, is eaſily and beautifully ſeen; as it riſes from the centre of the optic nerve through the vitreous humour to the cryſtalline. It is this point, where the artery enters the eye through the cineritious part of the optic nerve, (which is in part near the middle of the nerve,) which is without ſenſibility to light; as is ſhewn by fixing three papers, each of them about half an inch in diameter, againſt a wall about a foot diſtant from each other, about the height of the eye; and then looking at the middle one, with one eye, and retreating till you loſe ſight of one of the external papers. Now as the animal grows older, the artery becomes leſs viſible, and perhaps carries only a tranſparent fluid, and at length in ſome ſubjects I ſuppoſe ceaſes to be pervious; then it follows, that the cryſtalline lens, loſing ſome fluid, and gaining none, becomes dry, and in conſequence opake; for the ſame reaſon, that wet or oiled paper is more tranſparent than when it is dry, as explained in Claſs I. 1. 4. 1. The want of moiſture in the cornea of old people, when the exhalation becomes greater than the ſupply, is the cauſe of its want of tranſparency; and which like the cryſtalline gains rather a milky opacity. The ſame analogy may be uſed to explain the whiteneſs of the hair of old people, which loſes its pellucidity along with its moiſture. See Claſs I. 2. 2. 11.

M. M. Small electric ſhocks through the eye. A quarter of a grain of corroſive ſublimate of mercury diſſolved in brandy, or taken in a pill, twice a day for ſix weeks. Couching by depreſſion, or by extraction. The former of theſe operations is much to be preferred to the latter, though the latter is at this time ſo faſhionable, that a ſurgeon is almoſt compelled to uſe it, leſt he ſhould not be thought an expert operator. For depreſſing the cataract is attended with no pain, no danger, no confinement, and may be as readily repeated, if the crystalline should rise again to the centre of the eye. The extraction of the cataract is attended with considerable pain, with long confinement, generally with fever, always with inflammation, and frequently with irreparable injury to the iris, and consequent danger to the whole eye. Yet has this operation of extraction been trumpeted into universal fashion for no other reason but because it is difficult to perform, and therefore keeps the business in the hands of a few empyrics, who receive larger rewards, regardless of the hazard, which is encountered by the flattered patient.

A friend of mine returned yesterday from London after an absence of many weeks; he had a cataract in a proper state for the operation, and in spite of my earnest exhortation to the contrary, was prevailed upon to have it extracted rather than depressed. He was confined to his bed three weeks after the operation, and is now returned with the iris adhering on one side so as to make an oblong aperture; and which is nearly, if not totally, without contraction, and thus greatly impedes the little vision, which he possesses. Whereas I saw some patients couched by depression many years ago by a then celebrated empyric, Chevalier Taylor, who were not confined above a day or two, that the eye might gradually be accustomed to light, and who saw as well as by extraction, perhaps better, without either pain, or inflammation, or any hazard of losing the eye.

As the inflammation of the iris is probably owing to forcing the crystalline through the aperture of it in the operation of extracting it, could it not be done more safely by making the opening behind the iris and ciliary process into the vitreous humour? but the operation would still be more painful, more dangerous, and not more useful than that by depressing it.

14. *Innutritio ossium.* Innutrition of the bones. Not only the blood effused in vibices and petechiæ, or from bruises, as well as the blood

blood and new veſſels in inflamed parts, are reabſorbed by the increaſed action of the lymphatics; but the harder materials, which conſtitute the fangs of the firſt ſet of teeth, and the ends of exfoliating bones, and ſometimes the matter of chalk-ſtones in the gout, the coagulable lymph, which is depoſited on the lungs, or on the muſcles after inflammation of thoſe parts, and which frequently produces difficulty of breathing, and the pains of chronic rheumatiſm, and laſtly the earthy part of the living bones are diſſolved and abſorbed by the increaſed actions of this ſyſtem of veſſels. See Sect. XXXIII. 3. 1.

The earthy part of bones in this diſeaſe of the innutrition of them ſeems to ſuffer a ſolution, and reabſorption; while the ſecerning veſſels do not ſupply a ſufficient quantity of calcareous earth and phoſphoric acid, which conſtitute the ſubſtance of bones. As calcareous earth abounds every where, is the want of phoſphoric acid the remote cauſe? One cauſe of this malady is given in the Philoſophic Tranſactions, where the patient had been accuſtomed to drink large quantities of vinegar. Two caſes are deſcribed by Mr. Gouch. In one caſe, which I ſaw, a conſiderable quantity of calcareous earth, and afterwards of bone-aſhes, and of decoction of madder, and alſo of ſublimate of mercury, were given without effect. All the bones became ſoft, many of them broke, and the patient ſeemed to die from the want of being able to diſtend her cheſt owing to the ſoftneſs of the ribs.

M. M. Salt of urine, called ſal microcoſmicum, phoſphorated ſoda. Calcined hartſhorn. Bone-aſhes. Hard or petrifying water, as that of Matlock, or ſuch as is found in all limeſtone or marly countries. The calcareous earth in theſe waters might poſſibly be carried to the bones, as madder is known to colour them. Warm bath. Volatile or fixed alcali as a lotion on the ſpine, or eſſential oils.

The innutrition of the bones is often firſt to be perceived by the difficulty of breathing and palpitation of the heart on walking a little

faſter

faſter than uſual, which I ſuppoſe is owing to the ſoftneſs of the ends of the ribs adjoining to the ſternum; on which account they do not perfectly diſtend the cheſt, when they are raiſed by the pectoral and intercoſtal muſcles with greater force than uſual. After this the ſpine becomes curved both by the ſoftneſs of its vertebræ, and for the purpoſe of making room for the diſturbed heart. See Species 16 of this genus.

As theſe patients are pale and weak, there would ſeem to be a deficiency of oxygene in their blood, and in conſequence a deficiency of phoſphoric acid; which is probably produced by oxygene in the act of reſpiration.

Mr. Bonhome in the Chemical Annals, Auguſt, 1793, ſuppoſes the rickets to ariſe from the prevalence of vegetable or acetous acid, which is known to ſoften bones out of the body. Mr. Dettaen ſeems to have eſpouſed a ſimilar opinion, and both of them in conſequence give alcalies and teſtacea. If this theory was juſt, the ſoft bones of ſuch patients ſhould ſhew evident marks of ſuch acidity after death; which I believe has not been obſerved. Nor is it analogous to other animal facts, that nutritious fluids ſecreted by the fineſt veſſels of the body ſhould be ſo little animalized, as to retain acetous or vegetable acidity.

The ſucceſs attending the following caſe in ſo ſhort a time as a fortnight I aſcribed principally to the uſe of the warm bath; in which the patient continued for full half an hour every night, in the degree of heat, which was moſt grateful to her ſenſation, which might be I ſuppoſe about 94. Miſs ——, about ten years of age, and very tall and thin, has laboured under palpitation of her heart, and difficult breathing on the leaſt exerciſe, with occaſional violent dry cough, for a year or more, with dry lips, little appetite either for food or drink, and dry ſkin, with cold extremities. She has at times been occaſionally worſe, and been relieved in ſome degree by the bark. She began to bend forwards, and to lift up her ſhoulders. The former ſeemed

 owing

owing to a beginning curvature of the ſpine, the latter was probably cauſed to facilitate her difficult reſpiration.

M. M. She uſed the warm bath, as above related; which by its warmth might increaſe the irritability of the ſmalleſt ſeries of veſſels, and by ſupplying more moiſture to the blood might probably tend to carry further the materials, which form calcareous or bony particles, or to convey them in more dilute ſolution. She took twice a day twenty grains of extract of bark, twenty grains of ſoda phoſphorata, and ten grains of chalk, and ten of calcined hartſhorn mixed into a powder with ten drops of laudanum; with fleſh food both to dinner and ſupper; and port wine and water inſtead of the ſmall beer, ſhe had been accuſtomed to; ſhe lay on a ſofa frequently in a day, and occaſionally uſed a neck-ſwing

15. *Rachitis*. Rickets. The head is large, protuberant chiefly on the forepart. The ſmaller joints are ſwelled; the ribs depreſſed; the belly tumid, with other parts emaciated. This diſeaſe from the innutrition or ſoftneſs of the bones aroſe about two centuries ago; ſeems to have been half a century in an increaſing or ſpreading ſtate; continued about half a century at its height, or greateſt diffuſion; and is now nearly vaniſhed: which gives reaſon to hope, that the ſmall-pox, meaſles, and venereal diſeaſe, which are all of modern production, and have already become milder, may in proceſs of time vaniſh from the earth, and perhaps be ſucceeded by new ones! See the preceding ſpecies.

16. *Spinæ diſtortio*. Diſtortion of the ſpine is another diſeaſe originating from the innutrition or ſoftneſs of the bones. I once ſaw a child about ſix years old with palpitation of heart, and quickneſs of reſpiration, which began to have a curvature of the ſpine; I then doubted, whether the palpitation and quick reſpiration were the cauſe or conſequence of the curvature of the ſpine; ſuſpecting either that nature

nature had bent the ſpine outwards to give room to the enlarged heart; or that the malformation of the cheſt had compreſſed and impeded the movements of the heart. But a few weeks ago on attending a young lady about ten years old, whoſe ſpine had lately began to be diſtorted, with very great difficulty and quickneſs of reſpiration, and alarming palpitation of the heart, I convinced myſelf, that the palpitation and difficult reſpiration were the effect of the change of the cavity of the cheſt from the diſtortion of the ſpine; and that the whole was therefore a diſeaſe of the innutrition or ſoftneſs of the bones.

For on directing her to lie down much in the day, and to take the bark, the diſtortion became leſs, and the palpitation and quick reſpiration became leſs at the ſame time. After this obſervation a neck-ſwing was directed, and ſhe took the bark, madder, and bone-aſhes; and ſhe continues to amend both in her ſhape and health.

Delicate young ladies are very liable to become awry at many boarding ſchools. This is occaſioned principally by their being obliged too long to preſerve an erect attitude, by ſitting on forms many hours together. To prevent this the ſchool-ſeats ſhould have either backs, on which they may occaſionally reſt themſelves; or deſks before them, on which they may occaſionally lean. This is a thing of greater conſequence than may appear to thoſe, who have not attended to it.

When the leaſt tendency to become awry is obſerved, they ſhould be adviſed to lie down on a bed or ſofa for an hour in the middle of the day for many months; which generally prevents the increaſe of this deformity by taking off for a time the preſſure on the ſpine of the back, and it at the ſame time tends to make them grow taller. Young perſons, when nicely meaſured, are found to be half an inch higher in a morning than at night; as is well known to thoſe, who inliſt very young men for ſoldiers. This is owing to the cartilages between the bones of the back becoming compreſſed by the weight of the head and ſhoulders on them during the day. It is the ſame preſſure

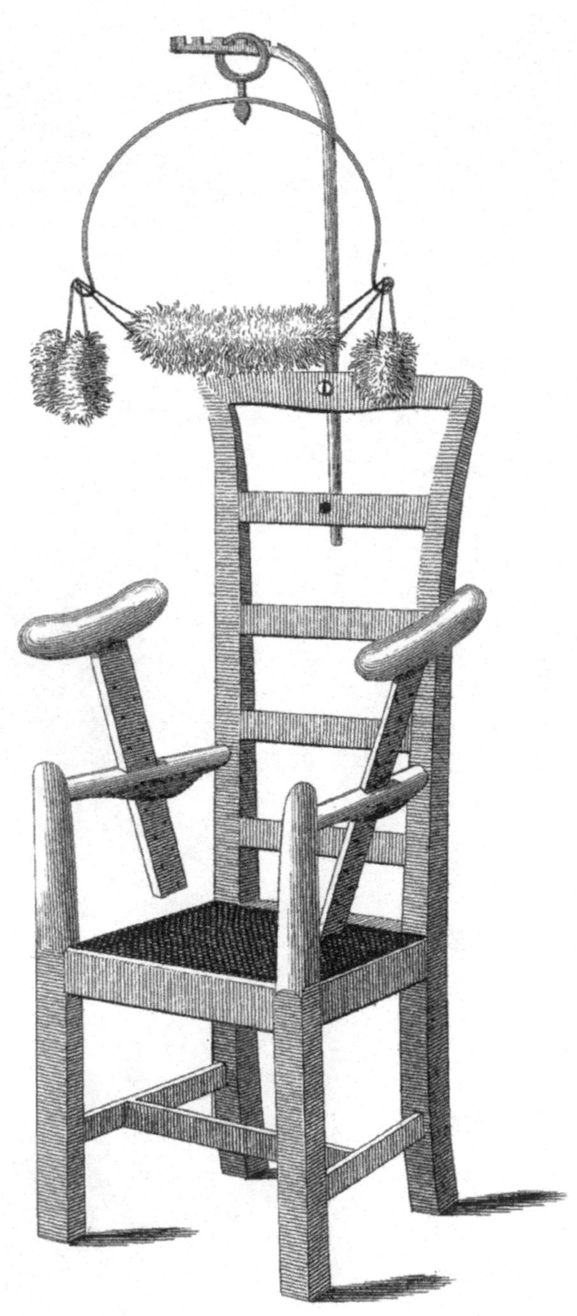

Barlow sculp.

preſſure which produces curvatures and diſtortions of the ſpine in growing children, where the bones are ſofter than uſual; and which may thus be relieved by an horizontal poſture for an hour in the middle of the day, or by being frequently allowed to lean on a chair, or to play on the ground on a carpet.

Young ladies ſhould alſo be directed, where two ſleep in a bed, to change every night, or every week, their ſides of the bed; which will prevent their tendency to ſleep always on the ſame ſide; which is not only liable to produce crookedneſs, but alſo to occaſion diſeaſes by the internal parts being ſo long kept in uniform contact as to grow together. For the ſame reaſon they ſhould not be allowed to ſit always on the ſame ſide of the fire or window, becauſe they will then be inclined too frequently to bend themſelves to one ſide.

Another great cauſe of injury to the ſhape of young ladies is from the preſſure of ſtays, or other tight bandages, which at the ſame time cauſe other diſeaſes by changing the form or ſituation of the internal parts. If a hard part of the ſtays, even a knot of the thread, with which they are ſewed together, is preſſed hard upon one ſide more than the other, the child bends from the ſide moſt painful, and thus occaſions a curvature of the ſpine. To counteract this effect ſuch ſtays, as have feweſt hard parts, and eſpecially ſuch as can be daily or weekly turned, are preferable to others.

Where frequent lying down on a ſofa in the day-time, and ſwinging frequently for a ſhort time by the hands or head, with looſe dreſs, do not relieve a beginning diſtortion of the back; recourſe may be had to a chair with ſtuffed moveable arms for the purpoſe of ſuſpending the weight of the body by cuſhions under the arm-pits, like reſting on crutches, or like the leading ſtrings of infants. From the top of the back of the ſame chair a curved ſteel bar may alſo project to ſuſpend the body occaſionally, or in part by the head, like the ſwing above mentioned. The uſe of this chair is more efficacious in ſtraightening the ſpine, than ſimply lying down horizontally; as it



not

not only takes off the preſſure of the head and ſhoulders from the ſpine, but at the ſame time the inferior parts of the body contribute to draw the ſpine ſtraight by their weight; or laſtly, recourſe may be had to a ſpinal machine firſt deſcribed in the Memoires of the academy of ſurgery in Paris, Vol. III. p. 600, by M. Le Vacher, and ſince made by Mr. Jones, at N° 6, North-ſtreet, Tottenham-court Road, London, which ſuſpends the head, and places the weight of it on the hips. This machine is capable of improvement by joints in the bar at the back of it, to permit the body to bend forwards without diminiſhing the extenſion of the ſpine.

The objections of this machine of M. Vacher, which is made by Mr. Jones, are firſt, that it is worn in the day-time, and has a very unſightly appearance. Mr. Jones has endeavoured to remedy this, by taking away the curved bar over the head, and ſubſtituting in its place a forked bar, riſing up behind each ear, with webs faſtened to it, which paſs under the chin and occiput. But this is not an improvement, but a deterioration of M. Vacher's machine, as it prevents the head from turning with facility to either ſide. Another objection is, that its being worn, when the muſcles of the back are in action, it is rather calculated to prevent the curvature of the ſpine from becoming greater, than to extend the ſpine, and diminiſh its curvature.

For this latter purpoſe I have made a ſteel bow, as deſcribed in the annexed plate, which receives the head longitudinally from the forehead to the occiput; having a fork furniſhed with a web to ſuſtain the chin, and another to ſuſtain the occiput. The ſummit of the bow is fixed by a ſwivel to the board going behind the head of the bed above the pillow. The bed is to be inclined from the head to the feet about twelve or ſixteen inches. Hence the patient would be conſtantly ſliding down during ſleep, unleſs ſupported by this bow, with webbed forks, covered alſo with fur, placed beneath the chin, and beneath the occiput. There are alſo proper webs lined with fur for the hands to take hold off occaſionally, and alſo to go under the arms. By theſe

Barlow sculp.

these means I ſhould hope great advantage from gradually extending the ſpine during the inactivity of the muſcles of the back; and that it may be done without diſturbing the ſleep of the patient, and if this ſhould happen, the bow is made to open by a joint at the ſummit of it, ſo as to be inſtantly diſengaged from the neck by the hand of the wearer. This bow I have not yet had opportunity to make uſe of, but it may be had from Mr. Harriſon, whiteſmith, Bridge-gate, Derby.

It will be from hence eaſily perceived, that all other methods of confining or directing the growth of young people ſhould be uſed with great ſkill; ſuch as back-boards, or bandages, or ſtocks for the feet; and that their application ſhould not be continued too long at a time, leſt worſe conſequences ſhould enſue, than the deformity they were deſigned to remove. To this may be added, that the ſtiff erect attitude taught by ſome modern dancing maſters does not contribute to the grace of perſon, but rather militates againſt it; as is well ſeen in one of the prints in Hogarth's Analyſis of Beauty; and is exemplifyed by the eaſy grace of ſome of the antient ſtatues, as of the Venus de Medici, and the Antinous, and in the works of ſome modern artiſts, as in a beautiful print of Hebe feeding an Eagle, painted by Hamilton, and engraved by Eginton, and many of the figures of Angelica Kauffman.

Where the bone of one of the vertebræ of the back has been ſwelled on both ſides of it, ſo as to become protuberant, iſſues near the ſwelled part have been found of great ſervice, as mentioned in Species 18 of this genus. This has induced me to propoſe in curvatures of the ſpine, to put an iſſue on the outſide of the curve, where it could be certainly aſcertained, as the bones on the convex ſide of the curve muſt be enlarged; in one caſe I thought this of ſervice, and recommend the further trial of it.

In the tendency to curvature of the ſpine, whatever ſtrengthens the general conſtitution is of ſervice; as the uſe of the cold bath in the

 ſummer

ſummer months. This however requires ſome reſtriction both in reſpect to the degree of coldneſs of the bath, the time of continuing in it, and the ſeaſon of the year. Common ſprings, which are of forty-eight degrees of heat, are too cold for tender conſtitutions, whether of children or adults, and frequently do them great and irreparable injury. The coldneſs of river water in the ſummer months, which is about ſixty-eight degrees, or that of Matlock, which is about ſixty-eight, or of Buxton, which is eighty-two, are much to be preferred. The time of continuing in the bath ſhould be but a minute or two, or not ſo long as to occaſion a trembling of the limbs from cold. In reſpect to the ſeaſon of the year, delicate children ſhould certainly only bathe in the ſummer months; as the going frequently into the cold air in winter will anſwer all the purpoſes of the cold bath.

17. *Claudicatio coxaria.* Lameneſs of the hip. A nodding of the thigh-bone is ſaid to be produced in feeble children by the ſoftneſs of the neck or upper part of that bone beneath the cartilage; which is naturally bent, and in this diſeaſe bends more downwards, or nods, by the preſſure of the body; and thus renders one leg apparently ſhorter than the other. In other caſes the end of the bone is protruded out of its ſocket, by inflammation or enlargement of the cartilages or ligaments of the joint, ſo that it reſts on ſome part of the edge of the acetabulum, which in time becomes filled up. When the legs are ſtraight, as in ſtanding erect, there is no verticillary motion in the knee-joint; all the motion then in turning out the toes further than nature deſigned, muſt be obtained by ſtraining in ſome degree this head of the thigh-bone, or the acetabulum, or cavity, in which it moves. This has induced me to believe, that this misfortune of the nodding of the head by the bone, or partial diſlocation of it, by which one leg becomes ſhorter than the other, is ſometimes occaſioned by making very young children ſtand in what are called ſtocks; that is

with

with their heels together, and their toes quite out. Whence the ſocket of the thigh-bone becomes inflamed and painful, or the neck of the bone is bent downward and outwards.

In this caſe there is no expectation of recovering the ſtraightneſs of the end of the bone; but theſe patients are liable to another misfortune, that is, to acquire afterwards a diſtortion of the ſpine; for as one leg is ſhorter than the other, they ſink on that ſide, and in conſequence bend the upper part of their bodies, as their ſhoulders, the contrary way, to balance themſelves; and then again the neck is bent back again towards the lame ſide, to preſerve the head perpendicular; and thus the figure becomes quite diſtorted like the letter S, owing originally to the deficiency of the length of one limb. The only way to prevent this curvature of the ſpine is for the child to wear a high-heeled ſhoe or patten on the lame foot, ſo as to ſupport that ſide on the ſame level with the other, and thus to prevent a greater deformity.

I have this day ſeen a young lady about twelve, who does not limp or waddle in walking; but nevertheleſs, when ſhe ſtands or ſits, ſhe ſinks down towards her right ſide, and turns out that toe more than the other. Hence, both as ſhe ſits and ſtands, ſhe bends her body to the right; whence her head would hang a little over her right ſhoulder: but to replace this perpendicularly, ſhe lifts up her left ſhoulder and contracts the muſcles on that ſide of the neck; which are therefore become thicker and ſtronger by their continued action; but there is not yet any very perceptible diſtortion of the ſpine.

As her right toe is turned outward rather more than natural, this ſhews the diſeaſe to be in the hip-joint; becauſe, when the limb is ſtretched out, the toe cannot turn horizontally in the leaſt without moving the end of the thigh-bone; although when the knee is bent, the toe can be turned through one third or half of a circle by the rotation of the tibia and fibula of the leg round each other. Hence



if

if children are ſet in ſtocks with their heels touching each other as they ſit, and are then made to riſe up, till they ſtand erect, the ſocket or head of the thigh-bone becomes injured, eſpecially in thoſe children, whoſe bones are ſoft; and a ſhortneſs of that limb ſucceeds either by the bending of the neck of the thigh-bone, or by its getting out of the acetabulum; and a conſequent riſing of one ſhoulder, and a curvature of the ſpine is produced from ſo diſtant a cauſe.

M. M. An elaſtic cuſhion made of curled hair ſhould be placed under the affected hip, whenever ſhe ſits; or ſhould be fitted to the part by means of drawers, ſo that ſhe cannot avoid ſitting on it. A neck-ſwing, and lying down in the day, ſhould be occaſionally uſed to prevent or remove any curvature of the ſpine. The reſt as in Species 13 and 15 of this genus.

18. *Spina protuberans.* Protuberant ſpine. One of the bones of the ſpine ſwells, and riſes above the reſt. This is not an uncommon diſeaſe, and belongs to the innutrition of the bones, as the bone muſt become ſoft before it ſwells; which ſoftneſs is owing to defect of the ſecretion of phoſphorated calcareous earth. The ſwelling of the bone compreſſes a part of the brain, called the ſpinal marrow, within the cavity of the back-bones; and in conſequence the lower limbs become paralytic, attended ſometimes with difficulty of emptying the bladder and rectum.

M. M. Iſſues put on each ſide of the prominent bone are of great effect, I ſuppoſe, by their ſtimulus; which excites into action more of the ſenſorial powers of irritation and ſenſation, and thus gives greater activity to the vaſcular ſyſtem in their vicinity. The methods recommended in diſtortion of the ſpine are alſo to be attended to.

19. *Spina bifida.* Divided ſpine, called alſo Hydrorachitis, as well

as the Hydrocephalus externus, are probably owing in part to a defect of ossification of the spine and cranium; and that the collection of fluid beneath them may originate from the general debility of the system; which affects both the secerning, and absorbent vessels.

A curious circumstance, which is affirmed to attend the spina bifida, is, that on compressing the tumor with the hand gently, the whole brain becomes affected, and the patient falls asleep. I suppose the same must happen on compressing the hydrocephalus externus? See Sect. XVIII. 20.

20. *Ossis palati defectus.* A defect of the bone of the palate, which frequently accompanies a division of the upper lip, occurs before nativity; and is owing to the deficient action of the secerning system, from whence the extremities are not completed. From a similar cause I have seen the point of the tongue deficient, and one joint of the two least fingers, and of the two least toes, in the same infant; who was otherwise a fine girl. See Sect. XXXIX. 4. 4.

The operation for the hare-lip is described by many surgical writers; but there is a person in London, who makes very ingenious artificial palates; which prevents that defect of speech, which attends this malformation. This factitious palate consists of a thin plate of silver of the shape and form of the roof of the mouth; from the front edge to the back edge of this silver plate four or five holes are made in a straight line large enough for a needle to pass through them; on the back of it is then sewed a piece of sponge; which when expanded with moisture is nearly as large as the silver plate. This sponge is slipped through the division of the bone of the palate, so as to lie above it, while the silver plate covers the aperture beneath, and is suspended by the expanding sponge. This is removed every night and washed, and returned into its place in the morning;

morning; on this account it is convenient to have five or fix of them, for the fake of cleanlinefs. I have been more particular in defcribing this invention, as I do not know the name, or place of refidence, of the maker.

ORDO

## ORDO II.

*Decreased Irritation.*

## GENUS III.

*The decreased Action of the Absorbent System.*

SOME decrease of heat attends these diseases, though in a less degree than those of the last genus, because the absorbent system of glands do not generate so much heat in their healthy state of action as the secerning system of glands, as explained in Class I. 1. 3.

### SPECIES.

1. *Mucus faucium frigidus.* Cold mucus from the throat. Much mucus, of rather a saline taste, and less inspissated than usual, is evacuated from the fauces by hawking, owing to the deficient absorption of the thinner parts of it. This becomes a habit in some elderly people, who are continually spitting it out of their mouths; and has probably been brought on by taking snuff, or smoking tobacco; which by frequently stimulating the fauces have at length rendered the absorbent vessels less excitable by the natural stimulus of the saline part of the secretion, which ought to be reabsorbed, as soon as secreted.

M. M. A few grains of powder of bark frequently put into the mouth, and gradually diffused over the fauces. A gargle of barley water.

2. *Sudor frigidus.* The cold dampneſs of the hands of ſome people is cauſed by the deficient abſorption of perſpirable matter; the clammy or viſcid feel of it is owing to the mucous part being left upon the ſkin. The coldneſs is produced both by the decreaſed action of the abſorbent ſyſtem, and by the evaporation of a greater quantity of the perſpirable matter into the air, which ought to have been abſorbed.

M. M. Waſh the hands in lime water, or with a ſmall quantity of volatile alcali in water.

3. *Catarrhus frigidus.* The thin diſcharge from the noſtrils in cold weather. The abſorbent veſſels become torpid by the diminution of external heat, ſooner than the ſecerning ones, which are longer kept warm by the circulating blood, from which they ſelect the fluid they ſecrete; whereas the abſorbent veſſels of the noſtrils drink up their fluids, namely the thin and ſaline part of the mucus, after it has been cooled by the atmoſphere. Hence the abſorbents ceaſing to act, and the ſecerning veſſels continuing ſome time longer to pour out the mucus, a copious thin diſcharge is produced, which trickles down the noſtrils in cold weather. This diſcharge is ſo acrid as to inflame the upper lip; which is owing to the neutral ſalts, with which it abounds, not being reabſorbed; ſo the tears in the fiſtula lacrymalis inflame the cheek. See Claſs I. 1. 2. 7.

4. *Expectoratio frigida.* Cold expectoration. Where the pulmonary abſorption is deficient, an habitual cough is produced, and a frequent expectoration of thin ſaline mucus; as is often ſeen in old enfeebled people. Though the ſtimulus of the ſaline fluid, which attends all ſecretions, is not ſufficient to excite the languid abſorbent veſſels to imbibe it; yet this ſaline part, together with the increaſed quantity of the whole of the ſecreted mucus, ſtimulates the branches of the bronchia, ſo as to induce an almoſt inceſſant cough to diſcharge it from the lungs. A ſingle grain of opium, or any other ſtimulant drug,

drug, as a wine-poſſet with ſpirit of hartſhorn, will cure this cold cough, and the cold catarrh of the preceding article, like a charm, by ſtimulating the torpid mouths of the abſorbents into action. Which has given riſe to an indiſcriminate and frequently pernicious uſe of the warm regimen in coughs and catarrhs of the warm or inflammatory kind, to the great injury of many.

M. M. Half a grain of opium night and morning promotes the abſorption of the more fluid and ſaline parts; and in conſequence thickens the mucus, and abates its acrimony. Warm diluent drink, wine-whey, with volatile alcali.

5. *Urina uberior pallida.* On being expoſed naked to cold air, or ſprinkled with cold water, a quantity of pale urine is ſoon diſcharged; for the abſorbents of the bladder become torpid by their ſympathy with thoſe of the ſkin; which are rendered quieſcent by the diminution of external heat; but the kidnies continue to ſecrete the urine, and as no part of it is abſorbed, it becomes copious and pale. This happens from a ſimilar cauſe in cold fits of agues; and in leſs degree to many debilitated conſtitutions, whoſe extremities are generally cold and pale. The great quantity of limpid water in hyſteric caſes, and in diabætes, belongs to Claſs I. 3. 1. 10. I. 3. 2. 6.

M. M. Tincture of cantharides, opium, alum, ſorbentia. Flannel ſhirt in cold weather. Animal food. Beer. Wine. Friction. Exerciſe. Fire.

6. *Diarrhœa frigida.* Liquid ſtools are produced by expoſing the body naked to cold air, or ſprinkling it with cold water, for the ſame reaſon as the laſt article.

But this diſeaſe is ſometimes of a dangerous nature; the inteſtinal abſorption being ſo impaired, that the aliment is ſaid to come away undiminiſhed in quantity, and almoſt unchanged by the powers of digeſtion, and is then called lientery.

The mucus of the rectum ſometimes comes away like pellucid hartſhorn jelly, and liquefies by heat like that, towards the end of inirritative fevers, which is owing to the thinner part of the mucus not being abſorbed, and thus reſembles the catarrh of ſome old people.

M. M. Opium, campechy wood, armenian bole. Bliſter. Flannel ſhirt in cold weather. Clyſters with opium. Friction on the bowels morning and night. Equitation twice a day.

7. *Fluor albus frigidus.* Cold fluor albus. In weak conſtitutions, where this diſcharge is pellucid and thin, it muſt proceed from want of abſorption of the mucous membrane of the vagina, or uterus, and not from an increaſed ſecretion. This I ſuſpect to be the moſt frequent kind of fluor albus; the former one deſcribed at Claſs I. 1. 2. 11. attends menſtruation, or is a diſcharge inſtead of it, and thus reſembles the venereal orgaſm of female quadrupeds. The diſcharge in this latter kind being more ſaline, is liable to excoriate the part, and thus produce ſmarting in making water; in its great degree it is difficult to cure.

M. M. Increaſe the evacuation by ſtool and by perſpiration, by taking rhubarb every night, about ſix or ten grains with one grain of opium for ſome months. Flannel ſhirt in winter. Balſam copaiva. Gum kino, bitters, chalybeates, friction over the whole ſkin with flannel morning and night. Partial cold bath, by ſprinkling the loins and thighs, or ſponging them with cold water. Mucilage, as iſinglaſs boiled in milk; blanc mange, hartſhorn jelly, are recommended by ſome. Tincture of cantharides ſometimes ſeems of ſervice given from ten to twenty drops or more, three or four times a day. A large plaſter of burgundy pitch and armenian bole, ſo as to cover the loins and lower part of the belly, is ſaid to have ſometimes ſucceeded by increaſing abſorption by its compreſſion in the manner of a bandage. A ſolution of metallic ſalts, as white vitriol,

vitriol, ſixty grains to a pint; or an infuſion of oak-bark may be injected into the vagina. Cold bath.

8. *Gonorrhœa frigida.* Cold gleet. Where the gleet is thin and pellucid, it muſt ariſe from the want of abſorption of the membranes of the urethra, rather than from an increaſed ſecretion from them. This I ſuppoſe to be a more common diſeaſe than that mentioned at Claſs I. 1. 2. 10.

M. M. Metallic injections, partial cold bath, internal method as in the fluor albus above deſcribed. Balſam of copaiva. Tincture of cantharides.

9. *Hepatis tumor.* The liver becomes enlarged from defect of the abſorption of mucus from its cells, as in anaſarca, eſpecially in feeble children; at the ſame time leſs bile is ſecreted from the torpid circulation in the vena portæ. And as the abſorbents, which reſume the thinner parts of the bile from the gall-bladder and hepatic ducts, are alſo torpid or quieſcent, the bile is more dilute, as well as in leſs quantity. From the obſtruction of the paſſage of the blood through the compreſſed vena porta theſe patients have tumid bellies, and pale bloated countenances; their paleneſs is probably owing to the deficiency of the quantity of red globules in the blood in conſequence of the inert ſtate of the bile.

Theſe ſymptoms in children are generally attended with worms, the dilute bile and the weak digeſtion not deſtroying them. In ſheep I have ſeen fleuke-worms in the gall-ducts themſelves among the dilute bile; which gall-ducts they eat through, and then produce ulcers, and the hectic fever, called the rot. See Claſs I. 1. 4. 10. and Article IV. 2. 6.

M. M. After a calomel purge, crude iron-filings are ſpecific in this diſeaſe in children, and the worms are deſtroyed by the returning

 acrimony

acrimony and quantity of the bile. A blifter on the region of the liver. Sorbentia, as worm-feed, fantonicum. Columbo. Bark.

10. *Chlorofis.* When the defect of the due action of both the abforbent and fecerning veffels of the liver affects women, and is attended with obftruction of the catamenia, it is called chlorofis; and is cured by the exhibition of fteel, which reftores by its fpecific ftimulus the abforbent power of the liver; and the menftruation, which was obftructed in confequence of debility, recurs.

Indigeftion, owing to torpor of the ftomach, and a confequent too great acidity of its contents, attend this difeafe; whence a defire of eating chalk, or marl. Sometimes a great quantity of pale urine is difcharged in a morning, which is owing to the inaction of the abforbents, which are diftributed on the neck of the bladder, during fleep. The fwelling of the ankles, which frequently attends chlorofis, is another effect of deficient action of the abforbent fyftem; and the pale countenance is occafioned by the deficient quantity of red globules of blood, caufed by the deficient quantity or acrimony of the bile, and confequent weaknefs of the circulation. The pulfe is fo quick in fome cafes of chlorofis, that, when attended with an accidental cough, it may be miftaken for pulmonary confumption. This quick pulfe is owing to the debility of the heart from the want of ftimulus occafioned by the deficiency of the quantity, and acrimony of the blood.

M. M. Steel. Bitters. Conftant moderate exercife. Friction with flannel all over the body and limbs night and morning. Rhubarb five grains, opium half a grain, every night. Flefh diet, with fmall beer, or wine and water. The difeafe continues fome months, but at length fubfides by the treatment above defcribed. A bath of about eighty degrees, as Buxton Bath, is of fervice; a colder bath may do great injury.

11. *Hydrocele.*

11. *Hydrocele.* Dropſy of the vagina teſtis. Dropſies have been divided into the incyſted and the diffuſed, meaning thoſe of the cellular membrane, the cells of which communicate with each other like a ſponge, and thoſe of any other cavity of the body. The collections of mucous fluids in the various cells and cavities of the body ariſe from the torpor of the abſorbent veſſels of thoſe parts. It is probable, that in dropſies attended with great thirſt the cutaneous abſorbents become paralytic firſt; and then from the great thirſt, which is thus occaſioned by the want of atmoſpheric moiſture, the abſorption of the fat enſues; as in fevers attended with great thirſt, the fat is quickly taken up. See Obeſitas I. 2. 3. 16. Some have believed, that the cellular and adipoſe membranes are different ones; as no fat is ever depoſited in the eye-lids or ſcrotum, both which places are very liable to be diſtended with the mucilaginous fluid of the anaſarca, and with air in Emphyſema. Sometimes a gradual abſorption of the accumulated fluid takes place, and the thinner parts being taken up, there remains a more viſcid fluid, or almoſt a ſolid in the part, as in ſome ſwelled legs, which can not eaſily be indented by the preſſure of the finger, and are called ſcorbutic. Sometimes the paralyſis of the abſorbents is completely removed, and the whole is again taken up into the circulation.

The Hydrocele is known by a tumor of the ſcrotum, which is without pain, gradually produced, with fluctuation, and a degree of pellucidity, when a candle is held behind it; it is the moſt ſimple incyſted dropſy, as it is not in general complicated with other diſeaſes, as aſcites with ſchirrous liver, and hydrocephalus internus, with general debility. The cure of this diſeaſe is effected by different ways; it conſiſts in diſcharging the water by an external aperture; and by ſo far inflaming the cyſt and teſticle, that they afterwards grow together, and thus prevent in future any ſecretion or effuſion of mucus; the diſeaſe is thus cured, not by the reviveſcence of the abſorbent power of the lymphatics, but by the prevention of ſecretion by the adheſion of the vagina to the teſtis. This I believe is performed with leſs pain, and is more certainly manageable by tapping, or difcharging the fluid by means of a trocar, and after the evacuation of it to fill the cyft with a mixture of wine and water for a few minutes till the neceffary degree of ftimulus is produced, and then to withdraw it; as recommended by Mr. Earle. See alfo Medical Commentaries by Dr. Duncan, for 1793.

12. *Hydrocephalus internus*, or dropfy of the ventricles of the brain, is fatal to many children, and fome adults. When this difeafe is lefs in quantity, it probably produces a fever, termed a nervous fever, and which is fometimes called a worm fever, according to the opinion of Dr. Gilchrift, in the Scots Medical effays. This fever is attended with great inirritability; as appears from the dilated pupils of the eyes, in which it correfponds with the dropfy of the brain. And the latter difeafe has its paroxyfms of quick pulfe, and in that refpect correfponds with other fevers with inirritability.

The hydrocephalus internus is diftinguifhed from apoplexy by its being attended with fever, and from nervous fever by the paroxyfms being very irregular, with perfect intermiffions many times in a day. In nervous fever the pain of the head generally affects the middle of the forehead; in hydrocephalus internus it is generally on one fide of the head. One of the earlieft criterions is the patient being uneafy on raifing his head from the pillow, and wifhing to lie down again immediately; which I fuppofe is owing to the preffure of the water on the larger trunks of the blood-veffels entering the cavity being more intolerable than on the fmaller ones; for if the larger trunks are compreffed, it muft inconvenience the branches alfo; but if fome of the fmall branches are compreffed only, the trunks are not fo immediately incommoded.

Blifters on the head, and mercurial ointment externally, with calomel internally, are principally recommended in this fatal difeafe. When the patient cannot bear to be raifed up in bed without great uneafinefs, uneaſineſs, it is a bad ſymptom. So I believe is deafneſs, which is commonly miſtaken for ſtupor. See Claſs I. 2. 5. 6. And when the dilatation of the pupil of either eye, or the ſquinting is very apparent, or the pupils of both eyes much dilated, it is generally fatal. As by ſtimulating one branch of lymphatics into inverted motion, another branch is liable to abſorb its fluid more haſtily; ſuppoſe ſtrong errhines, as common tobacco ſnuff to children, or one grain of turpeth mineral, (Hydrargyrus vitriolatus), mixed with ten or fifteen grains of ſugar, was gradually blown up the noſtrils? See Claſs I. 3. 2. 1. I have tried common ſnuff upon two children in this diſeaſe; one could not be made to ſneeze, and the other was too near death to receive advantage. When the mercurial preparations have produced ſalivation, I believe they may have been of ſervice, but I doubt their good effect otherwiſe. In one child I tried the tincture of Digitalis; but it was given with too timid a hand, and too late in the diſeaſe, to determine its effects. See Sect. XXIX. 5. 9.

As all the above remedies generally fail of ſucceſs, I think frequent, almoſt hourly, ſhocks of electricity from very ſmall charges might be paſſed through the head in all directions with probability of good event. And the uſe of the trephine, where the affected ſide can be diſtinguiſhed. See Strabiſmus, Claſs I. 2. 5. 4. When one eye is affected, does the diſeaſe exiſt in the ventricule of that ſide?

13. *Aſcites.* The dropſy of the cavity of the abdomen is known by a tenſe ſwelling of the belly; which does not ſound on being ſtruck like the tympany; and in which a fluctuation can be readily perceived by applying one hand expanded on one ſide, and ſtriking the tumour on the other.

Effuſions of water into large cavities, as into that of the abdomen or thorax, or into the ventricules of the brain or pericardium, are more difficult to be reabſorbed, than the effuſion of fluids into the cellular membrane; becauſe one part of this extenſive ſponge-like ſyſtem

 of

of cells, which connects all the ſolid parts of the body, may have its power of abſorption impaired, at the ſame time that ſome other part of it may ſtill retain that power, or perhaps poſſeſs it in an increaſed degree; and as all theſe cells communicate with each other, the fluid, which abounds in one part of it, can be transferred to another, and thus be reabſorbed into the circulation.

In the aſcites, cream of tartar has ſometimes been attended with ſucceſs; a dram or two drams are given every hour in a morning till it operates, and is to be repeated for ſeveral days; but the operation of tapping is generally applied to at laſt. Dr. Sims, in the Memoirs of the Medical Society of London, Vol. III. has lately propoſed, what he believes to be a more ſucceſsful method of performing this operation, by making a puncture with a lancet in the ſcar of the navel, and leaving it to diſcharge itſelf gradually for ſeveral days, without introducing a canula, which he thinks injurious both on account of the too ſudden emiſſion of the fluid, and the danger of wounding or ſtimulating the viſcera. This operation I have twice known performed with leſs inconvenience, and I believe with more benefit to the patient, than the common method.

After the patient has been tapped, ſome have tried injections into the cavity of the abdomen, but hitherto I believe with ill event. Nor are experiments of this kind very promiſing of ſucceſs. Firſt becauſe the patients are generally much debilitated, moſt frequently by ſpirituous potation, and have generally a diſeaſe of the liver, or of other viſcera. And ſecondly, becauſe the quantity of inflammation, neceſſary to prevent future ſecretion of mucus into the cavity of the abdomen, by uniting the peritoneum with the inteſtines or meſentery, as happens in the cure of the hydrocele, would I ſuppoſe generally deſtroy the patient, either immediately, or by the conſequence of ſuch adheſions.

This however is not the caſe in reſpect to the dropſy of the ovarium, or in the hydrocele.

14. *Hydrops*

14. *Hydrops thoracis.* The dropſy of the cheſt commences with loſs of fleſh, cold extremities, pale countenance, high coloured urine in ſmall quantity, and general debility, like many other dropſies. The patient next complains of numbneſs in the arms, eſpecially when elevated, with pain and difficulty of ſwallowing, and an abſolute impoſſibility of lying down for a few minutes, or with ſudden ſtarting from ſleep, with great difficulty of breathing and palpitation of his heart.

The numbneſs of the arms is probably owing more frequently to the increaſed action of the pectoral muſcles in reſpiration, whence they are leſs at liberty to perform other offices, than to the connexion of nerves mentioned in Sect. XXIX. 5. 2. The difficulty of ſwallowing is owing to the compreſſion of the œſophagus by the lymph in the cheſt; and the impoſſibility of breathing in an horizontal poſture originates from this, that if any parts of the lungs muſt be rendered uſeleſs, the inability of the extremities of them muſt be leſs inconvenient to reſpiration; ſince if the upper parts or larger trunks of the air-veſſels ſhould be rendered uſeleſs by the compreſſion of the accumulated lymph, the air could not gain admittance to the other parts, and the animal muſt immediately periſh.

If the pericardium is the principal ſeat of the diſeaſe, the pulſe is quick and irregular. If only the cavity of the thorax is hydropic, the pulſe is not quick nor irregular.

If one ſide is more affected than the other, the patient leans moſt that way, and has more numbneſs in that arm.

The hydrops thoracis is diſtinguiſhed from the anaſarca pulmonum, as the patient in the former cannot lie down half a minute; in the latter the difficulty of breathing, which occaſions him to riſe up, comes on more gradually; as the tranſition of the lymph in the cellular membranes from one part to another of it is ſlower, than that of the effuſed lymph in the cavity of the cheſt.

The hydrops thoracis is often complicated with fits of con-

 vulſive

vulſive breathing; and then it produces a diſeaſe for the time very ſimilar to the common periodic aſthma, which is perhaps owing to a temporary anaſarca of the lungs; or to an impaired venous abſorption in them. Theſe exacerbations of difficult breathing are attended with cold extremities, cold breath, cold tongue, upright poſture with the mouth open, and a deſire of cold air, and a quick, weak, intermittent pulſe, and contracted hands.

Theſe exacerbations recur ſometimes every two or three hours, and are relieved by opium, a grain every hour for two or three doſes, with ether about a dram in cold water; and ſeem to be a convulſion of the muſcles of reſpiration induced by the pain of the dyſpnea. As in Claſs III. 1. 1. 9.

M. M. A grain of dried ſquill, and a quarter of a grain of blue vitriol every hour for ſix or eight hours, unleſs it vomit or purge. A grain of opium. Bliſters. Calomel three grains every third day, with infuſion of ſenna. Bark. Chalybeates. Puncture in the ſide.

Can the fluctuation in the cheſt be heard by applying the ear to the ſide, as Hippocrates aſſerts? Can it be felt by the hand or by the patient before the diſeaſe is too great to admit of cure by the paracenteſis? Does this dropſy of the cheſt often come on after peripneumony? Is it ever cured by making the patient ſick by tincture of digitalis? Could it be cured, if on one ſide only, by the operation of puncture between the ribs, and afterwards by inflaming the cavity by the admiſſion of air for a time, like the cure of the hydrocele; the pleura afterwards adhering wholly to that lobe of the lungs, ſo as to prevent any future effuſion of mucus?

15. *Hydrops ovarii.* Dropſy of the ovary is another incyſted dropſy, which ſeldom admits of cure. It is diſtinguiſhed from aſcites by the tumour and pain, eſpecially at the beginning, occupying one ſide, and the fluctuation being leſs diſtinctly perceptible. When it happens to young ſubjects it is leſs liable to be miſtaken for aſcites.

It

It affects women of all ages, either married or virgins; and is produced by cold, fear, hunger, bad food, and other debilitating causes. I saw an elegant young lady, who was shortly to have been married to a sensible man, with great prospect of happiness; who, on being overturned in a chaise in the night, and obliged to walk two or three miles in wet, cold, and darkness, became much indisposed, and gradually afflicted with a swelling and pain on one side of the abdomen; which terminated in a dropsy of the ovary, and destroyed her in two or three years. Another young woman I recollect seeing, who was about seventeen, and being of the very inferior class of people, seemed to have been much weakened by the hardship of a cold floor, and little or no bed, with bad food; and who to these evils had to bear the unceasing obloquy of her neighbours, and the persecution of parish officers.

The following is abstracted from a letter of my friend Mr. Power, surgeon, at Bosworth in Leicestershire, on examining the body of an elderly lady who died of this disease, March 29, 1793. "On opening the abdomen I found a large cyst attached to the left ovarium by an elastic neck as thick as the little finger, and so callous as not to admit of being separated by scissars without considerable difficulty. The substance of the cyst had an appearance much resembling the gravid uterus near the full period of gestation, and was as thick. It had no attachment to the peritonæum, or any of the viscera, except by the hard callous neck I have mentioned; so that the blood must with difficulty have been circulated through it for some time. Its texture was extremely tender, being easily perforated with the finger, was of a livid red colour, and evidently in a sphacelated state. It contained about two gallons of a fluid of the colour of port wine, without any greater tenacity. It has fallen to my lot to have opened two other patients, whose deaths were occasioned by incysted dropsy of the ovarium. In one of these the ovarium was much enlarged with eight or ten cysts on its surface, but there was no adhesion formed by

any

any of the cyſts to any other part; nor had the ovarium formed any adheſion with the peritonæum, though in a very diſeaſed ſtate. In the other the diſeaſe was more ſimple, being only one cyſt, without any attachment but to the ovarium.

"As the ovarium is a part not neceſſary to life, and dropſies of this kind are ſo generally fatal in the end, I think I ſhall be induced, notwithſtanding the hazard attending wounds, which penetrate the cavity of the abdomen, to propoſe the extirpation of the diſeaſed part in the firſt caſe, which occurs to me, in which I can with preciſion ſay, that the ovarium is the ſeat of the diſeaſe, and the patient in other reſpects tolerably healthy; as the cavity of the abdomen is often opened in other caſes without bad conſequences."

An argument, which might further countenance the operation thus propoſed by Mr. Power, might be taken from the diſeaſe frequently affecting young perſons; from its being generally in theſe ſubjects local and primary; and not like the aſcites, produced or accompanied with other diſeaſed viſcera; and laſtly, as it is performed in adult quadrupeds, as old ſows, with ſafety, though by awkward operators.

16. *Anaſarca pulmonum.* The dropſy of the cellular membrane of the lungs is uſually connected with that of the other parts of the ſyſtem. As the cells of the whole cellular membrane communicate with each other, the mucaginous fluid, which remains in any part of it for want of due abſorption, ſinks down to the moſt depending cells; hence the legs ſwell, though the cauſe of the diſeaſe, the deficiency of abſorption, may be in other parts of the ſyſtem. The lungs however are an exception to this, ſince they are ſuſpended in the cavity of the thorax, and have in conſequence a depending part of their own.

The anaſarca of the lungs is known by the difficulty of reſpiration accompanied with ſwelled legs, and with a very irregular pulſe. This

 laſt

last circumstance has generally been ascribed to a dropsy at the same time existing in the pericardium, but is more probably owing to the difficult passage of the blood through the lungs; because I found on dissection, in one instance, that the most irregular pulse, which I ever attended to, was owing to very extensive adhesions of the lungs; insomuch that one lobe intirely adhered to the pleura; and secondly, because this kind of dropsy of the lungs is so certainly removed for a time along with the anasarca of the limbs by the use of digitalis.

This medicine, as well as emetic tartar, or squill, when given so as to produce sickness, or nausea, or perhaps even without producing either in any perceptible degree, by affecting the lymphatics of the stomach, so as either to invert their motion, or to weaken them, increases by reverse sympathy the action, and consequent absorbent power of these lymphatics, which open into the cellular membrane. But as those medicines seldom succeed in producing an absorption of those fluids, which stagnate in the larger cavities of the body, as in the abdomen, or chest, and do generally succeed in this difficulty of breathing with irregular pulse above described, I conclude that it is not owing to an effusion of lymph into the pericardium, but simply to an anasarca of the lungs.

M. M. Digitalis. See Art. V. 2. 1. Tobacco. Squill. Emetic tartar (antimonium tartarizatum). Then Sorbentia. Chalybeates. Opium half a grain twice a day. Raisin wine and water, or other wine and water, is preferred to the spirit and water, which these patients have generally been accustomed to.

The usual cause of anasarca is from a diseased liver, and hence it most frequently attends those, who have drank much fermented or spirituous liquors; but I suspect that there is another cause of anasarca, which originates from the brain; and which is more certainly fatal than that, which originates from a diseased liver. These patients, where the anasarca originates from, or commences in, the brain, have not other symptoms of diseased liver; have less difficulty of breathing

at

at the beginning; and hold themſelves more upright in their chair, and in walking. In this kind of dropſy I ſuſpect the digitalis has leſs or no effect; as it particularly increaſes the abſorption from the lungs.

17. *Obeſitas.* Corpulency may be called an anaſarca or dropſy of fat, ſince it muſt be owing to an analogous cauſe; that is, to the deficient abſorption of fat compared to the quantity ſecreted into the cells which contain it. See Claſs II. 1. 1. 4.

The method of getting free from too much fat without any injury to the conſtitution, conſiſts, firſt, in putting on a proper bandage on the belly, ſo that it can be tightened or relaxed with eaſe, as a tightiſh under waiſtcoat, with a double row of buttons. This is to compreſs the bowels and increaſe their abſorption, and it thus removes one principal cauſe of corpulency, which is the looſeneſs of the ſkin. Secondly, he ſhould omit one entire meal, as ſupper; by this long abſtinence from food the abſorbent ſyſtem will act on the mucus and fat with greater energy. Thirdly, he ſhould drink as little as he can with eaſe to his ſenſations; ſince, if the abſorbents of the ſtomach and bowels ſupply the blood with much, or perhaps too much, aqueous fluid, the abſorbents of the cellular membrane will act with leſs energy. Fourthly, he ſhould uſe much ſalt or ſalted meat, which will increaſe the perſpiration and make him thirſty; and if he bears this thirſt, the abſorption of his fat will be greatly increaſed, as appears in fevers and dropſies with thirſt; this I believe to be more efficacious than ſoap. Fifthly, he may uſe aerated alcaline water for his drink, which may be ſuppoſed to render the fat more fluid,—or he may take ſoap in large quantities, which will be decompoſed in the ſtomach. Sixthly, ſhort reſt, and conſtant exerciſe.

18. *Splenis tumor.* Swellings of the ſpleen, or in its vicinity, are frequently perceived by the hand in intermittents, which are called

called Ague-cakes, and ſeem owing to a deficiency of abſorption in the affected part.

Mr. Y——, a young man about twenty-five years of age, who lived intemperately, was ſeized with an obſtinate intermittent, which had become a continued fever with ſtrong pulſe, attended with daily remiſſion. A large hard tumour on the left ſide, on the region of the ſpleen, but extending much more downward, was ſo diſtinctly perceptible, that one ſeemed to get one's fingers under the edge of it, much like the feel of the brawn or ſhield on a boar's ſhoulder. He was repeatedly bled, and purged with calomel, had an emetic, and a bliſter on the part, without diminiſhing the tumour; after ſome time he took the Peruvian bark, and ſlight doſes of chalybeates, and thus became free from the fever, and went to Bath for ſeveral weeks, but the tumour remained. This tumour I examined every four or five years for above thirty years. His countenance was pale, and towards the end of his life he ſuffered much from ulcers on his legs, and died about ſixty, of general debility; like many others, who live intemperately in reſpect to the ingurgitation of fermented or ſpirituous liquors.

As this tumour commenced in the cold fit of an intermittent fever, and was not attended with pain, and continued ſo long without endangering his life, there is reaſon to believe it was ſimply occaſioned by deficient abſorption, and not by more energetic action of the veſſels which conſtitute the ſpleen. See Claſs II. 1. 2. 13.

M. M. Veneſection. Emetic, cathartic with calomel; then forbentia, chalybeates, Peruvian bark.

19. *Genu tumor albus.* White ſwelling of the knee, is owing to deficient abſorption of the lymphatics of the membranes including the joint, or capſular ligaments, and ſometimes perhaps of the gland which ſecretes the ſynovia; and the ends of the bones are probably affected in conſequence.



I ſaw

I ſaw an inſtance, where a cauſtic had been applied by an empiric on a large white ſwelling of the knee, and was told, that a fluid had been diſcharged from the joint, which became anchyloſed, and healed without loſs of the limb.

M. M. Repeated, bliſters on the part early in the diſeaſe are ſaid to cure it by promoting abſorption; ſaturnine ſolutions externally are recommended. Bark, animal charcoal, as burnt ſponge, opium in ſmall doſes. Friction with the hand.

20. *Bronchocele.* Swelled throat. An enlargement of the thyroid glands, ſaid to be frequent in mountainous countries, where river water is drank, which has its ſource from diſſolving ſnows. This idea is a very ancient one, but perhaps not on that account to be the more depended upon, as authors copy one another. Tumidum guttur quis miratur in alpibus, ſeems to have been a proverb in the time of Juvenal. The inferior people of Derby are much ſubject to this diſeaſe, but whether more ſo than other populous towns, I can not determine; certain it is, that they chiefly drink the water of the Derwent, which ariſes in a mountainous country, and is very frequently blackened as it paſſes through the moraſſes near its ſource; and is generally of a darker colour, and attended with a whiter foam, than the Trent, into which it falls; the greater quantity and whiteneſs of its froth I ſuppoſe may be owing to the viſcidity communicated to it by the colouring matter. The lower parts of the town of Derby might be eaſily ſupplied with ſpring water from St. Alkmond's well; or the whole of it from the abundant ſprings near Bowbridge; the water from which might be conveyed to the town in hollow bricks, or clay-pipes, at no very great expence, and might be received into frequent reſervoirs with pumps to them; or laid into the houſes.

M. M. Twenty grains of burnt ſponge with ten of nitre made with

with mucilage into lozenges, and permitted to dissolve slowly under the tongue twice a day, is asserted to cure in a few months; perhaps other animal charcoal, as candle-snuffs, might do the same.

I have directed in the early state of this disease a mixture of common salt and water to be held in the mouth, particularly under the tongue, for a few minutes, four or six times a day for many weeks, which has sometimes succeeded, the salt and water is then spit out again, or in part swallowed. Externally vinegar of squills has been applied, or a mercurial plaster, or fomentations of acetated ammoniac; or ether. Some empirics have applied caustics on the bronchocele, and sometimes, I have been told, with success; which should certainly be used where there is danger of suffocation from the bulk of it. One case I saw, and one I was well informed of, where the bronchocele was cured by burnt sponge, and a hectic fever supervened with colliquative sweats; but I do not know the final event of either of them.

De Haen affirms the cure of bronchocele to be effected by flowers of zinc, calcined egg-shells, and scarlet cloth burnt together in a close crucible, which was tried with success, as he assured me, by a late lamented physician, my friend, Dr. Small of Birmingham; who to the cultivation of modern sciences added the integrity of ancient manners; who in clearness of head, and benevolence of heart, had few equals, perhaps no superiors.

21. *Scrophula.* King's evil is known by tumours of the lymphatic glands, particularly of the neck. The upper lip, and division of the nostrils is swelled, with a florid countenance, a smooth skin, and a tumid abdomen. Cullen. The absorbed fluids in their course to the veins in the scrophula are arrested in the lymphatic or conglobate glands; which swell, and after a great length of time, inflame and suppurate. Materials of a peculiar kind, as the variolous and venereal matter, when absorbed in a wound, produce this torpor, and conse-

quent inflammation of thoſe lymphatic glands, where they firſt arrive, as in the axilla and groin. There is reaſon to ſuſpect, that the tonſils frequently become inflamed, and ſuppurate from the matter abſorbed from carious teeth; and I ſaw a young lady, who had both the axillary glands ſwelled, and which ſuppurated; which was believed to have been cauſed by her wearing a pair of new green gloves for one day, when ſhe had perſpired much, and was much exhauſted and fatigued by walking; the gloves were probably dyed in a ſolution of verditer.

Theſe indolent tumours of the lymphatic glands, which conſtitute the ſcrophula, originate from the inirritability of thoſe glands; which therefore ſooner fall into torpor after having been ſtimulated too violently by ſome poiſonous material; as the muſcles of enfeebled people ſooner become fatigued, and ceaſe to act, when exerted, than thoſe of ſtronger ones. On the ſame account theſe ſcrophulous glands are much longer in acquiring increaſe of motion, after having been ſtimulated into inactivity, and either remain years in a ſtate of indolence, or ſuppurate with difficulty, and ſometimes only partially.

The difference between ſcrophulous tumours, and thoſe before deſcribed, conſiſts in this; that in thoſe either glands of different kinds were diſeaſed, or the mouths only of the lymphatic glands were become torpid; whereas in ſcrophula the conglobate glands themſelves become tumid, and generally ſuppurate after a great length of time, when they acquire new ſenſibility. See Sect. XXXIX. 4. 5.

Theſe indolent tumours may be brought to ſuppurate ſometimes by paſſing electric ſhocks through them every day for two or three weeks, as I have witneſſed. It is probable, that the alternate application of ſnow or iced water to them, till they become painfully cold, and then of warm flannel or warm water, frequently repeated, might reſtore their irritability by accumulation of ſenſorial power; and thence either facilitate their diſperſion, or occaſion them to ſuppurate. See Claſs II. 1. 4. 13.

This

This disease is very frequent amongst the children of the poor in large towns, who are in general ill fed, ill lodged, and ill clothed; and who are further weakened by eating much salt with their scanty meal of insipid vegetable food, which is seldom of better quality than water gruel, with a little coarse bread in it. See diarrhœa of infants, Class I. 1. 2. 5. Scrophulous ulcers are difficult to heal, which is owing to the deficiency of absorption on their pale and flabby surfaces, and to the general inirritability of the system. See Class I. 1. 3. 13.

M. M. Plentiful diet of flesh-meat and vegetables with small-beer. Opium, from a quarter of a grain to half a grain twice a day. Sorbentia. Tincture of digitalis, thirty drops twice a day. Externally sea-bathing, or bathing in salt and water, one pound to three gallons, made warm. The application of Peruvian bark in fine powder, seven parts, and white lead, (cerussa) in fine powder one part, mixed together and applied on the ulcers in dry powder, by means of lint and a bandage, to be renewed every day. Or very fine powder of calamy alone, lapis calaminaris. If powder of manganese?

22. *Schirrus.* After the absorbent veins of a gland cease to perform their office, if the secerning arteries of it continue to act some time longer, the fluids are pushed forwards, and stagnate in the receptacles or capillary vessels of the gland; and the thinner part of them only being resumed by the absorbent system of the gland, a hard tumour gradually succeeds; which continues like a lifeless mass, till from some accidental violence it gains sensibility, and produces cancer, or suppurates. Of this kind are the schirrous glands of the breasts, of the lungs, of the mesentery, and the scrophulous tumours about the neck and the bronchocele.

Another seat of schirrus is in the membranous parts of the system, as of the rectum intestinum, the urethra, the gula or throat; and of this kind is the verucca or wart, and the clavus pedum, or corns on the toes. A wen sometimes arises on the back of the neck, and sometimes

ſometimes between the ſhoulders; and by diſtending the tendinous faſcia produces great and perpetual pain.

M. M. Mercurial ointment. Cover the part with oiled ſilk. Extirpation. Electric ſhocks through the tumour. An iſſue into the ſubſtance of the wen. Opium. Ether externally.

23. *Schirrus recti inteſtini.* Schirrus of the rectum. A ſchirrus frequently affects a canal, and by contracting its diameter becomes a painful and deplorable diſeaſe. The canals thus obſtructed are the rectum, the urethra, the throat, the gall-ducts, and probably the excretory ducts of the lymphatics, and of other glands.

The ſchirrus of the rectum is known by the patient having pain in the part, and being only able to part with liquid feces, and by the introduction of the finger; the ſwelled part of the inteſtine is ſometimes protruded downwards, and hangs like a valve, ſmooth and hard to the touch, with an aperture in the centre of it. See a paper on this ſubject by J. Sherwin. Memoirs of a London Medical Society, Vol. II. p. 9.

M. M. To take but little ſolid food. Aperient medicines. Introduce a candle ſmeared with mercurial ointment. Sponge-tent. Clyſters with forty drops of laudanum. Introduce a leathern canula, or gut, and then either a wooden maundril, or blow it up with air, ſo as to diſtend the contracted part as much as the patient can bear. Or ſpread mercurial plaſter on thick ſoft leather, and roll it up with the plaſter outwards to any thickneſs and length, which can be eaſily introduced and worn; or two or three ſuch pieces may be introduced after each other. The ſame may be uſed to compreſs bleeding internal piles. See Claſs I. 2. 1. 6.

24. *Schirrus urethræ.* Schirrus of the urethra. The paſſage becomes contracted by the thickened membrane, and the urine is forced through with great difficulty, and is thence liable to diſtend the canal behind

behind the ſtricture; till at length an aperture is made, and the urine forces its way into the cellular membrane, making large ſinuſes. This ſituation ſometimes continues many months, or even years, and ſo much matter is evacuated after making water, or at the ſame time, by the action of the muſcles in the vicinity of the ſinuſes, that it has been miſtaken for an increaſed ſecretion from the bladder, and has been erroneouſly termed a catarrh of the bladder. See a paper by Dr. R. W. Darwin in the Medical Memoirs.

M. M. Diſtend the part gradually by catgut bougies, which by their compreſſion will at the ſame time diminiſh the thickneſs of the membrane, or by bougies of elaſtic gum, or of horn boiled ſoft. The patient ſhould gain the habit of making water ſlowly, which is a matter of the utmoſt conſequence, as it prevents the diſtention, and conſequent rupture, of that part of the urethra, which is between the ſtricture and the neck of the bladder.

When there occurs an external ulcer in the perinæum, and the urine is in part diſcharged that way, the diſeaſe can not be miſtaken. Otherwiſe from the quantity of matter, it is generally ſuppoſed to come from the bladder, or proſtate gland; and the urine, which eſcapes from the ruptured urethra, mines its way amongſt the muſcles and membranes, and the patient dies tabid, owing to the want of an external orifice to diſcharge the matter. See Claſs II. 1. 4. 11.

25. *Schirrus œſophagi.* A ſchirrus of the throat contracts the paſſage ſo as to render the ſwallowing of ſolids impracticable, and of liquids difficult. It affects patients of all ages, but is probably moſt frequently produced by ſwallowing hard angular ſubſtances, when people have loſt their teeth; by which this membrane is over diſtended, or torn, or otherwiſe injured.

M. M. Put milk into a bladder tied to a canula or catheter; introduce it paſt the ſtricture, and preſs it into the ſtomach. Diſtend the ſtricture gradually by a ſponge-tent faſtened to the end of whale-

6 bone,

bone, or by a plug of wax, or a ſpermaceti candle, about two inches long; which might be introduced, and left there with a ſtring only fixed to it to hang out of the mouth, to keep it in its place, and to retract it by occaſionally; for which purpoſe the ſtring muſt be put through a catheter or hollow probang, when it is to be retracted. Or laſtly introduce a gut fixed to a pipe; and then diſtend it by blowing wind into it. The ſwallowing a bullet with a ſtring put through it, to retract it on the exhibition of an emetic, has alſo been propoſed. Externally mercurial ointment has been much recommended. Poultice. Oiled ſilk. Clyſters of broth. Warm bath of broth. Transfuſion of blood into a vein three or four ounces a day? See Claſs III. 1. 1. 15.

I directed a young woman about twenty-two years of age, to be fed with new milk put into a bladder, which was tied to a catheter, and introduced beyond the ſtricture in her throat; after a few days her ſpirits ſunk, and ſhe refuſed to uſe it further, and died. Above thirty years ago I propoſed to an old gentleman, whoſe throat was entirely impervious, to ſupply him with a few ounces of blood daily from an aſs, or from the human-animal, who is ſtill more patient and tractable, in the following manner. To fix a ſilver pipe about an inch long to each extremity of a chicken's gut, the part between the two ſilver ends to be meaſured by filling it with warm water; to put one end into the vein of a perſon hired for that purpoſe, ſo as to receive the blood returning from the extremity; and when the gut was quite full, and the blood running through the other ſilver end, to introduce that end into the vein of the patient upwards towards the heart, ſo as to admit no air along with the blood. And laſtly, to ſupport the gut and ſilver ends on a water plate, filled with water of ninety-eight degrees of heat, and to meaſure how many ounces of blood was introduced by paſſing the finger, ſo as to compreſs the gut, from the receiving pipe to the delivering pipe; and thence to determine how many gut-fulls were given from the healthy perſon to the patient. See Claſs IV. 2. 4. 11. L. Mr. —— conſidered a day on this propoſal, and then another day, and at length anſwered, that "he now found himſelf near the houſe of death; and that if he could return, he was now too old to have much enjoyment of life; and therefore he wiſhed rather to proceed to the end of that journey, which he was now ſo near, and which he muſt at all events ſoon go, than return for ſo ſhort a time." He lived but a few days afterwards, and ſeemed quite careleſs and eaſy about the matter.

26. *Lacteorum inirritabilitas.* Inirritability of the lacteals is deſcribed in Sect. XXVIII. under the name of paralyſis of the lacteals; but as the word paralyſis has generally been applied to the diſobedience of the muſcles to the power of volition, the name is here changed to inirritability of the lacteals, as more characteriſtic of the diſeaſe.

27. *Lymphaticorum inirritabilitas.* The inirritability of the cellular and cutaneous lymphatics is deſcribed in Sect. XXIX. 5. 1. and in Claſs I. 2. 3. 16. The inirritability of the cutaneous lymphatics generally accompanies anaſarca, and is the cauſe of the great thirſt in that malady. At the ſame time the cellular lymphatics act with greater energy, owing to the greater derivation of ſenſorial power to them in conſequence of the leſs expenditure of it by the cutaneous ones; and hence they abſorb the fat, and mucus, and alſo the thinner parts of the urine. Whence the great emaciation of the body, the muddy ſediment, and the ſmall quantity of water in this kind of dropſy.

## ORDO II.

*Decreased Irritation.*

## GENUS IV.

*With decreased Actions of other Cavities and Membranes.*

MANY of the diseases of this genus are attended with pain, and with cold extremities, both which cease on the exhibition of wine or opium; which shews, that they originate from deficient action of the affected organ. These pains are called nervous or spasmodic, are not attended with fever, but are frequently succeeded by convulsions and madness; both which belong to the class of volition. Some of them return at periods, and when these can be ascertained, a much less quantity of opium will prevent them, than is necessary to cure them, when they are begun; as the vessels are then torpid and inirritable from the want of sensorial power, till by their inaction it becomes again accumulated.

Our organs of sense properly so called are not liable to pain from the absence of their appropriated stimuli, as from darkness or silence; but the other senses, which may be more properly called appetites, as those by which we perceive heat, hunger, thirst, lust, want of fresh air, are affected with pain from the defect or absence of their accustomed stimuli, as well as with pleasure by the possession of them; it is probable that some of our glands, whose sense or appetite requires or receives something from the circulating blood, as the pancreas, liver, testes, prostate gland, may be affected with aching or pain, when they cannot acquire their appropriated fluid.

Wherever

Wherever this defect of ſtimulus occurs, a torpor or inaction of the organ enſues, as in the capillaries of the ſkin, when expoſed to cold; and in the glands, which ſecrete the gaſtric juice, when we are hungry. This torpor however, and concomitant pain, which is at firſt owing to defect of ſtimulus, is afterwards induced by other aſſociations or catenations, and conſtitutes the beginning of ague fits.

It muſt be further obſerved, that in the diſeaſes of pain without fever, the pain is frequently not felt in the part where the cauſe of the diſeaſe reſides; but is induced by ſympathy with a diſtant part, whoſe irritability or ſenſibility is greater or leſs than its own. Thus a ſtone at the neck of the bladder, if its ſtimulus is not very great, only induces the pain of ſtrangury at the glans penis. If its ſtimulus be greater, it then induces pain at the neck of the bladder. The concretions of bile, which are protruded into the neck of the gall-bladder, when the diſeaſe is not very great, produce pain at the other extremity of the bile-duct, which enters the duodenum immediately under the pit of the ſtomach; but, when the diſeaſe is great from the largeneſs of the bile-ſtone, the pain is felt in the region of the liver at the neck of the gall-bladder.

It appears from hence, that the pains enumerated in this genus are conſequences of the inactivity of the organ; and, as they do not occaſion other diſeaſes, ſhould be claſſed according to their proximate cauſe, which is defective irritation; there are nevertheleſs other pains from defect of ſtimulus, which produce convulſions, and belong to Claſs III. 1. 1.; and others, which produce pains of ſome diſtant part by aſſociation, and belong to Claſs IV. 2. 2.

## SPECIES.

1. *Sitis.* Thirſt. The ſenſes of thirſt and of hunger ſeem to have this connection, that the former is ſituated at the upper end, and the

latter at the lower end of the ſame canal. One about the pharinx, where the œſophagus opens into the mouth, and the other about the cardia ventriculi, where it opens into the ſtomach. The extremities of other canals have been ſhewn to poſſeſs correſpondent ſenſibilities, or irritabilities, as the two ends of the urethra, and of the common gall-duct. See IV. 2. 2. 2. and 4.

The membrane of the upper end of the gullet becomes torpid, and conſequently painful, when there is a deficiency of aqueous fluid in the general ſyſtem; it then wants its proper ſtimulus. In the ſame manner a want of the ſtimulus of more ſolid materials at the other end of the canal, which terminates in the ſtomach, produces hunger; as mentioned in Sect. XIV. 8. The proximate cauſes of both of them therefore conſiſt in deficient irritation, when they are conſidered as pains; becauſe theſe pains are in conſequence of the inactivity of the organ, according to the fifth law of animal cauſation. Sect. IV. 5. But when they are conſidered as deſires, namely of liquid or ſolid aliment, their proximate cauſe conſiſts in the pain of them, according to the ſixth law of animal cauſation. So the proximate cauſe of the pain of coldneſs is the inactivity of the organ, and perhaps the conſequent accumulation of ſenſorial power in it; but the pain itſelf, or the conſequent volition, is the proximate cauſe of the ſhuddering and gnaſhing the teeth in cold fits of intermittent fevers. See Claſs I. 2. 2. 1.

Thirſt may be divided into two varieties alluding to the remote cauſe of each, and may be termed ſitis calida, or warm thirſt, and ſitis frigida, or cold thirſt. The remote cauſe of the former ariſes from the diſſipation of the aqueous parts of our fluids by the increaſed ſecretion of perſpirable matter, or other evacuations. And hence it occurs in hot fits of fever, and after taking much wine, opium, ſpice, ſalt, or other drugs of the Art. incitantia or ſecernentia. The thirſt, which occurs about three hours after eating a couple of red herrings, to a perſon unaccuſtomed to ſalted meat, is of this kind; the increaſed action of the cutaneous veſſels diſſipates ſo much of our

4 fluids

fluids by insensible perspiration, as to require above two quarts of water to restore the fluidity of the blood, and to wash the salt out of the system. See Art. III. 2. 1.

M. M. Cold water. Vegetable acids. Warm bath.

The remote cause of sitis frigida, or cold thirst, is owing to the inaction of the cutaneous, pulmonary, urinary, and cellular absorbents; whence the blood is deprived of the great supply of moisture, which it ought to receive from the atmosphere, and from the cells of the cellular membrane, and from other cysts; this cause of thirst exists in dropsies, and in the cold fits of intermittents. The desire of fluids, like that of solids, is liable to acquire periods, and may therefore readily become diseased by indulgence in liquids grateful to the palate.

Of diseased thirst, the most common is either owing to defect of the action of the numerous absorbent vessels on the neck of the bladder, in which the patient makes much paleish water; or to the defective absorption of the skin and lungs, in which the patient makes but little water, and that high-coloured, and with sediment. In both the tongue and lips are liable to become very dry. The former in its greatest degree attends diabætes, and the latter anasarca.

M. M. Warm water, warm wine, warm bath. Opium. Cold bath. Iced water. Lemonade. Cyder.

2. *Esuries.* Hunger has been fancifully ascribed to the sides of the stomach rubbing against each other, and to the increased acidity of the gastric juice corroding the coats of it. If either of these were the cause of hunger, inflammation must occur, when they had continued some time; but, on the contrary, coldness and not heat are attendant on hunger; which evinces, that like thirst it is owing to the inactivity of the membrane, which is the seat of it; while the abundant nerves about the cardia ventriculi, and the pain of hunger being

 felt

felt in that part, gives great reaſon to conclude, that it is there ſituated.

The ſenſe of hunger as well as of thirſt is liable to acquire habits in reſpect to the times of its returning painfulneſs, as well as in reſpect to the quantity required to ſatiate its appetency, and hence may become diſeaſed by indulgence, as well as by want of its appropriate ſtimulus. Thoſe who have been accuſtomed to diſtend their ſtomach by large quantities of animal and vegetable food, and much potation, find a want of diſtention, when the ſtomach is empty, which occaſions faintneſs, and is miſtaken for hunger, but which does not appear to be the ſame ſenſation. I was well informed, that a woman near Lichfield, who eat much animal and vegetable food for a wager, affirmed, that ſince diſtending her ſtomach ſo much, ſhe had never felt herſelf ſatisfied with food; and had in general taken twice as much at a meal, as ſhe had been accuſtomed to, before ſhe eat ſo much for a wager.

3. *Nauſea ſicca.* Dry nauſea. Conſiſts in a quieſcence or torpor of the mucous or ſalivary glands, and precedes their inverted motions, deſcribed in nauſea humida, Claſs I. 3. 2. 3. In the ſame manner as ſickneſs of the ſtomach is a quieſcence of that organ preceding the action of vomiting, as explained in Sect. XXXV. 1. 3. This is ſometimes induced by diſagreeable drugs held in the mouth, at other times of diſguſtful ideas, and at other times by the aſſociation of theſe actions with thoſe of the ſtomach; and thus according to its different proximate cauſes may belong to this, or to the ſecond, or to the fourth claſs of diſeaſes.

M. M. Lemonade. Taſteful food. A bliſter. Warm bath.

4. *Ægritudo ventriculi.* Sickneſs of ſtomach is produced by the quieſcence or inactivity of that organ, as is explained in Sect. XXXV. 1. 3.

1. 3. It consists in the state between the usual peristaltic motions of that organ, in the digestion of our aliment, and the retrograde motions of it in vomiting; for it is evident, that the direct motions of it from the cardia to the pylorus must stop, before those in a contrary direction can commence. This sickness, like the nausea above described, is sometimes produced by disgustful ideas, as when nasty objects are seen, and nasty stories related, as well as by the exhaustion of the sensorial power by the stimulus of some emetic drugs, and by the defect of the production of it, as in enfeebled drunkards.

Sickness may likewise consist in the retrograde motions of the lymphatics of the stomach, which regurgitate into it the chyle or lymph, which they have lately absorbed, as in Class I. 3. 2. 3. It is probable, that these two kinds of sickness may be different sensations, though they have acquired but one name; as one of them attends hunger, and the other repletion; though either of them may possibly be induced by association with nauseous ideas.

M. M. A blister on the back. An emetic. Opium. Crude mercury. Covering the head in bed. See Sect. XXV. 16. Class IV. 1. 1. 2. and 3.

5. *Cardialgia.* Heartburn originates from the inactivity of the stomach, whence the aliment, instead of being subdued by digestion, and converted into chyle, runs into fermentation, producing acetous acid. Sometimes the gastric juice itself becomes so acid as to give pain to the upper orifice of the stomach; these acid contents of the stomach, on falling on a marble hearth, have been seen to produce an effervescence on it. The pain of heat at the upper end of the gullet, when any air is brought up from the fermenting contents of the stomach, is to be ascribed to the sympathy between these two extremities of the œsophagus rather than to the pungency of the carbonic gas, or fixed air; as the sensation in swallowing that kind of air in water is of a different kind. See Class I. 3. 1. 3. and IV. 2. 2. 5.

M. M.

M. M. This diseafe arifing from indigeftion is often very pertinacious, and afflicting; and attended with emaciation of the body from want of fufficient chyle. As the faliva fwallowed along with our food prevents its fermentation, as appears by the experiments of Pringle and Macbride, fome find confiderable relief by chewing parched wheat, or maftic, or a lock of wool, frequently in a day, when the pain occurs, and by fwallowing the faliva thus effufed; a temporary relief is often obtained from antiacids, as aerated alcaline water, Seltzer's water, calcareous earths, alcaline falts made into pills with foap, foap alone, tin, milk, bitters. More permanent ufe may be had from fuch drugs as check fermentation, as acid of vitriol; but ftill more permanent relief from fuch things as invigorate the digeftion, as a blifter on the back; a due quantity of vinous fpirit and water taken regularly. Steel. Temperance. A fleep after dinner. A waiftcoat made fo tight as flightly to comprefs the bowels and ftomach. A flannel fhirt in winter, not in fummer. A lefs quantity of potation of all kinds. Ten black pepper-corns fwallowed after dinner. Half a grain of opium twice a day, or a grain. The food fhould confift of fuch things as do not eafily ferment, as flefh, fhell-fifh, fea-bifcuit, toafted cheefe. I have feen toafted cheefe brought up from the ftomach 24 hours after it had been fwallowed, without apparently having undergone any chemical change. See Clafs II. 1. 3. 17. and IV. 1. 2. 13.

6. *Arthritis Ventriculi.* Sicknefs of the ftomach in gouty cafes is frequently a confequence of the torpor or inflammation of the liver, and then it continues many days or weeks. But when the patient is feized with great pain at the ftomach with the fenfation of coldnefs, which they have called an ice-bolt, this is a primary affection of the ftomach, and deftroys the patient in a few hours, owing to the torpor or inaction of that vifcus fo important to life.

The

This primary gout of the ſtomach, as it is a torpor of that viſcus, is attended with ſenſation of coldneſs, and with real defect of heat, in that part, and may thence be diſtinguiſhed from the pain occaſioned by the paſſage of a gall-ſtone into the duodenum, as well as by the weak pulſe, and cold extremities; to which muſt be added, that it affects thoſe only, who have been long afflicted with the gout, and much debilitated by its numerous attacks.

M.M. Opium. Vinous ſpirit. Volatile alcali. Spice. Warmth applied externally to the ſtomach by hot cloths or fomentation.

7. *Colica flatulenta.* The flatulent colic ariſes from the too great diſtention of the bowel by air, and conſequent pain. The cauſe of this diſeaſe is the inactivity or want of ſufficiently powerful contraction of the coats of the bowel, to carry forwards the gas given up by the fermenting aliment. It is without fever, and generally attended with cold extremities.

It is diſtinguiſhed, firſt, from the pain occaſioned by the paſſage of a gall-ſtone, as that is felt at the pit of the ſtomach, and this nearer the navel. Secondly, it is diſtinguiſhed from the colica ſaturnina, or colic from lead, as that ariſing from the torpor of the liver, or of ſome other viſcus, is attended with greater coldneſs, and with an aching pain; whereas the flatulent cholic being owing to diſtention of the muſcles of the bowel, the pain is more acute, and the coldneſs leſs. Thirdly, it is diſtinguiſhed from inflammation of the bowels, or ileus, as perpetual vomiting and fever attend this. Fourthly, it is diſtinguiſhed from cholera, becauſe that is accompanied with both vomiting and diarrhœa. And laſtly, from the colica epileptica, or hyſteric colic, as that is liable to alternate with convulſion, and ſometimes with inſanity; and returns by periods.

M. M. Spirit of wine and warm water, one ſpoonful of each. Opium one grain. Spice. Volatile alcali. Warm fomentation externally. Rhubarb.

8. *Colica ſaturnina.* Colic from lead. The pain is felt about the navel, is rather of an aching than acute kind at firſt, which increaſes after meals, and gradually becomes more permanent and more acute. It terminates in paralyſis, frequently of the muſcles of the arm, ſo that the hand hangs down, when the arm is extended horizontally. It is not attended with fever, or increaſe of heat. The ſeat of the diſeaſe is not well aſcertained, it probably affects ſome part of the liver, as a pale bluiſh countenance and deficiency of bile ſometimes attends or ſucceeds it, with conſequent anaſarca; but it ſeems to be cauſed immediately by a torpor of the inteſtine, whether this be a primary or ſecondary affection, as appears from the conſtipation of the bowels, which attends it; and is always produced in conſequence of the great ſtimulus of lead previouſly uſed either internally for a length of time, or externally on a large ſurface.

A delicate young girl, daughter of a dairy farmer, who kept his milk in leaden ciſterns, uſed to wipe off the cream from the edges of the lead with her finger; and frequently, as ſhe was fond of cream, licked it from her finger. She was ſeized with the ſaturnine colic, and ſemi-paralytic wriſts, and ſunk from general debility.

A feeble woman about 40 years of age ſprained her ancle, and bruiſed her leg and thigh; and applied by ill advice a ſolution of lead over the whole limb, as a fomentation and poultice for about a fortnight. She was then ſeized with the colica ſaturnina, loſt the uſe of her wriſts, and gradually ſunk under a general debility.

M. M. Firſt opium one or two grains, then a cathartic of ſenna, jalap, and oil, as ſoon as the pain is relieved. Oleum ricini. Alum. Oil of almonds. A bliſter on the navel. Warm bath. The ſtimulus of the opium, by reſtoring to the bowel its natural irritability in this caſe of painful torpor, aſſiſts the action of the cathartic.

9. *Tympanitis.* Tympany conſiſts in an elaſtic tumor of the abdomen,

domen, which ſounds on being ſtruck. It is generally attended with coſtiveneſs and emaciation. In one kind the air is ſaid to exiſt in the bowels, in which caſe the tumor is leſs equal, and becomes leſs tenſe and painful on the evacuation of air. In the other kind the air exiſts in the cavity of the abdomen, and ſometimes is in a few days exchanged for water, and the tympany becomes an aſcites.

Air may be diſtinguiſhed in the ſtomach of many people by the ſound on ſtriking it with the fingers, and comparing the ſound with that of a ſimilar percuſſion on other parts of the bowels; but towards the end of fevers, and eſpecially in the puerperal fever, a diſtention of the abdomen by air is generally a fatal ſymptom, though the eaſe, and often cheerfulneſs, of the patient vainly flatters the attendants.

M. M. In the former caſe a clyſter-pipe unarmed may be introduced, and left ſome time in the rectum, to take off the reſiſtance of the ſphincter, and thus diſcharge the air, as it is produced from the fermenting or putrefying aliment. For this purpoſe, in a diſeaſe ſomewhat ſimilar in horſes, a perforation is made into the rectum on one ſide of the ſphincter; through which fiſtula the air, which is produced in ſuch great exceſs from the quantity of vegetable food which they take, when their digeſtions are impaired, is perpetually evacuated. In both caſes alſo, balſams, eſſential oil, ſpice, bandage on the abdomen, and, to prevent the fermentation of the aliment, acid of vitriol, ſaliva. See Claſs I. 2. 4. 5.

10. *Hypochondriaſis.* The hypochondriac diſeaſe conſiſts in indigeſtion and conſequent flatulency, with anxiety or want of pleaſureable ſenſation. When the action of the ſtomach and bowels is impaired, much gas becomes generated by the fermenting or putreſcent aliment, and to this indigeſtion is catenated languor, coldneſs of the ſkin, and fear. For when the extremities are cold for too long a time in ſome weak conſtitutions, indigeſtion is produced by direct



ſympathy

ſympathy of the ſkin and the ſtomach, with conſequent heart-burn, and flatulency. The ſame occurs if the ſkin be made cold by fear, as in riding over dangerous roads in winter, and hence converſely fear is produced by indigeſtion or torpor of the ſtomach by aſſociation.

This diſeaſe is confounded with the fear of death, which is an inſanity, and therefore of a totally different nature. It is alſo confounded with the hyſteric diſeaſe, which conſiſts in the retrograde motions of the alimentary canal, and of ſome parts of the abſorbent ſyſtem.

The hypochondriaſis, like chloroſis, is ſometimes attended with very quick pulſe; which the patient ſeems to bear ſo eaſily in theſe two maladies, that if an accidental cough attends them, they may be miſtaken for pulmonary conſumption; which is not owing primarily to the debility of the heart, but to its direct ſympathy with the actions of the ſtomach.

M. M. Bliſter. A plaſter on the abdomen of Burgundy pitch. Opium a grain twice a day. Rhubarb ſix grains every night. Bark. Steel. Spice. Bath-water. Sieſta, or ſleep after dinner. Uniform hours of meals. No liquor ſtronger than ſmall beer, or wine and water. Gentle exerciſe on horſeback in the open air uniformly perſiſted in. See Cardialgia, I. 2. 4. 5.

11. *Cephalæa.* Head-ach frequently attends the cold paroxyſm of intermittents; afflicts inebriates the day after intoxication; and many people who remain too long in the cold bath. In all which caſes there is a general inaction of the whole ſyſtem, and as theſe membranes about the head have been more expoſed to the variations of heat and cold of the atmoſphere, they are more liable to become affected ſo far as to produce ſenſation, than other membranes; which are uſually covered either with clothes, or with muſcles, as mentioned in Sect. XXXIII. 2. 10.

 The

The promptitude of the membranes about the ſcalp to ſympathize with thoſe of other parts of the ſyſtem is ſo great, that this cephalæa without fever, or quickneſs of pulſe, is more frequently a ſecondary than a primary diſeaſe, and then belongs to Claſs IV. 1. 2. 11. The hemicrania, or partial head-ach, I believe to be almoſt always a diſeaſe from aſſociation; though it is not impoſſible, but a perſon may take cold on one ſide of the head only. As ſome people by ſitting always on the ſame ſide of the fire in winter are liable to render one ſide more tender than the other, and in conſequence more ſubject to pains, which have been erroneouſly termed rheumatic. See Claſs IV. 2. 1. 7. & 8.

M. M. The method of cure conſiſts in rendering the habit more robuſt, by gentle conſtant exerciſe in the open air, fleſh diet, ſmall beer at meals with one glaſs of wine, regular hours of reſt and riſing, and of meals. The cloathing about the head ſhould be warmer during ſleep than in the day; becauſe at that time people are more liable to take cold; that is, the membranous parts of it are more liable to become torpid. As explained in Sect. XVIII. 15. In reſpect to medicine, two drams of valerian root in powder three or four times a day are recommended by Fordyce. The bark. Steel in moderate quantities. An emetic. A bliſter. Opium, half a grain twice a day. Decayed teeth ſhould be extracted, particularly ſuch as either ache, or are uſeleſs. Cold bath between 60 and 70 degrees of heat. Warm bath of 94 or 98 degrees every day for half an hour during a month. See Claſs IV. 2. 2. 7. and 8.

A ſolution of arſenic, about the ſixteenth part of a grain, is reported to have great effect in this diſeaſe. It ſhould be taken thrice a day, if it produces no griping or ſickneſs, for two or three weeks. A medicine of this kind is ſold under the name of taſteleſs ague-drops; but a more certain method of aſcertaining the quantity is delivered in the ſubſequent materia medica, Art. IV. 2. 6.

12. *Odontalgia.*

12. *Odontalgia.* Tooth-ach. The pain has been erroneoufly fuppofed, where there is no inflammation, to be owing to fome acrid matter from a carious tooth ftimulating the membrane of the alveolar procefs into violent action and confequent pain; but the effect feems to have been miftaken for the caufe, and the decay of the tooth to have been occafioned by the torpor and confequent pain of the difeafed membrane.

Firft, becaufe the pain precedes the decay of the tooth in regard to time, and is liable to recur, frequently for years, without certainly being fucceeded at laft by a carious tooth, as I have repeatedly obferved.

Secondly, becaufe any ftimulant drug, as pyrethrum, or oil of cloves, applied to the tooth, or ether applied externally to the cheek, fo far from increafing the pain, as they would do if the pained membrane, already acted too ftrongly, that they frequently give immediate relief like a charm.

And thirdly, becaufe the torpor, or deficient action of the membrane, which includes the difeafed tooth, occafions the motions of the membranes moft connected with it, as thofe of the cheek and temples, to act with lefs than their natural energy; and hence a coldnefs of the cheek is perceived eafily by the hand of the patient, comparing it with the other cheek; and the pain of hemicrania is often produced in the temple of the affected fide.

This coldnefs of the cheek in common tooth-ach evinces, that the pain is not then caufed by inflammation; becaufe in all inflammations fo much heat is produced in the fecretions of new veffels and fluids, as to give heat to the parts in vicinity. And hence, as foon as the gum fwells and inflames along with the cheek, heat is produced, and the pain ceafes, owing to the increafed exertions of the torpid membrane, excited by the activity of the fenforial power of fenfation; which previoufly exifted in its paffive ftate in the painful torpid membrane. See Odontitis, Clafs II. 1. 4. 7. and IV. 2. 2. 8.

M. M. If the painful tooth be ſound, veneſection. Then a cathartic. Afterwards two grains of opium. Camphor and opium, one grain of each held in the mouth; or a drop or two of oil of cloves put on the painful tooth. Ether. If the tooth has a ſmall hole in it, it ſhould be widened within by an inſtrument, and then ſtopped with leaf-gold, or leaf-lead; but ſhould be extracted, if much decayed. It is probable that half a ſmall drop of a ſtrong ſolution of arſenic, put carefully into the hollow of a decayed aching tooth, would deſtroy the nerve without giving any additional pain; but this experiment requires great caution, leſt any of the ſolution ſhould touch the tongue or gums.

Much cold or much heat are equally injurious to the teeth, which are endued with a fine ſenſation of this univerſal fluid. The beſt method of preſerving them is by the daily uſe of a bruſh, which is not very hard, with warm water and fine charcoal duſt. A lump of charcoal ſhould be put a ſecond time into the fire till it is red hot, as ſoon as it becomes cool the external aſhes ſhould be blown off, and it ſhould be immediately reduced to fine powder in a mortar, and kept cloſe ſtopped in a phial. It takes away the bad ſmell from decayed teeth, by waſhing the mouth with this powder diffuſed in water immediately. The putrid ſmell of decaying ſtumps of teeth may be deſtroyed for a time by waſhing the mouth with a weak ſolution of alum in water. If the calcareous cruſt upon the teeth adheres very firmly, a fine powder of pumice-ſtone may be uſed occaſionally, or a tooth inſtrument.

Acid of ſea-ſalt, much diluted, may be uſed; but this very rarely, and with the greateſt caution, as in cleaning ſea-ſhells. When the gums are ſpongy, they ſhould be frequently pricked with a lancet. Should black ſpots in teeth be cut out? Does the enamel grow again when it has been perforated or abraded?

13. *Otalgia.* Ear-ach ſometimes continues many days without apparent inflammation, and is then frequently removed by filling the ear with laudanum, or with ether; or even with warm oil, or warm water. See Claſs II. 1. 4. 8. This pain of the ear, like hemicrania, is frequently the conſequence of aſſociation with a diſeaſed tooth; in that caſe the ether ſhould be applied to the cheek over the ſuſpected tooth, or a grain of opium and as much camphor mixed together and applied to the ſuſpected tooth. In this caſe the otalgia belongs to the fourth claſs of diſeaſes.

14. *Pleurodyne chronica*. Chronical pain of the ſide. Pains of the membranous parts, which are not attended with fever, have acquired the general name of rheumatic; which ſhould, nevertheleſs, be reſtricted to thoſe pains which exiſt only when the parts are in motion, and which have been left after inflammation of them; as deſcribed in Claſs I. 1. 3. 12. The pain of the ſide here mentioned affects many ladies, and may poſſibly have been owing to the preſſure of tight ſtays, which has weakened the action of the veſſels compoſing ſome membranous part, as, like the cold head-ach, it is attended with preſent debility; in one patient, a boy about ten years old, it was attended with daily convulſions, and was ſuppoſed to have originated from worms. The diſeaſe is very frequent, and generally withſtands the uſe of bliſters on the part; but in ſome caſes I have known it removed by electric ſhocks repeated every day for a fortnight through the affected ſide.

Pains of the ſide may be ſometimes occaſioned by the adheſion of the lungs to the pleura, after an inflammation of them; or to the adheſion of ſome abdominal viſcera to their cavity, or to each other; which alſo are more liable to affect ladies from the unnatural and ungraceful preſſure of tight ſtays, or by ſitting or lying too long in one poſture. But in theſe caſes the pain ſhould be more of the ſmarting, than of the dull kind.

M. M.

M. M. Ether. A bliſter. A plaſter of Burgundy pitch. An iſſue or ſeton on the part. Electric ſhocks. Friction on the part with oil and camphor. Looſe dreſs. Frequent change of poſture both in the day and night. Internally opium, valerian, bark.

15. *Sciatica frigida.* Cold ſciatica. The pain along the courſe of the ſciatic nerve, from the hip quite down to the top of the foot, when it is not attended with fever, is improperly termed either rheumatiſm or gout; as it occurs without inflammation, is attended with pain when the limb is at reſt; and as the pain attends the courſe of the nerve, and not the courſe of the muſcles, or of the faſcia, which contains them. The theory of Cotunnius, who believed it to be a dropſy of the ſheath of the nerve, which was compreſſed by the accumulated fluid, has not been confirmed by diſſection. The diſeaſe ſeems to conſiſt of a torpor of this ſheath of the nerve, and the pain ſeems to be in conſequence of this torpor. See Claſs II. 1. 2. 13.

M. M. Veneſection. A cathartic. And then one grain of calomel and one of opium every night for ten ſucceſſive nights. And a bliſter, at the ſame time, a little above the knee-joint on the outſide of the thigh, where the ſciatic nerve is not ſo deep ſeated. Warm bath. Cold bath. Cover the limb with oiled ſilk, or with a plaſter-bandage of emplaſtrum de minio.

16. *Lumbago frigida.* Cold lumbago. When no fever or inflammation attends this pain of the loins, and the pain exiſts without motion, it belongs to this genus of diſeaſes, and reſembles the pain of the loins in the cold fit of ague. As theſe membranes are extenſive, and more eaſily fall into quieſcence, either by ſympathy, or when they are primarily affected, this diſeaſe becomes very afflicting, and of great pertinacity. See Claſs II. 1. 2. 17.

M. M. Veneſection. A cathartic. Iſſues on the loins. Adheſive plaſter on the loins. Bliſter on the os ſacrum. Warm bath. Cold


bath.

bath. Remove to a warmer climate in the winter. Loose dress about the waist. Friction daily with oil and camphor.

17. *Hysteralgia frigida.* Cold pain of the uterus preceding or accompanying menstruation. It is attended with cold extremities, want of appetite, and other marks of general debility.

M. M. A clyster of half a pint of gruel, and 30 drops of laudanum; or a grain of opium and six grains of rhubarb every night. To sit over warm water, or go into a warm bath.

18. *Proctalgia frigida.* Cold pain at the bottom of the rectum previous to the tumor of the piles, which sometimes extends by sympathy to the loins; it seems to be similar to the pain at the beginning of menstruation, and is owing to the torpor or inirritability of the extremity of the alimentary canal, or to the obstruction of the blood in its passage through the liver, when that viscus is affected, and its consequent delay in the veins of the rectum, occasioning tumors of them, and dull sensations of pain.

M. M. Calomel. A cathartic. Spice. Clyster, with 30 drops of laudanum. Sitting over warm water. If chalybeates after evacuation? See Class I. 2. 3. 23. and I. 2. 1. 6.

19. *Vesicæ felleæ inirritabilitas.* The inirritability of the gall-bladder probably occasions one kind of *icterus*, or jaundice; which is owing to whatever obstructs the passage of bile into the duodenum. The jaundice of aged people, and which attends some fevers, is believed to be most frequently caused by an irritative palsy of the gall-bladder; on which account the bile is not pressed from the cyst by its contraction, as in a paralysis of the urinary bladder.

A thickening of the coats of the common bile-duct by inflammation or increased action of their vessels so as to prevent the passage of the bile into the intestine, in the same manner as the membrane, which lines

lines the noftrils, becomes thickened in catarrh fo as to prevent the paffage of air through them, is probably another frequent caufe of jaundice, efpecially of children; and generally ceafes in about a fortnight, like a common catarrh, without the aid of medicine; which has given rife to the character, which charms have obtained in fome countries for curing the jaundice of young people.

The fpiffitude of the bile is another caufe of jaundice, as mentioned in Clafs I. 1. 3. 8. This alfo in children is a difeafe of little danger, as the gall-ducts are diftenfible, and will the eafier admit of the exclufion of gall-ftones; but becomes a more ferious difeafe in proportion to the age of the patient, and his habits of life in refpect to fpirituous potation.

A fourth caufe of jaundice is the compreffion of the bile-duct by the enlargement of an inflamed or fchirrous liver; this attends thofe who have drank much fpirituous liquor, and is generally fucceeded by dropfy and death.

M. M. Repeated emetics. Mild cathartics. Warm bath. Electricity. Bitters. Then fteel, which, when the pain and inflammation is removed by evacuations, acts like a charm in removing the remainder of the inflammation, and by promoting the abforption of the new veffels or fluids; like the application of any acrid eye-water at the end of ophthalmia; and thus the thickened coats of the bile-duct become reduced, or the enlargement of the liver leffened, and a free paffage is again opened for the bile into the inteftine. Ether with yolk of egg is recommended, as having a tendency to diffolve infpiffated bile. And a decoction of madder is recommended for the fame purpofe; becaufe the bile of animals, whofe food was mixed with madder, was found always in a dilute ftate. Aerated alcaline water, or Seltzer's water. Raw cabbage, and other acrid vegetables, as water-creffes, muftard. Horfes are faid to be fubject to infpiffated bile, with yellow eyes, in the winter feafon, and to get well as foon as they feed on the fpring grafs.

The largeſt bile-ſtone I have ſeen was from a lady, who had parted with it ſome years before, and who had abſtained above ten years from all kinds of vegetable diet to prevent, as ſhe ſuppoſed, a colic of her ſtomach, which was probably a pain of the biliary duct; on reſuming the uſe of ſome vegetable diet, ſhe recovered a better ſtate of health, and formed no new bilious concretions.

A ſtrong aerated alcaline water is ſold by J. Schweppe, No. 8, King's-ſtreet, Holborn. See Claſs I. 1. 3. 10.

20. *Pelvis renalis inirritabilitas.* Inirritability of the pelvis of the kidney. When the nucleus of a ſtone, whether it be inſpiſſated mucus, or other matter, is formed in the extremity of any of the tubuli uriniferi, and being detached from thence falls into the pelvis of the kidney, it is liable to lodge there from the want of due irritability of the membrane; and in that ſituation increaſes by new appoſitions of indurated animal matter, in the ſame manner as the ſtone of the bladder. This is the general cauſe of hæmorrhage from the kidney; and of obtuſe pain in it on exerciſe; or of acute pain, when the ſtone advances into the ureter. See Claſs I. 1. 3. 9.

ORDO

## ORDO II.

*Decreased Irritation.*

## GENUS V.

*Decreased Action of the Organs of Sense.*

### SPECIES.

1. *Stultitia inirritabilis.* Folly from inirritability. Dulness of perception. When the motions of the fibrous extremities of the nerves of sense are too weak to excite sensation with sufficient quickness and vigour. The irritative ideas are nevertheless performed, though perhaps in a feeble manner, as such people do not run against a post, or walk into a well. There are three other kinds of folly; that from deficient sensation, from deficient volition, and from deficient association, as will be mentioned in their places. In delirium, reverie, and sleep, the power of perception is abolished from other causes.

2. *Visus imminutus.* Diminished vision. In our approach to old age our vision becomes imperfect, not only from the form of the cornea, which becomes less convex, and from its decreased transparency mentioned in Class I. 2. 3. 26.; but also from the decreased irritability of the optic nerve. Thus, in the inirritative or nervous fever, the pupil of the eye becomes dilated; which in this, as well as in the dropsy of the brain, is generally a fatal symptom. A part of the cornea as well as a part of the albuginea in these fevers is frequently seen during sleep; which is owing to the inirritability of the retina to light, or to the general paresis of muscular action, and in consequence

 to

to the lefs contraction of the fphincter of the eye, if it may be fo called, at that time.

There have been inftances of fome, who could not diftinguifh certain colours; and yet whofe eyes, in other refpects, were not imperfect. Philof. Tranfact. Which feems to have been owing to the want of irritability, or the inaptitude to action, of fome claffes of fibres which compofe the retina. Other permanent defects depend on the difeafed ftate of the external organ. Clafs I. 1. 3. 14. I. 2. 3. 25. IV. 2. 1. 11.

3. *Mufcæ volitantes.* Dark fpots appearing before the eyes, and changing their apparent place with the motions of the eyes, are owing to a temporary defect of irritability of thofe parts of the retina, which have been lately expofed to more luminous objects than the other parts of it, as explained in Sect. XL. 2. Hence dark fpots are feen on the bed-clothes by patients, when the optic nerve is become lefs irritable, as in fevers with great debility; and the patients are perpetually trying to pick them off with their fingers to difcover what they are; for thefe parts of the retina of weak people are fooner exhaufted by the ftimulus of bright colours, and are longer in regaining their irritability.

Other kinds of ocular fpectra, as the coloured ones, are alfo more liable to remain in the eyes of people debilitated by fevers, and to produce various hallucinations of fight. For after the contraction of a mufcle, the fibres of it continue in the laft fituation, till fome antagonift mufcles are exerted to retract them; whence, when any one is much exhaufted by exercife, or by want of fleep, or in fevers, it is eafier to let the fibres of the retina remain in their laft fituation, after having been ftimulated into contraction, than to exert any antagonift fibres to replace them.

As the optic nerves at their entrance into the eyes are each of them as thick as a crow-quill, it appears that a great quantity of fenforial



power

power is expended during the day in the perpetual activity of our ſenſe of viſion, beſides that uſed in the motions of the eye-balls and eye-lids; as much I ſuppoſe as is expended in the motions of our arms, which are ſupplied with nerves of about the ſame diameters. From hence we may conclude, that the light ſhould be kept from patients in fevers with debility, to prevent the unneceſſary exhauſtion of the ſenſorial power. And that on the ſame account their rooms ſhould be kept ſilent as well as dark; that they ſhould be at reſt in an horizontal poſture; and be cooled by a blaſt of cool air, or by waſhing them with cold water, whenever their ſkins are warmer than natural.

4. *Strabiſmus.* Squinting is generally owing to one eye being leſs perfect than the other; on which account the patient endeavours to hide the worſt eye in the ſhadow of the noſe, that his viſion by the other may not be confuſed. Calves, which have an hydatide with inſects incloſed in it in the frontal ſinus on one ſide, turn towards the affected ſide; becauſe the viſion on that ſide, by the preſſure of the hydatide, becomes leſs perfect; and the diſeaſe being recent, the animal turns round, expecting to get a more diſtinct view of objects.

In the hydrocephalus internus, where both eyes are not become inſenſible, the patient ſquints with only one eye, and views objects with the other, as in common ſtrabiſmus. In this caſe it may be known on which ſide the diſeaſe exiſts, and that it does not exiſt on both ſides of the brain; in ſuch circumſtances, as the patients I believe never recover as they are now treated, might it not be adviſeable to perforate the cranium over the ventricule of the affected ſide? which might at leaſt give room and ſtimulus to the affected part of the brain?

M. M. If the ſquinting has not been confirmed by long habit, and one eye be not much worſe than the other, a piece of gauze ſtretched

ſtretched on a circle of whale-bone, to cover the beſt eye in ſuch a manner as to reduce the diſtinctneſs of viſion of this eye to a ſimilar degree of imperfection with the other, ſhould be worn ſome hours every day. Or the better eye ſhould be totally darkened by a tin cup covered with black ſilk for ſome hours daily, by which means the better eye will be gradually weakened by the want of uſe, and the worſe eye will be gradually ſtrengthened by uſing it. Covering an inflamed eye in children for weeks together, is very liable to produce ſquinting, for the ſame reaſon.

5. *Amauroſis.* Gutta ſerena. Is a blindneſs from the inirritability of the optic nerve. It is generally eſteemed a palſy of the nerve, but ſhould rather be deemed the death of it, as paralyſis has generally been applied to a deprivation only of voluntary power. This is a diſeaſe of dark eyes only, as the cataract is a diſeaſe of light eyes only. At the commencement of this diſeaſe, very minute electric ſhocks ſhould be repeatedly paſſed through the eyes; ſuch as may be produced by putting one edge of a piece of ſilver the ſize of a half-crown piece beneath the tongue, and one edge of a piece of zinc of a ſimilar ſize between the upper lip and the gum, and then repeatedly bringing their exterior edges into contact, by which means very ſmall electric ſparks become viſible in the eyes. See additional note at the end of the firſt volume, p. 567. and Sect. XIV. 5.

M. M. Minute electric ſhocks. A grain of opium, and a quarter of a grain of corroſive ſublimate of mercury, twice a day for four or ſix weeks. Bliſter on the crown of the head.

6. *Auditus imminutus.* Diminiſhed hearing. Deafneſs is a frequent ſymptom in thoſe inflammatory or ſenſitive fevers with debility, which are generally called putrid; it attends the general ſtupor in thoſe fevers, and is rather eſteemed a ſalutary ſign, as during this ſtupor there is leſs expenditure of ſenſorial power.

In

In fevers of debility without inflammation, called nervous fevers, I ſuſpect deafneſs to be a bad ſymptom, ariſing like the dilated pupil from a partial paralyſis of the nerve of ſenſe. See Claſs IV. 2. 1. 15.

Nervous fevers are ſuppoſed by Dr. Gilchriſt to originate from a congeſtion of ſerum or water in ſome part of the brain, as many of the ſymptoms are ſo ſimilar to thoſe of hydrocephalus internus, in which a fluid is accumulated in the ventricules of the brain; on this idea the inactivity of the optic or auditory nerves in theſe fevers may ariſe from the compreſſion of the effuſed fluid; while the torpor attending putrid fever may depend on the meninges of the brain being thickened by inflammation, and thus compreſſing it; now the new veſſels, or the blood, which thickens inflamed parts, is more frequently reabſorbed, than the effuſed fluid from a cavity; and hence the ſtupor in one caſe is leſs dangerous than in the other.

In inflammatory or ſenſitive fevers with debility, deafneſs may ſometimes ariſe from a greater ſecretion and abſorption of the ear-wax, which is very ſimilar to the bile, and is liable to fill the meatus auditorius, when it is too viſcid, as bile obſtructs the gall-ducts.

M. M. In deafneſs without fever Dr. Darwin applied a cupping-glaſs on the ear with good effect, as deſcribed in Phil. Tranſ. Vol. LXIV. p. 348. Oil, ether, laudanum, dropped into the ears.

7. *Olfactus imminutus.* Inactivity of the ſenſe of ſmell. From our habits of truſting to the art of cookery, and not examining our food by the ſmell as other animals do, our ſenſe of ſmell is leſs perfect than theirs. See Sect. XVI. 5. Claſs IV. 2. 1. 16.

M. M. Mild errhines.

8. *Guſtus imminutus.* Want of taſte is very common in fevers, owing frequently to the dryneſs or ſcurf of the tongue, or external organ of that ſenſe, rather than to any injury of the nerves of taſte. See Claſs I. 1. 3. 1. IV. 2. 1. 16.

M. M Warm ſubacid liquids taken frequently.

9. *Tactus imminutus.* Numbneſs is frequently complained of in fevers, and in epilepſy, and the touch is ſometimes impaired by the dryneſs of the cuticle of the fingers. See Claſs IV. 2. 1. 16.

When the ſenſe of touch is impaired by the compreſſion of the nerve, as in ſitting long with one thigh croſſed over the other, the limb appears larger, when we touch it with our hands, which is to be aſcribed to the indiſtinctneſs of the ſenſation of touch, and may be explained in the ſame manner as the apparent largeneſs of objects ſeen through a miſt. In this laſt caſe the minute parts of an object, as ſuppoſe of a diſtant boy, are ſeen leſs diſtinctly, and therefore we inſtantly conceive them to be further from the eye, and in conſequence that the whole ſubtends a larger angle, and thus we believe the boy to be a man. So when any one's fingers are preſſed on a benumbed limb, the ſenſation produced is leſs than it ſhould be, judging from viſible circumſtances; we therefore conceive, that ſomething intervened between the object and the ſenſe, for it is felt as if a blanket was put between them; and that not being viſibly the caſe, we judge that the limb is ſwelled.

The ſenſe of touch is alſo liable to be deceived from the acquired habits of one part of it acting in the vicinity of another part of it. Thus if the middle finger be croſſed over either of the fingers next to it, and a nut be felt by the two ends of the fingers ſo croſſed at the ſame time, the nut appears as if it was two nuts. And laſtly, the ſenſe of touch is liable to be deceived by preconceived ideas; which we believe to be excited by external objects, even when we are awake. It has happened to me more than once, and I ſuppoſe to moſt others, to have put my hands into an empty baſon ſtanding in an obſcure corner of a room to waſh them, which I believed to contain cold water, and have inſtantly perceived a ſenſation of warmth, contrary to that which I expected to have felt.

In ſome paralytic affections, and in cold fits of ague, the ſenſation of touch has been much impaired, and yet that of heat has remained. See Sect. XIV. 6.

M. M. Friction

M. M. Friction alone, or with camphorated oil, warm bath. Ether. Volatile alcali and water. Internally ſpice, ſalt. Incitantia. Secernentia.

10. *Stupor.* The ſtupor, which occurs in fevers with debility, is generally eſteemed a favourable ſymptom; which may ariſe from the leſs expenditure of ſenſorial power already exiſting in the brain and nerves, as mentioned in ſpecies 6 of this genus. But if we ſuppoſe, that there is a continued production of ſenſorial power, or an accumulation of it in the torpid parts of the ſyſtem, which is not improbable, becauſe ſuch a production of it continues during ſleep, to which ſtupor is much allied, there is ſtill further reaſon for believing it to be a favourable ſymptom in inirritable fevers; and that much injury is often done by bliſters and other powerful ſtimuli to remove the ſtupor. See Sect. XII. 7. 8. and XXXIII. 1. 4.

Dr. Blane in his Croonian Lecture on muſcular motion for 1788, among many other ingenious obſervations and deductions, relates a curious experiment on ſalmon, and other fiſh, and which he repeated upon eels with ſimilar event.

"If a fiſh, immediately upon being taken out of the water, is ſtunned by a violent blow on the head, or by having the head cruſhed, the irritability and ſweetneſs of the muſcles will be preſerved much longer, than if it had been allowed to die with the organs of ſenſe entire. This is ſo well known to fiſhermen, that they put it in practice, in order to make them longer ſuſceptible of the operation called *crimping.* A ſalmon is one of the fiſh leaſt tenacious of life, inſomuch, that it will loſe all ſigns of life in leſs than half an hour after it is taken out of the water, if ſuffered to die without any farther injury; but if, immediately after being caught, it receives a violent blow on the head, the muſcles will ſhew viſible irritability for more than twelve hours afterwards."

Dr. Blane afterwards well remarks, that "in thoſe diſorders in which the exerciſe of the ſenſes is in a great meaſure deſtroyed, or ſuſpended, as in the hydrocephalus, and apoplectic palſy, it happens not uncommonly, that the appetite and digeſtion are better than in health."

ORDO

## ORDO III.

### *Retrograde Irritative Motions.*

## GENUS I.

### *Of the Alimentary Canal.*

THE retrograde motions of our ſyſtem originate either from defect of ſtimulus, or from defect of irritability. Thus ſickneſs is often induced by hunger, which is a want of ſtimulus; and from ipecacuanha, in which laſt caſe it would ſeem, that the ſickneſs was induced after the violence of the ſtimulus was abated, and the conſequent torpor had ſucceeded. Hence ſpice, opium, or food relieves ſickneſs.

The globus hyſtericus, ſalivation, diabætes, and other inverſions of motion attending hyſteric paroxyſms, ſeem to depend on the want of irritability of thoſe parts of the body, becauſe they are attended with cold extremities, and general debility, and are relieved by wine, opium, ſteel, and fleſh diet; that is, by any additional ſtimulus.

When the longitudinal muſcles are fatigued by long action, or are habitually weaker than natural, the antagoniſt muſcles replace the limb by ſtretching it in a contrary direction; and as theſe muſcles have had their actions aſſociated in ſynchronous tribes, their actions ceaſe together. But as the hollow muſcles propel the fluids, which they contain, by motions aſſociated in trains; when one ring is fatigued from its too great debility, and brought into retrograde action; the next ring, and the next, from its aſſociation in train falls into retrograde action. Which continue ſo long as they are excited to act, like the tremors of the hands of infirm people, ſo long as they endeavour to act.

act. Now as these hollow muscles are perpetually stimulated, these retrograde actions do not cease as the tremors of the longitudinal muscles, which are generally excited only by volition. Whence the retrograde motions of hollow muscles depend on two circumstances, in which they differ from the longitudinal muscles, namely, their motions being associated in trains, and their being subject to perpetual stimulus. For further elucidation of the cause of this curious source of diseases, see Sect. XXIX. 11. 5.

The fluids disgorged by the retrograde motions of the various vascular muscles may be distinguished, 1. From those, which are produced by secretion, by their not being attended by increase of heat, which always accompanies increased secretion. 2. They may be distinguished from those fluids, which are the consequence of deficient absorption, by their not possessing the saline acrimony, which those fluids possess; which inflames the skin or other membranes on which they fall; and which have a saline taste to the tongue. 3. They may be distinguished from those fluids, which are the consequence both of increased secretion and absorption, as these are attended with increase of warmth, and are inspissated by the abstraction of their aqueous parts. 4. Where chyle, or milk, are found in the feces or urine, or when other fluids, as matter, are translated from one part of the system to another, they have been the product of retrograde action of lymphatic or other canals. As explained in Sect. XXIX. 8.

## SPECIES.

1. *Ruminatio.* In the rumination of horned cattle the retograde motions of the œsophagus are visible to the eye, as they bring up the softened grass from their first stomach. The vegetable aliment in the first stomach of cattle, which have filled themselves too full of young clover, is liable to run into fermentation, and distend the stomach,

so

ſo as to preclude its exit, and frequently to deſtroy the animal. To diſcharge this air the farmers frequently make an opening into the ſtomach of the animal with ſucceſs. I was informed, I believe by the late Dr. Whytt of Edinburgh, that of twenty cows in this ſituation two had died; and that he directed a pint of gin or whiſky, mixed with an equal quantity of water, to be given to the other eighteen; all of which eructed immenſe quantities of air, and recovered.

There are hiſtories of ruminating men, and who have taken pleaſure in the act of chewing their food a ſecond time. Philoſ. Tranſact.

2. *Ructus.* Eructation. An inverted motion of the ſtomach excluding through its upper valve an elaſtic vapour generated by the fermentation of the aliment; which proceeds ſo haſtily, that the digeſtive power does not ſubdue it. This is ſometimes acquired by habit, ſo that ſome people can eruct when they pleaſe, and as long as they pleaſe; and there is gas enough generated to ſupply them for this purpoſe; for by Dr. Hale's experiments, an apple, and many other kinds of aliment, give up above ſix hundred times their own bulk of an elaſtic gas in fermentation. When people voluntarily eject the fixable air from their ſtomachs, the fermentation of the aliment proceeds the faſter; for ſtopping the veſſels, which contain new wines, retards their fermentation, and opening them again accelerates it; hence where the digeſtion is impaired, and the ſtomach ſomewhat diſtended with air, it is better to reſtrain than to encourage eructations, except the quantity makes it neceſſary. When wine is confined in bottles the fermentation ſtill proceeds ſlowly even for years, till all the ſugar is converted into ſpirit; but in the proceſs of digeſtion, the ſaccharine part is abſorbed in the form of chyle by the bibulous mouths of the numerous lacteals, before it has time to run into the vinous fermentation.

3. *Apepſia.* Indigeſtion. Water-qualm. A few mouthfuls of

the aliment are rejected at a time for some hours after meals. When the aliment has had time to ferment, and become acid, it produces cardialgia, or heart-burn. This disease is perhaps generally left after a slight inflammation of the stomach, called a surfeit, occasioned by drinking cold liquors, or eating cold vegetables, when heated with exercise. This inflammation of the stomach is frequently, I believe, at its commencement removed by a critical eruption on the face, which differs in its appearance as well as in its cause from the gutta rosea of drunkards, as the skin round the base of each eruption is less inflamed. See Class II. 1. 4. 7. This disease differs from Cardialgia, Class I. 2. 4. 5. in its being not uniformly attended with pain of the cardia ventriculi, and from its retrograde motions of a part of the stomach about the upper orifice of it. In the same manner as hysteria differs from hypochondriasis; the one consisting in the weakness and indigestion of the same portions of the alimentary canal, and the other in the inverted motions of some parts of it. This apepsia or water-qualm continues many years, even to old age; Mr. G—— of Lichfield suffered under this disease from his infancy; and, as he grew old, found relief only from repeated doses of opium.

M. M. A blister, rhubarb, a grain of opium twice a day. Soap, iron-powder. Tin-powder.

4. *Vomitus.* An inverted order of the motions of the stomach and œsophagus with their absorbent vessels, by which their contents are evacuated. In the act of vomiting less sensorial power is employed than in the usual peristaltic motion of the stomach, as explained in Sect. XXXV. 1. 3. Whence after the operation of an emetic the digestion becomes stronger by an accumulation of sensorial power during its decreased action. This decreased action of the stomach may be either induced by want of stimulus, as in the sickness which attends hunger; or it may be induced by temporary want of irritability, as in cold fits of fever; or from habitual want of irritability, as the vomiting

vomiting of enfeebled drunkards. Or laſtly, by having been previouſly too violently ſtimulated by an emetic drug, as by ipecacuanha.

M. M. A bliſter. An emetic. Opium. Warmth of a bed, covering the face for a while with the bed-clothes. Crude mercury. A poultice with opium or theriaca externally.

5. *Cholera.* When not only the ſtomach, as in the laſt article, but alſo the duodenum, and ilium, as low as the valve of the colon, have their motions inverted; and great quantities of bile are thus poured into the ſtomach; while at the ſame time ſome branches of the lacteals become retrograde, and diſgorge their contents into the upper part of the alimentary canal; and other branches of them diſgorge their contents into the lower parts of it beneath the valve of the colon; a vomiting and purging commence together, which is called cholera, as it is ſuppoſed to have its origin from increaſed ſecretion of bile; but I ſuppoſe more frequently ariſes from putrid food, or poiſonous drugs, as in the caſe narrated in Sect. XXV. 13. where other circumſtances of this diſeaſe are explained. See Claſs II. 1. 2. 11.

The cramps of the legs, which are liable to attend cholera, are explained in Claſs III. 1. 1. 15.

6. *Ileus.* Conſiſts in the inverted motions of the whole inteſtinal canal, from the mouth to the anus; and of the lacteals and abſorbents which ariſe from it. In this pitiable diſeaſe, through the valve of the colon, through the pylorus, the cardia, and the pharinx, are ejected, firſt, the contents of the ſtomach and inteſtines, with the excrement and even clyſters themſelves; then the fluid from the lacteals, which is now poured into the inteſtines by their retrograde motions, is thrown up by the mouth; and, laſtly, every fluid, which is abſorbed by the other lymphatic branches, from the cellular membrane, the ſkin, the bladder, and all other cavities of the body; and which

is then poured into the ſtomach or inteſtines by the retrograde motions of the lacteals; all which ſupply that amazing quantity of fluid, which is in this diſeaſe continually ejected by vomiting. See Sect. XXV. 15. for a further explanation of this diſeaſe.

M. M. Copious veneſection. Twenty grains of calomel in ſmall pills, or one grain of aloe every hour till ſtools are procured. Bliſters. Warm bath. Crude mercury. Clyſter of ice-water. Smear the ſkin all over with greaſe, as mentioned in Sect. XXV. 15.

As this malady is occaſioned ſometimes by an introſuſception of a part of the inteſtine into another part of it, eſpecially in children, could holding them up by their heels for a ſecond or two of time be of ſervice after veneſection? Or the exhibition of crude quickſilver two ounces every half hour, till a pound is taken, be particularly ſerviceable in this circumſtance? Or could half a pound, or a pound, of crude mercury be injected as a clyſter, the patient being elevated by the knees and thighs ſo as to have his head and ſhoulders much lower than his bottom, or even for a ſhort time held up by the heels? Could this alſo be of advantage in ſtrangulated hernia?

Where the diſeaſe is owing to ſtrangulated hernia, the part ſhould be ſprinkled with cold water, or iced water, or ſalt and water recently mixed, or moiſtened with ether. In caſes of ſtrangulated hernia, could acupuncture, or puncture with a capillary trocar, be uſed with ſafety and advantage to give exit to air contained in the ſtrangulated bowel? Or to ſtimulate it into action? It is not uncommon for baſhful men to conceal their being afflicted with a ſmall hernia, which is the cauſe of their death; this circumſtance ſhould therefore always be enquired into. Is the ſeat or cauſe of the ileus always below the valve of the colon, and that of the cholera above it? See Claſs II. 1. 2. 11.

7. *Globus hyſtericus.* Hyſteric ſuffocation is the perception of a globe rolling round in the abdomen, and aſcending to the ſtomach and

and throat, and there inducing ſtrangulation. It conſiſts of an ineffectual inverſion of the motions of the œſophagus, and other parts of the alimentary canal; nothing being rejected from the ſtomach.

M. M. Tincture of caſtor. Tinct. of opium of each 15 drops. See Hyſteria, Claſs I. 3. 1. 9.

8. *Vomendi conamen inane.* An ineffectual effort to vomit. It frequently occurs, when the ſtomach is empty, and in ſome caſes continues many hours; but as the lymphatics of the ſtomach are not inverted at the ſame time, there is no ſupply of materials to be ejected; it is ſometimes a ſymptom of hyſteria, but more frequently attends irregular epilepſies or reveries; which however may be diſtinguiſhed by their violence of exertion, for the exertions of hyſteric motions are feeble, as they are cauſed by debility; but thoſe of epilepſies, as they are uſed to relieve pain, are of the moſt violent kind; inſomuch that thoſe who have once ſeen theſe ineffectual efforts to vomit in ſome epilepſies, can never again miſtake them for ſymptoms of hyſteria. See a caſe in Sect. XIX. 2.

M. M. Bliſter. Opium. Crude mercury.

9. *Borborigmus.* A gurgling of the bowels proceeds from a partial invertion of the periſtaltic motions of them, by which the gas is brought into a ſuperior part of the bowel, and bubbles through the deſcending fluid, like air ruſhing into a bottle as the water is poured out of it. This is ſometimes a diſtreſſing ſymptom of the debility of the bowels joined with a partial inverſion of their motions. I attended a young lady about ſixteen, who was in other reſpects feeble; whoſe bowels almoſt inceſſantly made a gurgling noiſe ſo loud as to be heard at a conſiderable diſtance, and to attract the notice of all who were near her. As this noiſe never ceaſed a minute together for many hours in a day, it could not be produced by the uniform deſcent of

water, and aſcent of air through it, but there muſt have been alternately a retrograde movement of a part of the bowel, which muſt again have puſhed up the water above the air; or which might raiſe a part of the bowel, in which the fluid was lodged, alternately above and below another portion of it; which might readily happen in ſome of the curvatures of the ſmaller inteſtines, the air in which might be moved backward and forward like the air-bubble in a glaſs-level.

M. M. Eſſential oil. Ten corns of black pepper ſwallowed whole after dinner, that its effect might be ſlower and more permanent; a ſmall pipe occaſionally introduced into the rectum to facilitate the eſcape of the air. Crude mercury. See Claſs I. 2. 4. 9.

10. *Hyſteria.* The three laſt articles, together with the lymphatic diabætes, are the moſt common ſymptoms of the hyſteric diſeaſe; to which ſometimes is added the lymphatic ſalivation, and fits of ſyncope, or convulſion, with palpitation of the heart (which probably conſiſts of retrograde motions of it), and a great fear of dying. Which laſt circumſtance diſtinguiſhes theſe convulſions from the epileptic ones with greater certainty than any other ſingle ſymptom. The pale copious urine, cold ſkin, palpitation, and trembling, are the ſymptoms excited by great fear. Hence in hyſteric diſeaſes, when theſe ſymptoms occur, the fear, which has been uſually aſſociated with them, recurs at the ſame time, as in hypochondriaſis, Claſs I. 2. 4. 10. See Sect. XVI. 8. 1.

The convulſions which ſometimes attend the hyſteric diſeaſe, are exertions to relieve pain, either of ſome torpid, or of ſome retrograde organ; and in this reſpect they reſemble epileptic convulſions, except that they are ſeldom ſo violent as entirely to produce inſenſibility to external ſtimuli; for theſe weaker pains ceaſe before the total exhauſtion of ſenſorial power is produced, and the patient

ſinks

ſinks into imperfect ſyncope; whereas the true epilepſy generally terminates in temporary apoplexy, with perfect inſenſibility to external objects. Theſe convulſions are leſs to be dreaded than the epileptic ones, as they do not originate from ſo permanent a cauſe.

The great diſcharge of pale urine in this diſeaſe is owing to the inverted motions of the lymphatics, which ariſe about the neck of the bladder, as deſcribed in Sect. XXIX. 4. 5. And the lymphatic ſalivation ariſes from the inverted motions of the ſalivary lymphatics.

Hyſteria is diſtinguiſhed from hypochondriaſis, as in the latter there are no retrograde motions of the alimentary canal, but ſimply a debility or inirritability of it, with diſtention and flatulency. It is diſtinguiſhed from apepſia and cardialgia by there being nothing ejected from the ſtomach by the retrograde motions of it, or of the œſophagus.

M. M. Opium. Camphor. Aſſafœtida. Caſtor, with ſinapiſms externally; to which muſt be added a clyſter of cold water, or iced water; which, according to Monſ. Pomme, relieves theſe hyſteric ſymptoms inſtantaneouſly like a charm; which it may effect by checking the inverted motions of the inteſtinal canal by the torpor occaſioned by cold; or one end of the inteſtinal canal may become ſtrengthened, and regain its periſtaltic motion by reverſe ſympathy, when the other end is rendered torpid by ice-water. (Pomme des Affections Vaporeuſes, p. 25.) Theſe remove the preſent ſymptoms; and bark, ſteel, exerciſe, coldiſh bath, prevent their returns. See Art. VI. 2. 1.

11. *Hydrophobia.* Dread of water occaſioned by the bite of a mad dog, is a violent inverſion of the motions of the œſophagus on the contact or even approach of water or other fluids. The pharinx ſeems to have acquired the ſenſibility of the larinx in this diſeaſe, and

is

is as impatient to reject any fluid, which gets into it. Is not the cardia ventriculi the seat of this disease? As in cardialgia the pain is often felt in the pharinx, when the acid material stimulates the other end of the canal, which terminates in the stomach. As this fatal disease resembles tetanus, or locked jaw, in its tendency to convulsion from a distant wound, and affects some other parts by association, it is treated of in Class III. 1. 1. 15. and IV. 2. 1. 7.

ORDO

## ORDO III.

*Retrograde Irritative Motions.*

## GENUS II.

*Of the Abſorbent Syſtem.*

## SPECIES.

1. *Catarrhus lymphaticus.* Lymphatic catarrh. A periodical defluxion of a thin fluid from the noſtrils, for a few hours, occaſioned by the retrograde motions of their lymphatics; which may probably be ſupplied with fluid by the increaſed abſorption of ſome other lymphatic branches in their vicinity. It is diſtinguiſhed from that mucous diſcharge, which happens in froſty weather from decreaſed abſorption, becauſe it is leſs ſalt to the taſte; and from an increaſed ſecretion of mucus, becauſe it is neither ſo viſcid, nor is attended with heat of the part. This complaint is liable to recur at diurnal periods, like an intermittent fever, for weeks and months together, with great ſneezing and very copious diſcharge for an hour or two.

I have ſeen two of theſe caſes, both of which occurred in delicate women, and ſeemed an appendage to other hyſteric ſymptoms; whence I concluded, that the diſcharge was occaſioned by the inverted motions of the lymphatics of the noſtrils, like the pale urine in hyſteric caſes; and that they might receive this fluid from ſome other branches of lymphatic veſſels opening into the frontal or maxillary cavities in their vicinity.

Could ſuch a diſcharge be produced by ſtrong errhines, and excite an abſorption of the congeſtion of lymph in the dropſy of the brain?

2. *Salivatio lymphatica.* Lymphatic ſalivation. A copious expuition of a pellucid inſipid fluid, occaſioned by the retrograde motions of the lymphatics of the mouth. It is ſometimes periodical, and often attends the hyſteric diſeaſe, and nervous fevers; but is not accompanied with a ſaline taſte, or with heat of the mouth, or nauſea.

3. *Nauſea humida.* Moiſt nauſea conſiſts in a diſcharge of fluid, owing to the retrograde motions of the lymphatics about the fauces, without increaſe of heat, or ſaline taſte, together with ſome retrograde motions of the fauces or pharinx; along with this nauſea a ſickneſs generally precedes the act of vomiting; which may conſiſt of a ſimilar diſcharge of mucus or chyle into the ſtomach by the retrograde motions of the lymphatics or lacteals, which open into it. See Claſs I. 2. 4. 3. and I. 2. 4. 4.

M. M. Subacid liquids. Wine. Opium. A bliſter.

4. *Diarrhœa lymphatica.* Lymphatic diarrhœa. A quantity of mucus and lymph are poured into the inteſtines by the inverted motions of the inteſtinal lymphatics. The feces are leſs fetid and more liquid; and it ſometimes portends the commencement of a diabætes, or dropſy, or their temporary relief. This lymphatic diarrhœa ſometimes becomes chronical, in which the atmoſpheric moiſture, abſorbed by the cutaneous and pulmonary lymphatics, is poured into the inteſtines by the retrograde motions of the lacteals. See Section XXIX. 4. 6. where ſome caſes of this kind are related.

5. *Diarrhœa chylifera. cœliaca.* Chyliferous diarrhœa. The chyle drank up by the lacteals of the upper inteſtines is poured into the

the lower ones by the retrograde motions of their lacteals, and appears in the dejections. This circumstance occurs at the beginning of diarrhœa crapulosa, where the patient has taken and digested more aliment than the system can conveniently receive, and thus eliminates a part of it; as appears when there is curdled chyle in some of the dejections. See Sect. XXIX. 4. 7. It differs from the lymphatic diarrhœa, as the chyliferous diabætes differs from the aqueous and mucaginous diabætes.

6. *Diabætes.* By the retrograde motions of the urinary lymphatics, an immense quantity of fluid is poured into the bladder. It is either termed chyliferous, or aqueous, or mucaginous, from the nature of the fluid brought into the bladder; and is either a temporary disease, as in hysteric women, in the beginning of intoxication, in worm cases, or in those exposed to cold damp air, or to great fear, or anxiety, or in the commencement of some dropsies; or it becomes chronical.

When the urinary lymphatics invert their motions, and pour their refluent contents into the bladder, some other branch of the absorbent system acts with greater energy to supply this fluid. If it is the intestinal branch, the chyliferous diabætes is produced: if it is the cutaneous or pulmonary branch, the aqueous diabætes is produced: and if the cellular or cystic branches, the mucaginous diabætes. In the two last the urine is pellucid, and contains no sugar.

In dropsies the fluid is sometimes absorbed, and poured into the bladder by the retrograde motions of the urinary lymphatics, as during the exhibition of digitalis. In the beginning of the dropsies of infirm gouty patients, I have frequently observed, that they make a large quantity of water for one night, which relieves them for several days. In these cases the patient previously feels a fulness about the precordia, with difficult respiration, and symptoms similar to those of hysteria. Perhaps a previous defect of absorption takes place in some part of the body in thoſe hyſteric caſes, which are relieved by a copious diſcharge of pale urine. See Diabætes explained at large, Section XXIX. 4.

A diſcharge of blood ſometimes attends the diabætes, which was occaſionally a ſymptom of that diſeaſe in Mr. Brindley, the great navigable canal maker in this country. Which may be accounted for by the communication of a lymphatic branch with the gaſtric branch of the vena portarum, as diſcovered by J. F. Meckel. See Section XXVII. 2.

M. M. Alum. Earth of Alum. Cantharides. Calomel. Bark. Steel. Roſin. Opium. See Sect. XXIX. 4.

7. *Sudor lymphaticus.* Profuſe ſweats from the inverted motions of the cutaneous lymphatics, as in ſome fainting fits, and at the approach of death; and as perhaps in the ſudor anglicanus. See Sect. XXIX. 5. Theſe ſweats are glutinous to the touch, and without increaſed heat of the ſkin; if the part is not covered, the ſkin becomes cold from the evaporation of the fluid. Theſe ſweats without heat ſometimes occur in the act of vomiting, as in Sect. XXV. 9. and are probably the cauſe of the cold ſweaty hands of ſome people. As mentioned in Sect. XXIX. 4. 9. in the caſe of R. Davis, which he cured by frequent application of lime. Though it is poſſible, that cold ſweaty hands may alſo ariſe from the want of due abſorption of the perſpirable matter effuſed on them, and that the coldneſs may be owing to the greater evaporation in conſequence.

The acid ſweats deſcribed by Dr. Dobſon, which he obſerved in a diabætic patient, and aſcribes to the chyle effuſed on the ſkin, muſt be aſcribed to the retrograde action of the cutaneous lymphatics. See Sect. XXIX. 6.

8. *Sudor aſthmaticus.* The cold ſweats in this diſeaſe only cover the head, arms, and breaſt, and are frequently exceedingly profuſe.

Theſe ſweats are owing to the inverted motions of the cutaneous lymphatics of the upper part of the body, and at the ſame time the increaſed abſorption of the pulmonary abſorbents: hence theſe ſweats when profuſe relieve the preſent fit of aſthma. There is no other way to account for ſweats appearing on the upper parts of the body only, but by the fluid having been abſorbed by the lymphatic branch of the lungs, and effuſed on the ſkin by the retrograde movements of the cutaneous lymphatics; which join thoſe of the lungs before they enter into the venous circulation. For if they were occaſioned, as generally ſuppoſed, by the difficulty of the circulation of the blood through the lungs, the whole ſkin muſt be equally affected, both of the upper and lower parts of the body; for whatever could obſtruct the circulation in the upper part of the venous ſyſtem, muſt equally obſtruct it in the lower part of it. See Sect. XXIX. 6. In the convulſive aſthma theſe ſweats do not occur; hence they may be diſtinguiſhed; and might be called the hydropic aſthma, and the epileptic aſthma.

9. *Tranſlatio puris.* Tranſlation of matter from one part of the ſyſtem to another can only be explained from its being abſorbed by one branch of the lymphatic ſyſtem, and depoſited in a diſtant part by the retrograde motions of another branch; as mentioned Sect. XXIX. 7. 1. It is curious, that theſe tranſlations of matter are attended generally, I believe, with cold fits; for leſs heat is produced during the retrograde action of this part of the ſyſtem, as no ſecretion in the lymphatic glands of the affected branches can exiſt at the ſame time. Do any ineffectual retrograde motions occaſion the cold fits of agues? The time when the gout of the liver ceaſes, and the gout in the foot commences, is attended with a cold fit, as I have obſerved in two inſtances, which is difficult to explain, without ſuppoſing the new veſſels, or the matter produced on the inflamed liver, to be abſorbed,

and either eliminated by ſome retrograde motion, or carried to the newly inflamed part? See Claſs IV. 1. 2. 15.

10. *Tranſlatio lactis.* Tranſlation of milk to the bowels in puerperal fevers can only be explained by the milk being abſorbed by the pectoral branch of lymphatics, and carried to the bowels by the retrograde motions of the inteſtinal lymphatics or lacteals. See many inſtances of this in Sect. XXIX. 7. 4.

11. *Tranſlatio urinæ.* Tranſlation of urine. There is a curious caſe related in the Tranſaction of the College of Phyſicians at Philadelphia, Vol. I. p. 96. of a girl, who labouring under an iſcuria vomited her urine for many months; which could not be diſtinguiſhed from that which was at other times drawn off by the catheter. After having taken much opium, ſhe ſeems at length to have formed gravel, ſome of which was frequently brought up by vomiting. Dr. Senter aſcribes this to the retrograde motions of the lymphatics of the ſtomach, and the increaſed ones of thoſe of the bladder, and refers to thoſe of Sect. XXIX. of this work; which ſection was firſt publiſhed in 1780; and to Macquire's Dictionary of Chemiſtry, Art. Urine.

The patient above deſcribed ſometimes had a diſcharge of urine by the navel, and at other times by the rectum, and ſometimes by urinous ſweats.

ORDO

## ORDO III.

*Retrograde Irritative Motions.*

## GENUS III.

*Of the Sanguiferous Syſtem.*

### SPECIES.

1. *Capillarium motus retrogreſſus.* In microſcopic experiments it is uſual to ſee globules of blood regurgitate from the capillary veſſels again and again, before they paſs through them; and not only the mouths of the veins, which ariſe from theſe capillaries, are frequently ſeen by microſcopes to regurgitate ſome particles of blood during the ſtruggles of the animal; but a retrograde motion of the blood in the veins of theſe animals, from the very heart to the extremities of the limbs, is obſervable by intervals during the diſtreſſes of the dying creature. Haller, Elem. Phyſ. T. i. p. 216. See Section XXIX. 3. 8.

2. *Palpitatio cordis.* May not the ineffectual and weak unequal motions of the heart in hyſteric caſes be aſcribed to the retrograde motions of it, which continue for a ſhort time, or terminate in ſyncope? See Claſs IV. 3. 1. 6.

3. *Anhelatio*

3. *Anhelatio ſpaſmodica.* In ſome aſthmas may not the difficulty of reſpiration ariſe from the inverted action of the finer branches of the bronchia, or of the pulmonary artery or vein, like thoſe of the capillaries above deſcribed in No. 1. of this genus?

*The*

*The Orders and Genera of the Second Clafs of Difeafes.*

---

# CLASS II.

## DISEASES OF SENSATION.

### ORDO I.

*Increafed Senfation.*

### GENERA.

1. With increafed action of the mufcles.
2. With the production of new veffels by internal membranes or glands with fever.
3. With the production of new veffels by external membranes or glands with fever.
4. With the production of new veffels by internal membranes or glands without fever.
5. With the production of new veffels by external membranes or glands without fever.
6. With fever confequent to the production of new veffels or fluids.
7. With increafed action of the organs of fenfe.

### ORDO II.

*Decreafed Senfation.*

### GENERA.

1. With decreafed actions of the general fyftem.
2. With decreafed actions of particular organs.

### ORDO III.

*Retrograde Senfitive Motions.*

### GENERA.

1. Of the arterial fyftem.
2. Of the abforbent fyftem.
3. Of the excretory ducts.

## *The Orders, Genera, and Species, of the Second Clafs of Difeafes.*

## CLASS II.

### DISEASES OF SENSATION.

### ORDO I.

*Increafed Senfation.*

### GENUS I.

*With Increafed Action of the Mufcles.*

### SPECIES.

| | |
|---|---|
| 1. *Deglutitio.* | Deglutition. |
| 2. *Refpiratio.* | Refpiration. |
| 3. *Sternutatio.* | Sneezing. |
| 4. *Anhelitus.* | Panting. |
| 5. *Tuffis ebriorum.* | Cough of inebriates. |
| 6. *Syngultus.* | Hiccough. |
| 7. *Afthma humorale.* | Humoral afthma. |
| 8. *Nictitatio fenfitiva.* | Winking from pain. |
| 9. *Ofcitatio et pandiculatio.* | Yawning and ftretching. |
| 10. *Tenefmus.* | Tenefmus. |
| 11. *Stranguria.* | Strangury. |
| 12. *Parturitio.* | Parturition. |

GENUS.

## GENUS II.

*With the Production of new Vessels by internal Membranes or Glands, with Fever.*

### SPECIES.

| | |
|---|---|
| 1. *Febris sensitiva irritata.* | Sensitive irritated fever. |
| 2. *Ophthalmia interna.* | Inflammation of the eye. |
| 3. *Phrenitis.* | ——— of the brain. |
| 4. *Peripneumonia.* | ——— of the lungs. |
| ——— *trachealis.* | ——— the croup. |
| 5. *Pleuritis.* | ——— of the pleura. |
| 6. *Diaphragmitis.* | ——— of the diaphragm. |
| 7. *Carditis.* | ——— of the heart. |
| 8. *Peritonitis.* | ——— of the peritoneum. |
| 9. *Mesenteritis.* | ——— of the mesentery. |
| 10. *Gastritis.* | ——— of the stomach. |
| 11. *Enteritis.* | ——— of the bowels. |
| 12. *Hepatitis.* | ——— of the liver. |
| 13. *Splenitis.* | ——— of the spleen. |
| 14. *Nephritis.* | ——— of the kidney. |
| 15. *Cystitis.* | ——— of the bladder. |
| 16. *Hysteritis.* | ——— of the womb. |
| 17. *Lumbago sensitiva.* | ——— of the loins. |
| 18. *Ischias.* | ——— of the pelvis. |
| 19. *Paronychia interna.* | ——— beneath the nails. |

## GENUS III.

*With the Production of new Vessels by external Membranes or Glands, with Fever.*

### SPECIES.

| | |
|---|---|
| 1. *Febris sensitiva inirritata.* | Sensitive inirritated fever. |
| 2. *Erysipelas irritatum.* | Erysipelas irritated. |
| ——— *inirritatum.* | ——— inirritated. |
| ——— *sensitivum.* | ——— sensitive. |
| 3. *Tonsillitis interna.* | Angina internal. |
| ——— *superficialis.* | ——— superficial. |
| ——— *inirritata.* | ——— inirritated. |
| 4. *Parotitis suppurans.* | Mumps suppurative. |
| ——— *mutabilis.* | ——— mutable. |
| ——— *felina.* | ——— of cats. |
| 5. *Catarrhus sensitivus.* | Catarrh inflammatory. |
| 6. ——— *contagiosus.* | ——— contagious. |
| ——— *equinus et caninus.* | ——— among horses and dogs. |
| 7. *Peripneumonia superficialie.* | Superficial peripneumony. |
| 8. *Pertussis.* | Chin-cough. |
| 9. *Variola discreta.* | Small-pox distinct. |
| ——— *confluens.* | ——— confluent. |
| ——— *inoculata.* | ——— inoculated. |
| 10. *Rubeola irritata.* | Measles irritated. |
| ——— *inirritata.* | ——— inirritated. |
| 11. *Scarlatina mitis.* | Scarlet fever mild. |
| ——— *maligna.* | ——— malignant. |
| 12. *Miliaria sudatoria.* | Miliary fever sudatory. |

 *Miliaria*

| | |
|---|---|
| *Miliaria irritata.* | Miliary irritated. |
| ——— *inirritata.* | ——— inirritated. |
| 13. *Pestis.* | Plague. |
| —— *vaccina.* | ——— of horned cattle. |
| 14. *Pemphigus.* | Bladdery fever. |
| 15. *Varicella.* | Chicken-pox. |
| 16. *Urticaria.* | Nettle rash. |
| 17. *Aptha sensitiva.* | Thrush sensitive. |
| ——— *irritata.* | ——— irritated. |
| ——— *inirritata.* | ——— inirritated. |
| 18. *Dysenteria.* | Bloody flux. |
| 19. *Gastritis superficialis.* | Superficial inflam. of the stomach. |
| 20. *Enteritis superficialis.* | ——————— of the bowels. |

## GENUS IV.

*With the Production of new Vessels by internal Membranes or Glands, without Fever.*

### SPECIES.

| | |
|---|---|
| 1. *Ophthalmia superficialis.* | Ophthalmy superficial. |
| ——— *lymphatica.* | ——— lymphatic. |
| ——— *equina.* | ——— of horses. |
| 2. *Pterigion.* | Eye-wing. |
| 3. *Tarsitis palpebrarum.* | Red eyelids. |
| 4. *Hordeolum.* | Stye. |
| 5. *Paronychia superficialis.* | Whitlow. |
| 6. *Gutta rosea hepatica.* | Pimpled face hepatic. |
| ——— *stomatica.* | ——— stomatic. |
| ——— *hereditaria.* | ——— hereditary. |



7. *Odontitis.*

| | |
|---|---|
| 7. *Odontitis.* | Inflamed tooth. |
| 8. *Otitis.* | ——— ear |
| 9. *Fiſtula lacrymalis.* | Fiſtula lacrymalis. |
| 10. *Fiſtula in ano.* | Fiſtula in ano. |
| 11. *Hepatitis chronica.* | Chronical hepatitis. |
| 12. *Scrophula ſuppurans.* | Suppurating ſcrophula. |
| 13. *Scorbutus ſuppurans.* | Suppurating ſcurvy. |
| 14. *Schirrus ſuppurans.* | Suppurating ſchirrus. |
| 15. *Carcinoma.* | Cancer. |
| 16. *Arthrocele.* | Swelling of the joints. |
| 17. *Arthropuoſis.* | Suppuration of the joints. |
| 18. *Caries oſſium.* | Caries of the bones. |

## GENUS V.

*With the Production of new Veſſels by external Membranes or Glands, without Fever.*

### SPECIES.

| | |
|---|---|
| 1. *Gonorrhœa venerea.* | Clap. |
| 2. *Syphilis.* | Venereal diſeaſe. |
| 3. *Lepra.* | Leproſy. |
| 4. *Elephantiaſis.* | Elephantiaſis. |
| 5. *Framboeſia.* | Framboeſia. |
| 6. *Pſora.* | Itch. |
| 7. *Pſora ebriorum.* | Itch of drunkards. |
| 8. *Herpes.* | Herpes. |
| 9. *Zona ignea.* | Shingles. |
| 10. *Annulus repens.* | Ring-worm. |

11. *Tinea*

| | |
|---|---|
| 11. *Tinea capitis.* | Scald-head. |
| 12. *Crusta lactea.* | Milk-crust. |
| 13. *Trichoma.* | Plica polonica. |

## GENUS VI.

*With Fever consequent to the Production of new Vessels or Fluids.*

### SPECIES.

| | |
|---|---|
| 1. *Febris sensitiva.* | Sensitive fever. |
| 2. —— *a pure clauso.* | Fever from concealed matter. |
| 3. —— *a vomica.* | —— from vomica. |
| 4. —— *ab empyemate.* | —— from empyema. |
| 5. —— *mesenterica.* | —— mesenteric. |
| 6. —— *a pure aerato.* | —— from aerated matter. |
| 7. —— *a phthisi.* | —— from consumption. |
| 8. —— *scrophulosa.* | —— scrophulous. |
| 9. —— *ischiadica.* | —— from ischias. |
| 10. —— *arthropuodica.* | —— from joint-evil. |
| 11. —— *a pure contagioso.* | —— from contagious matter. |
| 12. —— *variolosa secundaria.* | —— secondary of small-pox. |
| 13. —— *carcinomatosa.* | —— cancarous. |
| 14. —— *venerea.* | —— venereal. |
| 15. —— *a sanie contagiosa.* | —— from contagious sanies. |
| 16. —— *puerpera.* | —— puerperal. |
| 17. —— *a sphacelo.* | —— from sphacelus. |

GENUS

## GENUS VII.

*With increaſed Action of the Organs of Senſe.*

### SPECIES.

| | |
|---|---|
| 1. *Delirium febrile.* | Delirium of fevers. |
| 2. ——— *maniacale.* | ———- maniacal. |
| 3. ——— *ebrietatis.* | ———- of drunkenneſs. |
| 4. *Somnium.* | Dreams. |
| 5. *Hallucinatio viſûs.* | Deception of ſight. |
| 6. ————- *auditus.* | ———— of hearing. |
| 7. *Rubor a calore.* | Bluſh from heat. |
| 8. —— *jucunditalis.* | —— from joy. |
| 9. *Priapiſmus amatorius.* | Amorous priapiſm. |
| 10. *Diſtentio mamularum.* | Diſtention of the nipples. |

## ORDO II.

*Decreaſed Senſation.*

## GENUS I.

*With decreaſed Action of the general Syſtem.*

### SPECIES.

| | |
|---|---|
| 1. *Stultitia inſenſibilis.* | Folly from inſenſibility. |
| 2. *Tædium vitæ.* | Irkſomeneſs of life. |
| 3. *Pareſis ſenſitiva,* | Senſitive debility. |

GENUS

## GENUS II.

*With decreaſed Actions of particular Organs.*

### SPECIES.

| | |
|---|---|
| 1. *Anorexia.* | Want of appetite. |
| 2. *Adipſia.* | Want of thirſt. |
| 3. *Impotentia.* | Impotence. |
| 4. *Sterilitas.* | Barrenneſs. |
| 5. *Inſenſibilitas artuum.* | Inſenſibility of the limbs. |
| 6. *Dyſuria inſenſitiva.* | Inſenſibility of the bladder. |
| 7. *Accumulatio alvina.* | Accumulation of feces. |

## ORDO III.

*Retrograde Senſitive Motions.*

## GENUS I.

*Of Excretory Ducts.*

### SPECIES.

| | |
|---|---|
| *Motus retrogreſſus.* | Retrograde motion. |
| 1. —— *ureterum.* | ———— of the ureters. |
| 2. —— *urethræ.* | ———— of the urethra. |
| 3. —— *ductus choledoci.* | ———— of the bile-duct. |

CLASS

# CLASS II.

## DISEASES OF SENSATION.

## ORDO I.

*Increased Sensation.*

## GENUS I.

*With increased Action of the Muscles.*

THE actions belonging to this genus are those which are immediately excited by the sensations of pain or pleasure, but which are neither followed by inflammation, nor by convulsion. The former of which belong to the subsequent genera of this order, and the latter to the class of voluntary motions.

The criterion between the actions, which are the immediate consequence of painful sensation, and convulsive actions properly so called, consists in the former having a tendency to dislodge the stimulating cause, which induces the painful sensation; and the latter being exerted for the purpose of expending the sensorial power, and thus dulling or destroying the general sensation of the system. See Class III. 1.

There is a degree of heat produced in the affected part by these sensitive actions without inflammation, but in much less quantity than when attended by inflammation; as in the latter there is a production of new vessels. See Sect. XXXIII. 2. 3.

Some of the species of this genus cannot properly be termed diseases in their natural state, but become so by their defect or excess, and are here inserted to facilitate the explanation of the others.

## SPECIES.

1. *Deglutitio.* Swallowing our food is immediately caused by the pleasureable sensation occasioned by its stimulus on the palate or fauces, and is acquired long before the nativity of the animal. Afterwards the pain of hunger previously produces the various voluntary exertions to procure the proper material, but the actions of masticating and of swallowing it are effected by the sensorial power of sensation; which appears by their not being always controulable by the will, as when children in vain attempt to swallow nauseous drugs. See Class IV. 1. 3. 1. The masticated food stimulates the palate, which is an organ of sense, into so much action, as to produce agreeable sensation; and the muscles subservient to deglutition are brought into action by the sensation thus produced. The pleasureable sensation is the proximate cause; the action of the fibres of the extremities of the nerves of taste is the remote cause; the sensorial power of irritation exciting these fibres of the nerves of taste into increased action is the pre-remote cause; the action of the muscles of deglutition is the proximate effect; the pushing the food into the stomach is the remote effect; and the nutrition of the body is the post-remote effect.

Though the muscles subservient to deglutition have their actions previously associated, so as to be excited into synchronous tribes or successive trains, either by volition, as when we swallow a disagreeable drug; or by sensation, as when we swallow agreeable food; or by irritation, as when we inattentively swallow our saliva; yet do all those three kinds of deglutition belong to the respective classes of volition, sensation, and irritation; because the first links of these tribes or trains of muscular action are excited by those sensorial powers, and the associated links, which accompany or succeed them, are excited by the combined powers either of volition, or of ſenſation, or of irritation, along with that of aſſociation.

2. *Reſpiratio.* Reſpiration is immediately cauſed by the ſenſorial power of ſenſation in conſequence of the baneful want of vital air; and not from the accumulation of blood in the lungs, as that might be carried on by inhaling azote alone, without the oxygenous part of the atmoſphere. The action of reſpiration is thus ſimilar to that of ſwallowing our food to appeaſe the pain of hunger; but the lungs being ſurrounded with air, their proper pabulum, no intermediate voluntary exertions are required, as in hunger, to obtain and prepare the wanted material.

Reſpiration is ſimilar to ſlow combuſtion; the oxygenous part of the atmoſphere is received through the moiſt membranes, which line the air-cells of the lungs, and uniting with the inflammable part of the blood generates an acid, probably the phoſphoric acid; a portion of carbonic acid is likewiſe produced in this proceſs; as appears by repeatedly breathing over lime-water, which then becomes turbid. See Botanic Garden, P. I. Canto I. l. 401. note.

3. *Sternutatio.* Sneezing conſiſts of muſcular actions produced by the ſenſorial faculty of ſenſation; and is an effort to diſlodge, by means of air forcibly impelled through the noſtrils, ſome material; which ſtimulates the membrane, which lines them, into too great action, and might thence injure the ſenſe of ſmell which is diffuſed on it.

In this operation the too great action of the veſſels of the membrane of the noſtrils is the remote cauſe; the ſenſation thence induced is the proximate cauſe; and the muſcular actions are the proximate effect.

This action of ſneezing frequently precedes common reſpiration in new-born children, but I believe not always; as like the latter it cannot have been previouſly acquired in the uterus.

It is produced in ſome people by ſudden light, as by looking up at the ſky in a morning, when they come out of a gloomy bed-chamber. It then becomes an aſſociate action, and belongs to Claſs IV. 1. 2. 2.

M. M. When it is exerted to exceſs it may be cured by ſnuffing ſtarch up the noſtrils. See Claſs I. 1. 2. 13.

4. *Anhelitus.* Panting. The quick and laborious breathing of running people, who are not accuſtomed to violent exerciſe, is occaſioned by the too great conflux of blood to the lungs. As the ſanguiferous ſyſtem, as well as the abſorbent ſyſtem, is furniſhed in many parts of its courſe with valves, which in general prevent the retrograde movement of their contained fluids; and as all theſe veſſels, in ſome part of their courſe, lie in contact with the muſcles, which are brought into action in running, it follows that the blood muſt be accelerated by the intermitted ſwelling of the bellies of the muſcles moving over them.

The difficulty of breathing, with which very fat people are immediately affected on exerciſe, is owing to the preſſure of the accumulated fat on the veins, arteries, and lymphatics; and which, by diſtending the ſkin, occaſions it to act as a tight bandage on the whole ſurface of the body. Hence when the muſcles are excited into quicker action, the progreſs of the blood in the veins, and of the lymph and chyle in the abſorbent ſyſtem, is urged on with much greater force, as under an artificial bandage on a limb, explained in Art. IV. 2. 10. and in Sect. XXXIII. 3. 2. Hence the circulation is inſtantly quickened to a great degree, and the difficulty of breathing is the conſequence of a more rapid circulation through the lungs. The increaſed ſecretion of the perſpirable matter is another conſequence of this rapid circulation; fat people, when at reſt, are believed to perſpire leſs than others, which may be gathered from their generally having more liquid ſtools, more and paler urine, and to their frequently taking leſs

food

food than many thin people; and laſtly, from the perſpiration of fat people being generally more inodorous than that of lean ones; but when corpulent people are put in motion, the ſweat ſtands in drops on their ſkins, and they "lard the ground" as they run. The increaſe of heat of corpulent people on exerciſe, is another conſequence of their more rapid circulation, and greater ſecretion. See Claſs I. 2. 3. 17.

Other cauſes of difficult or quick reſpiration will be treated of under Aſthma, Pertuſſis, Peripneumony, Tonſillitis.

5. *Tuſſis ebriorum.* Senſitive cough is an exertion of the muſcles uſed in expiration excited into more violent action by the ſenſorial power of ſenſation, in conſequence of ſomething which too powerfully ſtimulates the lungs. As the ſaline part of the ſecreted mucus, when the abſorption of it is impeded; or the too great viſcidity of it, when the abſorption is increaſed; or the too great quantity of the mucus, when the ſecretion is increaſed; or the inflammation of the membranes of the lungs; it is an effort to diſlodge any of theſe extraneous materials.

Of this kind is the cough which attends free-drinkers after a debauch; it conſiſts of many ſhort efforts to cough, with a frequent expuition of half a tea-ſpoonful of frothy mucus, and is attended with conſiderable thirſt. The thirſt is occaſioned by the previous diſſipation of the aqueous parts of the blood by ſenſible or inſenſible perſpiration; which was produced by the increaſed action of the cutaneous and pulmonary capillaries during the ſtimulus of the wine. In conſequence of this an increaſed abſorption commences to replace this moiſture, and the ſkin and mouth become dry, and the pulmonary mucus becomes inſpiſſated; which ſtimulates the bronchia, and is raiſed into froth by the ſucceſſive currents of air in evacuating it. This production of froth is called by ſome free-drinkers "ſpitting ſixpences" after a debauch. This ſubſequent thirſt, dry mouth, and viſcid

viſcid expectoration in ſome people ſucceeds the ſlighteſt degree of intoxication, of which it may be eſteemed a criterion. See Claſs IV. 2. 1. 8.

As coughs are not always attended with pain, the muſcular actions, which produce them, are ſometimes excited by the ſenſorial faculty of irritation, as in Claſs I. 1. 2. 8. I. 1. 3. 4. I. 1. 4. 3. I. 2. 3. 4. Coughs are alſo ſometimes convulſive, as in Claſs III. 1. 1. 10. and ſometimes ſympathetic, as Claſs IV. 2. 1. 7.

M. M. Veneſection, when the cough is attended with inflammation. Mucilages. Opium. Torpentia. Bliſter.

6. *Singultus.* Hiccough is an exertion of the muſcles uſed in inſpiration excited into more violent action by the ſenſorial power of ſenſation, in conſequence of ſomething which too powerfully ſtimulates the cardia ventriculi, or upper orifice of the ſtomach. As when ſolid food is too haſtily taken without ſufficient dilution. And is an effort to diſlodge that offenſive material, and puſh it to ſome leſs ſenſible part of the ſtomach, or into the middle of the contained aliment.

At the end of fatal fevers it may ariſe from the acrimony of the undigeſted aliment, or from a part of the ſtomach being already dead, and by its weight or coldneſs affecting the ſurviving part with diſagreeable ſenſation. The pain about the upper orifice of the ſtomach is the proximate cauſe, the too great or too little action of the fibres of this part of the ſtomach is the remote cauſe, the action of the muſcles uſed in inſpiration is the proximate effect, and the repercuſſion of the offending material is the remote effect.

Hiccough is ſometimes ſympathetic, occaſioned by the pain of gravel in the kidney or ureter, as in Claſs IV. 1. 1. 7. and is ſometimes a ſymptom of epilepſy or reverie, as in Sect. XIX. 2.

M. M. Oil of cinnamon from one drop gradually increaſed to ten, on ſugar, or on chalk. Opium. Bliſter. Emetic.

7. *Asthma humorale.* The humoral asthma probably consists in a temporary anasarca of the lungs, which may be owing to a temporary defect of lymphatic absorption. Its cause is nevertheless at present very obscure, since a temporary deficiency of venous absorption, at the extremities of the pulmonary or bronchial veins, might occasion a similar difficulty of respiration. See Abortio, Class I. 2. 1. 14. Or it might be supposed, that the lymph effused into the cavity of the chest might, by some additional heat during sleep, acquire an aerial form, and thus compress the lungs; and on this circumstance the relief, which these patients receive from cold air, would be readily accounted for.

The paroxysms attack the patient in his first sleep, when the circulation through the lungs in weak people wants the assistance of the voluntary power. Class I. 2. 1. 3. And hence the absorbents of the lungs are less able to fulfil the whole of their duty. And part of the thin mucus, which is secreted into the air-cells, remains there unabsorbed, and occasions the difficult respiration, which awakes the patient. And the violent exertions of the muscles of respiration, which succeed, are excited by the pain of suffocation, for the purpose of pushing forwards the blood through the compressed capillaries, and to promote the absorption of the effused lymph.

In this the humoral differs from the convulsive asthma, treated of in Class III. 1. 1. 10. as in that there is probably no accumulated fluid to be absorbed; and the violent respiration is only an exertion for the purpose of relieving pain, either in the lungs or in some distant part, as in other convulsions, or epilepsy; and in this respect the fits of humoral and convulsive asthma essentially differ from each other, contrary to the opinion expressed without sufficient consideration in Sect. XVIII. 15.

The patients in the paroxysms both of humoral and convulsive asthma find relief from cold air, as they generally rise out of bed, and open the window, and put out their heads; for the lungs are not

sensible

ſenſible to cold, and the ſenſe of ſuffocation is ſomewhat relieved by there being more oxygen contained in a given quantity of cold freſh air, than in the warm confined air of a cloſe bed-chamber.

I have ſeen humoral aſthma terminate in confirmed anaſarca, and deſtroy the patient, who had been an exceſſive drinker of ſpirituous potation. And M. Savage aſſerts, that this diſeaſe frequently terminates in diabetes; which ſeems to ſhew, that it is a temporary dropſy relieved by a great flow of urine. Add to this, that theſe paroxyſms of the aſthma are themſelves relieved by profuſe ſweats of the upper parts of the body, as explained in Claſs I. 3. 2. 8. which would countenance the idea of their being occaſioned by congeſtions of lymph in the lungs.

The congeſtion of lymph in the lungs from the defective abſorption of it is probably the remote cauſe of humoral aſthma; but the pain of ſuffocation is the immediate cauſe of the violent exertions in the paroxyſms. And whether this congeſtion of lymph in the air-cells of the lungs increaſes during our ſleep, as above ſuggeſted, or not; the pain of ſuffocation will be more and more diſtreſſing after ſome hours of ſleep, as the ſenſibility to internal ſtimuli increaſes during that time, as deſcribed in Sect. XVIII. 15. For the ſame reaſon many epileptic fits, and paroxyſms of the gout, occur during ſleep.

In two gouty caſes, complicated with jaundice, and pain, and ſickneſs, the patients had each of them a ſhivering fit, like the commencement of an ague, to the great alarm of their friends; both which commenced in the night, I ſuppoſe during their ſleep; and the conſequence was a ceſſation of the jaundice, and pain about the ſtomach, and ſickneſs; and inſtead of that the gout appeared in their extremities. In theſe caſes I conjecture, that there was a metaſtaſis not only of the diſeaſed action from the membranes of the liver to thoſe of the foot; but that ſome of the new veſſels, or new fluids, which were previouſly produced in the inflamed liver, were tranſlated

to

to the feet during the cold-fit, by the increased absorption of the hepatic lymphatics, and by the retrograde motions of those of the affected limbs.

This I think resembles in some respects a fit of humoral asthma, where stronger motions of the absorbent vessels of the lungs are excited, and retrograde ones of the correspondent cutaneous lymphatics; whence the violent sweats of the upper parts of the body only are produced; and for a time the patient becomes relieved by the metastasis and elimination of the offending material by sensitive exertion. For a further account of this intricate subject see Class III. 1. 1. 10.

M. M. To relieve the paroxysm a tea-spoonful of ether may be given mixed with water, with 10 drops of laudanum, to be repeated three or four times. Venesection. An emetic. A blister. Afterwards the Peruvian bark, with a grain of opium at night, and two or three of aloes. A flannel shirt in winter, but not in summer. Issues. Digitalis?

In this species of asthma, there is great reason to believe, that the respiration of an atmosphere, with an increased proportion of oxygen, will prove of great advantage; some well-observed and well-attested cases of which are published by Dr. Beddoes; as this purer air invigorates the circulation, and the whole system in consequence, perhaps not only by its stimulus, but by its supplying the material from which the sensorial power is extracted or fabricated. In spasmodic asthma, on the contrary, Dr. Ferriar has found undoubted benefit from an atmosphere mixed with hydrogen. See Sect. XVIII. 15. and Class III. 1. 1. 10.

8. *Nictitatio sensitiva.* Winking of the eyes is performed every minute, without our attention, for the purpose of diffusing the tears over them, which are poured into the eye a little above the external corner of it, and which are afterwards absorbed by the lacrymal points

above and below the internal corner of it. When this operation is performed without our attention, it is caused by the faculty of irritation, and belongs to Class I. 1. 4. 1. but when it is produced by a stronger stimulus of any extraneous material in the eye, so as to cause pain, the violent and frequent nictitation is caused by the faculty of sensation.

This disease is sometimes produced by the introversion of the edge of the lower eyelid, which bends the points of the hairs of the eyelash upon the ball of the eye, which perpetually stimulate it into painful sensation. This introversion of the eyelid is generally owing to a tumor of the cellular membrane below the edge of the eyelid, and though a very troublesome complaint may often be cured by the following simple means. A little common plaster spread on thin linen, about a quarter of an inch long, must be rolled up so as to be about the size of a crow-quill, this must be applied immediately below the eyelash on the outside of the eye; and must be kept on by another plaster over it. This will then act as a slight compression on the tumor under the eyelash, and will prevent the hairs from touching the eye-ball. In a week or two the compression will diminish the tumor it lies over, and cure this painful deformity.

9. *Oscitatio et pandiculatio.* Yawning and stretching of the limbs is produced either by a long inactivity of the muscles now brought into action, as sometimes happens after sleep, or after listening a long time to a dull narrative; or it is produced by a too long continued action of the antagonist muscles. In the former case there is an accumulation of sensorial power during the quiescence of the muscles now brought into action; which probably constitutes the pain or wearisomeness of a continued attitude. In the latter case there is an exhaustion of sensorial power in the muscles, which have lately been acting violently, and a consequent accumulation in the muscles, which are antagonists to them, and which were at rest.

These

These involuntary motions are often ſeen in paralytic limbs, which are at the ſame time completely diſobedient to the will; and are frequently obſervable in very young children; and from thence we may conclude, that theſe motions are learnt before nativity; as puppies are ſeen to open their mouths before the membranes are broken. See Sect. XVI. 2.

Where theſe motions are obſerved in limbs otherwiſe paralytic, it is an indication that electric ſhocks may be employed with advantage, as the excitability of the limb by irritation is not extinct, though it be diſobedient both to volition and ſenſation.

10. *Teneſmus* conſiſts in violent and frequent ineffectual efforts to diſcharge the contents of the rectum, owing to pain of the ſphincter. The pain is produced by indurated feces, or by ſome acrid material, as the acidity of indigeſted aliment; and the efforts are attended with mucus from the pained membrane. The feces muſt ſometimes be taken away by the end of a marrow-ſpoon, as cathartics and even clyſter will paſs without removing them. It is ſometimes cauſed by ſympathy with the urethra, when there is a ſtone at the neck of the bladder. See Claſs II. 2. 2. 7. and IV. 1. 2. 8.

M. M. Fomentation, an enema with mucilage and laudanum.

The common excluſion of the feces from the rectum is a proceſs ſimilar to this, except that the muſcles of the ſphincter ani, and thoſe of the abdomen, which act along with them by the combined powers of ſenſation and aſſociation, are in teneſmus excited by painful ſenſation, and in the latter by a ſenſation, which may in ſome inſtances be almoſt called pleaſurable, as relieving us from a painful one in the excluſion of the feces.

11. *Stranguria.* Strangury conſiſts in painful efforts to diſcharge the contents of the urinary bladder. It is generally owing to a ſtone

in the ſphincter of the bladder; or to the inflammation of the neck of it occaſioned by cantharides. It is ſometimes cauſed by ſympathy with the piles; and then is liable in women to occaſion convulſions, from the violence of the pain without inflammation. See Claſs IV. 2. 2. 2. and 3.

M. M. Fomentation clyſter with oil and laudanum, puſh the ſtone back with a bougie; if from cantharides give half a pint of warm water every ten minutes. Mucilage of gum arabic and tragacanth.

The natural evacuation of the urine is a proceſs ſimilar to this, except that the muſcular fibres of the bladder, and the muſcles of the abdomen, which act in concert with them by the combined powers of ſenſation and of aſſociation, are, in the former caſe of ſtrangury, excited into action by painful ſenſation; and in the latter by a ſenſation, which may almoſt be termed pleaſurable, as it relieves us from a previous uneaſy one.

The ejectio ſeminis is another proceſs in ſome reſpects ſimilar to ſtrangury, as belonging to the ſame ſenſible canal of the urethra, and by exciting into action the accelerator muſcles; but in the ſtrangury theſe muſcles are excited into action by painful ſenſation, and in the ejection of the ſemen by pleaſureable ſenſation.

12. *Parturitio.* Parturition is not a diſeaſe, it is a natural proceſs, but is more frequently unfortunate in high life than amongſt the middle claſs of females; which may be owing partly to fear, with which the prieſts of LUCINA are liable to inſpire the ladies of faſhion to induce them to lie in in town; and partly to the bad air of London, to which they purpoſely reſort.

There are however other cauſes, which render parturition more dangerous to the ladies of high life; ſuch as their greater general debility from neglect of energetic exerciſe, their inexperience of the variations of cold and heat, and their ſecluſion from freſh air. To which muſt be added, that great ſource of the deſtruction of female grace

8 and

and beauty, as well as of female health, the tight ſtays, and other bandages, with which they are generally tortured in their early years by the active folly of their friends, which by diſplacing many of the viſcera impedes their actions, and by compreſſing them together produces adheſions of one part to another, and affects even the form and aperture of the bones of the pelvis, through which the naſcent child muſt be protruded.

As parturition is a natural, not a morbid proceſs, no medicine ſhould be given, where there is no appearance of diſeaſe. The abſurd cuſtom of giving a powerful opiate without indication to all women, as ſoon as they are delivered, is, I make no doubt, frequently attended with injurious, and ſometimes with fatal conſequences. See Claſs II. 1. 2. 16.

Another thing very injurious to the child, is the tying and cutting the navel-ſtring too ſoon; which ſhould always be left till the child has not only repeatedly breathed, but till all pulſation in the cord ceaſes. As otherwiſe the child is much weaker than it ought to be; a part of the blood being left in the placenta, which ought to have been in the child; and at the ſame time the placenta does not ſo naturally collapſe, and withdraw itſelf from the ſides of the uterus, and is not therefore removed with ſo much ſafety and certainty. The folly of giving rue or rhubarb to new-born children, and the danger of feeding them with gruel inſtead of milk, is ſpoken of in Claſs I. 1. 2. 5. and II. 1. 2. 16.

ORDO

## ORDO I.

*Increased Sensation.*

## GENUS II.

*With the Production of new Vessels by internal Membranes or Glands, with Fever.*

IN the first class of diseases two kinds of fevers were described, one from excess, and the other from defect of irritation; and were in consequence termed irritative, and inirritative fevers. In this second class of diseases another kind of fever occurs, which is caused by excess of sensation, and termed in consequence Sensitive Fever. But there is no fever from defect of sensation, because the circulation is carried on in health without our consciousness, that is, without any sensation attending it.

But as excess of sensation may exist with excess or defect of irritation, two other kinds of fever arise from a combination of sensitive fever with the irritative, and inirritative ones. Making five kinds in all.

1. Irritative fever, described in Class I. 1. 1. 1.
2. Inirritative fever. Class I. 2. 1. 1.
3. Sensitive fever. Class II. 1. 6. 1.
4. Sensitive irritated fever. Class II. 1. 2. 1.
5. Sensitive inirritated fever. Class II. 1. 3. 1.

As the sensitive irritated fever attends all the diseases enumerated under the genus about to be described, it is placed at the head of it. And as the sensitive inirritated fever accompanies the greatest number

 of

of the species enumerated under the third genus of this order, it is placed at the head of them. And as the sensitive fever attends the diseases of the sixth genus, it is placed at the head of them. But as every febrile paroxysm consists of disordered tribes or trains of associated motions, it may be doubted, whether they ought not all to have been placed in the fourth class, amongst the diseases of association. See Class IV. 2. 4. 11.

All the subsequent species of this genus are attended with sensitive irritated fever; there are nevertheless some superficial inflammations, which affect the same situations without much fever, as the scrophulous ophthalmy and spurious peripneumony, which belong to other genera.

Inflammation is uniformly attended with the production or secretion of new fibres constituting new vessels; this therefore may be esteemed its essential character, or the criterion of its existence. The extension of the old vessels seems rather a consequence than a cause of the germination, or pullulation, of these new ones; for the old vessels may be enlarged, and excited with unusual energy, without any production of new ones, as in the blush of shame or of anger.

When these new vessels are formed, if they are not reabsorbed into the circulation, they secrete a new fluid called purulent matter; which generally opens itself a passage on the external skin, and produces an ulcer, which either gradually heals, or spreads, and is the cause of hectic fever; or they secrete contagious matter, which has the property of exciting the same kind of inflammation, and of producing the same kind of contagious matter, when inserted by inoculation into the skin of other persons. These contagious matters form ulcers, which either heal spontaneously, or by art; or continue to spread, and destroy the patient, by other kinds of hectic fever.

In this genus there is an increase of the sensorial power of irritation as well as of sensation; whence great arterial energy is produced, and the pulse becomes strong and full, as well as quick; and the

coats

coats of the arteries feel hard under the finger, being themſelves thickened and diſtended by inflammation. The blood drawn, eſpecially at the ſecond bleeding, is covered with a tough ſize; which is probably the mucus from the inflamed internal ſurface of the arteries, increaſed in quantity, and more coagulable than in its natural ſtate; the thinner part being more perfectly abſorbed by the increaſed action of the inflamed abſorbents. See Sect. XXXIII. 2. 2. This is rendered more probable, becauſe the hard feel of the pulſe, and the abundance of coagulable lymph commence, exiſt, and ceaſe together.

Great heat is produced from the new chemical combinations ariſing in the ſecretion of new fibres, and great pain from the diſtention of old ones, or from their increaſed action. The increaſed quantity of ſenſation from a topical inflammation or phlegmon is the immediate cauſe of the febris ſenſitiva irritata, or inflammatory fever; as when it ariſes from the pain of pleuriſy, or paronychia; but generally an irritative fever precedes this topical inflammation, which occurs during the hot fit of it; and then the irritative fever is changed into a ſenſitive irritated fever, by the additional cauſe of the ſenſorial power of ſenſation beſides that of irritation.

## SPECIES.

1. *Febris ſenſitiva irritata.* Senſitive irritated fever, or inflammatory fever. Phlegmaſia. A ſtrong full pulſe, with inflammation of the coats of the arteries, conſtitutes this diſeaſe. It originates from ſome topical inflammation, which, if the fever is not ſubdued, terminates in ſuppuration; and differs from irritative fever in reſpect to the painful ſenſation which accompanies it. For as pleaſurable ſenſation is the cauſe of the growth of the new veſſels, and diſtention of the old ones, in the natural enlargement of the body during our infancy;

fancy; so a painful sensation is the cause of the unnatural production of new vessels, and enlargement of old ones in inflammatory diseases.

When matter is thus formed in any internal viscus, or in the cellular membrane, as in the lungs or liver; so long as this abscess remains without admission of air, this inflammatory fever is liable to continue, receiving only temporary relief by bleeding or emetics, or cathartics; till the patient, after a month, or two, or three, expires. But, if air be admitted to these internal abscesses, this kind of fever is changed into a hectic fever in a single day. It also sometimes happens, that when the abscess remains unopened to the air, if the matter has become putrid, that hectic fever supervenes, with colliquative sweats, or diarrhœa; the matter in both cases is sometimes absorbed, and the sides of the abscess grow together again without an external aperture. See Class II. 1. 4. 1. and 2. Another termination of inflammation is in gangrene, but this belongs to the inflammation of the external skin; as the production of purulent matter belongs to inflammation of the internal or mucous membranes. Thus when the external skin is the seat of inflammation, as in erythema, or erysipelas, and produces sensitive irritated fever, no collection of purulent matter can be formed; but a material oozes out, and lies upon the surface, like that in the confluent small-pox, and the cuticle at length peels off, or gangrene supervenes. It must be noted, that these kinds of inflammation can exist together; and some parts of the cellular membrane may suppurate at the same time that the external skin is affected with erythema, or erysipelas.

M. M. Venesection. Cathartics. Diluents. Cool air. Torpentia. Cold Bath? See Sect. XII. 6.

The increased arterial action in this sensitive irritated fever is not simply owing to the increased irritability of the arterial system, or to the stimulus of the distention of the vessels, but also to the increased acrimony or pungency of the blood; which has now so far changed its nature as to become more fluid, more denſe, and to be loaded with coagulable lymph. Hence it becomes neceſſary not only to leſſen the quantity of blood by veneſection and by cathartics, but alſo to dilute its acrimony, or pungency, by the introduction of aqueous and mucilaginous fluids, ſuch as barley water, cream and water, ſugar and water, weak broths; to which may be added ſo much of ſome vegetable eſſential oil, as may render them grateful to the ſtomach, and thus promote their abſorption, as by infuſing parſley or cellery and turneps in the broth; or by balm, mint, or ſage teas.

The following ſpecies of this genus only diſtinguiſh the ſituation of the part previouſly inflamed, and which is the remote cauſe of the ſenſitive irritated, or inflammatory fever, which attends it.

2. *Ophthalmia interna.* Inflammation of the eye is attended with the production of new veſſels, which ſpread over the tunica adjunctiva, and over the cornea; theſe new veſſels are eaſily ſeen, as they lie on a white ground, and give ocular demonſtration of their production in inflammation. When this inflammation of the cornea ſuppurates, it is liable to leave little ulcers, which may be ſeen beneath the ſurface in the form of little excavations; and as theſe heal, they are liable to be covered with an opake ſcar. This ſcar, in ſome months or years, is liable to wear away, and become tranſparent, without the aſſiſtance of any poliſhing powder, as of very finely levigated glaſs, as ſome have recommended. But when the cornea is affected through all its thickneſs, the return of its tranſparency becomes hopeleſs. See Claſs I. 1. 3. 14.

In violent degrees of ophthalmy the internal parts, as the retina, optic artery, iris, ciliary proceſs, become inflamed, as well as the external ones; hence the leaſt light admitted to the eye occaſions intolerable pain. This curious circumſtance cannot be owing to the action of light on the inflamed veſſels of the cornea; it therefore ſhews, that the extremity of the optic nerve or retina is alſo rendered more exquiſitely

exquiſitely ſenſible to light, by partaking of the inflammation; and I have been told, that red colours are in theſe caſes ſometimes painfully perceived even in perfect darkneſs. This ſhews that the retina is excited into motion by the ſtimulus of light; and that, when it is inflamed, theſe motions give great pain, like thoſe of other inflamed parts, as the muſcles, or membranes. And ſecondly, that the ideas of colours conſiſt in the motions of the retina; which ideas occaſion pain, when the extremity of the moving nerve is inflamed.

M. M. Veneſection. Cathartics. Diluents. Torpentia. Frequently moiſten the eye with cold water by means of a rag. Cool airy room. Darkneſs. When the inflammation begins to decline, white vitriol gr. vi. in an ounce of water is more efficacious to moiſten the eye than ſolutions of lead. Tincture of opium diluted. New veſſels from the inflamed tunica adnata frequently ſpread like a fly's wing upon the tranſparent cornea, which is then called Pterigium. To ſtop the growth of this, the principal veſſels ſhould be cut through with a lancet. When the inflammation begins to decline, after due evacuation any ſtimulating material put into the eye increaſes the abſorption, which ſoon removes the new red veſſels; which has given riſe to a hundred famous eye-waters, and eye-doctors; if theſe ſtimulating materials are uſed too ſoon, the inflammation is increaſed by them. See Sect. XXXII. 2. 10.

There is another ophthalmia, which attends weak children, and is generally eſteemed a ſymptom of ſcrophula, as deſcribed in Claſs II. 1. 5. 3. and another, which is of venereal origin, mentioned in Claſs II. 1. 5. 2. both which may be termed ophthalmia ſuperficialis.

3. *Phrenitis.* Inflammation of the brain is attended with intolerance of light and ſound; which ſhews, that the extremities of the nerves of thoſe ſenſes are at the ſame time inflamed; it is alſo attended with great pain of the head, with watchfulneſs, and furious de-

 lirium.

lirium. The violent efforts, theſe patients are ſaid ſometimes to exert, are owing to the increaſed ſecretion of ſenſorial power in the brain; as all other inflamed glands have a greater circulation of blood paſſing through them, and a greater ſecretion in conſequence of their peculiar fluids, as in the hepatitis much more bile is generated.

M. M. Veneſection. Cathartics. Torpentia. Foment the head with cold water for hours together. Or with warm water. Cool airy room. Afterwards cupping on the occiput. Leeches to the temples. When the patient is weakened a bliſter on the head, and after further exhauſtion five or ſix drops of tincture of opium.

4. *Peripneumonia.* Inflammation of the lungs. The pulſe is not always hard, ſometimes ſoft; which is probably owing to a degree of ſickneſs or inaction of the ſtomach; with dull pain of the cheſt; reſpiration conſtantly difficult, ſometimes with erect poſture; the face bloated and purpliſh; cough generally with moiſt expectoration, often ſtained with blood.

When the difficulty of reſpiration is very great, the patient is not able to cough; in this ſituation, after copious bleeding, the cough is liable to return, and is ſo far a favourable ſymptom, as it ſhews ſome abatement of the inflammation.

A peripneumony frequently occurs in the chin-cough, and deſtroys the patient, except immediate recourſe be had to the lancet, or to four or five leeches; when blood cannot be otherwiſe taken.

The peripneumony is very fatal to young children, eſpecially as I believe it is frequently miſtaken for a ſpaſmodic aſthma, or for the croup, or cynanche trachealis of Cullen. Both which, however, when they occur, require immediate veneſection by the lancet or by leeches, as well as the peripneumony.

The croup is an inflammation of the upper part, and the peripneumony of the lower part of the ſame organ, viz. the trachea or windpipe. See Claſs I. 1. 3. 4. But as the inflammation is ſeldom I ſuppoſe

pose confined to the upper part of the trachea only, but exists at the same time in other parts of the lungs, and as no inflammation of the tonsils is generally perceptible, the uncouth name of cynanche trachealis should be changed for *peripneumonia trachialis*. The method of cure consists in immediate and repeated bleeding. A vomit. A grain of calomel or other mild cathartic. Bathing in subtepid water, and in breathing over the steam of warm water, with or without a little vinegar in it. And lastly, by keeping the child raised high in bed.

Inflammation of the lungs is also liable to occur in the measles, and must be attacked by venesection at any time of the disease; otherwise either a present death, or an incurable consumption, is the consequence.

The peripneumony is frequently combined with inflammation of the pleura, and sometimes with that of the diaphragm; either of these may generally be distinguished, not only by the pain which attends inflammation of these membranes, but by inspecting the naked chest, and observing whether the patient breathes more by elevating the ribs, or by depressing the diaphragm.

A crisis happens in children about the sixth day with much pale urine, which must be waited for after evacuations have been used, as far as can be done with safety; in this situation the warm bath twice a day, and small blisters repeatedly in succession, are of peculiar service.

After the termination of peripneumony a collection of coagulable lymph is frequently left in the cavity of the chest unabsorbed; or a common anasarca of the lungs occurs from the present inaction of the absorbent vessels, which had previously been excited too violently. This difficulty of breathing is cured or relieved by the exhibition of digitalis. See Art. IV. 2. 8.

M. M. The lancet is the anchor of hope in this disease; which must be repeated four or five times, or as often as the fever and difficulty of breathing increase, which is generally in the evening; antimonials,

timonials, diluents, repeated fmall blifters about the cheft, mucilage, pediluvium, warm bath. Is a decoction of feneka-root of ufe? Do not neutral falts increafe the tendency to cough by their ftimulus, as they increafe the heat of urine in gonorrhœa? Children in every kind of difficult breathing, from whatever caufe, fhould be kept as upright in bed as may be, and continually watched; fince, if they flip down, they are liable to be immediately fuffocated. After the patient is greatly debilitated, fo that no further evacuation can be admitted, and the difficult breathing and cough continue, I have given four or five drops of tincture of opium, that is, about a quarter of a grain of folid opium, with great advantage, and I believe in feveral cafes I have faved the patient. A greater quantity of opium in this ftate of debility cannot be ufed without hazarding the life of the perfon. This fmall quantity of an opiate fhould be given about fix in the evening, or before the accefs of the evening paroxyfm, and repeated three or four nights, or longer.

There is a peripneumony with weak pulfe, which may be termed *peripneumonia inirritata*, as defcribed in Sect. XXVII. 2. which belongs to this place. See alfo Superficial Peripneumony, Clafs II. 1. 3. 7.

5. *Pleuritis*. Pleurify. Inflammation of the pleura, with hard pulfe, pain chiefly of the fide, pungent, particularly increafed during infpiration; lying on either fide uneafy, the cough very painful, dry at the beginning, afterwards moift, often bloody.

One caufe of pleurify is probably a previous adhefion of the lungs to a part of the pleura, which envelops them. This in many cafes has been produced in infancy, by fuffering children to lie too long on one fide. Or by placing them uniformly on one fide of a fire, or window, to which they will be liable always to bend themfelves.

When matter is produced during peripneumony or pleurify in one fide of the cheft, fo long as it is a concealed vomica, the fever continues,

tinues, if the diſeaſe be great, for many weeks, and even months; and requires occaſional veneſection, till the patient ſinks under the inflammatory or ſenſitive irritated fever. But if air be admitted, by a part of the abſceſs opening itſelf a way into the air-veſſels of the lungs, a hectic fever, with colliquitive ſweats or diarrhœa, ſupervenes, and frequently deſtroys the patient; or the abſceſs heals the lungs adhering to the pleura.

M. M. The lancet muſt be uſed copiouſly, and repeated as often as the pain and difficult reſpiration increaſe. A bliſter on the pained part. Antimonial preparations. Diluents. Cool air. Do neutral ſalts increaſe the tendency to cough? Pediluvium or ſemicupium frequently repeated.

6. *Diaphragmitis.* Inflammation of the diaphragm. Pain round the lower ribs as if girt with a cord. Difficult reſpiration performed only by elevating the ribs and in an erect poſture. The corners of the mouth frequently retracted into a diſagreeable ſmile, called riſus Sardonicus.

Thoſe animals, which are furniſhed with clavicles, or collar-bones, not only uſe their foremoſt feet as hands, as men, monkies, cats, mice, ſquirrels, &c. but elevate their ribs in reſpiration as well as depreſs the diaphragm for the purpoſe of enlarging the cavity of the cheſt. Hence an inflammation of the diaphragm is ſudden death to thoſe animals, as horſes and dogs, which can only breathe by depreſſing the diaphragm; and is I ſuppoſe the cauſe of the ſudden death of horſes that are over-worked; whereas, in the human animal, when the diaphragm is inflamed, ſo as to render its motions impoſſible from the pain they occaſion, reſpiration can be carried on, though in a leſs perfect manner, by the intercoſtal muſcles in the elevation of the ribs. In pleuriſy the ribs are kept motionleſs, and the reſpiration is performed by the diaphragm, as may be readily ſeen on inſpecting the naked cheſt, and which is generally a bad ſymptom; in in the diaphragmitis the ribs are alternately elevated, and depreſſed, but the lower part of the belly is not ſeen to move.

M. M. As in pleuriſy and peripneumony. When the patient becomes delirious, and ſmiles diſagreeably by intervals, and is become ſo weak, that evacuations by the lancet could be uſed no further, and I have almoſt deſpaired of my patient, I have found in two or three inſtances, that about five or ſix drops of tinct. thebaic. given an hour before the evening exacerbation, has had the happieſt effect, and cured the patient in this caſe, as well as in common peripneumony; it muſt be repeated two or three evenings, ſee Claſs II. 1. 2. 4. as the exacerbation of the fever and difficult reſpiration and delirium generally increaſe towards night.

The ſtimulus of this ſmall quantity of opium on a patient previouſly ſo much debilitated, acts by increaſing the exertion of the abſorbent veſſels, in the ſame manner as a ſolution of opium, or any other ſtimulant, put on an inflamed eye after the veſſels are previouſly emptied by evacuations, ſtimulates the abſorbent ſyſtem, ſo as to cauſe the remaining new veſſels to be immediately reabſorbed. Which ſame ſtimulants would have increaſed the inflammation, if they had been applied before the evacuations. See Claſs II. 1. 2. 2. Sect. XXXIII. 3. 1. When the ſanguiferous ſyſtem is full of blood, the abſorbents cannot act ſo powerfully, as the progreſs of their contents is oppoſed by the previous fulneſs of the blood-veſſels; whence ſtimulants in that caſe increaſe the action of the ſecerning ſyſtem more than of the abſorbent one; but after copious evacuation this reſiſtance to the progreſs of the abſorbed fluids is removed; and when ſtimulants are then applied, they increaſe the action of the abſorbent ſyſtem more than that of the ſecerning one. Hence opium given in the commencement of inflammatory diſeaſes deſtroys the patient; and cures them, if given in very ſmall doſes at the end of inflammatory diſeaſes.

7. *Carditis*

7. *Carditis.* Inflammation of the heart is attended with unequal intermitting pulse, palpitation, pain in the middle of the sternum, and constant vomiting. It cannot certainly be distinguished from peripneumony, and is perhaps always combined with it.

8. *Peritonitis.* Inflammation of the peritonæum is known by pain all over the abdomen, which is increased on erecting the body. It has probably most frequently a rheumatic origin. See Class II. 1. 2. 17.

9. *Mesenteritis.* Inflammation of the mesentery is attended with pains like colic, and with curdled or chyle-like stools. It is a very frequent and dangerous disease, as the production of matter more readily takes place in it than in any other viscus. The consequence of which, after a hard labour, is probably the puerperal fever, and in scrophulous habits a fatal purulent fever, or hopeless consumption.

M. M. Venesection. Warm bath. Emollient clysters.

10. *Gastritis.* In inflammation of the stomach the pulse is generally soft, probably occasioned by the sickness which attends it. The pain and heat of the stomach is increased by whatever is swallowed, with immediate rejection of it. Hiccough.

This disease may be occasioned by acrid or indigestible matters taken into the stomach, which may chemically or mechanically injure its interior coat. There is however a slighter species of inflammation of this viscus, and perhaps of all others, which is unattended by much fever; and which is sometimes induced by drinking cold water, or eating cold insipid food, as raw turnips, when the person has been much heated and fatigued by exercise. For when the sensorial power has been diminished by great exertion, and the stomach has become less irritable by having been previously stimulated by much heat, it sooner becomes quiescent by the application of cold.

In conſequence of this ſlight inflammation of the ſtomach an eruption of the face frequently enſues by the ſenſitive aſſociation of this viſcus with the ſkin, which is called a ſurfeit. See Claſs IV. 1. 2. 13. and II. 1. 4. 6. and II. 1. 3. 19.

M. M. Veneſection. Warm bath. Bliſter. Anodyne clyſters. Almond ſoap. See Claſs II. 1. 3. 17.

11. *Enteritis*. Inflammation of the bowels is often attended with ſoft pulſe, probably owing to the concomitant ſickneſs; which prevents ſometimes the early uſe of the lancet, to the deſtruction of the patient. At other times it is attended with ſtrong and full pulſe like other inflammations of internal membranes. Can the ſeat of the diſeaſe being higher or lower in the inteſtinal canal, that is, above or below the valve of the colon, produce this difference of pulſe by the greater ſympathy of one part of the bowels with the ſtomach than another? In enteritis with ſtrong pulſe the pain is great about the navel, with vomiting, and the greateſt difficulty in procuring a ſtool. In the other, the pain and fever is leſs, without vomiting, and with diarrhœa. Whence it appears, that the enteritis with hard quick pulſe differs from Ileus, deſcribed in Claſs I. 3. 1. 6. only in the exiſtence of fever in the former and not the latter, the other ſymptoms generally correſponding; and, ſecondly, that the enteritis with ſofter quick pulſe, differs from the cholera deſcribed in Claſs I. 3. 1. 5. only in the exiſtence of fever in the former, and not the latter, the other ſymptoms being in general ſimilar. See Claſs II. 1. 3. 20.

Inflammation of the bowels ſometimes is owing to extraneous indigeſtible ſubſtances, as plum-ſtones, eſpecially of the damaſin, which has ſharp ends. Sometimes to an introſuſception of one part of the inteſtine into another, and very frequently to a ſtrangulated hermia or rupture. In reſpect to the firſt, I knew an inſtance where a damaſin ſtone, after a long period of time, found its way out of the body near the groin. I knew another child, who vomited ſome damaſin ſtones, which had lain for near twenty hours, and given great

great pain about the navel, by the exhibition of an emetic given in repeated doſes for about an hour. The ſwallowing of plum-ſtones in large quantities, and even of cherry-ſtones, is annually fatal to many children. In reſpect to the introſuſception and hermia, ſee Ileus, Claſs I. 3. 1. 6.

M. M. Repeated veneſection. Calomel from ten to twenty grains given in ſmall pills as in Ileus; theſe means uſed early in the diſeaſe generally ſucceed. After theſe evacuations a bliſter contributes to ſtop the vomiting. Warm bath. Crude mercury. Aloes one grain-pill every hour will frequently ſtay in the ſtomach. Glauber's ſalt diſſolved in pepper-mint water given by repeated ſpoonfuls.

When the patient is much reduced, opium in very ſmall doſes may be given, as a quarter of a grain, as recommended in pleuriſy. If the pain ſuddenly ceaſes, and the patient continues to vomit up whatever is given him, it is generally fatal; as it indicates, that a mortification of the bowel is already formed. Some authors have adviſed to join cathartic medicines with an opiate in inflammation of the bowels, as recommended in colica ſaturnina. This may ſucceed in ſlighter caſes, but is a dangerous practice in general; ſince, if the obſtruction be not removed by the evacuation, the ſtimulus of the opium is liable to increaſe the action of the veſſels, and produce mortification of the bowel, as I think I have ſeen more than once.

12. *Hepatitis.* Inflammation of the liver is attended with ſtrong quick pulſe; tenſion and pain of the right ſide; often pungent as in pleuriſy, oftner dull. A pain is ſaid to affect the clavicle, and top of the right ſhoulder; with difficulty in lying on the left ſide; difficult reſpiration; dry cough; vomiting; hiccough.

There is another hepatitis mentioned by authors, in which the fever, and other ſymptoms, are wanting, or are leſs violent; as deſcribed in Claſs II. 1. 4. 11. and which is probably ſometimes relieved

 by

by eruptions of the face; as in thoſe who are habituated to the intemperate uſe of fermented liquors.

M. M. Hepatic inflammation is very liable to terminate in ſuppuration, and the patient is deſtroyed by the continuance of a fever with ſizy blood, but without night-ſweats, or diarrhœa, as in other unopened abſceſſes. Whence copious and repeated veneſection is required early in the diſeaſe, with repeated doſes of calomel, and cathartics. Warm bath. Towards the end of the diſeaſe ſmall doſes of opium before the evening paroxyſms, and laſtly the Peruvian bark, and chalybeate wine, at firſt in ſmall doſes, as 20 drops twice a day, and afterwards, if neceſſary, in larger. See Art. IV. 2. 6.

Mrs. C. a lady in the laſt month of her pregnancy, was ſeized with violent hepatitis, with ſymptoms both of peripneumony and of pleuriſy, for it ſeldom happens in violent inflammations, that one viſcus alone is affected; ſhe wanted then about a fortnight of her delivery, and after frequent veneſection, with gentle cathartics, with fomentation or warm bath, ſhe recovered and was ſafely delivered, and both herſelf and child did well. Rheumatic and eruptive fevers are more liable to induce abortion.

13. *Splenitis.* Inflammation of the ſpleen commences with tenſion, heat, and tumour of the left ſide, and with pain, which is increaſed by preſſure. A caſe is deſcribed in Claſs I. 2. 3. 18. where a tumid ſpleen, attended with fever, terminated in ſchirrus of that viſcus.

14. *Nephritis.* Inflammation of the kidney ſeems to be of two kinds; each of them attended with different ſymptoms, and different modes of termination. One of them I ſuppoſe to be an inflammation of the external membrane of the kidney, ariſing from general cauſes of inflammation, and accompanied with pain in the loins without vomiting; and the other to conſiſt in an inflammation of the interior

parts of the kidney, occaſioned by the ſtimulus of gravel in the pelvis of it, which is attended with perpetual vomiting, with pain along the courſe of the ureter, and retraction of the teſtis on that ſide, or numbneſs of the thigh.

The former of theſe kinds of nephritis is diſtinguiſhed from lumbago by its ſituation being more exactly on the region of the kidney, and by its not being extended beyond that part; after three or four days I believe this inflammation is liable to change place; and that a herpes or eryſipelas, called zona, or ſhingles, breaks out about the loins in its ſtead; at other times it is cured by a cathartic with calomel, with or without previous veneſection.

The other kind of nephritis, or inflammation of the interior part of the kidney, generally ariſes from the pain occaſioned by the ſtimulus of a ſtone entering the ureter from the pelvis of the kidney; and which ceaſes when the ſtone is protruded forwards into the bladder; or when it is returned into the pelvis of the kidney by the retrograde action of the ureter. The kidney is nevertheleſs inflamed more frequently, though in a leſs degree, from other cauſes; eſpecially from the intemperate ingurgitation of ale, or other fermented or ſpirituous liquors. This leſs degree of inflammation is the cauſe of gravel, as that before mentioned is the effect of it. The mucus ſecreted to lubricate the internal ſurface of the uriniferous tubes of the kidney becomes ſecreted in greater quantity, when theſe veſſels are inflamed; and, as the correſpondent abſorbent veſſels act more energetically at the ſame time, the abſorption of its more fluid parts is more powerfully effected; on both theſe accounts the mucus becomes both changed in quality and more indurated. And in this manner ſtones are produced on almoſt every mucous membrane of the body; as in the lungs, bowels, and even in the pericordium, as ſome writers have affirmed. See Claſs I. 1. 3. 9.

M. M. Veneſection. Ten grains of calomel given in ſmall pills. then infuſion of ſena with oil. Warm bath. Then opium a grain and

and half. See Clafs I. 1. 3. 9. for a further account of the method of cure.

15. *Cyftitis.* Inflammation of the bladder is attended with tumor and pain of the lower part of the belly; with difficult and painful micturition; and tenefmus. It generally is produced by the exiftence of a large ftone in the bladder, when in a great degree; or is produced by common caufes, when in a flighter degree.

The ftone in the bladder is generally formed in the kidney, and paffing down the ureter into the bladder becomes there gradually increafed in fize; and this moft frequently by the appofition of concentric fpheres, as may be feen by fawing fome of the harder calculi through the middle, and polifhing one furface. Thefe new concretions fuperinduced on the nucleus, which defcended from the kidney, as defcribed in Clafs I. 1. 3. 9. and in the preceding article of this genus, is not owing to the microcofmic falt, which is often feen to adhere to the fides of chamber-pots, as this is foluble in warm water, but to the mucus of the bladder, as it rolls along the internal furface of it. Now when the bladder is flightly inflamed, this mucus of its internal furface is fecreted in greater quantity, and is more indurated by the abforption of its more liquid part at the inftant of fecretion, as explained in Clafs I. 1. 3. 9. and II. 1. 2. 14. and thus the ftimulus and pain of a ftone in the bladder contributes to its enlargement by inflaming the interior coat of it.

M. M. Venefection. Warm bath. Diluents. Anodyne clyfters. See Clafs I. 1. 3. 9.

16. *Hyfteritis.* Inflammation of the womb is accompanied with heat, tenfion, tumor, and pain of the lower belly. The os uteri painful to the touch. Vomiting. This difeafe is generally produced by improper management in the delivery of pregnant women. I knew an unfortunate cafe, where the placenta was left till the next day;

 and

and then an unſkilful accoucheur introduced his hand, and forcibly tore it away; the conſequence was a moſt violent inflammatory fever, with hard throbbing pulſe, great pain, very ſizy blood, and the death of the patient. Some accoucheurs have had a practice of introducing their hand into the uterus immediately after the birth of the child, to take away the placenta; which they ſaid was to ſave time. Many women I believe have been victims to this unnatural practice.

Others have received injury, where inflammation has been beginning, by the univerſal practice of giving a large doſe of opium immediately on delivery, without any indication of its propriety; which, though a proper and uſeful medicine, where the patient is too feeble, when given in a ſmall doſe, as 10 drops of tincture of opium, or half a grain of ſolid opium, muſt do a proportionate injury, when it is given improperly; and as delivery is a natural proceſs, it is certainly more wiſe to give no medicines, except there be ſome morbid ſymptom, which requires it; and which has only been introduced into cuſtom by the ill-employed activity of the Prieſts or Prieſteſſes of LUCINA; like the concomitant nonſenſe of cramming rue or rheubarb into the mouth of the unfortunate young ſtranger, who is thus ſoon made to experience the evils of life. See Claſs II. 1. 1. 12. and I. 1. 2. 5. Juſt ſo ſome over-wiſe beldames force young ducks and turkeys, as ſoon as they are hatched, to ſwallow a pepper-corn.

M. M. Veneſection repeatedly; diluents; fomentation; the patient ſhould be frequently raiſed up in bed for a ſhort time, to give opportunity of diſcharge to the putrid lochia; mucilaginous clyſters. See Febris Puerpera.

17. *Lumbago ſenſitiva.* Senſitive lumbago. When the extenſive membranes, or ligaments, which cover the muſcles of the back are torpid,

torpid, as in the cold paroxyſm of ague, they are attended with pain in conſequence of the inaction of the veſſels, which compoſe them. When this inaction continues without a conſequent renewal or increaſe of activity, the diſeaſe becomes chronical, and forms the lumbago frigida, or irritativa, deſcribed in Claſs I. 2. 4. 16. But when this cold fit or torpor of theſe membranes, or ligaments or muſcles of the back, is ſucceeded by a hot fit, and conſequent inflammation, a violent inflammatory fever, with great pain, occurs, preventing the erect poſture of the body; and the affected part is liable to ſuppurate, in which caſe a very dangerous ulcer is formed, and a part of one of the vertebræ is generally found carious, and the patient ſinks after a long time under the hectic fever occaſioned by the aerated or oxygenated matter.

This diſeaſe bears no greater analogy to rheumatiſm than the inflammation of the pleura, or any other membranous inflammation; and has therefore unjuſtly been arranged under that name. It is diſtinguiſhed from nephritis, as it is ſeldom attended with vomiting, I ſuppoſe never, except the ureter happens to be inflamed at the ſame time.

The pain ſometimes extends on the outſide of the thigh from the hip to the ankle, heel, or toes, and is then called ſciatica; and has been thought to conſiſt in an inflammation of the theca, or covering of the ſciatic nerve, as the pain ſometimes ſo exactly attends the principal branches of that nerve. See Claſs I. 2. 4. 15. 16.

M. M. Veneſection repeatedly; calomel; gentle cathartics; diluents; warm bath; poultice on the back, conſiſting of camomile flowers, turpentine, ſoap, and opium; a burgundy-pitch plaſter. A debility of the inferior limbs from the torpor of the muſcles, which had previouſly been too much excited, frequently occurs at the end of this diſeaſe; in this caſe electricity, and iſſues on each ſide of the lumber vertebræ, are recommended. See Claſs I. 2. 4. 16.

18. *Iſchias.* The iſchias conſiſts of inflammatory fever, with great pain about the pelvis, the os coccigis, and the heads of the thigh-bones, preventing the patient from walking or ſtanding erect, with increaſe of pain on going to ſtool. This malady, as well as the preceding, has been aſcribed to rheumatiſm; with which it ſeems to bear no greater analogy, than the inflammations of any other membranes.

The patients are left feeble, and ſometimes lame after this diſeaſe; which is alſo ſometimes accompanied with great flow of urine, owing to the defective abſorption of its aqueous parts; and with conſequent thirſt occaſioned by the want of ſo much fluid being returned into the circulation; a lodgment of fæces in the rectum ſometimes occurs after this complaint from the leſſened ſenſibility of it. See Claſs I. 2. 4. 15.

M. M. Veneſection; gentle cathartics; diluents; fomentation; poultice with camomile flowers, turpentine, ſoap, and opium; afterwards the bark. See Claſs I. 1. 3. 5.

When this inflammation terminates in ſuppuration the matter generally can be felt to fluctuate in the groin, or near the top of the thigh. In this circumſtance, my friend Mr. Bent, Surgeon near Newcaſtle in Staffordſhire, propoſes to tap the abſceſs by means of a trocar, and thus as often as neceſſary to diſcharge the matter without admitting the air. Might a weak injection of wine and water, as in the hydrocele, be uſed with great caution to inflame the walls of the abſceſs, and cauſe them to unite? See Claſs II. 1. 6. 9.

19. *Paronychia interna.* Inflammation beneath the finger-nail. The pain occaſioned by the inflammatory action and tumor of parts bound down between the nail on one ſide and the bone on the other, neither of which will yield, is ſaid to occaſion ſo much pain as to produce immediate delirium, and even death, except the parts are di-

vided by a deep incision; which must pass quite through the periosteum, as the inflammation is said generally to exist beneath it. This disease is thus resembled by the process of toothing in young children; where an extraneous body lodged beneath the periosteum induces pain and fever, and sometimes delirium, and requires to be set at liberty by the lancet.

ORDO

## ORDO I.

*Increased Sensation.*

## GENUS III.

*With the Production of new Vessels by external Membranes or Glands, with Fever.*

THE diseases of this genus are perhaps all productive of contagious matter; or which becomes so by its exposure to the air, either through the cuticle, or by immediate contact with it; such are the matters of the small-pox and measles. The purulent matter formed on parts covered from the air by thicker membranes or muscles, as in the preceding genus, does not induce fever, and cannot therefore be called contagious; but it acquires this property of producing fever in a few hours, after the abscess has been opened, so as to admit the air to its surface, and may then be said to consist of contagious miasmata. This kind of contagious matter only induces fever, but does not produce other matter with properties similar to its own; and in this respect it differs from the contagious miasmata of small-pox or measles, but resembles those which have their origin in crowded jails; for these produce fever only, which frequently destroys the patient; but do not produce other matters similar to themselves; as appears from none of those, who died of the jail-fever, caught at the famous black assizes at Oxford, at the beginning of this century, having infected their physicians or attendants.

If indeed the matter has continued so long as to become putrid, and thus to have given out air from a part of it, it acquires the power of producing fever; in the same manner as if the ulcer had been opened, and exposed to the common air; instances of which are not

 unfrequent.

unfrequent. And from these circumstances it seems probable, that the matters secreted by the new vessels formed in all kinds of phlegmons, or pustles, are not contagious, till they have acquired something from the atmosphere, or from the gas produced by putrefaction; which will account for some phenomena in the lues venerea, cancer, and of other contagious secretions on the skin without fever, to be mentioned hereafter. See Class II. 1. 4. 14.

The theory of contagion has been perplexed by comparing it with fermenting liquors; but the contagious material is shewn in Section XXXIII. to be produced like other secreted matters by certain animal motions of the terminations of the vessels. Hence a new kind of gland is formed at the terminations of the vessels in the eruptions of the small-pox; the animal motions of which produce from the blood variolous matter; as other glands produce bile or saliva. Now if some of this matter is introduced beneath the cuticle of a healthy person, or enters the circulation, and excites the extremities of the blood-vessels into those kinds of diseased motions, by which it was itself produced, either by irritation or association, these diseased motions of the extremities of the vessels will produce other similar contagious matter. See Sect. XXXIII. 2. 5. and 9. Hence contagion seems to be propagated two ways; one, by the stimulus of contagious matter applied to the part, which by an unknown law of nature excites the stimulated vessels to produce a similar matter; as in venereal ulcers, which thus continue to spread; or as when variolous matter is inserted beneath the cuticle; or when it is supposed to be absorbed, and diffused over the body mixed with the blood, and applied in that manner to the cutaneous glands. The other way, by which contagion seems to be diffused, is by some distant parts sympathizing or imitating the motions of the part first affected; as the stomach and skin in the eruptions of the inoculated small-pox, or in the bite of a mad dog; as treated of in Sect. XXII. 3. 3.

In some of the diseases of this genus the pulse is strong, full, and hard,

hard, conſtituting the ſenſitive irritated fever, as deſcribed in the preceding genus; as in one kind of eryſipelas, which requires repeated veneſection. In others the arterial action is ſometimes moderate, ſo as to conſtitute the ſenſitive fever, as in the inoculated ſmall-pox; where the action of the arteries is neither increaſed by the ſenſorial power of irritation, as in the ſenſitive irritated fever; nor decreaſed by the defect of that power, as in the ſenſitive inirritated fever. But in the greateſt number of the diſeaſes of this genus the arterial action is greatly diminiſhed in reſpect to ſtrength, and conſequently the frequency of pulſation is proportionally increaſed, as explained in Sect. XXXII. 2. 1. Which is owing to the deficiency of the ſenſorial power of irritation joined with the increaſe of that of ſenſation, and thus conſtitutes the ſenſitive inirritated fever; as in Scarlatina with gangrenous tonſils.

From this great debility of the action of the arteries, there appears to be leſs of the coagulable lymph or mucus ſecreted on their internal ſurfaces; whence there is not only a defect of that buff or ſize upon the blood, which is ſeen on the ſurface of that, which is drawn in the ſenſitive irritated fever; but the blood, as it cools, when it has been drawn into a baſon, ſcarcely coagulates; and is ſaid to be diſſolved, and is by ſome ſuppoſed to be in a ſtate of actual putrefaction. See Sect. XXXIII. 1. 3. where the truth of this idea is controverted. But in the fevers of both this genus and the preceding one great heat is produced from the chemical combinations in the ſecretions of new veſſels and fluids, and pain or uneaſineſs from the diſtention of the old ones; till towards the termination of the diſeaſe ſenſation ceaſes, as well as irritation, with the mortification of the affected parts, and the death of the patient.

Dyſenteria, as well as tonſillitis and aphtha, are enumerated amongſt the diſeaſes of external membranes, becauſe they are expoſed either to the atmoſpheric air, which is breathed, and ſwallowed with our food and ſaliva; or they are expoſed to the inflammable air, or hydrogen, which is generated in the inteſtines; both which contribute to produce or promote the contagious quality of theſe fluids; as mentioned in Claſs II. 1. 5.

It is not ſpeaking accurate language, if we ſay, that in the diſeaſes of this genus the fever is contagious; ſince it is the material produced by the external membranes, which is contagious, after it has been expoſed to air; while the fever is the conſequence of this contagious matter, and not the cauſe of it. As appears from the inoculated ſmall-pox, in which the fever does not commence, till after ſuppuration has taken place in the inoculated arm, and from the diſeaſes of the fifth genus of this order, where contagion exiſts without fever. See Claſs II. 1. 5. and II. 1. 3. 18.

## SPECIES.

1. *Febris ſenſitiva inirritata.* Senſitive inirritated fever. Typhus gravior. Putrid malignant fever. Jail fever. The immediate cauſe of this diſeaſe is the increaſe of the ſenſorial power of ſenſation, joined with the decreaſe of the ſenſorial power of irritation; that is, it conſiſts in the febris ſenſitiva joined with the febris inirritativa of Claſs I. 2. 1. 1. as the febris ſenſitiva irritata of the preceding genus conſiſts of the febris ſenſitiva joined with the febris irritativa of Claſs I. 1. 1. 1. In both which the word irritata, and inirritata, are deſigned to expreſs more or leſs irritation than the natural quantity; and the ſame when applied to ſome of the diſeaſes of this genus.

This fever is frequently accompanied with topical inflammation, which is liable, if the arterial ſtrength is not ſupported, to end in ſphacelus; and as mortified parts, ſuch as ſloughs of the throat, if they adhere to living parts, ſoon become putrid from the warmth and moiſture of their ſituation; theſe fevers have been termed putrid, and have been thought to owe their cauſe to what is only their conſe-

quence.

quence. In hot climates this fever is frequently induced by the exhalations of ſtagnating lakes or marſhes, which abound with animal ſubſtances; but which in colder countries produce fevers with debility only, as the quartan ague, without inflammation.

The ſenſitive inirritated, or malignant, fever is alſo frequently produced by the putrid exhalations and ſtagnant air in priſons; but perhaps moſt frequently by contact or near approach of the perſons, who have reſided in them. Theſe cauſes of malignant fevers contributed to produce, and to ſupport for a while, the ſeptic and antiſeptic theory of them; ſee Sect. XXXIII. 1. 3. The vibices or bruiſes, and petechiæ or purples, were believed to be owing to the diſſolved ſtate of the blood by its incipient putrefaction; but hydroſtatical experiments have been made, which ſhew the ſizy blood of the patient in ſenſitive irritated or inflammatory fever, with ſtrong pulſe, is more fluid, while it is warm, than this uncoagulable blood taken in this ſenſitive inirritated, or malignant fever; from whence it is inferred, that theſe petechiæ, and vibices, are owing to the deficient power of abſorption in the terminations of the veins. See Claſs I. 2. 1. 5.

This ſenſitive inirritated fever, or typhus gravior, is diſtinguiſhed from the inirritative fever, or typhus mitior, in the early ſtages of it, by the colour of the ſkin; which in the latter is paler, with leſs heat, owing to the leſs violent action of the capillaries; in this it is higher coloured, and hotter, from the greater energy of the capillary action in the production of new veſſels. In the more advanced ſtate petechiæ, and the production of contagious matter from inflamed membranes, as the aphthæ of the mouth, or ulcers of the throat, diſtinguiſhes this fever from the former. Delirium, and dilated pupils of the eyes, are more frequent in nervous fevers; and ſtupor with deafneſs more frequent attendants on malignant fevers. See Claſs I. 2. 5. 6.

There is another criterion diſcernible by the touch of an experienced finger; and that is, the coat of the artery in inflammatory fevers, both thoſe attended with ſtrength of pulſation, and theſe with weak pulſation, feels harder, or more like a cord; for the coats of the arteries in theſe fevers are themſelves inflamed, and are conſequently turgid with blood, and thence are leſs eaſily compreſſed, though their pulſations are nevertheleſs weak: when the artery is large or full with an inflamed coat, it is called hard; and when ſmall or empty with an inflamed coat, it is called ſharp, by many writers.

M. M. The indications of cure conſiſt, 1. In procuring a regurgitation of any offenſive material, which may be lodged in the long mouths of the lacteals or lymphatics, or in their tumid glands. 2. To excite the ſyſtem into neceſſary action by the repeated exhibition of nutrientia, ſorbentia, and incitantia; and to preſerve the due evacuation of the bowels. 3. To prevent any unneceſſary expenditure of ſenſorial power. 4. To prevent the formation of ulcers, or to promote the abſorption in them, for the purpoſe of healing them.

1. One ounce of wine of ipecacuanha, or about ten grains of the powder, ſhould be given as an emetic. After a few hours three or four grains of calomel ſhould be given in a little mucilage, or conſerve. Where ſomething ſwallowed into the ſtomach is the cauſe of the fever, it is liable to be arreſted by the lymphatic glands, as the matter of the ſmall-pox inoculated in the arm is liable to be ſtopped by the axillary lymphatic gland; in this ſituation it may continue a day or two, or longer, and may be regurgitated during the operation of an emetic or cathartic into the ſtomach or bowel, as evidently happens on the exhibition of calomel, as explained in Sect. XXIX. 7. 2. For this reaſon an emetic and cathartic, with veneſection, if indicated by the hardneſs and fulneſs of the pulſe, will very frequently remove fevers, if exhibited on the firſt, ſecond, or even third day.

2. Wine and opium, in ſmall doſes repeated frequently, but ſo that not the leaſt degree of intoxication follows, for in that caſe a greater

greater degree of debility is produced from the expenditure of ſenſorial power in unneceſſary motions. Many weak patients have been thus ſtimulated to death. See Sect. XII. 7. 8. The Peruvian bark ſhould be given alſo in repeated doſes in ſuch quantity only as may ſtrengthen digeſtion, not impede it. For theſe purpoſes two ounces of wine, or of ale, or cyder, ſhould be given every ſix hours; and two ounces of decoction of bark, with two drachms of the tincture of bark, and ſix drops of tincture of opium, ſhould be given alſo every ſix hours alternately; that is, each of them four times in twenty-four hours. As much rhubarb as may induce a daily evacuation, ſhould be given to remove the colluvies of indigeſted materials from the bowels; which might otherwiſe increaſe the diſtreſs of the patient by the air it gives out in putrefaction, or by producing a diarrhœa by its acrimony; the putridity of the evacuations are in conſequence of the total inability of the digeſtive powers; and their delay in the inteſtines, to the inactivity of that canal in reſpect to its periſtaltic motions.

The quantities of wine or beer and opium, and bark, above mentioned, may be increaſed by degrees, if the patient ſeems refreſhed by them; and if the pulſe becomes ſlower on their exhibition; but this with caution, as I have ſeen irrecoverable miſchief done by greater quantities both of opium, wine, and bark, in this kind of fever; in which their uſe is to ſtrengthen the digeſtion of the weak patient, rather than to ſtop the paroxyſms of fever; but when they are adminiſtered in intermittents, much larger quantities are neceſſary.

The ſtimulus of ſmall bliſters applied in ſucceſſion, one every three or four days, when the patient becomes weak, is of great ſervice by ſtrengthening digeſtion, and by preventing the coldneſs of the extremities, owing to the ſympathy of the ſkin with the ſtomach, and of one part of the ſkin with another.

In reſpect to nutriment, the patient ſhould be ſupplied with wine and water, with toaſted bread, and ſugar or ſpice in it; or with ſago with wine; freſh broth with turnips, cellery, parſley; fruit; new milk. Tea with cream and ſugar; bread pudding, with lemon juice and ſugar; chicken, fiſh, or whatever is grateful to the palate of the ſick perſon, in ſmall quantity repeated frequently; with ſmall beer, cyder and water, or wine and water, for drink, which may be acidulated with acid of vitriol in ſmall quantities.

3. All unneceſſary motions are to be checked, or prevented. Hence horizontal poſture, obſcure room, ſilence, cool air. All the parts of the ſkin, which feel too hot to the hand, ſhould be expoſed to a current of cool air, or bathed with cold water, whether there are eruptions on it or not. Waſh the patient twice a day with cold vinegar and water, or cold ſalt and water, or cold water alone, by means of a ſponge. If ſome parts are too cold, as the extremities, while other parts are too hot, as the face or breaſt, cover the cold parts with flannel, and cool the hot parts by a current of cool air, or bathing them as above.

4. For the healing of ulcers, if in the mouth, ſolution of alum in water about 40 grains to an ounce, or of blue vitriol in water, one grain or two to an ounce may be uſed to touch them with three or four times a day. Of theſe perhaps a ſolution of alum is to be preferred, as it inſtantly takes away the ſtench from ulcers I ſuppoſe by combining with the volatile alcali which attends it. For this purpoſe a ſolution of alum of an ounce to a pint of water ſhould be frequently injected by means of a ſyringe into the mouth. If there are ulcers on the external ſkin, fine powder of bark ſeven parts, and ceruſſa in fine powder one part, ſhould be mixed, and applied dry on the ſore, and kept on by lint, and a bandage.

As ſloughs in the mouth are frequently produced by the previous dryneſs of the membranes, which line it, this dryneſs ſhould be prevented by frequently moiſtening them, which may be effected by injection with a ſyringe, or by a moiſt ſponge, or laſtly in the following manner. Place a glaſs of wine and water, or of milk and ſugar, on a table by the bedſide, a little above the level of the mouth of the patient; then, having previouſly moiſtened a long piece of narrow liſting, or cloth, or flannel, with the ſame liquor, leave one end of it in the glaſs, and introduce the other into the mouth of the patient; which will thus be ſupplied with a conſtant oozing of the fluid through the cloth, which acts as a capillary ſyphon.

The viſcid phlegm, which adheres to the tongue, ſhould be coagulated by ſome auſtere acid, as by lemon-juice evaporated to half its quantity, or by crab-juice; and then it may be ſcraped off by a knife, or rubbed off by flannel, or a ſage leaf dipped in vinegar, or in ſalt and water.

2. *Eryſipelas*, St. Anthony's fire, may be divided into three kinds, which differ in their method of cure, the irritated, the inirritated, and the ſenſitive eryſipelas.

*Eryſipelas irritatum* is attended with increaſe of irritation beſides increaſe of ſenſation; that is, with ſtrong, hard, and full pulſe, which requires frequent veneſection, like other inflammations with arterial ſtrength. It is diſtinguiſhed from the phlegmonic inflammations of the laſt genus by its ſituation on the external habit, and by the redneſs, heat, and tumour not being diſtinctly circumſcribed; ſo that the eye or finger cannot exactly trace the extent of them.

When the external ſkin is the ſeat of inflammation, and produces ſenſitive irritated fever, no collection of matter is formed, as when a phlegmon is ſituated in the cellular membrane beneath the ſkin; but the cuticle riſes as beneath a bliſter-plaſter, and becomes ruptured; and a yellow material oozes out, and becomes inſpiſſated, and lies upon its ſurface; as is ſeen in this kind of eryſipelas, and in the confluent ſmall-pox; or if the new veſſels are reabſorbed the cuticle peels off in ſcales. This difference of the termination of eryſipelatous and phlegmonic inflammation ſeems to be owing in part to the leſs diſtenſibility of the cuticle than of the cellular membrane, and in part to the ready exhalation of the thinner parts of the secreted fluids through its pores.

This eryſipelas is generally preceded by a fever for two or three days before the eruption, which is liable to appear in ſome places, as it declines in others; and ſeems frequently to ariſe from a previous ſcratch or injury of the ſkin; and is attended ſometimes with inflamflammation of the cellular membrane beneath the ſkin; whence a real phlegmon and collection of matter becomes joined to the eryſipelas, and either occaſions or increaſes the irritated fever, which attends it.

There is a greater ſympathy between the external ſkin and the meninges of the brain, than between the cellular membrane and thoſe meninges; whence eryſipelas is more liable to be preceded or attended, or ſucceeded, by delirium than internal phlegmons. I except the mumps, or parotitis, deſcribed below; which is properly an external gland, as its excretory duct opens into the air. When pain of the head or delirium precedes the cutaneous eruption of the face, there is ſome reaſon to believe, that the primary diſeaſe is a torpor of the meninges of the brain; and that the ſucceeding violent action is transferred to the ſkin of the face by ſenſitive aſſociation; and that a ſimilar ſympathy occurs between ſome internal membranes and the ſkin over them, when eryſipelas appears on other parts of the body. If this circumſtance ſhould be ſupported by further evidence, this diſeaſe ſhould be removed into Claſs IV. along with the rheumatiſm and gout. See Claſs IV. 1. 2. 17.

This ſuppoſed retropulſion of eryſipelas on the brain from the frequent appearance of delirium, has prevented the free uſe of the lancet early in this diſeaſe to the deſtruction of many; as it has prevented the ſubduing of the general inflammation, and thus has in the end produced the particular one on the brain. Mr. B——, a delicate gentleman about ſixty, had an eryſipelas beginning near one ear, and extending by degrees over the whole head, with hard, full, and ſtrong pulſe;

pulſe; blood was taken from him four or five times in conſiderable quantity, with gentle cathartics, with calomel, diluents, and cool air, and he recovered without any ſigns of delirium, or inflammation of the meninges of the brain. Mr. W——, a ſtrong corpulent man of inferior life, had eryſipelas over his whole head, with ſtrong hard pulſe: he was not evacuated early in the diſeaſe through the timidity of his apothecary, and died delirious. Mrs. F—— had eryſipelas on the face, without either ſtrong or weak pulſe; that is, with ſenſitive fever alone, without ſuperabundance or deficiency of irritation; and recovered without any but natural evacuations. From theſe three caſes of eryſipelas on the head it appears, that the evacuations by the lancet muſt be uſed with courage, where the degree of inflammation requires it; but not where this degree of inflammation is ſmall, nor in the eryſipelas attended with inirritation, as deſcribed below.

M. M. Veneſection repeated according to the degree of inflammation. An emetic. Calomel three grains every other night. Cool air. Diluents, emetic tartar in ſmall doſes, as a quarter of a grain every ſix hours. Tea, weak broth, gruel, lemonade, neutral ſalts. See Sect. XII. 6.

Such external applications as carry away the heat of the ſkin may be of ſervice, as cold water, cold flour, ſnow, ether. Becauſe theſe applications impede the exertions of the ſecerning veſſels, which are now in too great action; but any applications of the ſtimulant kind, as ſolutions of lead, iron, copper, or of alum, uſed early in the diſeaſe, muſt be injurious; as they ſtimulate the ſecerning veſſels, as well as the abſorbent veſſels, into greater action; exactly as occurs when ſtimulant eye-waters are uſed too ſoon in ophthalmy. See Claſs II. 1. 2. 2. But as the cuticle peels off in this caſe after the inflammation ceaſes, it differs from ophthalmy; and ſtimulant applications are not indicated at all, except where ſymptoms of gangrene appear. For as a new cuticle is formed under the old one, as under a bliſter, a bliſter, the ſerous fluid between them is a defence to the new cuticle, and ſhould dry into a ſcab by exhalation rather than be reabſorbed. Hence we ſee how greaſy or oily applications, and even how moiſt ones, are injurious in eryſipelas; becauſe they prevent the exhalation of the ſerous effuſion between the old and new cuticle, and thus retard the formation of the latter.

*Eryſipelas inirritatum* differs from the former in its being attended with weak pulſe, and other ſymptoms of ſenſitive inirritated fever. The feet and legs are particularly liable to this eryſipelas, which precedes or attends the ſphacelus or mortification of thoſe parts. A great and long coldneſs firſt affects the limb, and the eryſipelas on the ſkin ſeems to occur in conſequence of the previous torpor of the interior membranes. As this generally attends old age, it becomes more dangerous in proportion to the age, and alſo to the habitual intemperance of the patient in reſpect to the uſe of fermented or ſpirituous liquor.

When the former kind, or irritated eryſipelas, continues long, the patient becomes ſo weakened as to be liable to all the ſymptoms of this inirritated eryſipelas; eſpecially where the meninges of the brain are primarily affected. As in that caſe, after two or three efforts have been made to remove the returning periods of torpor of the meninges to the external ſkin, thoſe meninges become inflamed themſelves, and the patient ſinks under the diſeaſe; in a manner ſimilar to that in old gouty patients, where the torpor of the liver or ſtomach is relieved by aſſociation of the inflammation of the membranes of the feet, and then of other joints, and laſtly the power of aſſociation ceaſing to act, but the exceſs of ſenſation continuing, the liver or ſtomach remains torpid, or become themſelves inflamed, and the patient is deſtroyed.

M. M. Where there exiſts a beginning gangrene of the extremities, the Peruvian bark, and wine, and opium, are to be given in large quantities; ſo as to ſtrengthen the patient, but not to intoxicate, or or to impede his digeſtion of aliment, as mentioned in the firſt ſpecies of this genus. Claſs II. 1. 2. 1. But where the brain is inflamed or oppreſſed, which is known either by delirium, with quick pulſe; or by ſtupor, and ſlow reſpiration with ſlow pulſe; other means muſt be applied. Such as, firſt, a fomentation on the head with warm water, with or without aromatic herbs, or ſalt in it, ſhould be continued for an hour or two at a time, and frequently repeated. A bliſter may alſo be applied on the head, and the fomentation nevertheleſs occaſionally repeated. Internally very gentle ſtimulants, as camphor one grain or two in infuſion of valerian. Wine and water or ſmall beer, weak broth. An enema. Six grains of rhubarb and one of calomel. Afterwards five drops of tincture of opium, which may be repeated every ſix hours, if it ſeems of ſervice. Might the head be bathed for a minute with cold water? or with ether? or vinegar?

*Eryſipelas ſenſitivum* is a third ſpecies, differing only in the kind of fever which attends it, which is ſimply inflammatory, or ſenſitive, without either exceſs of irritation, as in the firſt variety; or the defect of irritation, as in the ſecond variety: all theſe kinds of eryſipelas are liable to return by periods in ſome people, who have paſſed the middle of life, as at periods of a lunation, or two lunations, or at the equinoxes. When theſe periods of eryſipelas happen to women, they ſeem to ſupply the place of the receding catamenia; when to men, I have ſometimes believed them to be aſſociated with a torpor of the liver; as they generally occur in thoſe who have drank vinous ſpirit exceſſively, though not approbriouſly; and that hence they ſupply the place of periodical piles, or gout, or gutta roſea.

M. M. As the fever requires no management, the diſeaſe takes its progreſs ſafely, like a moderate paroxyſm of the gout; but in this caſe, as in ſome of the former, the eryſipelas does not appear to be a primary diſeaſe, and ſhould perhaps be removed to the Claſs of Aſſociation.

3. *Tonſillitis.*

3. *Tonsillitis.* Inflammation of the tonsils. The uncouth term Cynanche has been used for diseases so dissimilar, that I have divided them into Tonsillitis and Parotitis; and hope to be excused for adding a Greek termination to a Latin word, as one of those languages may justly be considered as a dialect of the other. By tonsillitis the inflammation of the tonsils is principally to be understood; but as all inflammations generally spread further than the part first affected; so, when the summit of the windpipe is also much inflamed, it may be termed tonsillitis trachealis, or croup. See Class I. 1. 3. 4. and II. 1. 2. 4.; and when the summit of the gullet is much inflamed along with the tonsil, it may be called tonsillitis pharyngea, as described in Dr. Cullen's Nosologia, Genus X. p. 92. The inflammation of the tonsils may be divided into three kinds, which require different methods of cure.

*Tonsillitis interna.* Inflammation of the internal tonsil. When the swelling is so considerable as to produce difficulty of breathing, the size of the tonsil should be diminished by cutting it with a proper lancet, which may either give exit to the matter it contains, or may make it less by discharging a part of the blood. This kind of angina is frequently attended with irritated fever besides the sensitive one, which accompanies all inflammation, and sometimes requires venesection. An emetic should be given early in the disease, as by its inducing the retrograde action of the vessels about the fauces during the nausea it occasions, it may eliminate the very cause of the inflammation; which may have been taken up by the absorbents, and still continue in the mouths of the lymphatics or their glands. The patient should then be induced to swallow some aperient liquid, an infusion of senna, so as to induce three or four evacuations. Gargles of all kinds are rather hurtful, as the action of using them is liable to give pain to the inflamed parts; but the patients find great relief from frequently holding warm water in their mouths, and putting it out again, or by syringing warm water into the mouth, as this acts like

like a warm bath or fomentation to the inflamed part. Laſtly, ſome mild ſtimulant, as a weak ſolution of ſalt and water, or of white vitriol and water, may be uſed to waſh the fauces with in the decline of the diſeaſe, to expedite the abſorption of the new veſſels, if neceſſary, as recommended in ophthalmy.

*Tonſillitis ſuperficialis.* Inflammation of the ſurface of the tonſils. As the tonſils and parts in their vicinity are covered with a membrane, which, though expoſed to currents of air, is nevertheleſs conſtantly kept moiſt by mucus and ſaliva, and is liable to diſeaſes of its ſurface like other mucous membranes, as well as to ſuppuration of the internal ſubſtance of the gland; the inflammation of its ſurface is ſucceeded by ſmall elevated puſtules with matter in them, which ſoon diſappears, and the parts either readily heal, or ulcers covered with ſloughs are left on the ſurface.

This diſeaſe is generally attended with only ſenſitive fever, and therefore is of no danger, and may be diſtinguiſhed with great certainty from the dangerous inflammation or gangrene of the tonſils at the height of the ſmall-pox, or ſcarlet fever, by its not being attended with other ſymptoms of thoſe diſeaſes. One emetic and a gentle cathartic is generally ſufficient; and the frequent ſwallowing of weak broth, or gruel, both without ſalt in them, relieves the patient, and abſolves the cure. When theſe tumours of the tonſils frequently return I have ſometimes ſuſpected them to originate from the abſorption of putrid matter from decaying teeth. See Claſs I. 2. 3. 21. and II. 2. 2. 1.

*Tonſillitis inirritata.* Inflammation of the tonſils with ſenſitive inirritated fever is a ſymptom only of contagious fever, whether attended with ſcarlet eruption, or with confluent ſmall-pox, or otherwiſe. The matter of contagion is generally diffuſed, not diſſolved in the air; and as this is breathed over the mucaginous ſurface of the tonſils, the contagious atoms are liable to be arreſted by the tonſil; which therefore becomes the neſt of the future diſeaſe, like the inflamed circle round the inoculated puncture of the arm in supposititious small-pox. This swelling is liable to suffocate the patient in small-pox, and to become gangrenous in scarlet fever, and some other contagious fevers, which have been received in this manner. The existence of inflammation of the tonsil previous to the scarlet eruption, as the arm inflames in the inoculated small-pox, and suppurates before the variolous eruption, should be a criterion of the scarlet fever being taken in this manner.

M. M. All the means which strengthen the patient, as in the sensitive inirritated fever, Class II. 1. 2. 1. As it is liable to continue a whole lunation or more, great attention should be used to nourish the patient with acidulous and vinous panada, broth with vegetables boiled in it, sugar, cream, beer; all which given frequently will contribute much to moisten, clean, and heal the ulcuscles, or sloughs, of the throat; warm water and wine, or acid of lemon, should be frequently applied to the tonsils by means of a syringe, or by means of a capillary syphon, as described in Class II. 1. 2. 1. A slight solution of blue vitriol, as two grains to an ounce, or a solution of sugar of lead of about six grains to an ounce, may be of service; especially the latter, applied to the edges of the sloughs, drop by drop by means of a small glass tube, or small crow-quill with the end cut off, or by a camel's-hair pencil or sponge; to the end of either of which a drop will conveniently hang by capillary attraction; as solutions of lead evidently impede the progress of erysipelas on the exterior skin, when it is attended with feeble pulse. Yet a solution of alum injected frequently by a syringe is perhaps to be preferred, as it immediately removes the fetor of the breath, which must much injure the patient by its being perpetually received into the lungs by respiration.

4. *Parotitis.* Mumps, or branks, is a contagious inflammation of the parotis and maxillary glands, and has generally been classed under

the word Cynanche or Angina, to which it bears no analogy. It divides itself into two kinds, which differ in the degree of fever which attends them, and in the method of cure.

*Parotitis ſuppurans.* The ſuppurating mumps is to be diſtinguiſhed by the acuteneſs of the pain, and the ſenſitive, irritated, or inflammatory fever, which attends it.

M. M. Veneſection. Cathartic with calomel three or four grains repeatedly. Cool air, diluents. This antiphlogiſtic treatment is to be continued no longer than is neceſſary to relieve the violence of the pain, as the diſeaſe is attended with contagion, and muſt run through a certain time, like other fevers with contagion.

*Parotitis mutabilis.* Mutable parotitis. A ſenſitive fever only, or a ſenſitive irritated fever, generally attends this kind. And when the tumor of the parotis and maxillary glands ſubſides, a new ſwelling occurs in ſome diſtant part of the ſyſtem; as happens to the hands and feet, at the commencement of the ſecondary fever of the ſmall-pox, when the tumor of the face ſubſides. This new ſwelling in the parotitis mutabilis is liable to affect the teſtes in men, and form a painful tumor, which ſhould be prevented from ſuppuration by very cautious means, if the violence of the pain threaten ſuch a termination; as by bathing the part with coldiſh water for a time, veneſection, a cathartic; or by a bliſter on the perinæum, or ſcrotum, or a poultice.

When women are affected with this complaint, after the ſwelling of the parotis and maxillary glands ſubſides, a tumor with pain is liable to affect their breaſts; which, however, I have never ſeen terminate in ſuppuration.

On the retroceſſion of the tumor of the teſtes above deſcribed, and I ſuppoſe of that of the breaſts in women, a delirium of the calm kind is very liable to occur; which in ſome caſes has been the firſt ſymptom which has alarmed the friends of the patient; and it has thence been difficult to diſcover the cauſe of it without much inquiry; the

 previous

previous ſymptoms having been ſo ſlight as not to have occaſioned any complaints. In this delirium, if the pulſe will bear it, veneſection ſhould be uſed, and three or four grains of calomel, with fomentation of the head with warm water for an hour together every three or four hours.

Though this diſeaſe generally terminates favourably, conſidering the numbers attacked by it, when it is epidemic, yet it is dangerous at other times in every part of its progreſs. Sometimes the parotis or maxillary glands ſuppurate, producing ulcers which are difficult to cure, and frequently deſtroy the patient, where there was a previous ſcrophulous tendency. The teſtis in men is alſo liable to ſuppurate with great pain, long confinement, and much danger; and laſtly the affection of the brain is fatal to many.

Mr. W. W. had a ſwelled throat, which after a few days ſubſided. He became delirious or ſtupid, in which ſtate he was dying when I ſaw him; and his friends aſcribed his death to a coup de ſoleil, which he was ſaid to have received ſome months before, when he was abroad.

Mr. A. B. had a ſwelling of the throat, which after a few days ſubſided. When I ſaw him he had great ſtupor, with ſlow breathing, and partial delirium. On fomenting his head with warm water for an hour theſe ſymptoms of ſtupor were greatly leſſened, and his oppreſſed breathing gradually ceaſed, and he recovered in one day.

Mr. C. D. I found walking about the houſe in a calm delirium without ſtupor; and not without much inquiry of his friends could get the previous hiſtory of the diſeaſe; which had been attended with parotitis, and ſwelled teſtis, previous to the delirium. A few ounces of blood were taken away, a gentle cathartic was directed, and his head fomented with warm water for an hour, with a ſmall bliſter on the back, and he recovered in two or three days.

Mr. D. D. came down from London in the coach alone, ſo that no previous hiſtory could be obtained. He was walking about the houſe

houſe in a calm delirium, but could give no ſenſible anſwers to any thing which was propoſed to him. His pulſe was weak and quick. Cordials, a bliſter, the bark, were in vain exhibited, and he died in two or three days.

Mr. F. F. came from London in the ſame manner in the coach. He was mildly delirious with conſiderable ſtupor, and moderate pulſe, and could give no account of himſelf. He continued in a kind of cataleptic ſtupor, ſo that he would remain for hours in any poſture he was placed, either in his chair, or in bed; and did not attempt to ſpeak for about a fortnight; and then gradually recovered. Theſe two laſt caſes are not related as being certainly owing to parotitis, but as they might probably have that origin.

The parotitis ſuppurans, or mumps with irritated fever, is at times epidemic among cats, and may be called *parotitis felina*; as I have reaſon to believe from the ſwellings under the jaws, which frequently ſuppurate, and are very fatal to thoſe animals. In the village of Haywood, in Staffordſhire, I remember a whole breed of Perſian cats, with long white hair, was deſtroyed by this malady, along with almoſt all the common cats of the neighbourhood; and as the parotitis or mumps had not long before prevailed amongſt human beings in that part of the country, I recollect being inclined to believe, that the cats received the infection from mankind; though in all other contagious diſeaſes, except the rabies canina can be ſo called, no different genera of animals naturally communicate infection to each other; and I am informed, that vain efforts have been made to communicate the ſmall-pox and meaſles to ſome quadrupeds by inoculation. A diſeaſe of the head and neck deſtroyed almoſt all the cats in Weſtphalia. Savage, Noſol. Claſs X. Art. 30. 8.

5. *Catarrhus ſenſitivus* conſiſts of an inflammation of the membrane, which lines the noſtrils and fauces. It is attended with ſenſitive fever alone, and is cured by the ſteam of warm water externally,

and

and by diluents internally, with moderate venesection and gentle cathartics. This may be termed catarrhus sensitivus, to distinguish it from the catarrhus contagiosus, and is in common language called a violent cold in the head; it differs from the catarrhus calidus, or warm catarrh, of Class I. 1. 2. 7. in the production of new vessels, or inflammation of the membrane, and the consequent more purulent appearance of the discharge.

Rancedo catarrhalis, or catarrhal hoarseness, is a frequent symptom of this disease; and is occasioned by the pain or soreness which attends the thickened and inflamed membranes of the larynx; which prevents the muscles of vocallity from sufficiently contracting the aperture of it. It ceases with the inflammation, or may be relieved by the steam of warm water alone, or of water and vinegar, or of water and ether. See Paralytic Hoarseness, Class III. 2. 1. 4.

6. *Catarrhus contagiosus.* This malady attacks so many at the same time, and spreads gradually over so great an extent of country, that there can be no doubt but that it is disseminated by the atmosphere. In the year 1782 the sun was for many weeks obscured by a dry fog, and appeared red as through a common mist. The material, which thus rendered the air muddy, probably caused the epidemic catarrh, which prevailed in that year, and which began far in the north, and extended itself over all Europe. See Botanic Garden, Vol. II. note on Chunda, and Vol. I. Canto IV. line 294, note; and was supposed to have been thrown out of a volcano, which much displaced the country of Iceland.

In many instances there was reason to believe, that this disease became contagious, as well as epidemic; that is, that one person might receive it from another, as well as by the general unsalutary influence of the atmosphere. This is difficult to comprehend, but may be conceived by considering the increase of contagious matter in the small-pox. In that disease one particle of contagious matter stimulates the

the ſkin of the arm in inoculation into morbid action ſo as to produce a thouſand particles ſimilar to itſelf; the ſame thing occurs in catarrh, a few deleterious atoms ſtimulate the mucous membrane of the noſtrils into morbid actions, which produce a thouſand other particles ſimilar to themſelves. Theſe contagious particles diffuſed in the air muſt have conſiſted of animal matter, otherwiſe how could an animal body by being ſtimulated by them produce ſimilar particles? Could they then have had a volcanic origin, or muſt they not rather have been blown from putrid marſhes full of animal matter? But the greateſt part of the ſolid earth has been made from animal and vegetable recrements, which may be diſperſed by volcanos.—Future diſcoveries muſt anſwer theſe queſtions.

As the ſenſitive fever attending theſe epidemic catarrhs is ſeldom either much irritated or inirritated, veneſection is not always either clearly indicated or forbid; but as thoſe who have died of theſe catarrhs have generally had inflamed livers, with conſequent ſuppuration in them, veneſection is adviſeable, wherever the cough and fever are greater than common, ſo as to render the uſe of the lancet in the leaſt dubious. And in ſome caſes a ſecond bleeding was neceſſary, and a mild cathartic or two with four grains of calomel; with mucilaginous ſubacid diluents; and warm ſteam occaſionally to alleviate the cough, finiſhed the cure.

The catarrhus contagioſus is a frequent diſeaſe amongſt horſes and dogs; it ſeems firſt to be diſſeminated amongſt theſe animals by miaſmata diffuſed in the atmoſphere, becauſe ſo many of them receive it at the ſame time; and afterwards to be communicable from one horſe or dog to another by contagion, as above deſcribed. Theſe epidemic or contagious catarrhs more frequently occur amongſt dogs and horſes than amongſt men; which is probably owing to the greater extenſion and ſenſibility of the mucous membrane, which covers the organ of ſmell, and is diffuſed over their wide noſtrils,

and

and their large maxillary and frontal cavities. And to this circumstance may be aſcribed the greater fatality of it to theſe animals.

In reſpect to horſes, I ſuſpect the fever at the beginning to be of the ſenſitive, irritated, or inflammatory kind, becauſe there is ſo great a diſcharge of purulent mucus; and that therefore they will bear once bleeding early in the diſeaſe; and alſo one mild purgative, conſiſting of about half an ounce of aloe, and as much white hard ſoap, mixed together. They ſhould be turned out to graſs both day and night for the benefit of pure air, unleſs the weather be too cold (and in that caſe they ſhould be kept in an open airy ſtable, without being tied), that they may hang down their heads to facilitate the diſcharge of the mucus from their noſtrils. Graſs ſhould be offered them, or other freſh vegetables, as carrots and potatoes, with maſhes of malt, or of oats, and with plenty of freſh warm or cold water frequently in a day. When ſymptoms of debility appear, which may be known by the coldneſs of the ears or other extremities, or when ſloughs can be ſeen on the membrane which lines the noſtrils, a drink conſiſting of a pint of ale with half an ounce of tincture of opium in it, given every ſix hours, is likely to be of great utility.

In dogs I believe the catarrh is generally joined with ſymptoms of debility early in the diſeaſe. Theſe animals ſhould be permitted to go about in the open air, and ſhould have conſtant acceſs to freſh water. The uſe of being as much as may be in the air is evident, becauſe all the air which they breathe paſſes twice over the putrid ſloughs of the mortified parts of the membrane which lines the noſtrils, and the maxillary and frontal cavities; that is, both during inſpiration and expiration; and muſt therefore be loaded with contagious particles. Freſh new milk, and freſh broth, ſhould be given them very frequently, and they ſhould be ſuffered to go amongſt the graſs, which they ſometimes eat for the purpoſe of an emetic; and if poſſible ſhould have acceſs to a running ſtream of water. As the

contagious

contagious mucus of the noſtrils, both of theſe animals and of horſes, generally drops into the water they attempt to drink. Bits of raw fleſh, if the dog will eat them, are preferred to cooked meat; and from five to ten drops of tincture of opium may be given with advantage, when ſymptoms of debility are evident, according to the ſize of the dog, every ſix hours. If ſloughs can be ſeen in the noſtrils, they ſhould be moiſtened twice a day, both in horſes and dogs, with a ſolution of ſugar of lead, or of alum, by means of a ſponge fixed on a bit of whale bone, or by a ſyringe. The lotion may be made by diſſolving half an ounce of ſugar of lead in a pint of water.

Ancient philoſophers ſeem to have believed, that the contagious miaſmata in their warm climates affected horſes and dogs previous to mankind. If thoſe contagious particles were ſuppoſed to be diffuſed amongſt the heavy inflammable air, or carbonated hydrogen, of putrid marſhes, as theſe animals hold their heads down lower to the ground, they may be ſuppoſed to have received them ſooner than men. And though men and quadrupeds might receive a diſeaſe from the ſame ſource of marſh-putrefaction, they might not afterwards be able to infect each other, though they might infect other animals of the ſame genus; as the new contagious matter generated in their own bodies might not be preciſely ſimilar to that received; as happened in the jail-fever at Oxford, where thoſe who took the contagion and died, did not infect others.

> On mules and dogs the infection firſt began,
> And, laſt, the vengeful arrows fix'd on man.
>
> POPE's Homer's Iliad, I.

7. *Peripneumonia ſuperficialis.* The ſuperficial or ſpurious peripneumony conſiſts in an inflammation of the membrane, which lines the bronchia, and bears the ſame analogy to the true peripneumony, as the inflammations of other membranes do to that of the parenchyma, or ſubſtantial parts of the viſcus, which they ſurround. It affects elderly people, and frequently occaſions their death; and

exists at the end of the true peripneumony, or along with it; when the lancet has not been used sufficiently to cure by reabsorbing the inflamed parts, or what is termed by resolution.

M. M. Diluents, mucilage, antimonials, warmish air constantly changed, venesection once, perhaps twice, if the pulse will bear it. Oily volatile draughts. Balsams? Neutral salts increase the tendency to cough. Blisters in succession about the chest. Warm bath. Mild purgatives. Very weak chicken broth without salt in it. Boiled onions. One grain of calomel every night for a week. From five drops to ten of tincture of opium at six every night, when the patient becomes weak. Digitalis? See Class II. 1. 6. 7.

8. *Pertussis.* Tussis convulsiva. Chin-cough resembles peripneumonia superficialis in its consisting in an inflammation of the membrane which lines the air-vessels of the lungs; but differs in the circumstance of its being contagious; and is on that account of very long duration; as the whole of the lungs are probably not infected at the same time, but the contagious inflammation continues gradually to creep on the membrane. It may in this respect be compared to the ulcers in the pulmonary consumption; but it differs in this, that in chin-cough some branches of the bronchia heal, as others become inflamed.

This complaint is not usually classed amongst febrile disorders, but a sensitive fever may generally be perceived to attend it during some part of the day, especially in weak patients. And a peripneumony very frequently supervenes, and destroys great numbers of children, except the lancet or four or six leeches be immediately and repeatedly used. When the child has permanent difficulty of breathing, which continues between the coughing fits: unless blood be taken from it, it dies in two, three, or four days of the inflammation of the lungs. During this permanent difficulty of breathing the hooping-cough abates, or quite ceases, and returns again after once

or

or twice bleeding; which is then a good ſymptom, as the child now poſſeſſing the power to cough ſhews the difficulty of breathing to be abated. I dwell longer upon this, becauſe many loſe their lives from the difficulty there is in bleeding young children; where the apothecary is old or clumſy, or is not furniſhed with a very ſharp and fine-pointed lancet. In this diſtreſſing ſituation the application of four leeches to one of the child's legs, the wounds made by which ſhould continue to bleed an hour or two, is a ſuccedaneum; and ſaves the patient, if repeated once or twice according to the difficulty of the reſpiration.

The chin-cough ſeems to reſemble the gonorrhœa venerea in ſeveral circumſtances. They are both received by infection, are both diſeaſes of the mucous membrane, are both generally cured in four or ſix weeks without medicine. If ulcers in the cellular membrane under the mucous membrane occur, they are of a phagedenic kind, and deſtroy the patient in both diſeaſes, if no medicine be adminiſtered.

Hence the cure ſhould be ſimilar in both theſe diſeaſes; firſt general evacuations and diluents, then, after a week or two, I have believed the following pills of great advantage. The doſe for a child of about three years old was one ſixth part of a grain of calomel, one ſixth part of a grain of opium, and two grains of rhubarb, to be taken twice a day.

The opium promotes abſorption from the mucous membrane, and hence contributes to heal it. The mercury prevents ulcers from being formed under the mucous membrane, or cures them, as in the lues venerea; and the rhubarb is neceſſary to keep the bowels open.

M. M. Antimonial vomits frequently repeated. Mild cathartics. Cool air. Tincture of cantharides, or repeated bliſters; afterwards opiates in ſmall doſes, and the bark. Warm bath frequently uſed. The ſteam of warm water with a little vinegar in it may be inhaled twice a day. Could the breathing of carbonic acid gas mixed with atmoſpheric air be of ſervice? Copious veneſection, when a difficulty of breathing continues between the fits of coughing; otherwiſe the cough and the expectoration ceaſe, and the patient is deſtroyed. Ulcers of the lungs ſometimes ſupervene, and the phthiſis pulmonalis in a few weeks terminates in death. Where the cough continues after ſome weeks without much of the hooping, and a ſenſitive fever daily ſupervenes, ſo as to reſemble hectic fever from ulcers of the lungs; change of air for a week or fortnight acts as a charm, and reſtores the patient beyond the hopes of the phyſician.

Young children ſhould lie with their heads and ſhoulders raiſed; and ſhould be conſtantly watched day and night; that when the cough occurs, they may be held up eaſily, ſo as to ſtand upon their feet bending a little forwards; or nicely ſupported in that poſture which they ſeem to put themſelves into. A bow of whalebone, about the ſize of the bow of a key, is very uſeful to extract the phlegm out of the mouths of infants at the time of their coughing; as an handkerchief, if applied at the time of their quick inſpirations after long holding their breath, is dangerous, and may ſuffocate the patient in an inſtant, as I believe has ſometimes happened.

9. *Variola diſcreta.* The ſmall-pox is well divided by Sydenham into diſtinct and confluent. The former conſiſts of diſtinct puſtules, which appear on the fourth day of the fever, are circumſcribed and turgid; the fever ceaſing when the eruption is complete. Headach, pain in the loins, vomiting frequently, and convulſive fits ſometimes, precede the eruption.

The diſtinct ſmall-pox is attended with ſenſitive fever only, when very mild, as in moſt inoculated patients; or with ſenſitive irritated fever, when the diſeaſe is greater: the danger in this kind of ſmall-pox is owing either to the tumor and ſoreneſs of the throat about the height, or eighth day of the eruption; or to the violence of the ſecondary fever. For, firſt, as the natural diſeaſe is generally taken

by

by particles of the duſt of the contagious matter dried and floating in the air, theſe are liable to be arreſted by the mucus about the throat and tonſils in their paſſage to the lungs, or to the ſtomach, when they are previouſly mixed with ſaliva in the mouth. Hence the throat inflames like the arm in inoculated patients; and this increaſing, as the diſeaſe advances, deſtroys the patient about the height.

Secondly, all thoſe upon the face and head come out about the ſame time, namely, about one day before thoſe on the hands, and two before thoſe in the trunk; and thence, when the head is very full, a danger ariſes from the ſecondary fever, which is a purulent, not a variolous fever; for as the matter from all theſe of the face and head is reabſorbed at the ſame time, the patient is deſtroyed by the violence of this purulent fever; which in the diſtinct ſmall-pox can only be abated by veneſection and cathartics; but in the confluent ſmall-pox requires cordials and opiates, as it is attended with arterial debility. See Sect. XXXV. 1. and XXXIII. 2. 10.

When the puſtules on the face recede, the face ſwells; and when thoſe of the hands recede, the hands ſwell; and the ſame of the feet in ſucceſſion. Theſe ſwellings ſeem to be owing to the abſorption of variolous matter, which by its ſtimulus excites the cutaneous veſſels to ſecrete more lymph, or ſerum, or mucus, exactly as happens by the ſtimulus of a bliſter. Now, as a bliſter ſometimes produces ſtrangury many hours after it has riſen; it is plain, that a part of the cantharides is abſorbed, and carried to the neck of the bladder; whether it enters the circulation, or is carried thither by retrograde movements of the urinary branch of lymphatics; and by parity of reaſoning the variolous matter is abſorbed, and ſwells the face and hands by its ſtimulus.

*Variola confluens*. The confluent ſmall-pox conſiſts of numerous puſtules, which appear on the third day of the fever, flow together, are irregularly circumſcribed, flaccid, and little elevated; the fever continuing

continuing after the eruption is complete; convulſions do not precede this kind of ſmall-pox, and are ſo far to be eſteemed a favourable ſymptom.

The confluent ſmall-pox is attended with ſenſitive inirritated fever, or inflammation with arterial debility; whence the danger of this diſeaſe is owing to the general tendency to gangrene, with petechiæ, or purple ſpots, and hæmorrhages; beſides the two ſources of danger from the tumor of the throat about the height, or eleventh day of the eruption, and the purulent fever after that time; which are generally much more to be dreaded in this than in the diſtinct ſmall-pox deſcribed above.

M. M. The method of treatment muſt vary with the degree and kind of fever. Veneſection may be uſed in the diſtinct ſmall-pox early in the diſeaſe, according to the ſtrength or hardneſs of the pulſe; and perhaps on the firſt day of the confluent ſmall-pox, and even of the plague, before the ſenſorial power is exhauſted by the violence of the arterial action? Cold air, and even waſhing or bathing in cold water, is a powerful means in perhaps all eruptive diſeaſes attended with fever; as the quantity of eruption depends on the quantity of the fever, and the activity of the cutaneous veſſels; which may be judged of by the heat produced on the ſkin; and which latter is immediately abated by expoſure to external cold. Mercurial purges, as three grains of calomel repeated every day during the eruptive fever, ſo as to induce three or four ſtools, contribute to abate inflammation; and is believed by ſome to have a ſpecific effect on the variolous, as it is ſuppoſed to have on the venereal contagion.

It has been ſaid, that opening the pock and taking out the matter has not abated the ſecondary fever; but as I had conceived, that the pits, or marks left after the ſmall-pox, were owing to the acrimony of the matter beneath the hard ſcabs, which not being able to exhale eroded the ſkin, and produced ulcers, I directed the faces of

 two

two patients in the confluent ſmall-pox to be covered with cerate early in the diſeaſe, which was daily renewed; and I was induced to think, that they had much leſs of the ſecondary fever, and were ſo little marked, that one of them, who was a young lady, almoſt entirely preſerved her beauty. Perhaps mercurial plaſters, or cerates, made without turpentine in them, might have been more efficacious in preventing the marks, and eſpecially if applied early in the diſeaſe, even on the firſt day of the eruption, and renewed daily. For it appears from the experiments of Van Woenſel, that calomel or ſublimate corroſive, triturated with variolous matter, incapacitates it from giving the diſeaſe by inoculation. Calomel or ſublimate given as an alterative for ten days before inoculation, and till the eruptive fever commences, is ſaid with certainty to render the diſeaſe mild by the ſame author. Exper. on Mercury by Van Woenſel, tranſlated by Dr. Fowle, Saliſbury.

*Variola inoculata.* The world is much indebted to the great diſcoverer of the good effects of inoculation, whoſe name is unknown; and our own country to Lady Wortley Montague for its introduction into this part of Europe. By inſerting the variolous contagion into the arm, it is not received by the tonſils, as generally happens, I ſuppoſe, in the natural ſmall-pox; whence there is no dangerous ſwelling of the throat, and as the puſtules are generally few and diſtinct, there is ſeldom any ſecondary fever; whence thoſe two ſources of danger are precluded; hence when the throat in inoculated ſmall-pox is much inflamed and ſwelled, there is reaſon to believe, that the diſeaſe had been previouſly taken by the tonſils in the natural way.—Which alſo, I ſuppoſe, has generally happened, where the confluent kind of ſmall-pox has occurred on inoculation.

I have known two inſtances, and have heard of others, where the natural ſmall-pox began fourteen days after the contagion had been received; one of theſe inſtances was of a countryman, who went to a market town many miles from his home, where he ſaw a perſon

in

in the ſmall-pox, and on returning the fever commenced that day fortnight: the other was of a child, whom the ignorant mother carried to another child ill of the ſmall-pox, on purpoſe to communicate the diſeaſe to it; and the variolous fever began on the fourteenth day from that time. So that in both theſe caſes fever commenced in half a lunation after the contagion was received. In the inoculated ſmall-pox the fever generally commences on the ſeventh day, or after a quarter of a lunation; and on this circumſtance probably depends the greater mildneſs of the latter. The reaſon of which is difficult to comprehend; but ſuppoſing the facts to be generally as above related, the ſlower progreſs of the contagion indicates a greater inirritability of the ſyſtem, and in conſequence a tendency to malignant rather than to inflammatory fever. This difference of the time between the reception of the infection and the fever in the natural and artificial ſmall-pox may nevertheleſs depend on its being inſerted into a different ſeries of veſſels; or to ſome unknown effect of lunar periods. It is a ſubject of great curioſity, and deſerves further inveſtigation.

When the inoculated ſmall-pox is given under all the moſt favourable circumſtances I believe leſs than one in a thouſand miſcarry, which may be aſcribed to ſome unavoidable accident, ſuch as the patient having previouſly received the infection, or being about to be ill of ſome other diſeaſe. Thoſe which have lately miſcarried under inoculation, as far as has come to my knowledge, have been chiefly children at the breaſt; for in theſe the habit of living in the air has been confirmed by ſo ſhort a time, that it is much eaſier deſtroyed, than when theſe habits of life have been eſtabliſhed by more frequent repetition. See Sect. XVII. 3. Thus it appears from the bills of mortality kept in the great cities of London, Paris, and Vienna, that out of every thouſand children above three hundred and fifty die under two years old. (Kirkpatrick on Inoculation.) Whence a ſtrong reaſon againſt our hazarding inoculation before that age is paſſed,

eſpecially

efpecially in crowded towns; except where the vicinity of the natural contagion renders it neceffary, or the convenience of inoculating a whole family at a time; as it then becomes better to venture the lefs favourable circumftances of the age of the patient, or the chance of the pain from toothing, than to rifk the infection in the natural way.

The moft favourable method confifts in, firft, for a week before inoculation, reftraining the patients from all kinds of fermented or fpirituous liquor, and from animal food; and by giving them from one grain to three or four of calomel every other day for three times. But if the patients be in any the leaft danger of taking the natural infection, the inoculation had better be immediately performed, and this abftinence then began; and two or three gentle purges with calomel fhould be given, one immediately, and on alternate days. Thefe cathartics fhould not induce more than two or three ftools. I have feen two inftances of a confluent fmall-pox in inoculation following a violent purging induced by too large a dofe of calomel.

Secondly, the matter ufed for inoculation fhould be in a fmall quantity, and warm, and fluid. Hence it is beft when it can be recently taken from a patient in the difeafe; or otherwife it may be diluted with part of a drop of warm water, fince its fluidity is likely to occafion its immediate abforption; and the wound fhould be made as fmall and fuperficial as poffible, as otherwife ulcers have been fuppofed fometimes to enfue with fubaxillary abfceffes. Add to this, that the making two punctures either on the fame, or one on each arm, fecures the fuccefs of the operation in refpect to communicating the infection.

Thirdly, at the time of the fever or eruption the application of cool air to thofe parts of the fkin, which are too warm, or appear red, or are covered with what is termed a rafh, fhould be ufed freely, as well as during the whole difeafe. And at the fame time, if the feet or hands are colder than natural, theſe ſhould be covered with flannel. See Claſs IV. 2. 2. 10.

10. *Rubeola irritata, morbilli.* The meaſles commence with ſneezing, red eyes, dry hoarſe cough, and is attended with ſenſitive irritated fever. On the fourth day, or a little later, ſmall thick eruptions appear, ſcarcely eminent above the ſkin, and, after three days, changing into very ſmall branny ſcales.

As the contagious material of the ſmall-pox may be ſuppoſed to be diffuſed in the air like a fine dry powder, and mixing with the ſaliva in the mouth to infect the tonſils in its paſſage to the ſtomach; ſo the contagious material of the meaſles may be ſuppoſed to be more completely diſſolved in the air, and thus to impart its poiſon to the membrane of the noſtrils, which covers the ſenſe of ſmell; whence a catarrh with ſneezing uſhers in the fever; the termination of the naſal duct of the lacrymal ſac is ſubject to the ſame ſtimulus and inflammation, and affects by ſympathy the lacrymal glands, occaſioning a great flow of tears. See Sect. XVI. 8. And the redneſs of the eye and eyelids is produced in conſequence of the tears being in ſo great quantity, that the ſaline part of them is not entirely reabſorbed. See Sect. XXIV. 2. 8.

The contagion of the meaſles, if it be taken a ſufficient time before inoculation, ſo that the eruption may commence before the variolous fever comes on, ſtops the progreſs of the ſmall-pox in the inoculated wound, and delays it till the meaſle-fever has finiſhed its career. See Sect. XXXIII. 2. 9.

The meaſles are uſually attended with inflammatory fever with ſtrong pulſe, and bear the lancet in every ſtage of the diſeaſe. In the early periods of it, veneſection renders the fever and cough leſs; and, if any ſymptoms of peripneumony occur, is repeatedly neceſſary; and at the decline of the diſeaſe, if a cough be left after the eruption

8 has

has ceafed, and the fubfequent branny fcales are falling off, venefection fhould be immediately ufed; which prevents the danger of confumption. At this time alfo change of air is of material confequence, and often removes the cough like a charm, as mentioned in a fimilar fituation at the end of the chin-cough.

*Rubeola inirritata.* Meafles with inirritated fever, or with weak pulfe, has been fpoken of by fome writers. See London Med. Obferv. Vol. IV. Art. XI. It has alfo been faid to have been attended with fore throat. Edinb. Effays, Vol. V. Art. II. Could the fcarlet fever have been miftaken for the meafles? or might one of them have fucceeded the other, as in the meafles and fmall-pox mentioned in Sect. XXXIII. 2. 9.?

From what has been faid, it is probable that inoculation might difarm the meafles as much as the fmall-pox, by preventing the catarrh, and frequent pulmonary inflammation, which attends this difeafe; both of which are probably the confequence of the immediate application of the contagious miafmata to thefe membranes. Some attempts have been made, but a difficulty feems to arife in giving the difeafe; the blood, I conjecture, would not infect, nor the tears; perhaps the mucous difcharge from the noftrils might fucceed; or a drop of warm water put on the eruptions, and fcraped off again with the edge of a lancet; or if the branny fcales were collected, and moiftened with a little warm water? Further experiments on this fubject would be worthy the public attention.

11. *Scarlatina mitis.* The fcarlet fever exifts with all degrees of virulence, from a flea-bite to the plague. The infectious material of this difeafe, like that of the fmall-pox, I fuppofe to be diffufed, not diffolved, in the air; on which account I fufpect, that it requires a much nearer approach to the fick, for a well perfon to receive the infection, than in the meafles; the contagion of which I believe to be more volatile, or diffufible in the atmofphere. But as the

 contagious

contagious miaſmata of ſmall-pox and ſcarlet fever are ſuppoſed to be more fixed, they may remain for a longer time in clothes or furniture; as a thread dipped in variolous matter has given the diſeaſe by inoculation after having been expoſed many days to the air, and after having been kept many months in a phial. This alſo accounts for the ſlow or ſporadic progreſs of the ſcarlet fever, as it infects others at but a very ſmall diſtance from the ſick; and does not produce a quantity of pus-like matter, like the ſmall-pox, which can adhere to the clothes of the attendants, and when dried is liable to be ſhook off in the form of powder, and thus propagate the infection.

This contagious powder of the ſmall-pox, and of the ſcarlet fever, becomes mixed with ſaliva in the mouth, and is thus carried to the tonſils, the mucus of which arreſts ſome particles of this deleterious material; while other parts of it are carried into the ſtomach, and are probably decompoſed by the power of digeſtion; as ſeems to happen to the venom of the viper, when taken into the ſtomach. Our perception of bad taſtes in our mouths, at the ſame time that we perceive diſagreeable odours to our noſtrils, when we inhale very bad air, occaſions us to ſpit out our ſaliva; and thus, in ſome inſtances, to preſerve ourſelves from infection. This has been ſuppoſed to originate from the ſympathy between the organs of taſte and ſmell; but any one who goes into a ſick room cloſe ſhut up, or into a crowded aſſembly-room, or tea-room, which is not ſufficiently ventilated, may eaſily mix the bad air with the ſaliva on his tongue ſo as to taſte it; as I have myſelf frequently attended to.

Hence it appears that theſe heavy infectious matters are more liable to mix with the ſaliva, and inflame the tonſils, and that either before or at the commencement of the fever; and this is what generally happens in the ſcarlet fever, always I ſuppoſe in the malignant kind, and very frequently in the mild kind. But as this infection may be taken by other means, as by the ſkin, it alſo happens in the moſt mild kind, that there is no inflammation of the tonſils at all; in the ſame

ſame manner as there is generally no inflammation of the tonſils in the inoculated ſmall-pox.

In the mild ſcarlatina on the fourth day of the fever the face ſwells a little, at the ſame time a florid redneſs appears on various parts of the ſkin, in large blotches, at length coaleſcing, and after three days changing into branny ſcales.

M. M. Cool air. Fruit. Lemonade. Milk and water.

*Scarlatina maligna.* The malignant ſcarlet fever begins with inflamed tonſils; which are ſucceeded by dark drab-coloured ſloughs three or five lines in diameter, flat, or beneath the ſurrounding ſurface; and which conceal beneath them ſpreading gangrenous ulcers. The ſwellings of the tonſils are ſenſible to the eye and touch externally, and have an elaſtic rather than an œdematous feel, like parts in the vicinity of gangrenes. The pulſe is very quick and weak, with delirium, and the patient generally dies in a few days; or if he recovers, it is by ſlow degrees, and attended with anaſarca.

M. M. A vomit once. Wine. Beer. Cyder. Opium. Bark, in ſmall repeated doſes. Small ſucceſſive bliſters, if the extremities are cooler than natural. Cool air on the hot parts of the ſkin, the cool extremities being at the ſame time covered. Iced lemonade. Broth. Cuſtards. Milk. Jellies. Bread pudding. Chicken. Touch the ulcers with a dry ſponge to abſorb the contagious matter, and then with a ſponge filled with vinegar, with or without ſugar of lead diſſolved in it, about ſix grains to an ounce; or with a very little blue vitriol diſſolved in it, as a grain to an ounce; but nothing ſo inſtantaneouſly corrects the putrid ſmell of ulcers as a ſolution of alum, about half an ounce to a pint of water, which ſhould be a little warmiſh, and injected into the fauces gently by means of a ſyringe. Theſe ſhould be repeated frequently in a day, if it can be done eaſily, and without fatigue to the child. A little powder of bark taken frequently into the mouth, as a grain or two, that it may mix with the ſaliva,

ſaliva, and thus frequently ſtimulate the dying tonſils. Could a warm bath made of decoction of bark, or a cold fomentation with it, be of ſervice? Could oxygene gas mixed with common air ſtimulate the languid ſyſtem? Small electric ſhocks through the tonſils every hour? ether frequently applied externally to the ſwelled tonſils?

As this diſeaſe is attended with the greateſt degree of debility, and as ſtimulant medicines, if given in quantity, ſo as to produce more than natural warmth, contribute to expend the already too much exhauſted ſenſorial power; it appears, that there is nothing ſo neceſſary to be nicely attended to, as to prevent any unneceſſary motions of the ſyſtem; this is beſt accompliſhed by the application of cold to thoſe parts of the ſkin, which are in the leaſt too hot. And ſecondly, that the exhibition of the bark in ſuch quantity, as not to oppreſs the ſtomach and injure digeſtion, is next to be attended to, as not being liable to increaſe the actions of the ſyſtem beyond their natural quantity; and that opium and wine ſhould be given with the greateſt caution, in very ſmall repeated quantity, and ſo managed as to prevent, if poſſible, the cold fits of fever; which probably occur twice in 25 hours, obeying the lunations like the tides, as mentioned in Sect. XXXII. 6. that is, I ſuppoſe, the cold periods, and conſequent exacerbations of fever, in this malignant ſcarlatina, occur twice in a lunar day; which is about ten minutes leſs than 25 hours; ſo that if the commencement of one cold fit be marked, the commencement of the next may be expected, if not diſturbed by the exhibition of wine or opium, or the application of bliſters, to occur in about twelve hours and a half from the commencement of the former; or if not prevented by large doſes of the bark.

No one could do an act more beneficial to ſociety, or glorious to himſelf, than by teaching mankind how to inoculate this fatal diſeaſe; and thus to deprive it of its malignity. Matter might be taken from the ulcers in the throat, which would probably convey the contagion. Or warm water might be put on the eruption, and ſcraped off

off again by the edge of a lancet. These experiments could be attended with no danger, and should be tried for the public benefit, and the honour of medical science.

12. *Miliaria.* Miliary fever. An eruption produced by the warmth, and more particularly by the stimulus of the points of the wool in flannel or blankets applied to the skin, has been frequently observed; which, by cool dress, and bed-clothes without flannel, has soon ceased. See Class I. 1. 2. 3. This, which may be called *miliaria sudatoria*, has been confounded with other miliary fevers, and has made the existence of the latter doubted. Two kinds of eruptions I have seen formerly attended with fever, but did not sufficiently mark their progress, which I conceived to be miliary eruptions, one with arterial strength, or with sensitive irritated fever, and the other with arterial debility, or with sensitive inirritated fever.

In the former of these, or *miliaria irritata*, the eruptions were distinct and larger than the small-pox, and the fever was not subdued without two or three venesections, and repeated cathartics with calomel.

The latter, or *miliaria inirritata*, was attended with great arterial debility; and during the course of the fever pellucid points appeared within the skin, particularly on the soft parts of the fingers. And, in one patient, whom I esteemed near her end, I well recollect to have observed round pellucid globules, like what are often seen on vines in hot-houses, no larger than the smallest pins' heads, adhere to her neck and bosom; which were hard to the touch, but were easily rubbed off. These diseases, if they are allied, do not differ more than the kinds of small-pox; but require many further observations.

The eruption so often seen on children in the cradle, and called by the nurses red-gum, and which is attended with some degree of fever, I suspect to be produced by too great warmth, and the contact of flannel

nel

nel next their tender ſkins, like the miliaria ſudatoria; and like that requires cool air, cool clothes, and linen next their ſkin.

13. *Peſtis.* The plague, like other diſeaſes of this claſs, ſeems to be ſometimes mild, and ſometimes malignant; according to the teſtimony of different writers. It is ſaid to be attended with inflammation, with the greateſt arterial debility, and to be very contagious, attended at an uncertain time of the fever with buboes and carbuncles. Some authors affirm, that the contagion of the plague may be repeatedly received, ſo as to produce the diſeaſe; but as this is contrary to the general analogy of all contagious diſeaſes, which are attended with fever, and which cure themſelves ſpontaneouſly; there is reaſon to ſuſpect, that where it has been ſuppoſed to have been repeatedly received, that ſome other fever with arterial debility has been miſtaken for it, as has probably univerſally been the caſe, when the ſmall-pox has been ſaid to have been twice experienced.

M. M. Veneſection has been recommended by ſome writers on the firſt day, where the inflammation was ſuppoſed to be attended with ſufficient arterial ſtrength, which might perhaps ſometimes happen, as the bubo ſeems to be a ſuppuration; but the carbuncle, or anthrax, is a gangrene of the part, and ſhews the greateſt debility of circulation. Whence all the means before enumerated in this genus of diſeaſes to ſupport the powers of life are to be adminiſtered. Currents of cold air, cold water, ice, externally on the hot parts of the ſkin.

The methods of preventing the ſpreading of this diſeaſe have been much canvaſſed, and ſeem to conſiſt in preventing all congregations of the people, as in churches, or play-houſes; and to remove the ſick into tents on ſome airy common by the ſide of a river, and ſupply them with freſh food, both animal and vegetable; with beer and wine in proper quantities, and to encourage thoſe who can, daily to waſh both their clothes and themſelves.

The

The *pestis vaccina*, or disease amongst the cows, which afflicted this island about half a century ago, seems to have been a contagious fever with great arterial debility; as in some of them in the latter stage of the disease, an emphysema could often be felt in some parts, which evinced a considerable progress of gangrene beneath the skin. In the sensitive inirritated fevers of these animals, I suppose about sixty grains of opium, with two ounces of extract of oak-bark, every six hours, would supply them with an efficacious medicine; to which might be added thirty grains of vitriol of iron, if any tendency to bloody urine should appear, to which this animal is liable. The method of preventing the infection from spreading, if it should ever again gain access to this island, would be immediately to obtain an order from government to prevent any cattle from being removed, which were found within five miles of the place supposed to be infected, for a few days; till the certainty of the existence of the pestilence could be ascertained, by a committee of medical people. As soon as this was ascertained, all the cattle within five miles of the place should be immediately slaughtered, and consumed within the circumscribed district; and their hides put into lime-water before proper inspectors.

14. *Pemphigus* is a contagious disease attended with bladdery eruptions appearing on the second or third day, as large as filberts, which remain many days, and then effuse a thin ichor. It seems to be either of a mild kind with sensitive fever only, of which I have seen two instances, or with irritated, or with inirritated fever, as appears from the observations of M. Salabert. See Medical Comment. by Dr. Duncan, Decad. II. Vol. VI.

15. *Varicella*. Chicken-pox is accompanied with sensitive fever, pustules break out after a mild fever like the small-pox, seldom suppurate, and generally terminate in scales without scars. I once saw a

lady, who miſcarryed during this diſeaſe, though all her children had it as ſlightly as uſual. It ſometimes leaves ſcars or marks on the ſkin. This diſeaſe has been miſtaken for the ſmall-pox, and inoculated for it; and then the ſmall-pox has been ſuppoſed to happen twice to the ſame perſon. See Tranſ. of the College London. It is probable that the pemphigus and urticaria, as well as this diſeaſe, have formerly been diſeaſes of more danger; which the habit of innumerable generations may have rendered mild, and will in proceſs of time annihilate. In the ſame manner as the ſmall-pox, venereal diſeaſe, and rickets, ſeem to become milder or leſs in quantity every half century. While at the ſame time it is not improbable, that other new diſeaſes may ariſe, and for a ſeaſon thin mankind!

16. *Urticaria.* Nettle-raſh begins with mild ſenſitive fever, which is ſometimes ſcarcely perceptible. Hence this eruption has been thought of two ſorts, one with and the other without fever. On the ſecond day red ſpots, like parts ſtung with nettles, are ſeen; which almoſt vaniſh during the day, and recur in the evening with the fever, ſucceeded in a few days by very minute ſcales. See Tranſ. of the College, London.

17. *Aphtha.* Thruſh. It has been doubted, whether aphtha or thruſh, which conſiſts of ulcers in the mouth, ſhould be enumerated amongſt febrile diſeaſes; and whether theſe ulcers are always ſymptomatic, or the conſequence rather than the cauſe of the fevers which attend them. The tongue becomes rather ſwelled; its colour and that of the fauces purpliſh; ſloughs or ulcers appear firſt on the throat and edges of the tongue, and at length over the whole mouth. Theſe ſloughs are whitiſh, ſometimes diſtinct, often coaleſcing, and remain an uncertain time. Cullen. I ſhall conciſely mention four caſes of aphtha, but do not pretend to determine whether they were all of them ſymptomatic or original diſeaſes.

*Aphtha ſenſitiva.* A lady during pregnancy was frequently ſeized with ulcers on her tongue and cheeks, or other parts of the mouth, without much apparent fever; which continued two or three weeks, and returned almoſt every month. The thruſh in the mouths of young children ſeems to be a ſimilar diſeaſe. Theſe ulcers reſemble thoſe produced in the ſea-ſcurvy, and have probably for their cauſe an increaſed action of the ſecerning ſyſtem from increaſed ſenſation, with a decreaſed action of the abſorbent ſyſtem from decreaſed irritation. See Claſs I. 2. 1. 14.

M. M. Solutions of alum, of blue vitriol. Powder of bark taken frequently into the mouth in very ſmall quantity. See Claſs II. 1. 3. 1.

*Aphtha irritata.* Inflammatory aphtha. A caſe of this kind is related under the title of ſuppurative rheumatiſm. Claſs IV. 2. 1. 16.

*Aphtha inirritata.* Sloughs or ulcers of the mouth, attended with ſenſitive fever with great arterial debility. They ſeem to ſpread downwards from the throat into the ſtomach, and probably through the whole inteſtinal canal, beginning their courſe with cardialgia, and terminating it with teneſmus; and might perhaps be called an eryſipelas of this mucous membrane.

M. M. Cool air. A ſmall bliſter on the back. Bark. Wine. Opium in ſmall repeated quantities. Soap neutralizes the gaſtric acid without efferveſcence, and thus relieves the pain of cardialgia, where the ſtomach is affected. Milk alſo deſtroys a part of this acid. Infuſion of ſage leaves two ounces, almond ſoap from five grains to ten, with ſugar and cream, is generally both agreeable and uſeful to theſe patients. See I. 2. 4. 5.

Where the ſtomach may be ſuppoſed to be excoriated by poiſons containing acid, as ſublimate of mercury or arſenic; or if it be otherwiſe inflamed, or very ſenſible to the ſtimulus of the gaſtric acid; or where it abounds with acid of any kind, as in cardialgia; the exhibi-

 tion

tion of ſoap is perhaps a preferable manner of giving alcali than any other, as it decompoſes in the ſtomach without effervefcence; while the cauſtic alcali is too acrid to be adminiſtered in ſuch caſes, and the mild alcali produces carbonic gas. If a drop of acid of vitriol be put on cap paper, it will be long before it deſtroys the paper; but if a drop of mild alcali be added, a ſudden effervefcence ariſes, and the paper is inſtantly deſtroyed by the eſcape of the fixed air; in the ſame manner as lumps of ſolid lime are broken into powder by the eſcape of the ſteam produced from the water, which is poured on them. This ſhews why a ſucceſſion of acid and of alcaline cauſtics ſooner deſtroys a part, than either of them applied ſeparately.

18. *Dyſenteria.* Bloody-flux is attended with ſenſitive fever generally with arterial debility; with frequent mucous or bloody ſtools; which contain contagious matter produced by the membranes of the inteſtines; the alimentary excrement being nevertheleſs retained; with griping pains and teneſmus.

M. M. Emetics. Antimonials. Peruvian bark. Opium and calomel of each a grain every night. Bolus armeniæ. Earth of alum. Chalk. Calcined hartſhorn. Mucilage. Bee's wax mixt with yolk of egg. Cerated glaſs of antimony. Warm bath. Flannel clothing next to the ſkin. Large clyſters with opium. With ipecacuanha, with ſmoke of tobacco? Two dyſenteric patients in the ſame ward of the infirmary at Edinburgh quarrelled, and whipped each other with horſewhips a long time, and were both much better after it, owing perhaps to the exertion of ſo much of the ſenſorial power of volition; which, like real inſanity, added excitement to the whole ſyſtem.

The prevention of this contagion muſt conſiſt principally in ventilation and cleanlineſs; hence the patients ſhould be removed into cottages diſtant from each other, or into tents; and their fæces buried as ſoon as may be; or conveyed into a running ſtream; and themſelves ſhould

ſhould be waſhed with cold or warm water after every evacuation. For the contagious matter conſiſts in the múcous or purulent diſcharge from the membrane which lines the inteſtines; and not from the febrile perſpiration, or breath of the patients. For the fever is only the conſequence and not the cauſe of contagion; as appears from Genus the Fifth of this Order, where contagion exiſts without fever.

19. *Gaſtritis ſuperficialis.* Superficial inflammation of the ſtomach. An eryſipelatous inflammation of the ſtomach is mentioned by Dr. Cullen from his own obſervations; which is diſtinguiſhed from the inflammatory gaſtritis by leſs pain and fever, and by an eryſipelatous redneſs about the fauces. Does this diſeaſe belong to aphtha?

20. *Enteritis ſuperficialis.* Superficial inflammation of the bowels is alſo mentioned by Dr. Cullen from his own obſervation under the name of enteritis erythematica; and is ſaid to be attended with leſs pain and fever, without vomiting, and with diarrhoea. May not this diſeaſe be referred to aphtha, or to dyſentery?

ORDO

## ORDO I.

*Increaſed Senſation.*

## GENUS IV.

*With the Production of new Veſſels by internal Membranes or Glands, without Fever.*

WHERE inflammation is produced in a ſmall part, which has not great natural ſenſibility, the additional ſenſation does not produce an increaſed action of the arterial ſyſtem; that is, the aſſociated motions which are employed in the circulation of the blood, thoſe for inſtance of the heart, arteries, glands, capillaries, and their correſpondent veins, are not thrown into increaſed action by ſo ſmall an addition of the ſenſorial power of ſenſation. But when parts, which naturally poſſeſs more ſenſibility, become inflamed, the quantity of the ſenſorial power of ſenſation becomes ſo much increaſed, as to affect the aſſociated motions belonging to the circulation, occaſioning them to proceed with greater frequency; that is, a fever is induced. This is well exemplified in the internal and ſuperficial paronychia, one of which is attended with great pain and fever, and the other with little pain and no fever. See Claſs II. 1. 2. 19. and II. 1. 4. 5.

From hence it appears, that the ſenſitive fever is an accidental conſequence of the topical phlegmon, or inflammation, and not a cauſe of it; that it is often injurious, but never ſalutary; and ſhould therefore always be extinguiſhed, as ſoon as may be, either by the lancet and cathartics, and diluents, and cold air, when it is of the irritated kind; or by the bark, opium, cool air, and nutrientia, when it is of the inirritated kind.

SPECIES.

## SPECIES.

1. *Ophthalmia ſuperficialis.* As the membranes, which cover the eye, are excluded from the air about one third part of the twenty-four hours; and are moiſtened by perpetual nictitation during the other ſixteen; they may be conſidered as internal membranes; and from the analogy of their inflammation to that of other internal membranes, it is arranged under this genus; whilſt the tonſillitis is eſteemed an inflammation of an external membrane, becauſe currents of air are perpetually paſſing both day and night over the fauces.

The ſuperficial ophthalmy has generally been eſteemed a ſymptom of ſcrophula, when it recurs frequently in young perſons; but is probably only a concomitant of that diſeaſe, as a ſymptom of general debility; ramifications of new red veſſels, and of enlarged old ones, are ſpread over the white part of the eye; and it is attended with leſs heat, leſs pain, and leſs intolerance of light than the ophthalmia interna, deſcribed in Claſs II. 1. 2. 2. It occurs in thoſe of feeble circulation, eſpecially children of a ſcrophulous tendency, and ſeems to ariſe from a previous torpor of the veſſels of the tunica albuginea from their being expoſed to cold air; and from this torpor being more liable to occur in habits, which are naturally inirritable; and therefore more readily fall into quieſcence by a ſmaller deduction of the ſtimulus of heat, than would affect ſtronger or more irritable habits; the conſequence of this torpor is increaſed action, which produces pain in the eye, and that induces inflammation by the acquiſition of the additional ſenſorial power of ſenſation.

*Ophthalmia lymphatica* is a kind of anaſarca of the tunica adnata; in this the veſſels over the ſclerotica, or white part of the eye, riſe conſiderably above the cornea, which they ſurround, are leſs red than in the ophthalmia ſuperficialis, and appear to be ſwelled by an accumulation

of

of lymph rather than of blood; it is probably owing to the temporary obſtruction of a branch of the lymphatic ſyſtem.

M. M. If the pain be great, veneſection by leeches on the temple, or cutting the temporal artery, and one purge with three or four grains of calomel ſhould be premiſed. Then the Peruvian bark twice a day. Opium from a quarter to half a grain twice a day for ſome weeks. Bathe the eye frequently with cold water alone, or with cold water, to a pint of which is added half an ounce of ſalt. White vitriol ſix grains diſſolved in one ounce of water; a drop or two to be put between the eyelids twice a day. Take very ſmall electric ſparks from the eyes every day for a fortnight. Bathe the whole head with ſalt and water made warm every night for ſome months. Send ſuch children to a ſchool near the ſea for the convenience of ſea-bathing for many months annually; ſuch ſchools are to be found in or near Liverpool.

When a child is afflicted with an inflamed eye of this kind, he ſhould always ſit with his back to the window or candle; but it is generally not neceſſary to cover it, or if the uneaſy ſenſation of light makes this proper, the cover ſhould ſtand off from the eye, ſo as not much to exclude the cool air from it. As covering an eye unneceſſarily is liable to make that eye weaker than the other, from its not being ſufficiently uſed, and thence to produce a ſquinting for ever afterwards.

Nevertheleſs, when the pain is great, a poultice muſt be applied to keep the eyes moiſt, or a piece of oiled ſilk bound lightly over them. Or thus, boil an egg till it is hard, cut it longitudinally into two hemiſpheres, take out the yolk, ſew the backs of the two hollow hemiſpheres of the white to a ribbon, and bind them over the eyes every night on going to bed; which, if nicely fitted on, will keep the eyes moiſt without any diſagreeable preſſure. See Claſs I. 1. 3. 14.

*Ophthalmia*

*Ophthalmia equina.* An inflammation of this kind is liable to affect the eyes of horses; one cause of which is owing to a silly custom of cutting the hair out of horses' ears; by which they are not only liable to take cold at the ear, but grass seeds are liable to fall into their ears from the high racks in stables; and in both cases the eye becomes inflamed by sympathy. I once directed the temporal artery of a horse to be opened, who had frequent returns of an inflamed eye; and I believed it was of essential service to him; it is probable that the artery was afterwards contracted in the wounded part, and that thence less blood was derived to the eye: the hæmorrhage was stopped by two persons alternately keeping their fingers on the orifice, and afterwards by a long bandage of broad tape.

2. *Pterigion.* Eye-wing. A spot of inflammation sometimes begins on the inside of the lower eyelid, or on the tunica albuginea, and spreads an intertexture of red vessels from it, as from a center, which extend on the white part of the eye, and have the appearance of the wing of a fly, from whence its name.

M. M. Cut the ramifications of vessels again and again with the point of a lancet close to the center of inflammation.

3. *Tarsitis palpebrarum.* Inflammation of the edges of the eyelids. This is a disease of the glands, which produce the hairs of the eyelashes, and is frequently the cause of their falling off. After this inflammation a hard scar-like ridge is left on the edge of the eyelid, which scratches and inflames the eyeball, and becomes a very troublesome disease.

The Turkish ladies are said to colour the edge of the eyelash with crude antimony in very fine powder, which not only gives lustre to the eye, as a diamond set on a black foil, but may prevent extraneous light from being reflected from these edges into the eye, and thus serve the purpose of the black feathers about the eyes of swans, deſcribed in Sect. XXXIX. 5. 1. and may alſo prevent the edges of the eyelids from being inflamed by the frequent ſtimulus of tears on them. Black lead in fine powder might be better for all theſe purpoſes than antimony, and might be put on with a camel's hair bruſh.

M. M. Mercurial ointment ſmeared at night on the edges of the eyelids. Burnt alum ſixty grains, hog's greaſe half an ounce, well rubbed into an ointment to be ſmeared on them in the night. Cold water frequently in the day. See Claſs II. 1. 1. 8.

4. *Hordeolum.* Stye. This inflammation begins either on or near the edges of the eyelids, or in the looſe ſkin of them, and is ſometimes very ſlow either in coming to ſuppuration or in diſperſing. The ſkin beneath the lower eyelid is the moſt frequent ſeat of this tumor, which ſometimes never ſuppurates at all, but becomes an incyſted tumor: for as this ſkin is very looſe for the purpoſe of admitting great motion to the eyelid, the abſorbent power of the veins ſeems particularly weak in this part; whence when any perſon is weakened by fatigue or otherwiſe, a darker ſhade of colour is ſeen beneath the eyes; which is owing to a leſs energetic action of the abſorbent terminations of the veins, whence the currents of dark or venous blood are delayed in them. This dark ſhade beneath the eyes, when it is permanent, is a ſymptom of habitual debility, or inirritability of the circulating ſyſtem. See Claſs I. 2. 2. 2.

M. M. Smear the tumors with mercurial ointment, moiſten them frequently with ether. To promote their ſuppuration they may be wounded with a lancet, or ſlit down the middle, or they may be cut out. A cauſtic leaves a large ſcar.

*Paronychia ſuperficialis.* Whitlow. An inflammation about the roots of the nail beneath the ſkin, which ſuppurates without fever, and ſometimes deſtroys the nail; which is however gradually reproduced.

duced. This kind of abſceſs, though not itſelf dangerous, has given opportunity for the inoculation of venereal matter in the hands of accoucheurs, and of putrid matter from the diſſection of diſeaſed bodies; and has thus been the cauſe of diſeaſe and death. When putrid matter has been thus abſorbed from a dead body, a livid line from the finger to the ſwelled gland in the axilla is ſaid to be viſible; which ſhews the inflammation of the abſorbent veſſel along its whole courſe to the lymphatic gland; and death has generally been the conſequence.

M. M. In the common paronychia a poultice is generally ſufficient. In the abſorption of putrid matter rub the whole hand and arm with mercurial ointment three or four times a day, or perpetually. Could the ſwelled axillary gland be exſected? In the abſorption of venereal matter the uſual methods of cure in ſyphilis muſt be adminiſtered, as in Claſs II. 1. 5. 1.

6. *Gutta roſea.* The roſy drop on the face is of three kinds. Firſt, the *gutta roſea hepatica*, or the red pimples on the faces of drunkards, which are probably a kind of criſis, or vicarious inflammation, which ſucceeds, or prevents, a torpor of the membranes of the liver. This and the ſucceeding ſpecies properly belong to Claſs IV. 1. 2. 14.

Secondly, the pimpled face in conſequence of drinking cold water, or eating cold turnips, or other inſipid food, when much heated with exerciſe; which probably ariſes from the ſympathy between the ſkin of the face and the ſtomach; and may be called the *gutta roſea ſtomatica.* Which is diſtinguiſhed from the former by the habits of the patient in reſpect to drinking; by the colour of the eruptions being leſs deep; and by the patient continuing generally to be troubled with ſome degree of apepſia. See Claſs I. 3. 1. 3. I knew a lady, who had long been afflicted with pain about the region of the ſtomach; and, on drinking half a pint of vinegar, as a medicine, ſhe had a breaking



out

out commenced on her face; which remained, and ſhe became free from the pain about the ſtomach. Was this a ſtomachic, or an hepatic diſeaſe?

Thirdly, there is a red face, which conſiſts of ſmaller pimples than thoſe above mentioned; and which is leſs liable to ſuppurate; and which ſeems to be hereditary, or at leaſt has no apparent cauſe like thoſe above mentioned; which may be termed *gutta roſea hereditaria*, or puncta roſea.

Mrs. S. had a pimpled face, which I believe aroſe from potation of ale. She applied alum in a poultice to it, and had ſoon a paralytic ſtroke, which diſabled her on one ſide, and terminated in her death.

Mrs. L. had a red pimpled face, which ſeemed to have been derived from her mother, who had probably acquired it by vinous potation; ſhe applied a quack remedy to it, which I believe was a ſolution of lead, and was ſeized with epileptic fits, which terminated in palſy, and deſtroyed her. This ſhews the danger of uſing white paint on the face, which is called biſmuth, but is in reality white lead or ceruſſa.

Mr. Y—— had acquired the gutta roſea on his noſe, and applied a ſaturnine ſolution on it for a few nights, and was then ſeized with paralyſis on one ſide of his face; which however he gradually recovered, and has ſince acquired the gutta roſea on other parts of his face.

Theſe fatal effects were probably cauſed by the diſagreeable ſenſation of an inflamed liver, which uſed before to be relieved of the ſympathetic action and conſequent inflammation of the ſkin of the face, which was now prevented by the ſtronger ſtimulus of the application of calx of lead. The manner in which diſagreeable ſenſations induce epilepſy and palſy is treated of in Claſs III. In ſome caſes where habitual diſcharges, or eruptions, or ulcers are ſtopped, a torpor of the ſyſtem may follow, owing to the want of the accuſtomed

accuſtomed quantity of ſenſation or irritation. See Claſs I. 1. 2. 9. and II. 1. 5. 6. In both theſe ſituations ſome other ſtimulus ſhould be uſed to ſupply the place of that which is taken away; which may either be perpetual, as an iſſue; or periodical, as a cathartic repeated once a fortnight or month.

Miſs W. an elegant young lady of about twenty, applied a mercurial lotion to her face, which was covered with very ſmall red points; which ſeemed to have been not acquired by any known or avoidable means; ſhe was ſeized with inflammation of her liver, and after repeated bleeding and cathartics recovered, and in a few weeks the eruption appeared as before.

M. M. Five grains of calomel once a month, with a cathartic, five grains of rhubarb and a quarter of a grain of emetic tartar every night for many weeks. With this preparation mercurial plaſters, made without turpentine, and applied every night, and taken off every morning, will ſometimes ſucceed, and may be uſed with ſafety. But bliſtering the face all over the eruption, beginning with a part, ſucceeds better than any other means, as I have more than once experienced.—Something like this is mentioned in the Letters of Lady Mary Wortley Montague, who bliſtered her face with balſam of Mecca.

Mrs. F. had for many years had a diſagreeably looking eruption on her chin, after a cathartic with calomel, ſhe was adviſed to bliſter her whole chin; on the healing of the bliſter a few eruptions again appeared, which ceaſed on the application of a ſecond bliſter. She took rhubarb five grains, and emetic tartar a quarter of a grain every night for many weeks.

Miſs L. a young lady about eighteen, had tried variety of advice for pimples over the greateſt part of her face in vain. She took the above medicines internally, and bliſtered her face by degrees all over and became quite beautiful. A ſpot or two now and then appeared, and on this account ſhe frequently ſlept with parts of her face covered with mercurial plaſter, made without turpentine, which was held on by a paſteboard maſk, and taken off in the mornings; if any part of the plaſter adhered, a little butter or oil deſtroyed the adheſion.

7. *Odontitis.* Inflammatory tooth-ach is occaſioned by inflammation of the membranes of the tooth, or a caries of the bone itſelf. The gum ſometimes ſuppurates, otherwiſe a ſwelling of the cheek ſucceeds by aſſociation, and thus the violence of the pain in the membranes of the tooth is relieved, and frequently cured; and when this happens the diſeaſe properly belongs to Claſs IV. as it ſo far reſembles the tranſlations of morbid actions in the gout and rheumatiſm.

At other times the tooth dies without caries, eſpecially in people about ſixty years of age, or before; and then it ſtimulates its involving membrane, like any other extraneous ſubſtance. The membrane then becomes inflamed and thickened, occaſioning ſome pain, and the tooth riſes upwards above the reſt, and is gradually puſhed out whole and undecayed; on its riſing up a pus-like mucus is ſeen diſcharged from the gum, which ſurrounds it; and the gum ſeems to have left the tooth, as the fangs or roots of it are in part naked.

M. M. Where the tooth is ſound it can only be ſaved by evacuations by veneſection, and a cathartic; and after its operation two grains of opium, a bliſter may alſo be uſed behind the ear, and ether applied to the cheek externally. In ſlighter caſes two grains of opium with or without as much camphor may be held in the mouth, and ſuffered to diſſolve near the affected tooth, and be gradually ſwallowed. See Claſs I. 2. 4. 12. Odontalgia may be diſtinguiſhed from otitis by the application of cold water to the affected tooth; for as the pain of common tooth-ach is owing to torpor, whatever decreaſes ſtimulus adds to the torpor and conſequent pain; whereas the pain of an inflamed tooth being ceaſed by the increaſed action of the membranes of it is in ſome meaſure alleviated by the application of cold.

8. *Otitis.* Inflammation and conſequent ſuppuration of ſome membranes of the internal ear frequently occur in children, who ſleep in cold rooms, or near a cold wall, without a night-cap. If the bones are affected, they come out in a long proceſs of time, and the child remains deaf of that ear. But in this caſe there is generally a fever attends this inflammation; and it then belongs to another genus.

M. M. A warmer night-cap. Warmiſh water ſhould be gently ſyringed into the ear to keep it clean twice a day; and if it does not heal in a week, a little ſpirit of wine ſhould be added; firſt about a fourth part, and it ſhould be gradually increaſed to half rectified ſpirit and half water: if it continues long to diſcharge matter with a very putrid ſmell, the bones are injured, and will in time find their exit, during which time the ear ſhould be kept clean by filling it with a weaker mixture of ſpirit of wine and water; or a ſolution of alum in water; which may be poured into the ear, as the head is inclined, and ſhook out again by turning the head, two or three times morning and evening. See Claſs II. 1. 4. 10.

9. *Fiſtula lacrymalis.* The lacrymal ſack, with its puncta lacrymalia and naſal duct, are liable to be deſtroyed by ſuppuration without fever; the tears then run over the eyelids, and inflame the edges of them, and the cheeks, by their perpetual moiſture, and ſaline acrimony.

M. M. By a nice ſurgical operation a new aperture is to be made from the internal corner of the eye into the noſtril, and a ſilver tube introduced, which ſupplies the defect by admitting the tears to paſs again into the noſtril. See Melanges de Chirurgie par M. Pouteau; who thinks he has improved this operation.

10. *Fistula in ano.* A mucous discharge from the anus, called by some white piles, or matter from a suppurated pile, has been mistaken for the matter from a concealed fistula. A bit of cotton wool applied to the fundament to receive the matter, and renewed twice a day for a week or two, should always be used before examination with the probe. The probe of an unskilful empyric sometimes does more harm in the loose cellular membrane of these parts than the original ulcer, by making a fistula he did not find. The cure of a fistula in ano of those, who have been much addicted to drinking spirituous liquor, or who have a tendency to pulmonary consumption, is frequently of dangerous consequence, and is succeeded by ulcers of the lungs, and death.

M. M. Ward's paste, or 20 black pepper-corns taken after each meal twice a day; the pepper-corns should be cut each into two or three pieces. The late Dr. Monro of Edinburgh asserted in his lectures, that he had known a fistula in ano cured by injecting first a mixture of rectified spirit of wine and water; and by gradually increasing the strength of it, till the patient could bear rectified spirit alone; by the daily use of which at length the sides of the fistula became callous, and ceased to discharge, though the cavity was left. A French surgeon has lately affirmed, that a wire of lead put in at the external opening of the ulcer, and brought through the rectum, and twisted together, will gradually wear itself through the gut, and thus effect a cure without much pain. The ends of the leaden wire must be twisted more and more as it becomes loose. Or, lastly, it must be laid open by the knife.

11. *Fistula urethræ.* Where a stricture of the urethra exists, from whatever cause, the patient, in forcing the stream of urine through the structure, distends the urethra behind it; which after a time is liable to burst, and to become perforated; and some of the urine is pushed into the cellular membrane, occasioning fistulas, which sometimes

times have large ſurfaces producing much matter, which is preſſed out at the time of making water, and has been miſtaken for a catarrh of the bladder; theſe fiſtulas ſometimes acquire an external opening in the perinæum, and part of the urine is diſcharged that way.

Can this matter be diſtinguiſhed from mucus of the bladder by the criterion delivered in Claſs II. 1. 6. 6?

M. M. The perpetual uſe of bougies, either of catgut or of coartchouc. The latter may be had at No. 37, Red-lion ſtreet, Holborn, London. The former are eaſily made, by moiſtening the catgut, and keeping it ſtretched till dry, and then rounding one end with a pen-knife. The uſe of a warm bath every day for near an hour, at the heat of 94 or 96 degrees, for two or three months, I knew to be uncommonly ſucceſsful in one caſe; the extenſive fiſtulas completely healing. The patient ſhould introduce a bougie always before he makes water, and endeavour to make it as ſlowly as poſſible. See Claſs I. 2. 3. 24.

12. *Hepatitis chronica.* Chronical inflammation of the liver. A collection of matter in the liver has frequently been found on diſſection, which was not ſuſpected in the living ſubject. Though there may have been no certain ſigns of ſuch a collection of matter, owing to the inſenſibility of the internal parts of this viſcus; which has thus neither been attended with pain, nor induced any fever; yet there may be in ſome caſes reaſon to ſuſpect the exiſtence of ſuch an abſceſs; either from a ſenſe of fulneſs in the right hypochondre, or from tranſient pains ſometimes felt there, or from pain on preſſure, or from lying on the left ſide, and ſometimes from a degree of ſenſitive fever attending it.

Dr. Saunders ſuſpects the acute hepatitis to exiſt in the inflammation of the hepatic artery, and the chronical one in that of the vena portarum. Treatiſe on the Liver. Robinſon. London.

13. *Scrophula ſuppurans.* Suppurating ſcrophula. The indolent tumors of the lymphatic glands are liable, after a long time, to regain their ſenſibility; and then, owing to their former torpor, an increaſed action of the veſſels, beyond what is natural, with inflammation, is the conſequence of their new life, and ſuppuration ſucceeds. This cure of ſcrophula generally happens about puberty, when a new energy pervades the whole ſyſtem, and unfolds the glands and organs of reproduction.

M. M. See Claſs I. 2. 3. 21. Where ſcrophulous ulcers about the neck are difficult to heal, Dr. Beddoes was informed, in Ireland, that an empyric had had ſome ſucceſs by inflaming them by an application of wood ſorrel, oxalis acetoſella, the leaves of which are bruiſed in a mortar, and applied on the ulcers for two or three days, and then ſome more lenient application is uſed.

A poor boy, about twelve years old, had a large ſcrophulous ulcer on one ſide of the cheſt beneath the clavicle, and another under his jaw; he was directed, about three weeks ago, to procure a pound of dry oak-bark from the tanners, and to reduce it to fine powder, and to add to it one ounce of white lead in fine powder, and to cover the ulcers daily with it, keeping it on by brown paper and a bandage. He came to me a few minutes ago, to ſhew me that both the ulcers are quite healed. The conſtant application of linen rags, moiſtened with a ſolution of an ounce of ſugar of lead in a pint of water, I think I have ſeen equally efficacious.

14. *Scorbutus ſuppurans.* In the ſea-ſcurvy there exiſts an inactivity of venous abſorption, whence vibices and petechiæ, and ſometimes ulcers. As the column of blood preſſing on the origins of the veins of the lower extremities, when the body is erect, oppoſes the aſcent of the blood in them, they are more frequently liable to become enlarged, and to produce varixes, or vibices, or, laſtly, ulcers about the legs, than on the upper parts of the body. The expoſure to

cold

cold is believed to be another cause of ulcers on the extremities; as happens to many of the poor in winter at Lisbon, who sleep in the open air, without stockings, on the steps of their churches or palaces. See Class I. 2. 1. 15.

M. M. A bandage spread with plaster to cover the whole limb tight. Rags dipped in a solution of sugar of lead. A warm flannel stocking or roller. White lead and oak bark, both in fine powder. Horizontal rest.

15. *Scirrhus suppurans.* When a scirrhus affects any gland of no great extent or sensibility, it is, after a long period of time, liable to suppurate without inducing fever, like the indolent tumors of the conglobate or lymphatic glands above mentioned; whence collections of matter are often found after death both in men and other animals; as in the liver of swine, which have been fed with the grounds of fermented mixtures in the distilleries. Another termination of scirrhus is in cancer, as described below. See Class I. 2. 3. 22.

16. *Carcinoma.* Cancer. When a schirrous tumor regains its sensibility by nature, or by any accidental hurt, new vessels shoot amongst the yet insensible parts of it, and a new secretion takes place of a very injurious material. This cancerous matter is absorbed, and induces swelling of the neighbouring lymphatic glands; which also become schirrous, and afterwards cancerous.

This cancerous matter does not seem to acquire its malignant or contagious quality, till the cancer becomes an open ulcer; and the matter secreted in it is thus exposed to the air. Then it evidently becomes contagious, because it not only produces hectic fever, like common matter in ulcers open to the air; but it also, as it becomes absorbed, swells the lymphatic glands in its vicinity; as those of the axilla, when the open cancer is on the breast. See Class II. 1. 3.

Hence exsection before the cancer is open is generally a cure; but after the matter has been expofed to the air, it is feldom of fervice; as the neighbouring lymphatic glands are already infected. I have obferved fome of thefe patients after the operation to have had difeafed livers, which might either have previoufly exifted, or have been produced by the fear or anxiety attending the operation.

Erofion with arfenic, after the cancer is become an open ulcer, has generally no better effect than exfection, but has been fuccefsful before ulceration. The beft manner of ufing arfenic, is by mixing one grain with a dram of lapis calaminaris, and ftrewing on the cancer fome of the powder every day, till the whole is deftroyed.

Cancers on the face are faid to arife from the periofteum, and that unlefs this be deftroyed by the knife, or by cauftics, the cancer certainly recurs. After the cancer becomes an open ulcer of fome extent, a purulent fever fupervenes, as from other open ulcers, and gradually deftroys the patient. See Clafs II. 1. 6. 13.

Two very interefting cafes have been lately publifhed by Dr. Ewart, of Bath, in which carbonic acid gas, or fixed air, was kept conftantly in contact with the open cancerous ulcers of the breaft; which then healed like other common ulcers. This is rather to be afcribed to the exclufion of oxygen, than to any fpecific virtue in the carbonic acid. As in common ulcers the matter does not induce hectic fever, till it has been expofed to the air, and then probably united with oxygen.

The manner of applying the fixed air, is by including the cancer in one half or hemifphere of a large bladder; the edges are made to adhere to the fkin by adhefive plafter, or perhaps a mixture of one part of honey with about twenty parts of carpenter's glue might better fuit fome tender fkins. The bladder is then kept conftantly filled with carbonic acid gas, by means of a pipe in the neck of it; and the matter let out at a fmall aperture beneath.

17. *Arthrocele.*

17. *Arthrocele.* Swelling of the joints ſeems to have its remote cauſe in the ſoftneſs of the bones, for they could not ſwell unleſs they were previouſly ſoftened, ſee Claſs I. 2. 2. 12. The epiphyſes, or ends of the bones, being naturally of a looſer texture, are moſt liable to this diſeaſe, and perhaps the cartilages and capſular ligaments may alſo become inflamed and ſwelled along with the heads of the bones. This malady is liable to diſtort the fingers and knees, and is uſually called gout or rheumatiſm; the former of which is liable to diſable the fingers by chalk-ſtones, and thence to have ſomewhat a ſimilar appearance. But the arthrocele, or ſwelling of the joints, affects people who have not been intemperate in the uſe of fermented or ſpirituous liquors; or who have not previouſly had a regular gout in their feet; and in both theſe circumſtances differs from the gout. Nor does it accord with the inflammatory rheumatiſm, as it is not attended with fever, and becauſe the tumors of the joints never entirely ſubſide. The pain or ſenſibility, which the bones acquire, when they are inflamed, may be owing to the new veſſels, which ſhoot in them in their ſoft ſtate, as well as to the diſtention of the old ones.

M. M. Half a grain of opium twice a day, gradually increaſed to a grain, but not further, for many months. Thirty grains of powder of bark twice a day for many months. Ten grains of bone-aſhes, or calcined hartſhorn, twice a day, with decoction of madder? Soda phoſphorata?

18. *Arthropuoſis.* Joint-evil. This differs from the former, as that never ſuppurates; theſe ulcers of the joints are generally eſteemed to ariſe from ſcrophula; but as ſcrophula is a diſeaſe of the lymphatic or abſorbent ſyſtem, and this conſiſts in the ſuppuration of the membranes, or glands, or cartilages about the joints, there does not ſeem a ſufficient analogy to authorize their arrangement under the ſame name.

The white ſwelling of the knee, when it ſuppurates, comes under this

this ſpecies, with variety of other ulcers attended with carious bones.

19. *Caries oſſium.* A caries of the bones may be termed a ſuppuration of them; it differs from the above, as it generally is occaſioned by ſome external injury, as in decaying teeth; or by venereal virus, as in nodes on the tibia; or by other matter derived to the bone in malignant fevers; and is not confined to the ends of them.

The ſeparation of the dead bone from the living is a work of ſome time. See Sect. XXXIII. 3. 1.

ORDO

## ORDO I.

*Increased Sensation.*

## GENUS V.

*With the Production of new Vessels by external Membranes or Glands, without Fever.*

THE ulcers, or eruptions, which are formed on the external skin, or on the mouth or throat, or on the air-cells of the lungs, or on the intestines, all of which are more or less exposed to the contact of the atmospheric air, which we breathe, and which in some proportion we swallow with our food and saliva; or to the contact of the inflammable air, or hydrogen, which is set at liberty by the putrefying aliment in the intestines, or by putrefying matter in large abscesses; all of them produce contagious matter; which, on being inoculated into the skin of another person, will produce fever, or a similar disease.

In some cases even the matter formed beneath the skin becomes in some degree contagious, at least so much so as to produce fever of the hectic or malignant kind, as soon as it has pierced through the skin, and has thus gained access to some kind of air; as the fresh puss of a common abscess; or the putrid pus of an abscess, which has been long confined; or of cancerous ulcers.

From this analogy there is reason to suspect, that the matter of all contagious diseases, whether with or without fever, is not infectious till it has acquired something from the air; which, by oxygenating the secreted matter, may probably produce a new acid. And secondly, that in hectic fever a part of the purulent matter is absorbed; or



acts

acts on the ſurface of the ulcer; as variolous matter affects the inoculated part of the arm. And that hectic fever is therefore cauſed by the matter of an open ulcer; and not by the ſenſation in the ulcer independent of the aerated pus, which lies on it. Which may account for the venereal matter from buboes not giving the infection, according to the experiments of the late Mr. Hunter, and for ſome other phenomena of contagion. See Variola diſcreta, Claſs II. 1. 3. 9.

## SPECIES.

1. *Gonorrhœa venerea.* A pus-like contagious material diſcharged from the urethra after impure cohabitation, with ſmarting or heat on making water; which begins at the external extremity of the urethra, to which the contagious matter is applied, and where it has acceſs to the air.

M. M. In this ſtate of the venereal diſeaſe once veneſection, with mild cathartics of ſenna and manna, with mucilage, as almond emulſion, and gum arabic, taken for two or three weeks, abſolve the cure. Is camphor of uſe to relieve the ardor urinæ? Do balſams increaſe or leſſen the heat of urine? Neutral ſalts certainly increaſe the ſmarting in making water, by increaſing the acrimony of the urine.

Can the diſcharge from the urethra be ſoon ſtopped by ſaturnine injections, or mercurial ones, or with ſolution of blue vitriol, at firſt very dilute, and gradually made ſtronger? And at the ſame time leſt the ſyphilis, or general diſeaſe, ſhould ſupervene, the patient might take a quarter of a grain of corroſive ſublimate of mercury twice a day, as directed below?

2. *Syphilis.* Venereal diſeaſe. The contagion ſhews itſelf in ulcers on the part firſt inoculated, as chancres; ulcers on the tonſils ſucceed,

ſucceed, with eruption on the ſkin, eſpecially about the roots of the hair; afterwards on other parts of the ſkin, terminating in dry ſcabs; and laſtly, with pain and ſwelling of the bones.

The corona veneris, or crown of Venus, conſiſts of the eruptions at the roots of the hair appearing moſt round the forehead; which is occaſioned by this part being more expoſed to the air; which we obſerved, at the beginning of this genus, either produces or increaſes the virulence of contagious matter. But it is difficult to conceive from this hiſtory, why the throat ſhould be firſt affected; as it cannot be ſuppoſed, that the diſeaſe is ſo often taken by the ſaliva, like the ſmall-pox, though this may ſometimes occur, perhaps very often. The connection between the genitals in men and the throat, is treated of in Claſs IV. 1. 2. 7. Hydrophobia.

M. M. A quarter of a grain of corroſive ſublimate of mercury, taken thrice a day for five or ſix weeks, made into a pill with bread-crumbs, or diſſolved in a ſpoonful of brandy and water, is a very efficacious and almoſt certain cure. When it does not ſucceed, it is owing either to the drug being bad, or to its having precipitated from the brandy, or from its being ſpoiled in the pill by long keeping. Opium contributes much to expedite the cure both of the ſimple gonorrhœa, and of venereal ulcers, by increaſing abſorption both from the mucous membrane, and from the ſurface of ulcers.

3. *Lepra.* Leproſy. Leproſy of the Greeks. The ſkin is rough with white branny ſcales, which are full of chinks; often moiſt beneath, and itching. The ſcales on the head or arms of ſome drinking people are a diſeaſe of this kind. The perſpirable matter deſigned for the purpoſe of lubricating the external ſkin is ſecreted in this diſeaſe in a too viſcid ſtate, owing to the inflammation of the ſubcutaneous veſſels; and, as the abſorbents act too ſtrongly at the ſame time, a viſcid mucus is left adhering to the ſurface of the ſkin.

In the leproſy of the Jews, deſcribed in the thirteenth and four-



teenth

teenth chapters of Leviticus, the depreffion of the fore beneath the furface of the fkin, and the hairs in it becoming white, feem to have been the principal circumftances, which the prieft was directed to attend to for the purpofe of afcertaining the difeafe.

M. M. Effence of antimony from 20 drops to 100 twice or thrice a day, with half a pint of decoction of elm-bark; or tincture of cantharides from 20 to 60 drops four times a day; or fublimate of mercury, with much diluting fluid. Acid of vitriol? Perhaps the cure chiefly depends on much dilution with water, from two to four pints a day, in which elm-bark, or pine-buds, or juniper-tops, may be boiled. Bath or Buxton water drank in large quantities. Warm bath. Oil-fkin bound on the part to confine the perfpirable matter. Ointment of tar and fuet; or poultice for two or three days, and then cerate with lapis calaminaris. Diet of raifins and bread. Abftinence from wine, beer, and all fpirits.

4. *Elephantiafis.* Leprofy of the Arabs. A contagious difeafe; the fkin is thickened, wrinkled, rough, unctuous, deftitute of hair, without any fenfation of touch in the extremities of the limbs; the face deformed with tubercles; the voice hoarfe, and with a nafal tone. Cullen.

5. *Framboefia.* Yaws is faid to be contagious and hereditary. It principally affects the negroes in the Weft Indies. Edinb. Effays, Vol. VI.

6. *Pfory.* Itch. A contagious prurient eruption. There are two kinds of itch, that which appears between the fingers, and under the joints of the knees and elbows; and that which feldom is feen in thefe places, but all over the other parts of the body. The latter is feldom thought to be the itch, as it does not eafily infect even a bedfellow, and refifts the ufual means of cure by brimftone.

If

If the itch be cured too haſtily by rubbing mercurial or arſenical preparations over the whole body, or on too great a part of it, many bad ſymptoms are produced; as weakneſs of digeſtion, with pale bloated countenance, and tendency to dropſy. I have twice ſeen St. Vitus's dance occur from the uſe of a mercurial girdle; and once a ſwelled liver. I have alſo ſeen a ſwelled ſpleen and ſwelled legs from the external uſe of arſenic in the cure of the itch. And very numerous and large phlegmons commonly ſucceed the too haſty cure of it by other means.

There does not appear a ſtrict analogy between the haſty cure of the itch, and the retroceſſion of the puſtles in the ſecondary fever of the ſmall-pox; becauſe in that the abſorption of the matter is evinced by the ſwelling of the face and hands, as the puſtles recede, as explained in Claſs II. 1. 3. 9. Variola diſcreta. And a fever is produced by this abſorption; neither of which happen, when the puſtles of the itch are deſtroyed by mercury or arſenic.

Nor can theſe inconveniences, which occur on the too haſty cure of the itch, be explained by thoſe which follow the cure of ſome kinds of gutta roſea, Claſs II. 1. 4. 6. as in thoſe the eruptions on the face were an aſſociated diſeaſe with inflammation of the liver or ſtomach, which they were accuſtomed to relieve; whereas the itch is not known to have had any previous catenation with other diſeaſes.

In the itch there exiſts not only great irritation in the production of the puſtles, but great ſenſation is cauſed by their acrimony afterwards; inſomuch that the pain of itching, without the interrupted ſmarting occaſioned by ſcratching, would be intolerable. This great excitement of the two ſenſorial powers of irritation and ſenſation is ſo great, when the puſtles are diffuſed over the whole ſurface of the body, that a torpor ſucceeds the ſudden ceaſing of it; which affects thoſe parts of the ſyſtem, which were moſt catenated with the new motions of the

 ſkin,

ſkin, as the ſtomach, whence indigeſtion and flatulency; or which are generally moſt liable to fall into torpor, as the numerous glands, which form the liver. Whence the diſeaſes conſequent to the haſty cure of the itch are diſeaſes of debility, as tumid viſcera, œdematous ſwellings, and St. Vitus's dance, which is a debility of aſſociation. In the ſame manner indigeſtion, with green evacuations, are ſaid to follow an injudicious application of ceruſſa to ſtop too haſtily the exſudation behind the ears of children, Claſs I. 1. 2. 9. And dropſies are liable to ſucceed the cure of old ulcers of the legs, which have long ſtimulated the ſyſtem.

M. M. The ſize of a large pea, of an ointment conſiſting of one part of white precipitate of mercury to ſix parts of hogs' lard well triturated together, to be rubbed on a part of the body every night, and waſhed off with ſoap and water next morning, till every part is cleared; with lac ſulphuris twenty grains to be taken every morning inwardly. Warm ſaline bath, with white vitriol in it. Flowers of ſulphur mixed with thick gruel, with hogs fat. With either of which the body may be ſmeared all over.

7. *Pſora ebriorum.* Elderly people, who have been much addicted to ſpirituous drinks, as beer, wine, or alcohol, are liable to an eruption all over their bodies; which is attended with very afflicting itching, and which they probably propagate from one part of their bodies to another with their own nails by ſcratching themſelves. I ſaw fatal effects in one ſuch patient, by a too extenſive uſe of a ſolution of lead; the eruption diſappeared, he became dropſical, and died; I ſuppoſe from the too ſuddenly ceaſing of the great ſtimulus cauſed by the eruptions over the whole ſkin, as in the preceding article.

M. M. The patient ſhould gradually accuſtom himſelf to half his uſual quantity of vinous potation. The warm bath, with one pound of ſalt to every three gallons. Mercurial ointments on ſmall parts of the

the ſkin at a time. A grain of opium at night inſtead of the uſual potation of wine or beer.

8. *Herpes.* Herpes conſiſts of gregarious ſpreading excoriations, which are ſucceeded by branny ſcales or ſcabs. In this diſeaſe there appears to be a deficient abſorption of the ſubcutaneous mucus, as well as inflammation and increaſed ſecretion of it. For the fluid not only excoriates the parts in its vicinity by its acrimony, but is very ſaline to the taſte, as ſome of theſe patients have aſſured me; I believe this kind of eruption, as well as the tinea, and perhaps all other cutaneous eruption, is liable to be inoculated in other parts of the body by the finger-nails of the patients in ſcratching themſelves.

It is liable to affect the hands, and to return at diſtant periods; and is probably a ſecondary diſeaſe, as well as the zona ignea, or ſhingles, deſcribed below.

M. M. Poultice the eruption with bread and milk, or raw carrots grated, for two or three whole days, to dilute or receive the diſcharged fluid, and abate the inflammation; then cover the parts with freſh cerate mixed with lapis calaminaris. On the parts not excoriated mercurial ointment, made of one part of white calx of mercury and ſix of hogs' fat. Internally, after veneſection, gentle repeated cathartics. Laſtly, the bark. Acid of vitriol. Bolus Armeniæ, or teſtacia. Antimonials. Decoction of interior bark of elm.

9. *Zona ignea.* Shingles. This eruption has been thought a ſpecies of herpes by ſome writers, and by others a ſpecies of eryſipelas. Yellow or livid veſicles appear, producing a corroſive ichor, which is ſometimes attended with a degree of fever. It is ſaid to infeſt ſometimes the thorax and ribs, but its moſt general ſituation is on the ſmall of the back, over one kidney, extending forward over the courſe of one of the ureters.

There is reaſon to ſuſpect, that this alſo is a ſecondary or ſympathetic

thetic diseafe, as well as the preceding one; but future obfervations are required, before it can be removed to the fourth clafs, or difeafes of affociation. In three patients I have been induced to believe, that the eruption on the loins was a tranflation of inflammation from the external membrane of the kidney to the fkin. They had, for a day or two before the appearance of the eruption, complained of a dull pain on the region of one kidney, but without vomiting; by which it was diftinguifhed from nephritis interna, or gravel; and without pain down the outfide of the thigh, by which it was diftinguifhed from fciatica. In other fituations the fhingles may fympathize with other internal membranes, as in a cafe publifhed by Dr. Ruffel (De Tabe Glandulari), where the retroceffion of the fhingles was fucceeded by a ferious dyfpnæa.

M. M. Venefection, if the pulfe is ftrong. Calomel three or four grains, very mild repeated cathartics. Poultice for a few days, then cerate of lapis calaminaris, as in herpes. A grain of emetic tartar diffolved in a pint of water, and taken fo as to empty the ftomach and inteftines, is faid much to haften the cure; compreffes foaked in a faturnine folution are recommended externally on the eruption; and cerate where there are ulcerations. Defanet's Surgical Journal, Vol. II. p. 378. If this be a vicarious difeafe, it fhould continue half a lunation; left, on its ceafing, the bad habits of motion of the primary difeafe fhould not have been fo perfectly diffevered, but that they may recur.

10. *Annulus repens.* Ring-worm. A prurient eruption formed in a circle, affecting children, and would feem to be the work of infects, according to the theory of Linnæus, who afcribes the itch and dyfentery to microfcopic animalcula. Thefe animalcula are probably the effect, and not the caufe, of thefe eruptions; as they are to be feen in all putrefcent animal fluids. The annular propagation of the ringworm, and its continuing to enlarge its periphery, is well accounted

 for

for by the acrimony of the ichor or ſaline fluid eroding the ſkin in its vicinity.

M. M. Cover the eruption daily with ink. With white mercurial ointment, as deſcribed above in herpes. With ſolution of white vitriol ten grains to an ounce. Theſe metallic calces ſtimulate the abſorbents into ſtronger action, whence the fluid has its ſaline part reabſorbed, and that before it has acceſs to the air, which probably adds to its acrimony by oxygenating it, and thus producing a new acid.

11. *Tinea.* Scald head. This contagious eruption affects the roots of the hair, and is generally moſt virulent around the edges of the hair on the back part of the head; as the corona veneris appears moſt on the edges of the hair on the forepart of the head; for in theſe parts the eruption about the roots of the hair is moſt expoſed to the external air, by which its acrimony or noxious quality is increaſed.

The abſorption of the matter thus oxygenated ſwells the lymphatics of the neck by its ſtimulus, occaſioning many little hard lumps beneath the ſeat of the eruption; when this happens, the ſooner it is cured the better, leſt the larger lymphatics of the neck ſhould become affected.

M. M. The art of curing theſe eruptions conſiſts, firſt, in abating the inflammation, and conſequent ſecretion of a noxious material. Secondly, to prevent its acceſs to the air, which ſo much increaſes its acrimony. And thirdly, to promote the abſorption of it, before it has been expoſed to the air; for theſe purpoſes veneſection once, and gentle cathartics, which promote abſorption by emptying the blood-veſſels. Next poultices and fomentations, with warm water, abate inflammation by diluting the ſaline acrimony of the ſecreted fluid, and abating the painful ſenſation. Afterwards cerate joined with ſome metallic calx, as of zinc or lead, or ſolution of lead, mercury,

cury, or copper, or iron, which may ſtimulate the abſorbent ſyſtem into ſtronger action.

Cover the ſhaved head with tar and ſuet, and a bladder; this, by keeping the air from the ſecreted fluid, much contributes to its mildneſs, and the ſtimulus of the tar increaſes its abſorption. See the three preceding ſpecies of this genus.

12. *Cruſta lactea.* Milk-cruſt is a milder diſeaſe than tinea, affecting the face as well as the hairy ſcalp of very young children. It is not infectious, nor liable to ſwell the lymphatics in its vicinity like the tinea.

M. M. Cover the eruption with cerate made with lapis calaminaris, to be renewed every day. Mix one grain of emetic tartar with forty grains of chalk, and divide into eight papers, one to be taken twice a day, or with magneſia alba, if ſtools are wanted. The child ſhould be kept cool and much in the air.

13. *Trichoma.* Plica polonica. A contagious diſeaſe, in which the hair is ſaid to become alive and bleed, forming inextricable knots or plaits of great length, like the fabled head of Meduſa, with intolerable pain, ſo as to confine the ſufferer on his bed for years.

ORDO

## ORDO I.

*Increased Sensation.*

## GENUS VI.

*With Fever consequent to the Production of new Vessels or Fluids.*

### SPECIES.

1. *Febris sensitiva.* Sensitive fever, when unmixed with either irritative or inirritative fever, may be distinguished from either of them by the less comparative diminution of muscular strength; or in other words, from its being attended with less diminution of the sensorial power of irritation. An example of unmixed sensitive fever may generally be taken from the pulmonary consumption; in this disease patients are seen to walk about with ease, and to do all the common offices of life for weeks, and even months, with a pulse of 120 strokes in a minute; while in other fevers, whether irritated or inirritated, with a pulse of this frequency, the patient generally lies upon the bed, and exerts no muscular efforts without difficulty.

The cause of this curious phenomenon is thus to be understood; in the sensitive fever a new sensorial power, viz. that of sensation, is superadded to that of irritation; which in other fevers alone carries on the increased circulation. Whence the power of irritation is not much more exhausted than in health; and those muscular motions, which are produced in consequence of it, as those which are exerted in keeping the body upright in walking, riding, and in the performance of many customary actions, are little impaired. For an account of the irritated ſenſitive fever, ſee Claſs II. 1. 2. 1.; for the inirritated ſenſitive fever, Claſs II. 1. 3. 1. IV. 2. 4. 11.

2. *Febris a pure clauſo.* Fever from incloſed matter is generally of the irritated ſenſitive kind, and continues for many weeks, and even months, after the abſceſs is formed; but is diſtinguiſhed from the fever from aerated matter in open ulcers, becauſe there are ſeldom any night-ſweats, or colliquative diarrhœa in this, as in the latter. The pulſe is alſo harder, and requires occaſional veneſection, and cathartics, to abate the inflammatory fever; which is liable to increaſe again every three or four days, till at length, unleſs the matter has an exit, it deſtroys the patient. In this fever the matter, not having been expoſed to the air, has not acquired oxygenation; in which a new acid, or ſome other noxious property, is produced; which acts like contagion on the conſtitution inducing fever-fits, called hectic fever, which terminate with ſweats or diarrhœa; whereas the matter in the cloſed abſceſs is either not abſorbed, or does not ſo affect the circulation as to produce diurnal or hectic fever-fits; but the ſtimulus of the abſceſs excites ſo much ſenſation as to induce perpetual pyrexia, or inflammatory fever, without ſuch marked remiſſions. Neverthleſs there ſometimes is no fever produced, when the matter is lodged in a part of little ſenſibility, as in the liver; yet a white puslike ſediment in thoſe caſes exiſts I believe generally in the urine, with occaſional wandering pains about the region of the liver or cheſt.

3. *Vomica.* An abſceſs in the lungs is ſometimes produced after peripneumony, the cough and ſhortneſs of breath continue in leſs degree, with difficulty in lying on the well ſide, and with ſenſitive irritated fever, as explained in the preceding article.

The occaſional increaſe of fever, with hard pulſe and ſizy blood, in theſe patients, is probably owing to the inflammation of the walls of

6 the

the vomica; as it is attended with difficulty of breathing, and requires venesection. Mr. B——, a child about seven years old, lived about five weeks in this situation, with a pulse from 150 to 170 in a minute, without sweats, or diarrhœa, or sediment in his water, except mucus occasionally; and took sufficient nourishment during the whole time. The blood taken was always covered with a strong cupped size, and on his death three or four pints of matter were found in one side of the chest; which had probably, but lately, been effused from a vomica. This child was frequently induced to swing, both in a reciprocating and in a rotatory swing, without any apparent absorption of matter; in both these swings he expressed pleasure, and did not appear to be vertiginous.

M. M. Repeated emetics. Digitalis? Perseverance in rotatory swinging. See Class II. 1. 6. 7.

Mr. I. had laboured some months under a vomica after a peripneumony, he was at length taken with a catarrh, which was in some degree endemic in March 1795, which occasioned him to sneeze much, during which a copious hæmorrhage from the lungs occurred, and he spit up at the same time half a pint of very fetid matter, and recovered. Hence errhines may be occasionally used with advantage.

4. *Empyema.* When the matter from an abscess in the lungs finds its way into the cavity of the chest, it is called an empyema. A servant man, after a violent peripneumony, was seized with symptoms of empyema, and it was determined, after some time, to perform the operation; this was explained to him, and the usual means were employed by his friends to encourage him, "by advising him not to be afraid." By which good advice he conceived so much fear, that he ran away early next morning, and returned in about a week quite well. Did the great fear promote the absorption of the matter, like the sickness occasioned by digitalis? Fear renders the external skin pale; by this continued decrease of the action of the absorbents

of the ſkin might not thoſe of the lungs be excited into greater activity? and thus produce increaſed pulmonary abſorption by reverſe ſympathy, as it produces pale urine, and even ſtools, by direct ſympathy?

M. M. Digitalis?

5. *Febris Meſenterica.* Fever from matter formed in the meſentery is probably more frequent than is ſuſpected. It commences with pain in the bowels, with irritated ſenſitive fever; and continues many weeks, and even months, requiring occaſional veneſection, and mild cathartics; till at length the continuance of the pyrexia, or inflammatory fever, deſtroys the patient. This is an affection of the lymphatic glands, and properly belongs to ſcrophula; but as the matter is not expoſed to the air, no hectic fever, properly ſo called, is induced.

6. *Febris a pure aerato.* Fever from aerated matter. A great collection of matter often continues a long time, and is ſometimes totally abſorbed, even from venereal buboes, without producing any diſorder in the arterial ſyſtem. At length, if it becomes putrid by its delay, and one part of the matter thus becomes aerated by the air given out by the other part; or if the ulcer has been opened, ſo that any part of it has been expoſed to the air for but one day, a hectic fever is produced. Whence the utility ariſes of opening large abſceſſes by ſetons, as in that caſe little or no hectic fever is induced; becauſe the matter is ſqueezed out by the ſide of the ſpongy threads of cotton, and little or no air is admitted; or by tapping the abſceſs with a trocar, as mentioned in iſchias, Claſs II. 1. 2. 18.

In this fever the pulſe is about 120 in a minute, and its acceſs is generally in an evening, and ſometimes about noon alſo, with ſweats or purging towards morning, or urine with pus-like ſediment; and the patients bear this fever better than any other with ſo quick a pulſe;

pulse; and lastly, when all the matter from a concealed ulcer is absorbed, or when an open ulcer is healed, the hectic fever ceases. Here the absorbed matter is supposed to produce the fever, and the diarrhoea, sweats, or copious muddy urine, to be simply the consequence of increased secretion, and not to consist of the purulent matter, which was supposed to be absorbed from the ulcer. See Sudor calidus, Class I. 1. 2. 3.

The action of the air on ulcers, as we have already shewn, increases the acrimony of the purulent matter, and even converts it into a weaker kind of contagious matter; that is, to a material inducing fever. This was ascribed to the union of the azotic part of the atmosphere with the effused pus in Sect. XXVIII. 2. but by contemplating more numerous facts and analogies, I am now induced to believe, that it is by the union of oxygen with it; first, because oxygen so greedily unites with other animal substances, as the blood, that it will pass through a moist bladder to combine with it, according to the experiment of Dr. Priestley. Secondly, because the poisons of venomous creatures are supposed to be acids of different kinds, and are probably formed by the contact of air after their secretion. And lastly, because the contagious matter from other ulcers, as in itch, or small-pox, are formed on external membranes, and are probably combinations of animal matter and oxygen, producing other new acids; but further experiments must determine this question.

It was thought a subject of consequence by the Æsculapian Society at Edinburgh, to find a criterion which should distinguish pus from mucus, for the purpose of more certainly discovering the presence of ulcers in pulmonary diseases, or in the urinary passages. For this purpose that society offered their first gold medal, which was conferred on the late Mr. Charles Darwin, in the year 1778, for his experiments on this subject. From which he deduces the following conclusions:

"1. Pus

"1. Pus and mucus are both ſoluble in the vitriolic acid, though in very different proportions, pus being much the leſs ſoluble.

2. The addition of water to either of theſe compounds decompoſes it; the mucus thus ſeparated, either ſwims on the mixture, or forms large flocci in it; whereas the pus falls to the bottom, and forms on agitation a uniform turbid mixture.

3. Pus is diffuſible through a diluted vitriolic acid, though mucus is not; the ſame occurs with water, or a ſolution of ſea ſalt.

4. Nitrous acid diſſolves both pus and mucus; water added to the ſolution of pus produces a precipitate; and the fluid above becomes clear and green; while water and the ſolution of mucus form a dirty coloured fluid.

5. Alkaline lixivium diſſolves (though ſometimes with difficulty) mucus, and generally pus.

6. Water precipitates pus from ſuch a ſolution, but does not mucus.

7. Where alkaline lixivium does not diſſolve pus, it ſtill diſtinguiſhes it from mucus; as it then prevents its diffuſion through water.

8. Coagulable lymph is neither ſoluble in diluted nor concentrated vitriolic acid.

9. Water produces no change on a ſolution of ſerum in alkaline lixivium, until after long ſtanding, and then only a very ſlight ſediment appears.

10. Corroſive ſublimate coagulates mucus, but does not pus.

From the above experiments it appears, that ſtrong vitriolic acid and water, diluted vitriolic acid, and cauſtic alkaline lixivium and water will ſerve to diſtinguiſh pus from mucus; that the vitriolic acid can ſeparate it from coagulable lymph, and alkaline lixivium from ſerum.

And hence, when a perſon has any expectorated material, the com-

composition of which he wishes to ascertain, let him dissolve it in vitriolic acid, and in caustic alkaline lixivium; and then add pure water to both solutions: and if there is a fair precipitation in each, he may be assured that some pus is present. If in neither a precipitation occurs, it is a certain test, that the material is entirely mucus. If the material cannot be made to dissolve in alkaline lixivium by time and trituration, we have also reason to believe that it is pus." Experiments on Pus and Mucus. Cadell. London.

7. *Phthisis pulmonalis.* In pulmonary consumption the fever is generally supposed to be the consequence of the stimulus of absorbed matter circulating in the blood-vessels, and not simply of its stimulus on their extremities in the surface of the ulcers; as mentioned in Class II. 1. 5. and Class II. 1. 3. 9. The ulcers are probably sometimes occasioned by the putrid acrimony of effused blood remaining in the air-cells of the lungs after an hæmoptoe. See Class I. 2. 1. 9. The remote cause of consumption is ingeniously ascribed by Dr. Beddoes to the hyper-oxygenation of the blood, as mentioned Section XXVIII. 2.

As the patients liable to consumption are of the inirritable temperament, as appears by the large pupils of their eyes; there is reason to believe, that the hæmoptoe is immediately occasioned by the deficient absorption of the blood at the extremities of the bronchial vein; and that one difficulty of healing the ulcers is occasioned by the deficient absorption of the fluids effused into them. See Sect. XXX. 1. and 2.

The difficulty of healing pulmonary ulcers may be owing, as its remote cause, to the incessant motion of all the parts of the lungs; whence no scab, or indurated mucus, can be formed so as to adhere on them. Whence these naked ulcers are perpetually exposed to the action of the air on their surfaces, converting their mild purulent matter into a contagious ichor; which not only prevents them from healing,

healing, but by its action on their circumferences, like the matter of itch or tinea, contributes to ſpread them wider. See the preceding article, and Sect. XXXIII. 2. 7. where the pulmonary phthiſis is ſuppoſed to be infectious.

This acidifying principle is found in all the metallic calces, as in lapis calaminaris, which is a calciform ore of zinc; and in ceruſſa, which is a calx of lead; two materials which are powerful in healing excoriations, and ulcers, in a ſhort time by their external application. How then does it happen, that the oxygen in the atmoſphere ſhould prevent pulmonary ulcers from healing, and even induce them to ſpread wider; and yet in its combination with metals, it ſhould facilitate their healing? The healing of ulcers conſiſts in promoting the abſorption of the fluids effuſed into them, as treated of in Section XXXIII. 3. 2. Oxygen in combination with metals, when applied in certain quantity, produces this effect by its ſtimulus; and the metallic oxydes not being decompoſed by their contact with animal matter, no new acid, or contagious material, is produced. So that the combined oxygen, when applied to an ulcer, ſimply I ſuppoſe promotes abſorption in it, like the application of other materials of the articles ſorbentia or incitantia, if applied externally; as opium, bark, alum. But in the pulmonary ulcers, which cannot protect themſelves from the air by forming a ſcab, the uncombined oxygen of the atmoſphere unites with the purulent matter, converting it into a contagious ichor; which by infection, not by eroſion, enlarges the ulcers, as in the itch or tinea; which might hence, according to Dr. Beddoes's ingenious theory of conſumption, be induced to heal, if expoſed to an atmoſphere deprived of a part of its oxygen. This I hope future experiments will confirm, and that the pneumatic medicine will alleviate the evils of mankind in many other, as well as in this moſt fatal malady.

M. M. Firſt, the reſpiration of air lowered by an additional quantity of azote, or mixed with ſome proportion of hydrogen, or of carbonic

bonic acid air, may be tried; as described in a late publication of Dr. Beddoes on the medicinal use of factitious airs. Johnson, London. Or lastly, by breathing a mixture of one tenth part of hydro-carbonate mixed with common air, according to the discovery of Mr. Watt, which has a double advantage in these cases, of diluting the oxygen of the atmospheric air, and inducing sickness, which increases pulmonary absorption, as mentioned below. An atmosphere diluted with fixed air (carbonic acid) might be readily procured by setting tubs of new wort, or fermenting beer, in the parlour and lodging-room of the patient. For it is not acids floating in the air, but the oxygen or acidifying principle, which injures or enlarges pulmonary ulcers by combining with the purulent matter.

Another easy method of adding carbonic acid gas to the air of a room, would be by means of an apparatus invented by Mr. Watt, and sold by Bolton and Watt at Birmingham, as described in Dr. Beddoes' Treatise on Pneumatic Medicine. Johnson, London. It consists of an iron pot, with an arm projecting, and a method of letting water drop by slow degrees on chalk, which is to be put into the iron pot, and exposed to a moderate degree of heat over a common fire. By occasionally adding more and more chalk, carbonic acid gas might be carried through a tin pipe from the arm of the iron pot to any part of the room near the patient, or from an adjoining room. In the same manner a diffusion of solution of flowers of zinc might be produced and breathed by the patient, and would be likely much to contribute to the healing of pulmonary ulcers; as observed by Mr. Watt. See the treatise above mentioned.

Breathing over the vapour of caustic volatile alkali might easily be managed for many hours in a day; which might neutralize the acid poison formed on pulmonary ulcers by the contact of oxygen, and thus prevent its deleterious quality, as other acids become less caustic, when they are formed into neutral salts with alkalis. The volatile ſalt ſhould be put into a tin caniſter, with two pipes like horns from the top of it, one to ſuck the air from, and the other to admit it.

Secondly, the external ulcers in ſcrophulous habits are pale and flabby, and naturally diſinclined to heal, the depoſition of fluids in them being greater than the abſorption; theſe ulcers have their appearance immediately changed by the external application of metallic calxes, and the medicines of the article Sorbentia, ſuch as ceruſſa and the bark in fine powder, ſee Claſs I. 2. 3. 21. and are generally healed in a ſhort time by theſe means. Induced by theſe obſervations, I wiſhed to try the external application of ſuch powders to ulcers in the lungs, and conſtructed a box with a circulating bruſh in it, as deſcribed in the annexed plate; into this box two ounces of fine powder of Peruvian bark were put, and two drams of ceruſſa in fine powder; on whirling the central bruſh, part of this was raiſed into a cloud of powder, and the patient, applying his mouth to one of the tin pipes riſing out of the box, inhaled this powder twice a day into his lungs. I obſerved it did not produce any cough or uneaſineſs. This patient was in the laſt ſtage of conſumption, and was ſoon tired of the experiment, nor have I had ſuch patients as I wiſhed for the repetition of it. Perhaps a fine powder of manganeſe, or of the flowers of zinc, or of lapis calaminaris, might be thus applied to ulcers of the lungs with greater advantage? Perhaps air impregnated with flowers of zinc in their moſt comminuted ſtate, might be a better way of applying this powder to the lungs, as diſcovered by Mr. Watt. See Dr. Beddoes on Pneumatic Medicine. Johnſon.

Thirdly, as the healing of an ulcer conſiſts in producing a tendency to abſorption on its ſurface greater than the depoſition on it; ſee Sect. XXXIII. 3. 2. other modes of increaſing pulmonary abſorption, which are perhaps more manageable than the preceding ones, may be had recourſe to; ſuch as by producing frequent nauſea or ſickneſs. See Sect. XXIX. 5. 1. and Art. IV. 2. The great and ſudden abſorption

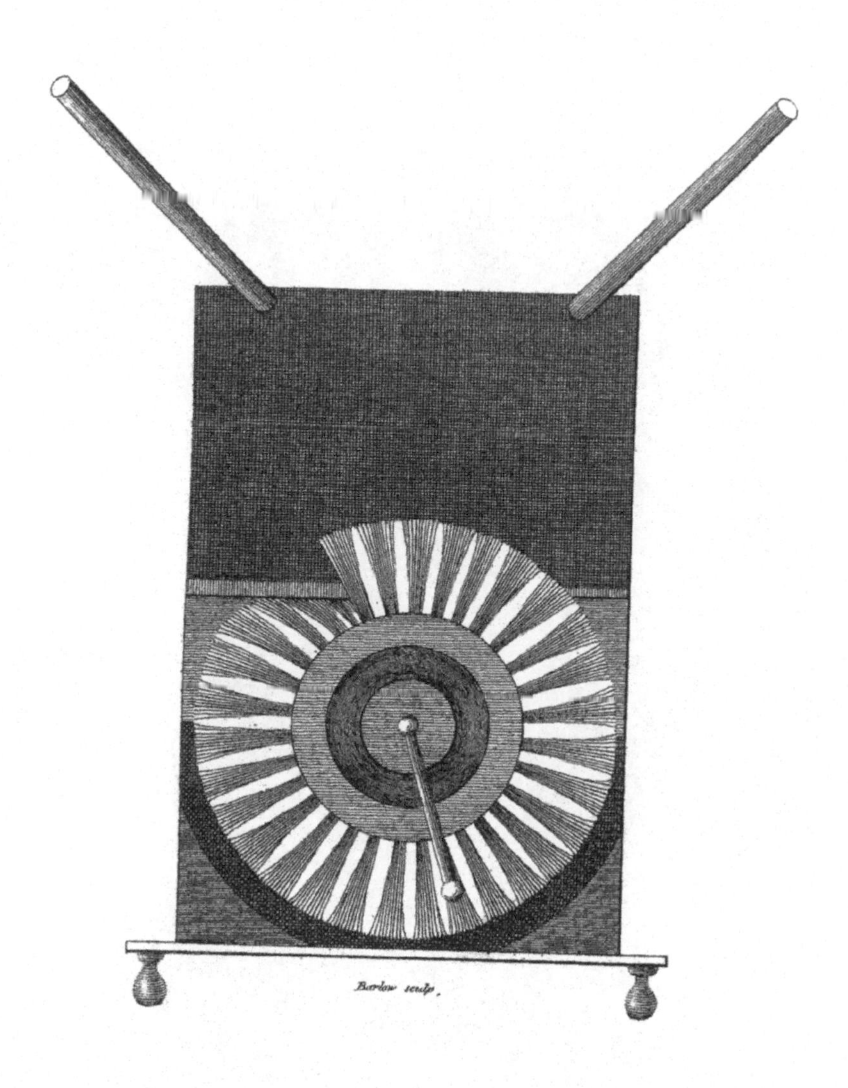


Published by J. Johnson St Pauls Ch...

ſorption of fluid from the lungs in the anaſarca pulmonum by the ſickneſs induced by the exhibition of digitalis, aſtoniſhes thoſe who have not before attended to it, by emptying the ſwelled limbs, and removing the difficulty of breathing in a few hours.

The moſt manageable method of uſing digitalis is by making a ſaturated tincture of it, by infuſing two ounces of the powder of the leaves in a mixture of four ounces of rectified ſpirit of wine, and four ounces of water. Of this from 30 to 60 drops, or upwards, from a two-ounce phial, are to be taken twice in the morning part of the day, and to be ſo managed as not to induce violent ſickneſs. If ſickneſs nevertheleſs comes on, the patient muſt for a day or two omit the medicine; and then begin it again in reduced doſes.

Mr. ——, a young man about twenty, with dark eyes, and large pupils, who had every ſymptom of pulmonary ulcers, I believed to have been cured by digitalis, and publiſhed the caſe in the Tranſactions of the College, Vol. III. But about two years afterwards I heard that he relapſed and died. Mr. L——, a corpulent man, who had for ſome weeks laboured under a cough with great expectoration, with quick pulſe, and difficulty of breathing, ſoon recovered by the uſe of digitalis taken twice a day; and though this caſe might probably be a peripneumonia notha, or catarrh, it is here related as ſhewing the power of pulmonary abſorption excited by the uſe of this drug.

Another method of inducing ſickneſs, and pulmonary abſorption in conſequence, is by ſailing on the ſea; by which many conſumptive patients have been ſaid to have received their cure; which has been erroneouſly aſcribed to ſea-air, inſtead of ſea-ſickneſs; whence many have been ſent to breathe the ſea-air on the coaſts, who might have done better in higher ſituations, where the air probably contains leſs oxygen gas, which is the heavieſt part of it. See a Letter from Dr. T. C. below.

A third method of inducing ſickneſs, and conſequent pulmonary

 abſorption,

abſorption, is by the vertigo occaſioned by ſwinging; which has lately been introduced into practice by Dr. Smith, (Eſſay on Pulmonary Conſumption), who obſerved that by ſwinging the hectic pulſe became ſlower, which is explained in Claſs IV. 2. 1. 10. The uſual way of reciprocating ſwinging, like the oſcillations of a pendulum, produces a degree of vertigo in thoſe, who are unuſed to it; but to give it greater effect, the patient ſhould be placed in a chair ſuſpended from the ceiling by two parallel cords in contact with each other, the chair ſhould then be forcibly revolved 20 or 40 times one way, and ſuffered to return ſpontaneouſly; which induces a degree of ſickneſs in moſt adult people, and is well worthy an exact and pertinacious trial, for an hour or two, three or four times a day for a month.

The common means of promoting abſorption in ulcers, and of thickening the matter in conſequence, by taking the bark and opium internally, or by metallic ſalts, as of mercury, ſteel, zinc, and copper, in ſmall quantities, have been repeatedly uſed in pulmonary conſumption; and may have relieved ſome of the ſymptoms. As mercury cures venereal ulcers, and as pulmonary ulcers reſemble them in their not having a diſpoſition to heal, and in their tendency to enlarge themſelves, there were hopes, from analogy, that it might have ſucceeded. Would a ſolution of gold in aqua regia be worth trying? When vinegar is applied to the lips, it renders them inſtantly pale, by promoting the venous abſorption; if the whole ſkin was moiſtened with warmiſh vinegar, would this promote venous abſorption in the lungs by their ſympathy with the ſkin? The very abſtemious diet on milk and vegetables alone is frequently injurious. Fleſh-meat once a day, with ſmall wine and water, or ſmall beer, is preferable. Half a grain of opium twice a day, or a grain, I believe to be of great uſe at the commencement of the diſeaſe, as appears from the ſubſequent caſe.

Miſs ——, a delicate young lady, of a conſumptive family, when ſhe was about eighteen, had frequent cough, with quick pulſe, a

8 pain

pain of her side, and the general appearances of a beginning consumption. She took about five drops of laudanum twice a day in a saline draught, which was increased gradually to ten. In a few weeks she recovered, was afterwards married, bore three or four children, and then became consumptive and died.

The following case of hereditary consumption is related by a physician of great ability and very extensive practice; and, as it is his own case, abounds with much nice observation and useful knowledge; and, as it has been attended with a favourable event, may give consolation to many, who are in a similar situation; and shews that Sydenham's recommendation of riding as a cure for consumption is not so totally ineffectual, as is now commonly believed.

" J. C. aged 27, with black hair, and a ruddy complexion, was subject to cough from the age of puberty, and occasionally to spitting of blood. His maternal grandfather died of consumption under thirty years of age, and his mother fell a victim to this disease, with which she had long been threatened, in her 43d year, and immediately after she ceased to have children. In the severe winter of 1783-4, he was much afflicted with cough; and being exposed to intense cold, in the month of February he was seized with peripneumony. The disease was violent and dangerous, and after repeated bleedings as well as blisterings, which he supported with difficulty, in about six weeks he was able to leave his bed. At this time the cough was severe, and the expectoration difficult. A fixed pain remained on the left side, where an issue was inserted; regular hectic came on every day about an hour after noon, and every night heat and restlessness took place, succeeded towards morning by general perspiration.

The patient, having formerly been subject to ague, was struck with the resemblance of the febrile paroxysm, with what he had experienced under that disease, and was willing to flatter himself it might be of the same nature. He therefore took bark in the interval of fever, but with an increase of his cough, and this requiring vene-section,

ſection, the blood was found highly inflammatory. The vaſt quantity of blood which he had loſt from time to time, produced a diſpoſition to fainting, when he reſumed the upright poſture, and he was therefore obliged to remain almoſt conſtantly in a recumbent poſition. Attempting to ride out in a carriage, he was ſurpriſed to find that he could ſit upright for a conſiderable time, while in motion, without inconvenience, though, on ſtopping the carriage, the diſpoſition to fainting returned.

At this time, having prolonged his ride beyond the uſual length, he one day got into an uneven road at the uſual period of the recurrence of the hectic paroxyſms, and that day he miſſed it altogether. This circumſtance led him to ride out daily in a carriage at the time the febrile acceſſion might be expected, and ſometimes by this means it was prevented, ſometimes deferred, and almoſt always mitigated.

This experience determined him to undertake a journey of ſome length, and Briſtol being, as is uſual in ſuch caſes, recommended, he ſet out on the 19th of April, and arrived there on the 2d of May. During the greater part of this journey (of 175 miles) his cough was ſevere, and being obliged to be bled three different times on the road, he was no longer able to ſit upright, but at very ſhort intervals, and was obliged to lie at length in the diagonal of a coach. The hectic paroxyſms were not interrupted during the journey, but they were irregular and indiſtinct, and the ſalutary effects of exerciſe, or rather of geſtation, were impreſſed on the patient's mind.

At Briſtol he ſtayed a month, but reaped no benefit. The weather was dry and the roads duſty; the water inſipid and inert. He attempted to ride on horſeback on the downs, but was not able to bear the fatigue for a diſtance of more than a hundred yards. The neceſſity of frequent bleedings kept down his ſtrength, and his hectic paroxyſms continued, though leſs ſevere. At this time, ſuſpecting that his cough was irritated by the weſt-winds bearing the vapour from

from the ſea, he reſolved to try the effects of an inland ſituation, and ſet off for Matlock in Derbyſhire.

During the journey he did not find the improvement he expected, but the nightly perſpirations began to diminiſh; and the extraordinary fatigue he experienced proceeded evidently from his travelling in a poſt-chaiſe, where he could not indulge in a recumbent poſition. The weather at Briſtol had been hot, and the earth arid and duſty. At Matlock, during the month of June 1784, there was almoſt a perpetual drizzle, the ſoil was wet, and the air moiſt and cold. Here, however, the patient's cough began to abate, and at intervals he found an opportunity of riding more or leſs on horſeback. From two or three hundred yards at a time, he got to ride a mile without ſtopping; and at length he was able to ſit on horſeback during a ride from Maſon's Bath to the village of Matlock along the Derwent, and round on the oppoſite banks, by the works of Mr. Arkwright, back to the houſe whence he ſtarted, a diſtance of five miles. On diſmounting, however, he was ſeized with diliquium, and ſoon after the ſtrength he had recovered was loſt by an attack of the hæmorrhoids of the moſt painful kind, and requiring much loſs of blood from the parts affected.

On reflection, it appeared that the only benefit received by the patient was during motion, and continued motion could better be obtained in the courſe of a journey than during his reſidence at any particular place. This, and other circumſtances of a private but painful nature, determined him to ſet out from Matlock on a journey to Scotland. The weather was now much improved, and during the journey he recruited his ſtrength. Though as yet he could not ſit upright at reſt for half an hour together without a diſpoſition to giddineſs, dimneſs of ſight, and deliquium, he was able to ſit upright under the motion of a poſt-chaiſe during a journey of from 40 to 70 miles daily, and his appetite began to improve. Still his cough continued,

tinued, and his hectic flushings, though the chills were much abated and very irregular.

The salutary effects of motion being now more striking than ever, he purchased a horse admirably adapted to a valetudinarian in Dumfriesshire, and being now able to sit on horseback for an hour together, he rode out several times a day. He fixed his residence for a few weeks at Moffat, a village at the foot of the mountains whence the Tweed, the Clyde, and the Annan, descend in different directions; a situation inland, dry, and healthy, and elevated about three hundred feet above the surface of the sea. Here his strength recovered daily, and he began to eat animal food, which for several months before he had not tasted. Persevering in exercise on horseback, he gradually increased the length of his rides, according to his strength, from four to twenty miles a day; and returning on horseback to Lancashire by the lakes of Cumberland, he arrived at Liverpool on the first of September, having rode the last day of his journey forty miles.

The two inferences of most importance to be drawn from this narrative, are, first, the extraordinary benefit derived from gestation in a carriage, and still more the mixture of gestation and exercise on horseback, in arresting or mitigating the hectic paroxysm; and secondly, that in the florid consumption, as Dr. Beddoes terms it, an elevated and inland air is in certain circumstances peculiarly salutary; while an atmosphere loaded with the spray of the sea is irritating and noxious. The benefit derived in this case from exercise on horseback, may lead us to doubt whether Sydenham's praise of this remedy be as much exaggerated as it has of late been supposed. Since the publication of Dr. C. Smyth on the effects of swinging in lowering the pulse in the hectic paroxysm, the subject of this narrative has repeated his experiments in a great variety of cases, and has confirmed them. He has also repeatedly seen the hectic paroxysm prevented,

vented, or cut ſhort, by external ablution of the naked body with tepid water.

So much was his power of digeſtion impaired or vitiated by the immenſe evacuations, and the long continued debility he underwent, that after the cough was removed, and indeed for ſeveral years after the period mentioned, he never could eat animal food without heat and fluſhing, with frequent pulſe and extreme drowſineſs. If this drowſineſs was encouraged, the fever ran high, and he awoke from diſturbed ſleep, wearied and depreſſed. If it was reſolutely reſiſted by gentle exerciſe, it went off in about an hour, as well as the increaſed frequency of the pulſe. This agitation was however ſuch as to incapacitate him during the afternoon for ſtudy of any kind. The ſame effects did not follow a meal of milk and vegetables, but under this diet his ſtrength did not recruit; whereas after the uſe of animal food it recovered rapidly, notwithſtanding the inconvenience already mentioned. For this inconvenience he at laſt found a remedy in the uſe of coffee immediately after dinner, recommended to him by his friend Dr. Percival. At firſt this remedy operated like a charm, but by frequent uſe, and indeed by abuſe, it no longer poſſeſſes its original efficacy.

Dr. Falconer, in his Diſſertation on the Influence of the Paſſions and Affections of the Mind on Health and Diſeaſe, ſuppoſes that the cheerfulneſs which attends hectic fever, the ever-ſpringing hope, which brightens the gloom of the conſumptive patient, increaſes the diſeaſed actions, and haſtens his doom. And hence he is led to enquire, whether the influence of fear might not be ſubſtituted in ſuch caſes to that of hope with advantage to the patient? This queſtion I ſhall not preſume to anſwer, but it leads me to ſay ſomething of the ſtate of the mind in the caſe juſt related.

The patient, being a phyſician, was not ignorant of his danger, which ſome melancholy circumſtances ſerved to impreſs on his mind. It has already been mentioned, that his mother and grandfather died of this diſeaſe. It may be added, that in the year preceding that on which he himſelf was attacked, a ſiſter of his was carried off by conſumption in her 17th year; that in the ſame winter in which he fell ill, two other ſiſters were ſeized with the ſame fatal diſorder, to which one of them fell a victim during his reſidence at Briſtol, and that the hope of bidding a laſt adieu to the other was the immediate cauſe of his journey to Scotland, a hope which, alas! was indulged in vain. The day on which he reached the end of his journey, her remains were committed to the duſt! It may be conjectured from theſe circumſtances, that whatever benefit may be derived from the apprehenſion of death, muſt in this caſe have been obtained. The expectation of this iſſue was indeed for ſome time ſo fixed that it ceaſed to produce much agitation; in conformity to that general law of our nature, by which almoſt all men ſubmit with compoſure to a fate that is foreſeen, and that appears inevitable. As however the progreſs of diſeaſe and debility ſeemed to be arreſted, the hope and the love of life revived, and produced, from time to time, the obſervations and the exertions already mentioned.

Wine and beer were rigorouſly abſtained from during ſix months of the above hiſtory; and all the blood which was taken was even to the laſt buffy." Feb. 3, 1795.

8. *Febris ſcrophuloſa.* The hectic fever occaſioned by ulcers of the lymphatic glands, when expoſed to the air, does not differ from that attending pulmonary conſumption, being accompanied with night-ſweats and occaſional diarrhœa.

M. M. The bark. Opium internally. Externally ceruſſa and bark in fine powder. Bandage. Sea-bathing. See Claſs I. 2. 3. 21. and II. 1. 4. 12.

9. *Febris iſchiadica.* A hectic fever from an open ulcer between the muſcles of the pelvis, which differs not from the preceding. If

 the

the matter in this ſituation lodges till part of it, I ſuppoſe, becomes putrid, and aerates the other part; or till it becomes abſorbed from ſome other circumſtance; a ſimilar hectic fever is produced, with night-ſweats, or diarrhœa.

Mrs. ——, after a lying in, had pain on one ſide of her loins, which extended to the internal part of the thigh on the ſame ſide. No fluctuation of matter could be felt; ſhe became hectic with copious night-ſweats, and occaſional diarrhœa, for four or five weeks; and recovered by, I ſuppoſe, the total abſorption of the matter, and the reunion of the walls of the abſceſs. See Claſs II. 1. 2. 18.

10. *Febris Arthropuodica.* Fever from the matter of diſeaſed joints. Does the matter from ſuppurating bones, which generally has a very putrid ſmell, produce hectic fever, or typhus? See Claſs II. 1. 4. 16.

11. *Febris a pure contagioſo.* Fever from contagious pus. When the contagious matters have been produced on the external habit, and in proceſs of time become abſorbed, a fever is produced in conſequence of this reabſorption; which differs with the previous irritability or inirritability, as well as with the ſenſibility of the patient.

12. *Febris varioloſa ſecundaria.* Secondary fever of ſmall-pox. In the diſtinct ſmall-pox the fever is of the ſenſitive irritated or inflammatory kind; in the confluent ſmall-pox it is of the ſenſitive inirritated kind, or typhus gravior. In both of them the ſwelling of the face, when the matter there begins to be abſorbed, and of the hands, when the matter there begins to be abſorbed, ſhew, that it ſtimulates the capillary veſſels or glands, occaſioning an increaſed ſecretion greater than the abſorbents can take up, like the action of the cantharides in a bliſter; now as the application of a bliſter on the ſkin frequently occaſions the ſtrangury, which ſhews, that ſome part of

the cantharides is abſorbed; there is reaſon to conclude, that a part of the matter of ſmall-pox is abſorbed, and thus produces the ſecondary fever. See Claſs II. 1. 3. 9. And not ſimply by its ſtimulus on the ſurface of the ulcers beneath the ſcabs. The exſudation of a yellow fluid from beneath the confluent eruptions on the face before the height is ſpoken of in Claſs II. 1. 3. 2.

The material thus abſorbed in the ſecondary fever of ſmall-pox differs from that of open ulcers, as it is only aerated through the elevated cuticle; and ſecondly, becauſe there is not a conſtant ſupply of freſh matter, when that already in the puſtules is exhauſted, either by abſorption, or by evaporation, or by its induration into a ſcab. Might not the covering the face aſſiduouſly and exactly with plaſters, as with cerate of calamy, or with minium plaſter, by precluding the air from the puſtules, prevent their contracting a contagious, or aceſcent, or fever-producing power? and the ſecondary fever be thus prevented entirely. If the matter in thoſe puſtules on the face in the confluent ſmall-pox were thus prevented from oxygenation, it is highly probable, both from this theory, and from the facts before mentioned, that the matter would not erode the ſkin beneath them, and by theſe means no marks or ſcars would ſucceed.

13. *Febris carcinomatoſa.* Fever from the matter of cancer. In a late publication the pain is ſaid to be relieved, and the fever cured, and the cancer eradicated, by the application of carbonic acid gas, or fixed air. See Claſs II. 1. 4. 16.

14. *Febris venerea.* From the abſorption of the matter from venereal ulcers and ſuppurating bones. See Syphilis, II. 1. 5. 2.

M. M. Any mercurial calx. Sarſaparilla? Mezereon?

15. *Febris a ſanie putrida.* Fever from putrid ſanies. When parts of the body are deſtroyed by external violence, as a bruiſe, or by mortification,

tification, a putrefaction ſoon ſucceeds; as they are kept in that degree of warmth and moiſture by their adheſion to the living parts of the body, which moſt forwards that proceſs. Thus the ſloughs of mortified parts of the tonſils give fetor to the breath in ſome fevers; the matter from putrefying teeth, or other ſuppurating bones, is particularly offenſive; and even the ſcurf, which adheres to the tongue, frequently acquires a bitter taſte from its incipient putridity. This material differs from thoſe before mentioned, as its deleterious property depends on a chemical rather than an animal proceſs.

16. *Febris puerpera.* Puerperal fever. It appears from ſome late diſſections, which have been publiſhed, of thoſe women who have died of the puerperal fever, that matter has been formed in the omentum, and found in the cavity of the abdomen, with ſome blood or ſanies. Theſe parts are ſuppoſed to have been injured by the exertions accompanying labour; and as matter in this viſcus may have been produced without much pain, this diſeaſe is not attended with arterial ſtrength and hard full pulſe like the inflammation of the uterus; and as the fever is of the inirritative or typhus kind, there is reaſon to believe, that the previous exhauſtion of the patient during labour may contribute to its production; as well as the abſorption of a material not purulent but putrid; which is formed by the delay of extravaſated or dead matter produced by the bruiſes of the omentum, or other viſcera, in the efforts of parturition, rather than by purulent matter, the conſequence of ſuppuration. The pulſe is generally about 120 when in bed and in the morning; and is increaſed to 134, or more, when the patient ſits up, or in the evening paroxyſm. The pulſe of all very weak patients increaſes in frequency when they ſit up; becauſe the expenditure of ſenſorial power neceſſary to preſerve an erect poſture deducts ſo much from their general ſtrength; and hence the pulſe becomes weaker, and in conſequence quicker. See Sect. XII. 1. 4.

In this fever time muſt be allowed for the abſorption of the matter. Very large and repeated quantities of the bark, by preventing ſufficient food from being taken, as bread, and wine, and water, I have thought has much injured the patient; for the bark is not here given as in intermittent fevers to prevent the paroxyſm, but ſimply to ſtrengthen the patient by increaſing the power of digeſtion. About two ounces of decoction of bark, with four drops of laudanum, and a dram of ſweet ſpirit of vitriol, once in ſix hours, and a glaſs of wine between thoſe times, with panada, or other food, I have thought of moſt advantage, with a ſmall bliſter occaſionally.

Where not only the ſtomach but alſo the bowels are much diſtended with air, ſo as to ſound on ſtriking them with the fingers, the caſe is always dangerous, generally hopeleſs; which is more ſo in proportion to the quickneſs of the pulſe. Where the bowels are diſtended two drops of oil of cinnamon ſhould be given in the panada three or four times a day.

17. *Febris a ſphacelo.* Fever from mortification. This fever from abſorption of putrid matter is of the inirritative or typhus kind. See the preceding article.

M. M. Opium and the bark are frequently given in too great quantity, ſo as to induce conſequent debility, and to oppreſs the power of digeſtion.

ORDO

## ORDO I.

*Increafed Senfation.*

## GENUS VII.

*With increafed Action of the Organs of Senfe.*

## SPECIES.

1. *Delirium febrile.* Paraphrofyne. The ideas in delirium confift of thofe excited by the fenfation of pleafure or pain, which precedes them, and the trains of other ideas affociated with thefe, and not of thofe excited by external irritations or by voluntary exertion. Hence the patients do not know the room which they inhabit, or the people who furround them; nor have they any voluntary exertion, where the delirium is complete; fo that their efforts in walking about a room or rifing from their bed are unfteady, and produced by their catenations with the immediate affections of pleafure or pain. See Section XXXIII. 1. 4.

By the above circumftances it is diftinguifhed from madnefs, in which the patients well know the perfons of their acquaintance, and the place where they are; and perform all the voluntary actions with fteadinefs and determination. See Sect. XXXIV. 2. 2.

Delirium is fometimes lefs complete, and then a new face and louder voice ftimulate the patient to attend to them for a few moments; and then they relapfe again into perfect delirium. At other times a delirium affects but one fenfe, and the perfon thinks he fees things which do not exift; and is at the fame time fenfible to the queftions which are afked him, and to the tafte of the food which is offered to him.

This

This partial delirium is termed an hallucination of the disordered organ; and may probably arise from the origin of one nerve of sense being more liable to inflammation than the others; that is, an exuberance of the sensorial power of sensation may affect it; which is therefore thrown into action by slighter sensitive catenations, without being obedient to external stimulus, or to the power of volition.

The perpetual flow of ideas in delirium is owing to the same circumstance, as of those in our dreams; namely, to the defect or paralysis of the voluntary power; as in hemiplagia, when one side of the body is paralytic, and thus expends less of the sensorial power, the limbs on the other side are in constant motion from the exuberance of it. Whence less sensorial power is exhausted in delirium, than at other times, as well as in sleep; and hence in fevers with great debility, it is perhaps, as well as the stupor, rather a favourable circumstance; and when removed by numerous blisters, the death of the patient often follows the recovery of his understanding. See Class I. 2. 5. 6. and I. 2. 5. 10.

Delirium in diseases from inirritability is sometimes preceded by a propensity to surprise. See Class I. 1. 5. 12.

M. M. Fomentations of the shaved head for an hour repeatedly. A blister on the head. Rising from bed. Wine and opium, and sometimes venesection in small quantity by cupping, if the strength of the arterial system will allow it.

2. *Delirium maniacale.* Maniacal delirium. There is another kind of delirium, described in Sect. XXXIII. 1. 4. which has the increase of pleasureable or painful sensation for its cause, without any diminution of the other sensorial powers; but as this excites the patient to the exertion of voluntary actions, for the purpose of obtaining the object of his pleasureable ideas, or avoiding the object of his painful ones, such as perpetual prayer, when it is of the religious kind, it belongs

to

to the insanities described in Class III. 1. 2. 1, and is more properly termed hallucinatio maniacalis.

3. *Dilirium ebrietatis.* The drunken delirium is in nothing different from the delirium attending fevers except in its cause, as from alcohol, or other poisons. When it is attended with an apoplectic stupor, the pulse is generally low; and venesection I believe sometimes destroys those, who would otherwise have recovered in a few hours.

M. M. Diluting liquids. An emetic.

4. *Somnium.* Dreams constitute the most complete kind of delirium. As in these no external irritations are attended to, and the power of volition is entirely suspended; so that the sensations of pleasure and pain, with their associations, alone excite the endless trains of our sleeping ideas; as explained in Sect. XVIII. on Sleep.

5. *Hallucinatio visûs.* Deception of sight. These visual hallucinations are perpetual in our dreams; and sometimes precede general delirium in fevers; and sometimes belong to reverie, and to insanity. See Class III. 1. 2. 1. and 2. and must be treated accordingly.

Other kinds of visual hallucinations occur by moon-light; when objects are not seen so distinctly as to produce the usual ideas associated with them, but appear to us exactly as they are seen. Thus the trunk of a tree appears a flat surface, instead of a cylinder as by day, and we are deceived and alarmed by seeing things as they really are seen. See Berkley on Vision.

6. *Hallucinatio auditûs.* Auricular deception frequently occurs in dreams, and sometimes precedes general delirium in fevers; and sometimes belongs to vertigo, and to reverie, and to insanity. See Sect. XX. 7. and Class III. 1. 2. 1. and 2.

7. *Rubor a calore.* The bluſh from heat is occaſioned by the increaſed action of the cutaneous veſſels in conſequence of the increaſed ſenſation of heat. See Claſs I. 1. 2. 1. and 3.

8. *Rubor jucunditatis.* The bluſh of joy is owing to the increaſed action of the capillary arteries, along with that of every moving veſſel in the body, from the increaſe of pleaſurable ſenſation.

9. *Priapiſmus amatorius.* Amatorial priapiſm. The blood is poured into the cells of the corpora cavernoſa much faſter than it can be reabſorbed by the vena penis, owing in this caſe to the pleaſurable ſenſation of love increaſing the arterial action. See Claſs I. 1. 4. 6.

10. *Diſtentio mamularum.* The teats of female animals, when they give ſuck, become rigid and erected, in the ſame manner as in the laſt article, from the pleaſurable ſenſation of the love of the mother to her offspring. Whence the teat may properly be called an organ of ſenſe. The nipples of men do the ſame when rubbed with the hand. See Claſs I. 1. 4. 7.

ORDO

## ORDO II.

*Decreaſed Senſation.*

## GENUS I.

*Of the General Syſtem.*

### SPECIES.

1. *Stultitia inſenſibilis.* Folly from inſenſibility. The pleaſure or pain generated in the ſyſtem is not ſufficient to promote the uſual activity either of the ſenſual or muſcular fibres.

2. *Tædium vitæ.* Ennui. Irkſomeneſs of life. The pain of lazineſs has been thought by ſome philoſophers to be that principle of action, which has excited all our induſtry, and diſtinguiſhed mankind from the brutes of the field. It is certain that, where the ennui exiſts, it is relieved by the exertions of our minds or bodies, as all other painful ſenſations are relieved; but it depends much upon our early habits, whether we become patient of lazineſs, or inclined to activity, during the remainder of our lives, as other animals do not appear to be affected with this malady; which is perhaps leſs owing to deficiency of pleaſurable ſenſation, than to the ſuperabundancy of voluntary power, which occaſions pain in the muſcles by its accumulation; as appears from the perpetual motions of a ſquirrel confined in a cage.

3. *Pareſis ſenſitiva.* Weakneſs of the whole ſyſtem from inſenſibility.

## ORDO II.

*Decreased Sensation.*

## GENUS II.

*Of Particular Organs.*

### SPECIES.

1. *Anorexia.* Want of appetite. Some elderly people, and those debilitated by fermented liquors, are liable to lose their appetite for animal food; which is probably in part owing to the deficiency of gastric acid, as well as to the general decay of the system: elderly people will go on years without animal food; but inebriates soon sink, when their digestion becomes so far impaired. Want of appetite is sometimes produced by the putrid matter from many decaying teeth being perpetually mixed with the saliva, and thence affecting the organ of taste, and greatly injuring the digestion.

M. M. Fine charcoal powder diffused in warm water held in the mouth frequently in a day, as in Class I. 1. 4. 4. or solution of alum in water. Extract the decayed teeth. An emetic. A blister. Chalybeates. Vitriolic acid. Bile of an ox inspissated, and made into pills; 20 grains to be taken before dinner and supper. Opium half a grain twice a day.

All the strength we possess is ultimately derived from the food, which we are able to digest; whence a total debility of the system frequently follows the want of appetite, and of the power

power of digeſtion. Some young ladies I have obſerved to fall into this general debility, ſo as but juſt to be able to walk about; which I have ſometimes aſcribed to their voluntary faſting, when they believed themſelves too plump; and who have thus loſt both their health and beauty by too great abſtinence, which could never be reſtored.

I have ſeen other caſes of what may be termed anorexia epileptica, in which a total loſs of appetite, and of the power of digeſtion, ſuddenly occurred along with epileptic fits. Miſs B. a girl about eighteen, apparently very healthy, and rather plump, was ſeized with fits, which were at firſt called hyſterical; they occurred at the end of menſtruation, and returned very frequently with total loſs of appetite. She was relieved by veneſection, bliſters, and opiates; her ſtrength diminiſhed, and after ſome returns of the fits, ſhe took to her bed, and has ſurvived 15 or 20 years; ſhe has in general eaten half a potato a day, and ſeldom ſpeaks, but retains her ſenſes, and had many years occaſional returns of convulſion. I have ſeen two ſimilar caſes, where the anorexia, or want of appetite, was in leſs degree; and but juſt ſo much food could be digeſted, as ſupplied them with ſufficient ſtrength to keep from the bed or ſofa for half the day. As well as I can recollect, all theſe patients were attended with weak pulſe, and cold pale ſkin; and received benefit by opium, from a quarter of a grain to a grain four times a day. See Claſs III. 1. 1. 7. and III. 1. 2. 1. and III. 1. 2. 20.

2. *Adipſia.* Want of thirſt. Several of the inferior people, as farmers wives, have a habit of not drinking with their dinner at all, or only take a ſpoonful or two of ale after it. I have frequently obſerved theſe to labour under bad digeſtion, and debility in conſequence; which I have aſcribed to the too great ſtimulus

of

of ſolid food undiluted, deſtroying in proceſs of time the irritability of the ſtomach.

3. *Impotentia* (ageneſia). Impotency much ſeldomer happens to the male ſex than ſterility to the female ſex. Sometimes a temporary impotence occurs from baſhfulneſs, or the interference of ſome voluntary exertion in the production of an effect, which ſhould be performed alone by pleaſurable ſenſation.

One, who was ſoon to be married to a lady of ſuperior condition to his own, expreſſed fear of not ſucceeding on the wedding night; he was adviſed to take a grain of opium before he went to bed, and to accuſtom himſelf to ſleep with a woman previouſly, but not to enjoy her, to take off his baſhfulneſs; which ſucceeded to his wiſh.

M. M. Chalybeates. Opium. Bark. Tincture of cantharides.

4. *Sterilitas*. Barrenneſs. One of the ancient medical writers aſſerts, that the female ſex become pregnant with moſt certainty at or near the time of menſtruation. This is not improbable, ſince theſe monthly periods ſeem to reſemble the monthly venereal orgaſm of ſome female quadrupeds, which become pregnant at thoſe times only; and hence the computation of pregnancy is not often erroneous, though taken from the laſt menſtruation. See Section XXXVI. 2. 3.

M. M. Opium a grain every night. Chalybeates in very ſmall doſes. Bark. Sea-bathing.

5. *Inſenſibilitas artuum*. As in ſome paralytic limbs. A great inſenſibility ſometimes accompanies the torpor of the ſkin in cold fits

fits of agues. Some parts have retained the ſenſe of heat, but not the ſenſe of touch. See Sect. XVI. 6.

M. M. Friction with flannel. A bliſter. Warmth.

6. *Dyſuria inſenſitiva.* Inſenſibility of the bladder. A difficulty or total inability to make water attends ſome fevers with great debility, owing to the inſenſibility or inirritability of the bladder. This is a dangerous but not always a fatal ſymptom.

M. M. Draw off the water with a catheter. Aſſiſt the patient in the excluſion of it by compreſſing the lower parts of the abdomen with the hands. Wine two ounces, Peruvian bark one dram in decoction, every three hours alternately. Balſam of copaiva. Oil of almonds, with as much camphor as can be diſſolved in it, applied as a liniment rubbed on the region of the bladder and perinæum, and repeated every four hours, was uſed in this diſeaſe with ſucceſs by Mr. Latham. Med. Comment. 1791, p. 213.

7. *Accumulatio alvina.* An accumulation of feces in the rectum, occaſioned by the torpor, or inſenſibility, of that bowel. But as liquids paſs by theſe accumulations, it differs from the conſtipatio alvi, which is owing to too great abſorption of the alimentary canal.

Old milk, and eſpecially when boiled, is liable to induce this kind of coſtiveneſs in ſome grown perſons; which is probably owing to their not poſſeſſing ſufficient gaſtric acid to curdle and digeſt it; for as both theſe proceſſes require gaſtric acid, it follows, that a greater quantity of it is neceſſary, than in the digeſtion of other aliments, which do not previouſly require being curdled. This ill digeſted milk not ſufficiently ſtimulating the rectum, remains till it becomes a too ſolid maſs. On this account milk ſeldom agrees with thoſe, who are ſubject to piles, by inducing coſtiveneſs and large ſtools.

M. M. Extract the hardened ſcybala by means of a marrow-ſpoon; or by a piece of wire, or of whale-bone bent into a bow, and introduced. Injections of oil. Caſtor oil, or oil of almonds, taken by the mouth. A large clyſter of ſmoak of tobacco. Six grains of rhubarb taken every night for many months. Aloes. An endeavour to eſtabliſh a habit of evacuation at a certain hour daily. See Claſs I. 1. 3. 5.

ORDO

## ORDO III.

### *Retrograde Senſitive Motions.*

## GENUS I.

### *Of Excretory Ducts.*

THE retrograde action of the œſophagus in ruminating animals, when they bring up the food from their firſt ſtomach for the purpoſe of a ſecond maſtication of it, may probably be cauſed by agreeable ſenſation; ſimilar to that which induces them to ſwallow it both before and after this ſecond maſtication; and then this retrograde action properly belongs to this place, and is erroneouſly put at the head of the order of irritative retrograde motions. Claſs I. 3. 1. 1.

### SPECIES.

1. *Ureterum motus retrogreſſus.* When a ſtone has advanced into the ureter from the pelvis of the kidney, it is ſometimes liable to be returned by the retrograde motion of that canal, and the patient obtains fallacious eaſe, till the ſtone is again puſhed into the ureter.

2. *Urethræ motus retrogreſſus.* There have been inſtances of bougies being carried up the urethra into the bladder moſt probably by an inverted motion of this canal; for which ſome have undergone an operation ſimilar to that for the extraction of a ſtone. A caſe is related in ſome medical publication, in which a catgut bougie was carried into the bladder, and after remaining many weeks, was voided

piece-meal in a ſemi-diſſolved ſtate. Another caſe is related of a French officer, who uſed a leaden bougie; which at length found its way into the bladder, and was, by injecting crude mercury, amalgamated and voided.

In the ſame manner the infection from a ſimple gonorrhœa is probably carried further along the courſe of the urethra; and ſmall ſtones frequently deſcend ſome way into the urethra, and are again carried up into the bladder by the inverted action of this canal.

3. *Ductus choledochi motus retrogreſſus.* The concretions of bile, called gall-ſtones, frequently enter the bile-duct, and give violent pain for ſome hours; and return again into the gall-bladder, by the retrograde action of this duct. May not oil be carried up this duct, when a gall-ſtone gives great pain, by its retrograde ſpaſmodic action? See Claſs I. 1. 3. 8.

M. M. Opium a grain and half.

*The Orders and Genera of the Third Clafs of Difeafes.*

# CLASS III.

DISEASES OF VOLITION.

## ORDO I.

*Increafed Volition.*

### GENERA.

1. With increafed actions of the mufcles.
2. With increafed actions of the organs of fenfe.

## ORDO II.

*Decreafed Volition.*

### GENERA.

1. With decreafed actions of the mufcles.
2. With decreafed actions of the organs of fenfe.

*The Orders, Genera, and Species, of the Third Class of Diseases.*

---

# CLASS III.

## DISEASES OF VOLITION.

## ORDO I.

*Increased Volition.*

## GENUS I.

*With Increased Actions of the Muscles.*

### SPECIES.

| | |
|---|---|
| 1. *Jactitatio.* | Restlessness. |
| 2. *Tremor febrilis.* | Febrile trembling. |
| 3. *Clamor.* | Screaming. |
| 4. *Risus.* | Laughter. |
| 5. *Convulsio.* | Convulsion. |
| ——— *debilis.* | ——— weak. |
| 6. ——— *dolorifica.* | ——— painful. |
| 7. *Epilepsia.* | Epilepsy. |
| 8. ——— *dolorifica.* | ——— painful. |
| 9. *Somnambulismus.* | Sleep-walking. |
| 10. *Asthma convulsivum.* | Asthma convulsive. |
| 11. ——— *dolorificum.* | ——— painful. |

12. *Stridor*

| | |
|---|---|
| 12. *Stridor dentium.* | Gnaſhing of the teeth. |
| 13. *Tetanus triſmus.* | Cramp of the jaw. |
| 14. ——— *dolorificus.* | ——— painful. |
| 15. *Hydrophobia.* | Dread of water. |

## GENUS II.

*With increaſed Actions of the Organs of Senſe.*

### SPECIES.

| | |
|---|---|
| 1. *Mania mutabilis.* | Mutable madneſs. |
| 2. *Studium inane.* | Reverie. |
| 3. *Vigilia.* | Watchfulneſs. |
| 4. *Erotomania.* | Sentimental love. |
| 5. *Amor ſui.* | Vanity. |
| 6. *Noſtalgia.* | Deſire of home. |
| 7. *Spes religioſa.* | Supeſtitious hope. |
| 8. *Superbia ſtemmatis.* | Pride of family. |
| 9. *Ambitio.* | Ambition. |
| 10. *Mæror.* | Grief. |
| 11. *Tædium vitæ.* | Irkſomeneſs of life. |
| 12. *Deſiderium pulchritudinis.* | Loſs of beauty. |
| 13. *Paupertatis timor.* | Fear of poverty. |
| 14. *Lethi timor.* | —— of death. |
| 15. *Orci timor.* | —— of hell. |
| 16. *Satyriaſis.* | Luſt. |
| 17. *Ira.* | Anger. |
| 18. *Rabies.* | Rage. |
| 19. *Citta.* | Depraved appetite. |

20. *Cacoſitia.*

20. *Cacositia.* Aversion to food.
21. *Syphilis imaginaria.* Imaginary pox.
22. *Psora imaginaria.* ——— itch.
23. *Tabes imaginaria.* ——— tabes.
24. *Sympathia aliena.* Pity.
25. *Educatio heroica.* Heroic education.

## ORDO II.

*Decreased Volition.*

### GENUS I.

*With decreased Actions of the Muscles.*

#### SPECIES.

1. *Lassitudo.* Fatigue.
2. *Vacillatio senilis.* See-saw of old age.
3. *Tremor senilis.* Tremor of old age.
4. *Brachiorum paralysis.* Palsy of the arms.
5. *Raucedo paralytica.* Paralytic hoarseness.
6. *Vesicæ urinariæ paralysis.* Palsy of the bladder.
7. *Recti paralysis.* Palsy of the rectum.
8. *Paresis voluntaria.* Voluntary debility.
9. *Catalepsis.* Catalepsy.
10. *Hemiplegia.* Palsy of one side.
11. *Paraplegia.* Palsy of the lower limbs.
12. *Somnus.* Sleep.
13. *Incubus.* Night-mare.

14. *Lethargus.*

| | |
|---|---|
| 14. *Lethargus.* | Lethargy. |
| 15. *Syncope epileptica.* | Epileptic fainting. |
| 16. *Apoplexia.* | Apoplexy. |
| 17. *Mors a frigore.* | Death from cold. |

## GENUS II.

*With decreaſed Actions of the Organs of Senſe.*

### SPECIES.

| | |
|---|---|
| 1. *Recollectionis jactura.* | Loſs of recollection. |
| 2. *Stultitia voluntaria.* | Voluntary folly. |
| 3. *Credulitas.* | Credulity. |

# CLASS III.

## DISEASES OF VOLITION.

## ORDO I.

*Increased Volition.*

## GENUS I.

*Increased Actions of the Muscles.*

WE now step forward to consider the diseases of volition, that superior faculty of the sensorium, which gives us the power of reason, and by its facility of action distinguishes mankind from brute animals; which has effected all that is great in the world, and superimposed the works of art on the situations of nature.

Pain is introduced into the system either by excess or defect of the action of the part. (Sect. IV. 5.) Both which circumstances seem to originate from the accumulation of sensorial power in the affected organ. Thus when the skin is exposed to great cold, the activity of the cutaneous vessels is diminished, and in consequence an accumulation of sensorial power obtains in them, because they are usually excited into incessant motion by the stimulus of heat, as explained in Sect. XII. 5. 2. Contrarywise, when the vessels of the skin are exposed to great heat, an excess of sensorial power is also produced in them, which is derived thither by the increase of stimulus above what is natural.

This accounts for the relief which is received in all kinds of pain by any violent exertions of our muscles or organs of sense; which may thus be in part aſcribed to the exhauſtion of the ſenſorial power by ſuch exertions. But this relief is in many caſes ſo inſtantaneous, that it ſeems nevertheleſs probable, that it is alſo in part owing to the different manner of progreſſion of the two ſenſorial powers of ſenſation and volition; one of them commencing at ſome extremity of the ſenſorium, and being propagated towards the central parts of it; and the other commencing in the central parts of the ſenſorium, and being propagated towards the extremities of it; as mentioned in Sect. XI. 2. 1.

Theſe violent voluntary exertions of our muſcles or ideas to relieve the ſenſation of pain conſtitute convulſions and madneſs; and are diſtinguiſhed from the muſcular actions owing to increaſed ſenſation, as in ſneezing, or coughing, or parturition, or ejectio ſeminis, becauſe they do not contribute to diſlodge the cauſe, but only to prevent the ſenſation of it. In two caſes of parturition, both of young women with their firſt child, I have ſeen general convulſions occur from exceſs of voluntary exertion, as above deſcribed, inſtead of the actions of particular muſcles, which ought to have been excited by ſenſation for the excluſion of the fetus. They both became inſenſible, and died after ſome hours; from one of them the fetus was extracted in vain. I have heard alſo of general convulſions being excited inſtead of the actions of the muſculi acceleratores in the ejectio ſeminis, which terminated fatally. See Claſs III. 1. 1. 7.

Theſe violent exertions are moſt frequently excited in conſequence of thoſe pains, which originate from defect of the action of the part. See Sect. XXXIV. 1. and 2. The pains from exceſs and defect of the action of the part are diſtinguiſhable from each other by the former being attended with increaſe of heat in the pained part, or of the whole body; while the latter not only exiſt without increaſe of heat in the pained part, but are generally attended with coldneſs of the extremities of the body.

As ſoon as theſe violent actions of our muſcular or ſenſual fibres for

for the purpose of relieving pain cease to be exerted, the pain recurs; whence the reciprocal contraction and relaxation of the muscles in convulsion, and the intervals of madness. Otherwise these violent exertions continue, till so great a part of the sensorial power is exhausted, that no more of it is excitable by the faculty of volition; and a temporary apoplexy succeeds, with snoring as in profound sleep; which so generally terminates epileptic fits.

When these voluntary exertions become so connected with certain disagreeable sensations, or with irritations, that the effort of the will cannot restrain them, they can no longer in common language be termed voluntary; but nevertheless belong to this class, as they are produced by excess of volition, and may still not improperly be called depraved voluntary actions. See Sect. XXXIV. 1. where many motions in common language termed involuntary are shewn to depend on excess of volition.

When these exertions from excess of volition, which in common language are termed involuntary motions, either of mind or body, are perpetually exerted in weak constitutions, the pulse becomes quick; which is occasioned by the too great expenditure of the sensorial power in these unceasing modes of activity. In the same manner as in very weak people in fevers, the pulse sometimes increases in frequency to 140 strokes in a minute, when the patients stand up or endeavour to walk; and subsides to 110, when they lie down again in their beds. Whence it appears, that when a very quick pulse accompanies convulsion or insanity, it simply indicates the weakness of the patient; that is, that the expenditure of sensorial power is too great for the supply of it. But if the strength of the patient is not previously exhausted, the exertions of the muscles are attended with temporary increase of circulation, the reciprocal swellings and elongations of their bellies push forwards the arterial blood, and promote the absorption of the venous blood; whence a temporary increase of secretion and of heat, and a stronger pulse.

## SPECIES.

1. *Jactitatio.* Restlessness. There is one kind of restlessness attending fevers, which consists in a frequent change of posture to relieve the uneasiness of the pressure of one part of the body upon another, when the sensibility of the system, or of some parts of it, is increased by inflammation, as in the lumbago; which may sometimes be distinguished in its early stage by the incessant desire of the patient to turn himself in bed. But there is another restlessness, which approaches towards writhing or contortions of the body, which is a voluntary effort to relieve pain; and may be esteemed a slighter kind of convulsion, not totally unrestrainable by opposite or counteracting volitions.

M. M. A blister. Opium. Warm bath.

2. *Tremor febrilis.* Reciprocal convulsions of the subcutaneous muscles, originating from the pain of the sense of heat, owing to defect of its usual stimulus, and consequent accumulation of sensorial power in it. The actual deficiency of heat may exist in one part of the body, and the pain of cold be felt most vividly in some other part associated with it by sensitive sympathy. So a chillness down the back is first attended to in ague-fits, though the disease perhaps commences with the torpor and consequent coldness of some internal viscus. But in whatever part of the system the defect of heat exists, or the sensation of it, the convulsions of the subcutaneous muscles exerted to relieve it are very general; and, if the pain is still greater, a chattering of the teeth is added, the more suddenly to exhaust the sensorial power, and because the teeth are very sensible to cold.

These convulsive motions are nevertheless restrainable by violent voluntary counteraction; and as their intervals are owing to the pain of of cold being for a time relieved by their exertion, they may be compared to laughter, except that there is no interval of pleaſure preceding each moment of pain in this as in the latter.

M. M. See I. 2. 2. 1.

3. *Clamor.* Screaming from pain. The talkative animals, as dogs, and ſwine, and children, ſcream moſt, when they are in pain, and even from fear; as they have uſed this kind of exertion from their birth moſt frequently and moſt forcibly; and can therefore ſooner exhauſt the accumulation of ſenſorial power in the affected muſcular or ſenſual organs by this mode of exertion; as deſcribed in Sect. XXXIV. 1. 3. This facility of relieving pain by ſcreaming is the ſource of laughter, as explained below.

4. *Riſus.* The pleaſurable ſenſations, which occaſion laughter, are perpetually paſſing into the bounds of pain; for pleaſure and pain are often produced by different degrees of the ſame ſtimulus; as warmth, light, aromatic or volatile odours, become painful by their exceſs; and the tickling on the ſoles of the feet in children is a painful ſenſation at the very time it produces laughter. When the pleaſurable ideas, which excite us to laugh, paſs into pain, we uſe ſome exertion, as a ſcream, to relieve the pain, but ſoon ſtop it again, as we are unwilling to loſe the pleaſure; and thus we repeatedly begin to ſcream, and ſtop again alternately. So that in laughing there are three ſtages, firſt of pleaſure, then pain, then an exertion to relieve that pain. See Sect. XXXIV. 1. 3.

Every one has been in a ſituation, where ſome ludicrous circumſtance has excited him to laugh; and at the ſame time a ſenſe of decorum has forbid the exertion of theſe interrupted ſcreams; and then the pain has become ſo violent, as to occaſion him to uſe ſome other great action, as biting his tongue, and pinching himſelf, in lieu of the reiterated ſcreams which conſtitute laughter.

5. *Convulſio.*

5. *Convulsio*. Convulsion. When the pains from defect or excess of motion are more distressing than those already described, and are not relievable by such partial exertions, as in screaming, or laughter, more general convulsions occur; which vary perhaps according to the situation of the pained part, or to some previous associations formed by the early habits of life. When these convulsive motions bend the body forwards, they are termed emprosthotonoi; when they bend it backward, they are termed opisthotonoi. They frequently succeed each other, but the opisthotonoi are generally more violent; as the muscles, which erect the body, and keep it erect, are naturally in more constant and more forcible action than their antagonists.

The causes of convulsion are very numerous, as from toothing in children, from worms or acidity in their bowels, from eruption of the distinct small-pox, and lastly, from breathing too long the air of an unventilated bed-room. Sir G. Baker, in the Transactions of the College, described this disease, and detected its cause; where many children in an orphan-house were crowded together in one chamber, without a chimney, and were almost all of them affected with convulsion; in the hospital at Dublin, many died of convulsions before the real cause was understood. See Dr. Beddoes's Guide to Self-preservation. In a large family, which I attended, where many female servants slept in one room, which they had contrived to render inaccessible to every blast of air; I saw four who were thus seized with convulsions, and who were believed to have been affected by sympathy from the first who fell ill. They were removed into more airy apartments, but were some weeks before they all regained their perfect health.

Convulsion is distinguished from epilepsy, as the patient does not intirely lose all perception during the paroxysm. Which only shews, that a less exhaustion of sensorial power renders tolerable the pains which cause convulsion, than those which cause epilepsy. The hysteric convulsions are distinguished from those, owing to other causes,

 by

by the presence of the expectation of death, which precedes and succeeds them, and generally by a flow of pale urine; these convulsions do not constantly attend the hysteric disease, but are occasionally superinduced by the disagreeable sensation arising from the torpor or inversion of a part of the alimentary canal. Whence the convulsion of laughter is frequently sufficient to restrain these hysteric pains, which accounts for the fits of laughter frequently attendant on this disease.

M. M. To remove the peculiar pain which excites the convulsions. Venesection. An emetic. A cathartic with calomel. Warm-bath. Opium in large quantities, beginning with smaller ones. Mercurial frictions. Electricity. Cold-bath in the paroxysm; or cold aspersion. See Memoirs of Med. Society, Lon. V. 3. p. 147. a paper by Dr. Currie.

*Convulsio debilis.* The convulsions of dying animals, as of those which are bleeding to death in the slaughter-house, are an effort to relieve painful sensation, either of the wound which occasions their death, or of faintness from want of due distention of the blood-vessels. Similar to this in a less degree is the subsultus tendinum, or starting of the tendons, in fevers with debility; these actions of the muscles are too weak to move the limb, but the belly of the acting muscles is seen to swell, and the tendon to be stretched. These weak convulsions, as they are occasioned by the disagreeable sensation of faintness from inanition, are symptoms of great general debility, and thence frequently precede the general convulsions of the act of dying. See a case of convulsion of a muscle of the arm, and of the fore-arm, without moving the bones to which they were attached, Sect. XVII. 1. 8. See twitchings of the face, Class. IV. 3. 2. 2.

6. *Convulsio dolorifica.* Raphania. Painful convulsion. In this disease the muscles of the arms and legs are exerted to relieve the

pains

pains left after the rheumatiſm in young and delicate people; it recurs once or twice a-day, and has been miſtaken for the chorea, or St. Vitus's dance; but differs from it, as the undue motions in that diſeaſe only occur, when the patient endeavours to exert the natural ones; are not attended with pain; and ceaſe, when he lies down without trying to move: the chorea, or dance of St. Vitus, is often introduced by the itch, this by the rheumatiſm.

It has alſo been improperly called nervous rheumatiſm; but is diſtinguiſhed from rheumatiſm, as the pains recur by periods once or twice a day; whereas in the chronic rheumatiſm they only occur on moving the affected muſcles. And by the warmth of a bed the pains of the chronic rheumatiſm are increaſed, as the muſcles or membranes then become more ſenſible to the ſtimulus of the extraneous mucaginous material depoſited under them. Whereas the pains of the raphania, or painful convulſion, commence with coldneſs of the part, or of the extremities. See Rheumatiſmus chronicus, Claſs I. 1. 3. 12.

The pains which accompany the contractions of the muſcles in this diſeaſe, ſeem to ariſe from the too great violence of thoſe contractions, as happens in the cramp of the calf of the leg; from which they differ in thoſe being fixed, and theſe being reiterated contractions. Thus theſe convulſions are generally of the lower limbs, and recur at periodical times from ſome uneaſy ſenſation from defect of action, like other periodic diſeaſes; and the convulſions of the limbs relieve the original uneaſy painful ſenſation, and then produce a greater pain from their own too vehement contractions. There is however another way of accounting for theſe pains, when they ſucceed the acute rheumatiſm; and that is by the coagulable lymph, which may be left ſtill unabſorbed on the membranes; and which may be in too ſmall quantity to affect them with pain in common muſcular exertions, but may produce great pain, when the bellies of the muſcles ſwell to a larger bulk in violent action.

 M. M. Vene-

M. M. Venesection. Calomel. Opium. Bark. One grain of calomel and one of opium for ten successive nights. A bandage spread with emplastrum de minio put tight on the affected part.

7. *Epilepsia* is originally induced, like other convulsions, by a voluntary exertion to relieve some pain. This pain is most frequently about the pit of the stomach, or termination of the bile-duct; and in some cases the torpor of the stomach, which probably occasioned the epileptic fits, remains afterwards, and produces a chronical anorexia; of which a case is related in Class II. 2. 2. 1. There are instances of its beginning in the heel, of which a case is published by Dr. Short, in the Med. Essays, Edinb. I once saw a child about ten years old, who frequently fell down in convulsions, as she was running about in play; on examination a wart was found on one ancle, which was ragged and inflamed; which was directed to be cut off, and the fits never recurred.

When epilepsy first commences, the patients are liable to utter one scream before they fall down; afterwards the convulsions so immediately follow the pain, which occasions them, that the patient does not recollect or seem sensible of the preceding pain. Thus in laughter, when it is not excessive, a person is not conscious of the pain, which so often recurs, and causes the successive screams or exertions of laughter, which give a temporary relief to it.

Epileptic fits frequently recur in sleep from the increase of sensibility at that time, explained in Sect. XVIII. 14. In two such cases, both of young women, one grain of opium given at night, and continued many months, had success; in one of them the opium was omitted twice at different times, and the fit recurred on both the nights. In the more violent case, described in Sect. XVIII. 15, opium had no effect.

Epileptic fits generally commence with setting the teeth, by which means the tongue is frequently wounded; and with rolling the eyeballs in every kind of direction; for the muscles which suspend the jaw, as well as thoſe which move the eyes, are in perpetual motion during our waking hours; and yet continue ſubſervient to volition; hence their more facile and forcible actions for the purpoſe of relieving pain by the exhauſtion of ſenſorial power. See Section XXXIV. 1. 4.

Epileptic convulſions are not attended with the fear of death, as in the hyſteric diſeaſe, and the urine is of a ſtraw colour. However it muſt be noted, that the diſagreeable ſenſations in hyſteric diſeaſes ſometimes are the cauſe of true epileptic convulſions, of ſyncope, and of madneſs.

The pain, which occaſions ſome fits of epilepſy, is felt for a time in a diſtant part of the ſyſtem, as in a toe or heel; and is ſaid by the patient gradually to aſcend to the head, before the general convulſions commence. This aſcending ſenſation has been called aura epileptica, and is ſaid to have been prevented from affecting the head by a tight bandage round the limb. In this malady the pain, probably of ſome torpid membrane, or diſeaſed tendon, is at firſt only ſo great as to induce ſlight ſpaſms of the muſcular fibres in its vicinity; which ſlight ſpaſms ceaſe on the numbneſs introduced by a tight bandage; when no bandage is applied, the pain gradually increaſes, till generally convulſions are exerted to relieve it. The courſe of a lymphatic, as when poiſonous matter is abſorbed; or of a nerve, as in the ſciatica, may, by the ſympathy exiſting between their extremities and origins, give an idea of the aſcent of an aura or vapour.

In difficult parturition it ſometimes happens, that general convulſions are excited to relieve the pain of labour, inſtead of the exertions of thoſe muſcles of the abdomen and diaphragm, which ought to forward the excluſion of the child. See Claſs III. 1. 1. That is, inſtead of the particular muſcular actions, which ought to be excited by ſenſation to remove the offending cauſe, general convulſions are produced by the power of volition, which ſtill the pain, as in common epilepſy, without removing the cauſe; and, as the parturition is not thus promoted, the convulſions continue, till the ſenſorial power is

 totally

totally exhausted, that is, till death. In patients afflicted with epilepsy from other causes, I have seen the most violent convulsions recur frequently during pregnancy without miscarriage, as they did not tend to forward the exclusion of the fetus.

M. M. Venesection. A large dose of opium. Delivery.

The later in life epileptic fits are first experienced, the more dangerous they may be esteemed in general; as in these cases the cause has generally been acquired by the habits of the patient, or by the decay of some part, and is thus probably in an increasing state. Whereas in children the changes in the system, as they advance to puberty, sometimes removes the cause. So in toothing, fits of convulsion with stupor frequently occur, and cease when the tooth advances; but this is not to be expected in advanced life. Sir ——, about sixty years of age, had only three teeth left in his upper jaw, a canine tooth, and one on each side of it. He was seized with epileptic fits, with pain commencing in these teeth. He was urged to have them extracted, which he delayed too long, till the fits were become habitual, and then had them extracted in vain, and in a few months sunk under the disease.

Mr. F——, who had lived intemperately, and had been occasionally affected with the gout, was suddenly seized with epileptic fits; the convulsions were succeeded by apoplectic snoring; from which he was, in about 20 minutes, disturbed by fresh convulsions, and had continued in this situation above four-and-twenty hours. About eight ounces of blood were then taken from him; and after having observed, that the apoplectic's torpor continued about 20 minutes, I directed him to be forcibly raised up in bed, after he had thus lain about fifteen minutes, to gain an interval between the termination of the sleep, and the renovation of convulsion. In this interval he was induced to swallow forty drops of laudanum. Twenty more were given him in the same manner in about half an hour, both which evidently shortened the convulsion fits, and the consequent stupor; he then took thirty more drops, which for the present removed the fits.

fits. He became rather insane the next day, and after about three more days lost the insanity, and recovered his usual state of health.

The case mentioned in Sect. XXVII. 2. where the patient was left after epileptic fits with a suffusion of blood beneath the tunica adjunctiva of the eye, was in almost every respect similar to the preceding, and submitted to the same treatment. Both of them suffered frequent relapses, which were relieved by the same means, and at length perished, I believe, by the epileptic fits.

In those patients, who have not been subject to epilepsy before they have arrived to about forty years of age, and who have been intemperate in respect to spirituous potation, I have been induced to believe, that the fits were occasioned by the pain of a diseased liver; and this became more probable in one of the above subjects, who had used means to repel eruptions on the face; and thus by some stimulant application had prevented an inflammation taking place on the skin of the face instead of on some part of the liver. Secondly, as in these cases insanity had repeatedly occurred, which could not be traced from an hereditary source; there is reason to believe, that this as well as the epileptic convulsions were caused by spirituous potation; and that this therefore is the original source both of epilepsy and of insanity in those families, which are afflicted with them. This idea however brings some consolation with it; as it may be inferred, that in a few sober generations these diseases may be eradicated, which otherwise destroy the family.

M. M. Venesection. Opium. Bark. Steel. Arsenic. Opium one grain twice a day for years together. See the preceding article.

8. *Epilepsia dolorifica.* Painful epilepsy. In the common epilepsy the convulsions are immediately induced, as soon as the disagreeable sensation, which causes them, commences; but in this the pain continues long with cold extremities, gradually increasing for two or three hours, till at length convulsions or madness come on; which terminate

terminate the daily paroxyſm, and ceaſe themſelves in a little time afterwards.

This diſeaſe ſometimes originates from a pain about the lower edge of the liver, ſometimes in the temple, and ſometimes in the pudendum; it recurs daily for five or ſix weeks, and then ceaſes for ſeveral months. The pain is owing to defect of action, that is, to the accumulation of ſenſorial power in the part, which probably ſympathizes with ſome other part, as explained in Sect. XXXV. 2. XII. 5. 3. and Claſs II. 1. 1. 11. and IV. 2. 2. 3.

It is the moſt painful malady that human nature is liable to!—See Sect. XXXIV. 1. 4.

Mrs. C—— was ſeized every day about the ſame hour with violent pain on the right ſide of her bowels about the ſituation of the lower edge of the liver, without fever, which increaſed for an hour or two, till it became totally intolerable. After violent ſcreaming ſhe fell into convulſions, which terminated ſometimes in fainting, with or without ſtertor, as in common epilepſy; at other times a tempory inſanity ſupervened; which continued about half an hour, and the fit ceaſed. Theſe paroxyſms had returned daily for two or three weeks, and were at length removed by large doſes of opium, like the fits of reverie or ſomnambulation. About half an hour before the expected return of the fit three or four grains of opium were exhibited, and then tincture of opium was given in warm brandy and water about 20 or 30 drops every half hour, till the eyes became ſomewhat inflamed, and the noſe began to itch, and by the ſharp movements of the patient, or quick ſpeech, an evident intoxication appeared; and then it generally happened that the pain ceaſed. But the effects of this large doſe of opium was ſucceeded by perpetual ſickneſs and efforts to vomit, with great general debility all the ſucceeding day.

The rationale of this temporary cure from the exhibition of opium and vinous ſpirit depends on the great expenditure of ſenſorial power in the increaſed actions of all the irritative motions, by the ſtimulus of ſuch large quantities of opium and vinous ſpirit; together with the

production

production of much ſenſation, and many movements of the organs of ſenſe or ideas in conſequence of that ſenſation; and laſtly, even the motions of the arterial ſyſtem become accelerated by this degree of intoxication, all which ſoon exhauſted ſo much ſenſorial power as to relieve the pain; which would otherwiſe have cauſed convulſions or inſanity, which are other means of expending ſenſorial power. The general debility on the ſucceeding day, and the particular debility of the ſtomach, attended in conſequence with ſickneſs and frequent efforts to vomit, were occaſioned by the ſyſtem having previouſly been ſo ſtrongly ſtimulated, and thoſe parts in particular on which the opium and wine more immediately acted. This ſickneſs continued ſo many hours as to break the catenation of motions, which had daily reproduced the paroxyſm; and thus it generally happened, that the whole diſeaſe ceaſed for ſome weeks or months from one great intoxication, a circumſtance not eaſily to be explained on any other theory.

The exceſs or defect of motion in any part of the ſyſtem occaſions the production of pain in that part, as in Sect. XII. 1. 6. This defect or exceſs of fibrous action is generally induced by exceſs or defect of the ſtimulus of objects external to the moving organ. But there is another ſource of exceſſive fibrous action, and conſequent pain, which is from exceſs of volition, which is liable to affect thoſe muſcles, that have weak antagoniſts; as thoſe which ſupport the under jaw, and cloſe the mouth in biting, and thoſe of the calf of the leg; which are thus liable to fixed or painful contractions, as in triſmus, or locked jaw, and in the cramp of the calf of the leg; and perhaps in ſome colics, as in that of Japan: theſe pains, from contraction ariſing from exceſs of volition in the part from the want of the counteraction of antagoniſt muſcles, may give occaſional cauſe to epileptic fits, and may be relieved in the ſame way, either by exciting irritative and ſenſitive motions by the ſtimulus of opium and wine; or by convulſions or inſanity, as deſcribed above, which are only different methods of exhauſting the general quantity of ſenſorial power.

Conſidering

Confidering the great refemblance between this kind of painful epilepfy and the colic of Japan, as defcribed by Kemfer; and that that difeafe was faid to be cured by acupuncture, or the prick of a needle; I directed fome very thin fteel needles to be made about three inches long, and of fuch a temper, that they would bend double rather than break; and wrapped wax thread over about half an inch of the blunt end for a handle. One of thefe needles, when the pain occurred, was pufhed about an inch into the painful part, and the pain inftantly ceafed; but I was not certain, whether the fear of the patient, or the ftimulus of the puncture, occafioned the ceffation of pain; and as the paroxyfm had continued fome weeks, and was then declining, the experiment was not tried again. The difeafe is faid to be very frequent in Japan, and its feat to be in the bowels, and that the acupuncture eliminates the air, which is fuppofed to diftend the bowel. But though the aperture thus made is too fmall to admit of the eduction of air; yet as the ftimulus of fo fmall a puncture may either excite a torpid part into action, or caufe a fpafmodic one to ceafe to act; and laftly, as no injury could be likely to enfue from fo fmall a perforation, I fhould be inclined at fome future time to give this a fairer trial in fimilar circumftances.

Another thing worth trial at the commencement of this deplorable difeafe would be electricity, by paffing ftrong fhocks through the painful part; which, whether the pain was owing to the inaction of that part, or of fome other membrane affociated with it, might ftimulate them into exertion; or into inactivity, if owing to fixed painful contraction.

And laftly, the cold bath, or afperfions with cold water on the affected part, according to the method of Dr. Currie in the Memoirs of a Med. Soc. London, V. iii. p. 147, might produce great effect at the commencement of the pain. Neverthelefs opium duly adminiftered, fo as to precede the expected paroxyfm, and in fuch dofes, given by degrees, as to induce intoxication, is principally to be depended upon in this deplorable malady. To which fhould be added,

that

that if venefection can be previoufly performed, even to but few ounces, the effect of the opium is much more certain; and ftill more fo, if there be time to premife a brifk cathartic, or even an emetic. The effect of increafed ftimulus is fo much greater after previous defect of ftimulus; and this is ftill of greater advantage where the caufe of the difeafe happens to confift in a material, which can be abforbed. See Art. IV. 2. 8.

M. M. Venefection. An emetic. A cathartic. Warm bath. Opium a grain every half hour. Wine. Spirit of wine. If the patient becomes intoxicated by the above means, the fit ceafes, and violent vomitings and debility fucceed on the fubfequent day, and prevent a return. Blifters or finapifms on the fmall of the leg, taken off when they give much pain, are of ufe in flighter convulfions. Acupuncture. Electricity. Afperfion with cold water on the painful part.

9. *Somnambulifmus.* Sleep-walking is a part of reverie, or ftudium inane, defcribed in Sect. XIX. In this malady the patients have only the general appearance of being afleep in refpect to their inattention to the ftimulus of external objects, but, like the epilepfies above defcribed, it confifts in voluntary exertions to relieve pain. The mufcles are fubfervient to the will, as appears by the patient's walking about, and fometimes doing the common offices of life. The ideas of the mind alfo are obedient to the will, becaufe their difcourfe is confiftent, though they anfwer imaginary queftions. The irritative ideas of external objects continue in this malady, becaufe the patients do not run againft the furniture of the room; and when they apply their volition to their organs of fenfe, they become fenfible of the objects they attend to, but not otherwife, as general fenfation is deftroyed by the violence of their voluntary exertions. At the fame time the fenfations of pleafure in confequence of ideas excited by volition are vividly experienced, and other ideas feem to be excited by thefe pleafurable fenfations, as appears in the cafe of Mafter A. Sect. XXXIV.

XXXIV. 3. 1. where a hiſtory of a hunting ſcene was voluntarily recalled, with all the pleaſurable ideas which attended it. In melancholy madneſs the patient is employed in voluntarily exciting one idea, with thoſe which are connected with it by voluntary aſſociations only, but not ſo violently as to exclude the ſtimuli of external objects. In reverie variety of ideas are occaſionally excited by volition, and thoſe which are connected with them either by ſenſitive or voluntary aſſociations, and that ſo violently as to exclude the ſtimuli of external objects. Theſe two ſituations of our ſenſual motions, or ideas, reſemble convulſion and epilepſy; as in the former the ſtimulus of external objects is ſtill perceived, but not in the latter. Whence this diſeaſe, ſo far from being connected with ſleep, though it has by univerſal miſtake acquired its name from it, ariſes from exceſs of volition, and not from a ſuſpenſion of it; and though, like other kinds of epilepſy, it often attacks the patients in their ſleep, yet thoſe two, whom I ſaw, were more frequently ſeized with it while awake, the ſleep-walking being a part of the reverie. See Sect. XIX. and XXXIV. 3. and Claſs II. 1. 7. 4. and III. 1. 2. 18.

M. M. Opium in large doſes before the expected paroxyſm.

10. *Aſthma convulſivum*. The fits of convulſive aſthma return at periods, and are attended with cold extremities, and ſo far reſemble the acceſs of an intermittent fever; but, as the lungs are not ſenſible to the pain of cold, a ſhivering does not ſucceed, but inſtead of it violent efforts of reſpiration; which have no tendency, as in the humoral aſthma, to diſlodge any offending material, but only to relieve the pain by exertion, like the ſhuddering in the beginning of ague-fits, as explained Claſs III. 1. 1. 2.

The inſenſibility of the lungs to cold is obſervable on going into froſty air from a warm room; the hands and face become painfully cold, but no ſuch ſenſation is excited in the lungs; which is another argument in favour of the exiſtence of a peculiar ſet of nerves for the

 purpoſe

purpose of perceiving the universal fluid matter of heat, in which all things are immersed. See Sect. XIV. 6. Yet are the lungs nevertheless very sensible to the deficiency of oxygen in the atmosphere, as all people experience, when they go into a room crowded with company and candles, and complain, that it is so close, they can scarcely breathe; and the same in some hot days in summer.

There are two diseases, which bear the name of asthma. The first is the torpor or inability of the minute vessels of the lungs, consisting of the terminations of the pulmonary and bronchial arteries and veins, and their attendant lymphatics; in this circumstance it resembles the difficulty of breathing, which attends cold bathing. If this continues long, a congestion of fluid in the air-cells succeeds, as the absorbent actions cease completely before the secerning ones; as explained in Class I. 1. 2. 3. And the coldness, which attends the inaction of these vessels, prevents the usual quantity of exhalation. Some fits cease before this congestion takes place, and in them no violent sweating nor any expuition of phlegm occurs. This is the humoral asthma, described at Class II. 1. 1. 7.

The second kind of asthma consists in the convulsive actions in consequence of the disagreeable sensations thus induced; which in some fits of asthma are very great, as appears in the violent efforts to raise the ribs, and to depress the diaphragm, by lifting the shoulders. These, so long as they contribute to remove the cause of the disease, are not properly convulsions, but exertions immediately caused by sensation; but in this kind of asthma they are only efforts to relieve pain, and are frequently preceded by other epileptic convulsions.

These two kinds of asthmas have so many resembling features, and are so frequently intermixed, that it often requires great attention to distinguish them; but as one of them is allied to anasarca, and the other to epilepsy, we shall acquire a clearer idea of them by comparing them with those disorders. A criterion of the humoral or hydropic asthma is, that it is relieved by copious sweats about the head and

and breaſt, which are to be aſcribed to the ſenſitive exertions of the pulmonary veſſels to relieve the pain occaſioned by the anaſarcous congeſtion in the air-cells; and which is effected by the increaſed abſorption of the mucus, and its elimination by the retrograde action of thoſe lymphatics of the ſkin, whoſe branches communicate with the pulmonary ones; and which partial ſweats do not eaſily admit of any other explanation. See Claſs I. 3. 2. 8. Another criterion of it is, that it is generally attended with ſwelled legs, or other ſymptoms of anaſarca. A criterion of the convulſive aſthma may be had from the abſence of theſe cold clammy ſweats of the upper part of the body only, and from the patient having occaſionally been ſubject to convulſions of the limbs, as in the common epilepſy.

It may thus frequently happen, that in the humoral aſthma ſome exertions of the lungs may occur, which may not contribute to diſcharge the anaſarcous lymph, but may be efforts ſimply to relieve pain; beſides thoſe efforts, which produce the increaſed abſorption and elimination of it; and thus we have a bodily diſeaſe reſembling in this circumſtance the reverie, in which both ſenſitive and voluntary motions are at the ſame time, or in ſucceſſion, excited for the purpoſe of relieving pain.

It may likewiſe ſometimes happen, that the diſagreeable ſenſation, occaſioned by the congeſtion of lymph in the air-cells in the humoral or hydropic aſthma, may induce voluntary convulſions of the reſpiratory organs only to relieve the pain, without any ſenſitive actions of the pulmonary abſorbents to abſorb and eliminate the congeſtion of ſerous fluid; and thus the ſame cauſe may occaſionally induce either the humoral or convulſive aſthma.

The humoral aſthma has but one remote cauſe, which is the torpor of the pulmonary veſſels, like that which occurs on going into the cold bath; or the want of abſorption of the pulmonary lymphatics to take up the lymph effuſed into the air-cell. Whereas the convulſive aſthma, like other convulſions, or epilepſies, may be occaſioned by pain in almoſt any remote part of the ſyſtem. But in ſome of the adult patients in this diſeaſe, as in many epilepſies, I have ſuſpected the remote cauſe to be a pain of the liver, or of the biliary ducts.

The aſthmas, which have been induced in conſequence of the receſs of eruptions, eſpecially of the leprous kind, countenance this opinion. One lady I knew, who for many years laboured under an aſthma, which ceaſed on her being afflicted with pain, ſwelling, and diſtortion of ſome of her large joints, which were eſteemed gouty, but perhaps erroneouſly. And a young man, whom I ſaw yeſterday, was ſeized with aſthma on the retroceſſion, or ceaſing of eruptions on his face.

The convulſive aſthma, as well as the hydropic, are more liable to return in hot weather; which may be occaſioned by the leſs quantity of oxygen exiſting in a given quantity of warm air, than of cold, which can be taken into the lungs at one inſpiration. They are both moſt liable to occur after the firſt ſleep, which is therefore a general criterion of aſthma. The cauſe of this is explained in Sect. XVIII. 15. and applies to both of them, as our ſenſibility to internal uneaſy ſenſation increaſes during ſleep.

When children are gaining teeth, long before they appear, the pain of the gums often induces convulſions. This pain is relieved in ſome by ſobbing and ſcreaming; but in others a laborious reſpiration is exerted to relieve the pain; and this conſtitutes the true aſthma convulſivum. In other children again general convulſions, or epileptic paroxyſms, are induced for this purpoſe; which, like other epilepſies, become eſtabliſhed by habit, and recur before the irritation has time to produce the painful ſenſation, which originally cauſed them.

The aſthma convulſivum is alſo ſometimes induced by worms, or by acidity in the ſtomachs of children, and by other painful ſenſations in adults; in whom it is generally called nervous aſthma, and is often joined with other epileptic ſymptoms.

This aſthma is diſtinguiſhed from the peripneumony, and from the

the croup, by the preſence of fever in the two latter. It is diſtinguiſhed from the humoral aſthma, as in that the patients are more liable to run to the cold air for relief, are more ſubject to cold extremities, and experience the returns of it more frequently after their firſt ſleep. It is diſtinguiſhed from the hydrops thoracis, as that has no intervals, and the patient ſits conſtantly upright, and the breath is colder; and, where the pericardium is affected, the pulſe is quick and unequal. See Hydrops Thoracis, I. 2. 3. 14.

M. M. Veneſection once. A cathartic with calomel once. Opium. Aſſafætida. Warm bath. If the cauſe can be detected, as in toothing or worms, it ſhould be removed. As this ſpecies of aſthma is ſo liable to recur during ſleep, like epileptic fits, as mentioned in Section XVIII. 15. there was reaſon to believe, that the reſpiration of an atmoſphere mixed with hydrogen, or any other innocuous air, which might dilute the oxygen, would be uſeful in preventing the paroxyſms by decreaſing the ſenſibility of the ſyſtem. This, I am informed by Dr. Beddoes, has been uſed with decided ſucceſs by Dr. Ferriar. See Claſs II. 1. 1. 7.

11. *Aſthma dolorificum.* Angina pectoris. The painful aſthma was firſt deſcribed by Dr. Heberden in the Tranſactions of the College; its principal ſymptoms conſiſt in a pain about the middle of the ſternum, or rather lower, on every increaſe of pulmonary or muſcular exertion, as in walking faſter than uſual, or going quick up a hill, or even up ſtairs; with great difficulty of breathing, ſo as to occaſion the patient inſtantly to ſtop. A pain in the arms about the inſertion of the tendon of the pectoral muſcle generally attends, and a deſire of reſting by hanging on a door or branch of a tree by the arms is ſometimes obſerved. Which is explained in Claſs I. 2. 3. 14. and in Sect. XXIX. 5. 2.

Theſe patients generally die ſuddenly; and on examining the thorax no certain cauſe, or ſeat, of the diſeaſe has been detected; ſome

have

have ſuppoſed the valves of the arteries, or of the heart, were imperfect; and others that the accumulation of fat about this viſcus or the lungs obſtructed their due action; but other obſervations do not accord with theſe ſuppoſitions.

Mr. W——, an elderly gentleman, was ſeized with aſthma during the hot part of laſt ſummer; he always waked from his firſt ſleep with difficult reſpiration, and pain in the middle of his ſternum, and after about an hour was enabled to ſleep again. As this had returned for about a fortnight, it appeared to me to be an aſthma complicated with the diſeaſe, which Dr. Heberden has called angina pectoris. It was treated by veneſection, a cathartic, and then by a grain of opium given at going to bed, with ether and tincture of opium when the pain or aſthma recurred, and laſtly with the bark, but was ſeveral days before it was perfectly ſubdued.

This led me to conceive, that in this painful aſthma the diaphragm, as well as the other muſcles of reſpiration, was thrown into convulſive action, and that the fibres of this muſcle not having proper antagoniſts, a painful fixed ſpaſm of it, like that of the muſcles in the calf of the leg in the cramp, might be the cauſe of death in the angina pectoris, which I have thence arranged under the name of painful aſthma, and leave for further inveſtigation.

From the hiſtory of the caſe of the late much lamented John Hunter, and from the appearances after death, the caſe ſeems to have been of this kind, complicated with vertigo and conſequent affection of the ſtomach. The remote cauſe ſeems to have ariſen from oſſifications of the coronary arteries; and the immediate cauſe of his death from fixed ſpaſm of the heart. Other hiſtories and diſſections are ſtill required to put this matter out of doubt; as it is poſſible, that either a fixed ſpaſm of the diaphragm, or of the heart, which are both furniſhed with but weak antagoniſts, may occaſion ſudden death; and theſe may conſtitute two diſtinct diſeaſes.

Four patients I have now in my recollection, all of whom I believed

lieved to labour under the angina pectoris in a great degree; which have all recovered, and have continued well three or four years by the use, as I believe, of issues on the inside of each thigh; which were at first large enough to contain two pease each, and afterwards but one. They took besides some slight antimonial medicine for a while, and were reduced to half the quantity or strength of their usual potation of fermented liquor.

The use of femoral issues in angina pectoris was first recommended by Dr. Macbride, physician at Dublin, Med. Observ. & Enquir. Vol. VI. And I was further induced to make trial of them, not only because the means which I had before used were inadequate, but from the ill effect I once observed upon the lungs, which succeeded the cure of a small sore beneath the knee; and argued conversely, that issues in the lower limbs might assist a difficult respiration.

Mrs. L——, about fifty, had a small sore place about the size of half a pea on the inside of the leg a little below the knee. It had discharged a pellucid fluid, which she called a ley-water, daily for fourteen years, with a great deal of pain; on which account she applied to a surgeon, who, by means of bandage and a saturnine application, soon healed the sore, unheedful of the consequences. In less than two months after this I saw her with great difficulty of breathing, which with universal anasarca soon destroyed her.

The theory of the double effect of issues, as above related, one in relieving by their presence the asthma dolorificum, and the other in producing by its cure an anasarca of the lungs, is not easy to explain. Some similar effects from cutaneous eruptions and from blisters are mentioned in Class I. 1. 2. 9. In these cases it seems probable, that the pain occasioned by issues, and perhaps the absorption of a small quantity of aerated purulent matter, stimulate the whole system into greater energy of action, and thus prevent the torpor which is the beginning of so many diseases. In confirmation of this effect of pain

on

on the ſyſtem, I remember the caſe of a lady of an ingenious and active mind, who, for many of the latter years of her life, was perpetually ſubject to great pains of her head from decaying teeth. When all her teeth were gone, ſhe became quite low ſpirited, and melancholy in the popular ſenſe of that word, and after a year or two became univerſally dropſical and died.

M. M. Iſſues in the thighs. Five grains of rhubarb, and one ſixth of a grain of emetic tartar every night for ſome months; with or without half a grain of opium. No ſtronger liquor than ſmall beer, or wine diluted with twice its quantity of water. Since I wrote the above I have ſeen two caſes of hydrops thoracis, attended with pain in the left arm, ſo as to be miſtaken for aſthma dolorificum, in which femoral iſſues, though applied early in the diſeaſe, had no effect.

12. *Stridor dentium.* The clattering of the teeth on going into cold water, or in the beginning of ague-fits, is an exertion along with the tremblings of the ſkin to relieve the pain of cold. The teeth and ſkin being more ſenſible to cold than the more internal parts, and more expoſed to it, is the reaſon that the muſcles, which ſerve them, are thrown into exertion from the pain of cold rather than thoſe of reſpiration, as in ſcreaming from more acute pain. Thus the poet,

> Put but your toes into cold water,
> Your correſpondent teeth will clatter.
>
> Prior.

In more acute pains the jaws are gnaſhed together with great vehemence, inſomuch that ſometimes the teeth are ſaid to have been broken by the force. See Sect. XXXIV. 1. 3. In theſe caſes ſomething ſhould be offered to the patient to bite, as a towel, otherwiſe they are liable to tear their own arms, or to bite their attendants, as I have witneſſed in the painful epilepſy.

13. *Tetanus*

13. *Tetanus trifmus.* Cramp. The tetanus confifts of a fixed fpafm of almoft all the mufcles of the body; but the trifmus, or locked jaw, is the moft frequent difeafe of this kind. It is generally believed to arife from fympathy with an injured tendon. In one cafe where it occurred in confequence of a broken ankle from a fall from a horfe, it was preceded by evident hydrophobia. Amputation was advifed, but not fubmitted to; two wounds were laid into one with fciffors, but the patient died about the feventh day from the accident. In this cafe the wounded tendon, like the wounds from the bite of a mad dog, did not produce the hydrophobia, and then the locked jaw, till feveral days after the accident.

I twice witneffed the locked jaw from a pain beneath the fternum, about the part where it is complained of in painful afthma, or angina pectoris, in the fame lady at fome years diftance of time. The laft time it had continued two days, and fhe wrote her mind, or expreffed herfelf by figns. On obferving a broken tooth, which made a fmall aperture into her mouth, I rolled up five grains of opium like a worm about an inch long, and introducing it over the broken tooth, pufhed it onward by means of a fmall crow-quill; as it diffolved I obferved fhe fwallowed her faliva, and in lefs than half an hour, fhe opened her mouth and converfed as ufual.

Men are taught to be afhamed of fcreaming from pain in their early years; hence they are prone to exert the mufcles of the jaws inftead, which they have learnt to exert frequently and violently from their infancy; whence the locked jaw. This and the following fpafm have no alternate relaxations, like the preceding ones; which is perhaps owing, firft, to the weaknefs of their antagonift mufcles, thofe which elevate the jaw being very ftrong for the purpofe of biting and mafticating hard fubftances, and for fupporting the under jaw, with very weak antagonift mufcles; and fecondly, to their not giving fufficient relief even for a moment to the pain, or its preceding irritation, which excited them.

M. M. Opium in very large quantities. Mercurial ointment ufed extenfively. Electricity. Cold bath. Dilate the wound, and fill it with lint moiftened with fpirit of turpentine; which inflames the wound, and cures or prevents the convulfions. See a cafe, Tranfact. of American Society, Vol. II. p. 227.

Wine in large quantities in one cafe was more fuccefsful than opium; it probably inflames more, which in this difeafe is defirable. Between two or three ounces of bark, and from a quart to three pints of wine a day, fucceeded better than opium. Ib.

14. *Tetanus dolorificus.* Painful cramp. This kind of fpafm moft frequently attacks the calf of the leg, or mufcles of the toes; it often precedes paroxyfms of gout, and appears towards the end of violent diarrhœa, and from indigeftion, or from acid diet. In thefe cafes it feems to fympathize with the bowels, but is alfo frequently produced by the pain of external cold, and to the too great previous extenfion of the mufcles, whence fome people get the cramp in the extenfor mufcles of the toes after walking down hill, and of thofe of the calf of the leg after walking up a fteep eminence. For the reafon why thefe cramps commence in fleep, fee Sect. XVIII. 15.

The mufcle in this difeafe contracts itfelf to relieve fome fmaller pain, either from irritation or affociation, and then falls into great pain itfelf, from the too great action of its own fibres. Hence any mufcle, by being too vehemently exerted, falls into cramp, as in fwimming too forcibly in water, which is painfully cold; and a fecondary pain is then induced by the too violent contraction of the mufcle; though the pain, which was the caufe of the contraction, ceafes. Which accounts for the continuance of the contraction, and diftinguifhes this difeafe from other convulfions, which are relaxed and exerted alternately. Hence whatever may be the caufe of the primary pain, which occafions the cramp of the calf of the leg, the fecondary one is relievable by ftanding up, and thus by the weight of the body on the toes forcibly extending the contracted muscles. For the cause, which induces these muscles of the calf of the leg to fail into more violent contraction than other spasmodic muscles, proceeds from the weakness of their antagonist muscles; as they are generally extended again after action by the weight of the body on the balls of the toes. See the preceding article.

M. M. Rub the legs with camphor dissolved in oil, and let the patient wear stockings in bed. If a foot-board be put at the bed's feet, and the bed be so inclined, that he will rest a little with his toes against the foot-board, that pressure is said to prevent the undue contractions of the musculi gastrocnemii, which constitute the calf of the leg. In gouty patients, or where the bowels are affected with acidity, half a grain of opium, and six grains of rhubarb, and six of chalk, every night. Flesh-meat to supper. A little very weak warm spirit and water may be taken for present relief, when these cramps are very troublesome to weak or gouty patients.

15. *Hydrophobia.* Dread of water generally attending canine madness. I was witness to a case, where this disease preceded the locked jaw from a wound in the ankle, occasioned by a fall from a horse; as mentioned in the preceding article. It came on about the sixth day after the accident; when the patient attempted to swallow fluids, he became convulsed all over from the pain of this attempt, and spurted them out of his mouth with violence. It is also said to happen in some hysterical cases. Hence it seems rather the immediate consequence of a pained tendon, than of a contagious poison. And is so far analogous to tetanus, according with the opinions of Doctor Rusch and Doctor Percival.

In other respects, as it is produced by the saliva of an enraged animal instilled into a wound, it would seem analogous to the poison of venomous animals. And from the manner of its access so long after the bite, and of its termination in a short time, it would

 seem

ſeem to reſemble the progreſs of contagious fevers. See Sect. XXII. 3. 3.

If the patient was bitten in a part, which could be totally cut away, as a finger, even after the hydrophobia appears, it is probable it might cure it; as I ſuſpect the cauſe ſtill remains in the wounded tendon, and not in a diffuſed infection tainting the blood. Hence there are generally uneaſy ſenſations, as cold or numbneſs, in the old cicatrix, before the hydrophobia commences. See a caſe in Medical Communications, Vol. II. p. 190.

If the diſeaſed tendon could be inflamed without cutting it out, as by cupping, or cauſtic, or bliſter after cupping, and this in the old wound long ſince healed, after the hydrophobia commences, might prevent the ſpaſms about the throat. As inflaming the teeth by the uſe of mercury is of uſe in ſome kinds of hemicrania. Put ſpirit of turpentine on the wound, waſh it well. See Claſs I. 3. 1. 11. IV. 1. 2. 7.

M. M. Wine, muſk, oil, internally. Opium, mercurial ointment, uſed extenſively. Mercurial fumigation. Turpeth mineral. To ſalivate the patient as ſoon as poſſible. Exſection or a cauſtic on the ſcar, even after the appearance of hydrophobia. Put a tight bandage on the limb above the ſcar of the old wound to benumb the pained tendon, however long the wound may have been healed. Could a hollow catheter of elaſtic gum, coartchouc, be introduced into the œſophagus by the mouth or noſtril, and liquid nouriſhment be thus conveyed into the ſtomach? See Deſault's Journal, Caſe I. where, in an ulcer of the mouth, ſuch a catheter was introduced by the noſtril, and kept in the œſophagus for a month, by which means the patient was nouriſhed and preſerved.

It is recommended by Dr. Bardſley to give oil internally by a ſimilar method contrived by Mr. John Hunter. He covered a probang with the ſkin of a ſmall eel, or the gut of a lamb or cat. It was tied up at one end above and below the ſponge, and a ſlit made above the upper

upper ligature; to the other end of the eel-ſkin or gut was fixed a bladder and pipe. The probang thus covered was introduced into the ſtomach, and the liquid food or medicine was put into the bladder and ſqueezed down through the eel-ſkin. Mem. of Society at Mancheſter. See Claſs I. 2. 3. 25.

Dr. Bardſley has endeavoured to prove, that dogs never experience the hydrophobia, or canine madneſs, without having been previouſly bitten or infected; and ſecondly, that the diſeaſe in this ſpecies of animal always ſhews itſelf in five or ſix weeks; and concludes from hence, that this dreadful malady might be annihilated by making all the dogs in Great Britain perform a kind of quarantine, by ſhutting them up for a certain number of weeks. Though the diſeaſe from the bite of the mad dog is perhaps more analogous to thoſe from the wounds inflicted by venomous animals than to thoſe from other contagious matter, yet theſe obſervations are well worthy further attention; which the author promiſes.

ORDO

## ORDO I.

*Increaſed Volition.*

## GENUS II.

*With increaſed Actions of the Organs of Senſe.*

In every ſpecies of madneſs there is a peculiar idea either of deſire or averſion, which is perpetually excited in the mind with all its connections. In ſome conſtitutions this is connected with pleaſurable ideas without the exertion of much muſcular action, in others it produces violent muſcular action to gain or avoid the object of it, in others it is attended with deſpair and inaction. Mania is the general word for the two former of theſe, and melancholia for the latter; but the ſpecies of them are as numerous as the deſires and averſions of mankind.

In the preſent age the pleaſurable inſanities are moſt frequently induced by ſuperſtitious hopes of heaven, by ſentimental love, and by perſonal vanity. The furious inſanities by pride, anger, revenge, ſuſpicion. And the melancholy ones by fear of poverty, fear of death, and fear of hell; with innumerable others.

> Quicquid agunt homines, votum, timor, ira, voluptas,
> Gaudia, diſcurſus, noſtri eſt farrago libelli.
>
> JUVEN. I. 85.

This idea, however, which induces madneſs or melancholy, is generally untrue; that is, the object is a miſtaken fact. As when a patient is perſuaded he has the itch, or venereal diſeaſe, of which he has no ſymptom, and becomes mad from the pain this idea occaſions. So that the object of madneſs is generally a delirious idea, and thence cannot

cannot be conquered by reaſon; becauſe it continues to be excited by painful ſenſation, which is a ſtronger ſtimulus than volition. Moſt frequently pain of body is the cauſe of convulſion, which is often however exchanged for madneſs; and a painful delirious idea is moſt frequently the cauſe of madneſs originally, but ſometimes of convulſion. Thus I have ſeen a young lady become convulſed from a fright, and die in a few days; and a temporary madneſs frequently terminates the paroxyſms of the epilepſia dolorifica, and an inſanity of greater permanence is frequently induced by the pains or bruiſes of parturition.

Where the patient is debilitated a quick pulſe ſometimes attends inſane people, which is nevertheleſs generally only a ſymptom of the debility, owing to the too great expenditure of ſenſorial power; or of the paucity of its production, as in inirritative, or in ſenſitive inirritated fever. See III. 1. 1.

But nevertheleſs where the quick pulſe is permanent, it ſhews the preſence of fever; and as the madneſs then generally ariſes from the diſagreeable ſenſations attending the fever, it is ſo far a good ſymptom; becauſe when the fever is cured, or ceaſes ſpontaneouſly, the inſanity moſt frequently vaniſhes at the ſame time.

The ſtimulus of ſo much volition ſupports inſane people under variety of hardſhips, and contributes to the cure of diſeaſes from debility, as ſometimes occurs towards the end of fevers. See Sect. XXXIV. 2. 5. And, on the ſame account, they bear large doſes of medicines to procure any operation on them; as emetics, and cathartics, which, before they produce their effect in inverting the motions of the ſtomach in vomiting, or of the abſorbents of the bowels in purging, muſt firſt weaken the natural actions of thoſe organs, as ſhewn in Sect. XXXV. 1. 13.

From theſe conſiderations it appears, that the indications of cure muſt conſiſt in removing the cauſe of the pain, whether it ariſes from a delirious idea, or from a real fact, or from bodily diſeaſe; or ſecondly,

condly, if this cannot be done, by relieving the pain in conſequence of ſuch idea or diſeaſe. The firſt is ſometimes effected by preſenting frequently in a day contrary ideas to ſhew the fallacy, or the too great eſtimation, of the painful ideas. 2dly. By change of place, and thus preſenting the ſtimulus of new objects, as a long journey. 3dly. By producing forgetfulneſs of the idea or object, which cauſes their pain; by removing all things which recal it to their minds; and avoiding all converſation on ſimilar ſubjects. For I ſuppoſe no diſeaſe of the mind is ſo perfectly cured by other means as by forgetfulneſs.

Secondly, the pain in conſequence of the ideas or bodily diſeaſes above deſcribed is to be removed, firſt, by evacuations, as veneſection, emetics, and cathartics; and then by large doſes of opium, or by the vertigo occaſioned by a circulating ſwing, or by a ſea-voyage, which, as they affect the organs of ſenſe as well as evacuate the ſtomach, may contribute to anſwer both indications of cure.

Where maniacs are outrageous, there can be no doubt but coercion is neceſſary; which may be done by means of a ſtraight waiſtcoat; which diſarms them without hurting them; and by tying a handkerchief round their ankles to prevent their eſcape. In others there can be no doubt, but that confinement retards rather than promotes their cure; which is forwarded by change of ideas in conſequence of change of place and of objects, as by travelling or ſailing.

The circumſtances which render confinement neceſſary, are firſt, if the lunatic is liable to injure others, which muſt be judged of by the outrage he has already committed. 2dly. If he is likely to injure himſelf; this alſo muſt be judged of by the deſpondency of his mind, if ſuch exiſts. 3dly. If he cannot take care of his affairs. Where none of theſe circumſtances exiſt, there ſhould be no confinement. For though the miſtaken idea continues to exiſt, yet if no actions are produced in conſequence of it, the patient cannot be called inſane, he can only be termed delirious. If every one, who poſſeſſes miſtaken ideas,

ideas, or who puts falſe eſtimates on things, was liable to confinement, I know not who of my readers might not tremble at the ſight of a madhouſe!

The moſt convenient diſtribution of inſanities will be into general, as mania mutabilis, ſtudium inane, and vigilia; and into partial inſanities. Theſe laſt again may be ſubdivided into deſires and averſions, many of which are ſucceeded by pleaſurable or painful ideas, by fury or dejection, according to the degree or violence of their exertions. Hence the analogy between the inſanities of the mind, and the convulſions of the muſcles deſcribed in the preceding genus, is curiouſly exact. The convulſions without ſtupor, are either juſt ſufficient to obliterate the pain, which occaſions them; or are ſucceeded by greater pain, as in the convulſio dolorifica. So the exertions in the mania mutabilis are either juſt ſufficient to allay the pain which occaſions them, and the patient dwells comparatively in a quiet ſtate; or thoſe exertions excite painful ideas, which are ſucceeded by furious diſcourſes, or outrageous actions. The ſtudium inane, or reverie, reſembles epilepſy, in which there is no ſenſibility to the ſtimuli of external objects. Vigilia, or watchfulneſs, may be compared to the general writhing of the body; which is juſt a ſufficient exertion to relieve the pain which occaſions it. Erotomania may be compared to triſmus, or other muſcular fixed ſpaſm, without much ſubſequent pain; and mæror to cramp of the muſcles of the leg, or other fixed ſpaſm with ſubſequent pain. All theſe coincidences contribute to ſhew, as explained in Sect. III. 5, that our ideas are motions of the immediate organs of ſenſe obeying the ſame laws as our muſcular motions.

The violence of action accompanying inſanity depends much on the education of the perſon; thoſe who have been proudly educated with unreſtrained paſſions, are liable to greater fury; and thoſe, whoſe education has been humble, to greater deſpondency. Where the delirious idea, above deſcribed, produces pleaſurable ſenſations, as in perſonal vanity or religious enthuſiaſm; it is almoſt a pity to ſnatch them from their fool's paradiſe, and reduce them again to the common lot of humanity; leſt they ſhould complain of their cure, like the patient deſcribed in Horace,

> ——Pol! me occidiſtis, amici,
> Non ſervaſtis, ait, cui ſic extorta voluptas,
> Et demptus per vim mentis gratiſſimus error!

The diſpoſition to inſanity, as well as to convulſion, is believed to be hereditary; and in conſequence to be induced in thoſe families from ſlighter cauſes than in others. Convulſions have been ſhewn to have been moſt frequently induced by pains owing to defect of ſtimulus, as the ſhuddering from cold, and not from pains from exceſs of ſtimulus, which are generally ſucceeded by inflammation. But inſanities are on the contrary generally induced by pains from exceſs of ſtimulus, as from the too violent actions of our ideas, as in common anger, which is an inſanity of ſhort duration; for inſanities generally, though not always, ariſe from pains of the organs of ſenſe; but convulſions generally, though not always, from pains of the membranes or glands. And it has been previouſly explained, that though the membrane and glands, as the ſtomach and ſkin, receive great pain from want of ſtimulus; yet that the organs of ſenſe, as the eye and ear, receive no pain from defect of ſtimulus.

Hence it follows, that the conſtitutions moſt liable to convulſion, are thoſe which moſt readily become torpid in ſome part of the ſyſtem, that is, which poſſeſs leſs irritability; and that thoſe moſt liable to inſanity, are ſuch as have exceſs of ſenſibility; and laſtly, that theſe two circumſtances generally exiſt in the ſame conſtitution; as explained in Sect. XXXI. 2. on Temperaments. Theſe obſervations explain why epilepſy and inſanity frequently ſucceed or reciprocate with each other, and why inirritable habits, as ſcrophulous ones, are liable to inſanity, of which I have known ſome inſtances.

In many cafes however there is no appearance of the difpofition to epilepfy or infanity of the parent being tranfmitted to the progeny. Firft, where the infanity has arifen from fome violent difappointment, and not from intemperance in the ufe of fpirituous liquors. Secondly, where the parent has acquired the infanity or epilepfy by habits of intoxication after the procreation of his children. Which habits I fuppofe to be the general caufe of the difpofition to infanity in this country. See Clafs III. 1. 1. 7.

As the difpofition to gout, dropfy, epilepfy, and infanity, appears to be produced by the intemperate ufe of fpirituous potation, and is in all of them hereditary; it feems probable, that this difpofition gradually increafes from generation to generation, in thofe families which continue for many generations to be intemperate in this refpect; till at length thefe difeafes are produced; that is, the irritability of the fyftem gradually is decreafed by this powerful ftimulus, and the fenfibility at the fame time increafed, as explained in Sect. XXXI. 1. and 2. This difpofition is communicated to the progeny, and becomes ftill increafed, if the fame ftimulus be continued, and fo on by a third and fourth generation; which accounts for the appearance of epilepfy in the children of fome families, where it was never known before to have exifted, and could not be afcribed to their own intemperance. A parity of reafoning fhews, that a few fober generations may gradually in the fame manner reftore a due degree of irritability to the family, and decreafe the excefs of fenfibility.

From hence it would appear probable, that fcrophula and dropfy are difeafes from inirritability; but that in epilepfy and infanity an excefs of fenfibility is added, and the two faulty temperaments are thus conjoined.

## SPECIES.

1. *Mania mutabilis.* Mutable madneſs. Where the patients are liable to miſtake ideas of ſenſation for thoſe from irritation, that is, imaginations for realities, if cured of one ſource of inſanity, they are liable in a few months to find another ſource in ſome new miſtaken or imaginary idea, and to act from this new idea. The idea belongs to delirium, when it is an imaginary or miſtaken one; but it is the voluntary actions exerted in conſequence of this miſtaken idea, which conſtitute inſanity.

In this diſeaſe the patient is liable carefully to conceal the object of his deſire or averſion. But a conſtant inordinate ſuſpicion of all people, and a careleſſneſs of cleanlineſs, and of decency, are generally concomitants of madneſs. Their deſigns cannot be counteracted, till you can inveſtigate the delirious idea or object of their inſanity; but as they are generally timid, they are therefore leſs to be dreaded.

Z. Z. called a young girl, one of his maid-ſervants, into the parlour, and, with cocked piſtols in his hands, ordered her to ſtrip herſelf naked; he then inſpected her with ſome attention, and diſmiſſed her untouched. Then he ſtripped two of his male ſervants in the ſame manner, to the great terror of the neighbourhood. After he was ſecured, with much difficulty he was perſuaded to tell me, that he had got the itch, and had examined ſome of his ſervants to find out from whom he had received it; though at the ſame time there was not a ſpot to be ſeen on his hands, or other parts. The outrages in conſequence of this falſe idea were in ſome meaſure to be aſcribed to the pride occaſioned by unreſtrained education, affluent wealth, and dignified family.

Madneſs is ſometimes produced by bodily pain, particularly I believe of a diſeaſed liver, like convulſion and epilepſy; at other times

 it

it is caused by very painful ideas occasioned by external circumstances, as of grief or disappointment; but the most frequent cause of insanity arises from the pain of some imaginary or mistaken idea; which may be termed hallucinatio maniacalis. This hallucination of one of the senses is often produced in an instant, and generally becomes gradually weakened in process of time, by the perpetual stimulus of external objects, or by the successions of other catenations of ideas, or by the operations of medicines; and when the maniacal hallucination ceases, or is forgotten, the violent exertions cease, which were in consequence of it, and the disease is cured.

Mr. ———, a clergyman, about forty years of age, who was rather a weak man, happened to be drinking wine in jocular company, and by accident swallowed a part of the seal of a letter, which he had just then received; one of his companions seeing him alarmed, cried out in humour, "It will seal your bowels up." He became melancholy from that instant, and in a day or two refused to swallow any kind of nourishment. On being pressed to give a reason for this refusal, he answered, he knew nothing would pass through him. A cathartic was given, which produced a great many evacuations, but he still persisted, that nothing passed through him; and though he was frightened into taking a little broth once or twice by threats, yet he soon ceased intirely to swallow any thing, and died in consequence of this insane idea.

Miss ——, a sensible and ingenious lady, about thirty, said she had seen an angel; who told her, that she need not eat, though all others were under the necessity of supporting their earthly existence by food. After fruitless persuasions to take food, she starved herself to death.— It was proposed to send an angel of an higher order to tell her, that now she must begin to eat and drink again; but it was not put into execution.

Mrs. ———, a lady between forty and fifty years of age, imagined that she heard a voice say to her one day, as she was at her toilet, "Repent,

"Repent, or you will be damned." From that moment ſhe became melancholy, and this hallucination affected her in greater or leſs degree for about two years; ſhe then recovered perfectly, and is now a cheerful old woman.

Mrs. ———, a farmer's wife, going up ſtairs to dreſs, found the curtains of her bed drawn, and on undrawing them, ſhe believed that ſhe ſaw the corpſe of her ſiſter, who was then ill at the diſtance of twenty miles; and became from that time inſane; and as her ſiſter died about the time, ſhe could not be produced to counteract the inſane hallucination, but ſhe perfectly recovered in a few months.

Mrs. ——, a moſt elegant, beautiful, and accompliſhed lady, about twenty-two years of age, had been married about two months to an elegant, poliſhed, and affluent young man, and it was well known to be a love-match on both ſides. She ſuddenly became melancholy, and yet not to ſo great a degree, but that ſhe could command herſelf to do the honours of her table with grace and apparent eaſe. After many days intreaty, ſhe at length told me, that ſhe thought her marrying her huſband had made him unhappy; and that this idea ſhe could not efface from her mind day or night. I withſtood her being confined, as ſome had adviſed, and propoſed a ſea-voyage to her, with expectation that the ſickneſs, as well as change of objects, might remove the inſane hallucination, by introducing other energetic ideas; this was not complied with, but ſhe travelled about England with her friends and her huſband for many months, and at length perfectly recovered, and is now I am informed in health and ſpirits.

Theſe caſes are related to ſhew the utility of endeavouring to inveſtigate the maniacal idea, or hallucination; as it may not only acquaint us with the probable deſigns of the patient, from whence may be deduced the neceſſity of confinement; but alſo may ſome time lead to the moſt effectual plan of cure.

I received

I received good information of the truth of the following caſe, which was publiſhed a few years ago in the newſpapers. A young farmer in Warwickſhire, finding his hedges broke, and the ſticks carried away during a froſty ſeaſon, determined to watch for the thief. He lay many cold hours under a hay-ſtack, and at length an old woman, like a witch in a play, approached, and began to pull up the hedge; he waited till ſhe had tied up her bottle of ſticks, and was carrying them off, that he might convict her of the theft, and then ſpringing from his concealment, he ſeized his prey with violent threats. After ſome altercation, in which her load was left upon the ground, ſhe kneeled upon her bottle of ſticks, and raiſing her arms to heaven beneath the bright moon then at the full, ſpoke to the farmer already ſhivering with cold, "Heaven grant, that thou never mayeſt know again the bleſſing to be warm." He complained of cold all the next day, and wore an upper coat, and in a few days another, and in a fortnight took to his bed, always ſaying nothing made him warm, he covered himſelf with very many blankets, and had a ſieve over his face, as he lay; and from this one inſane idea he kept his bed above twenty years for fear of the cold air, till at length he died.

M. M. As mania ariſes from pain either of our muſcles or organs of ſenſe, the arts of relieving pain muſt conſtitute the method of cure. See Sect. XXXIV. 3. 4. Veneſection. Vomits of from five grains to ten of emetic tartar, repeated every third morning for three or four times; with ſolution of gum-ammoniac, and ſoluble tartar, ſo as to purge gently every day. Afterwards warm bath for two or three hours a day. Opium in large doſes. Bark. Steel.

Dr. Binns gave two ſcruples (40 grains) of ſolid opium at a doſe, and twenty grains four hours afterwards; which reſtored the patient. Dr. Brandreth gave 400 drops of laudanum to a maniac in the greateſt poſſible furor, and in a few hours he became calm and rational. Med. Comment. for 1791, p. 384.

*Prognoſtic.*

*Prognostic.*

THE temporary quick pulse attending some maniacal cases is simply a symptom of debility, and is the consequence of too great exertions; but a permanent quick pulse shews the presence of fever, and is frequently a salutary sign; because, if the life of the patient be safe, when the fever ceases, the insanity generally vanishes along with it, as mentioned above. In this case the kind of fever must direct the method of curing the insanity; which must consist of moderate evacuations and diluents, if the pulse be strong; or by nutrientia, bark, and small doses of opium, if the pulse be weak.

Where the cause is of a temporary nature, as in puerperal insanity, there is reason to hope, that the disease will cease, when the bruises, or other painful sensations attending this state, are removed. In these cases the child should be brought frequently to the mother, and applied to her breast, if she will suffer it, and this whether she at first attends to it or not; as by a few trials it frequently excites the storgè, or maternal affection, and removes the insanity, as I have witnessed.

When the madness is occasioned by pain of the teeth, which I believe is no uncommon case, these must be extracted; and the cure follows the extinction of the pain. There is however some difficulty in detecting the delinquent tooth in this case, as in hemicrania, unless by its apparent decay, or by some previous information of its pain having been complained of; because the pain of the tooth ceases, as soon as the exertions of insanity commence.

When a person becomes insane, who has a family of small children to solicit his attention, the prognostic is very unfavourable; as it shews the maniacal hallucination to be more powerful than those ideas which generally interest us the most.

2. *Studium*

2. *Studium inane.* Reverie consists of violent voluntary exertions of ideas to relieve pain, with all the trains or tribes connected with them by sensations or associations. It frequently alternates with epileptic convulsions; with which it corresponds, in respect to the insensibility of the mind to the stimuli of external objects, in the same manner as madness corresponds with common convulsion, in the patient's possessing at the same time a sensibility of the stimuli of external objects.

Some have been reported to have been involved in reverie so perfectly, as not to have been disturbed by the discharge of a cannon; and others to have been insensible to torture, as the martyrs for religious opinions; but these seem more properly to belong to particular insanities than to reverie, like nostalgia and erotomania.

Reverie is distinguished from madness as described above; and from delirium, because the trains of ideas are kept consistent by the power of volition, as the person reasons and deliberates in it. Somnambulismus is a part of reverie, the latter consisting in the exertions of the locomotive muscles, and the former of the exertions of the organs of sense; see Class I. 1. 1. 9. and Sect. XIX. both which are mixed, or alternate with each other, for the purpose of relieving pain.

When the patients in reverie exert their volition on their organs of sense, they can occasionally perceive the stimuli of external objects, as explained in Sect. XIX. And in this case it resembles sometimes an hallucination of the senses, as there is a mixture of fact and imagination in their discourse; but may be thus distinguished: hallucinations of the senses are allied to delirium, and are attended generally with quick pulse, and other symptoms of great debility; but reverie is without fever, and generally alternates with convulsions; and so much intuitive analogy (see Sect. XVII. 3. 7.) is retained in its paroxysms, as to preserve a consistency in the trains of ideas.

Miss G——, whose case is related in Sect. III. 5. 8. said, as I once ſat by her, " My head is fallen off, ſee it is rolled to that corner of the room, and the little black dog is nibbling the noſe off." On my walking to the place which ſhe looked at, and returning, and aſſuring her that her noſe was unhurt, ſhe became pacified, though I was doubtful whether ſhe attended to me. See Claſs III. 1. 1. 9. and Claſs III. 1. 2. 2.

M. M. Large doſes of opium given before the expected paroxyſm, as in epilepſia dolorifica, Claſs III. 1. 1. 8.

The hallucinatio ſtudioſa, or falſe ideas in reverie, differ from maniacal hallucinations above deſcribed, as no inſane exertions ſucceed, and in the patients whom I have ſeen they have always been totally forgotten, when the paroxyſm was over.

Maſter ——, a ſchool-boy about twelve years old, after he came out of a convulſion fit and ſat up in bed, ſaid to me, " Don't you ſee my father ſtanding at the feet of the bed, he is come a long way on foot to ſee me." I anſwered, no: " What colour is his coat?" He replied, " A drab colour." " And what buttons?" " Metal ones," he anſwered, and added, " how ſadly his legs are ſwelled." In a few minutes he ſaid, with apparent ſurpriſe, " He is gone," and returned to his perfect mind. Other caſes are related in Sect. XIX. and XXXIV. 3. and in Claſs III. 1. 2. 2. with further obſervations on this kind of hallucination; which however is not the cauſe of reverie, but conſtitutes a part of it, the cauſe being generally ſome uneaſy ſenſation of the body.

3. *Vigilia.* Watchfulneſs conſiſts in the unceaſing exertion of volition; which is generally cauſed by ſome degree of pain either of mind or of body, or from defect of the uſual quantity of pleaſurable ſenſation; hence if thoſe, who are accuſtomed to wine at night, take tea inſtead, they cannot ſleep. The ſame happens from want of ſolid food for ſupper, to thoſe who are accuſtomed to uſe it; as in theſe caſes there is pain or defect of pleaſure in the ſtomach.

Sometimes

Sometimes the anxiety about ſleeping, that is the deſire to ſleep, prevents ſleep; which conſiſts in an abolition of deſire or will. This may ſo far be compared to the impediment of ſpeech deſcribed in Sect. XVII. 1. 10. as the interference of the will prevents the effect deſired.

Another ſource of watchfulneſs may be from the too great ſecretion of ſenſorial power in the brain, as in phrenzy, and as ſometimes happens from the exhibition of opium, and of wine; if the exhauſtion of ſenſorial power by the general actions of the ſyſtem occaſioned by the ſtimulus of theſe drugs can be ſuppoſed to be leſs than the increaſed ſecretion of it.

M. M. 1. Solid food to ſupper. Wine. Opium. Warm bath. 2. The patient ſhould be told that his want of ſleep is of no conſequence to his health. 3. Veneſection by cupping. Abſtinence from wine. 4. A bliſter by ſtimulating the ſkin, and rhubarb by ſtimulating the bowels, will ſometimes induce ſleep. Exerciſe. An uniform ſound, as of a pauſing drop of water, or the murmur of bees. Other means are deſcribed in Sect. XVIII. 20.

4. *Erotomania.* Sentimental love. Deſcribed in its exceſs by romance-writers and poets. As the object of love is beauty, and as our perception of beauty conſiſts in a recognition by the ſenſe of viſion of thoſe objects, which have before inſpired our love, by the pleaſure they have afforded to many of our ſenſes (Sect. XVI. 6); and as brute animals have leſs accuracy of their ſenſe of viſion than mankind (ib.); we ſee the reaſon why this kind of love is not frequently obſervable in the brute creation, except perhaps in ſome married birds, or in the affection of the mother to her offspring. Men, who have not had leiſure to cultivate their taſte for viſible objects, and who have not read the works of poets and romance-writers, are leſs liable to ſentimental love; and as ladies are educated rather with an idea of being choſen, than of chooſing; there are many men, and more women,

 who

who have not much of this insanity; and are therefore more easily induced to marry for convenience or interest, or from the flattery of one sex to the other.

In its fortunate gratification sentimental love is supposed to supply the purest source of human felicity; and from the suddenness with which many of those patients, described in Species I. of this genus, were seized with the maniacal hallucination, there is reason to believe, that the most violent sentimental love may be acquired in a moment of time, as represented by Shakespeare in the beginning of his Romeo and Juliet.

Some have endeavoured to make a distinction between beauty and grace, and have made them as it were rivals for the possession of the human heart; but grace may be defined beauty in action; for a sleeping beauty cannot be called graceful in whatever attitude she may recline; the muscles must be in action to produce a graceful attitude, and the limbs to produce a graceful motion. But though the object of love is beauty, yet the idea is nevertheless much enhanced by the imagination of the lover; which appears from this curious circumstance, that the lady of his passion seldom appears so beautiful to the lover after a few months separation, as his ideas had painted her in his absence; and there is, on that account, always a little disappointment felt for a minute at their next interview from this hallucination of his ideas.

This passion of love produces reverie in its first state, which exertion alleviates the pain of it, and by the assistance of hope converts it into pleasure. Then the lover seeks solitude, lest this agreeable reverie should be interrupted by external stimuli, as described by Virgil.

> Tantum inter densas, umbrosa cacumina, fagos
> Assidue veniebat, ibi hæc incondita solus
> Montibus et sylvis studio jactabat inani.

6 When

When the pain of love is ſo great, as not to be relieved by the exertions of reverie, as above deſcribed; as when it is miſplaced on an object, of which the lover cannot poſſeſs himſelf; it may ſtill be counteracted or conquered by the ſtoic philoſophy, which ſtrips all things of their ornaments, and inculcates "nil admirari." Of which leſſons may be found in the meditations of Marcus Antoninus. The maniacal idea is ſaid in ſome lovers to have been weakened by the action of other very energetic ideas; ſuch as have been occaſioned by the death of his favourite child, or by the burning of his houſe, or by his being ſhipwrecked. In thoſe caſes the violence of the new idea for a while expends ſo much ſenſorial power as to prevent the exertion of the maniacal one; and new catenations ſucceed. On this theory the lover's leap, ſo celebrated by poets, might effect a cure, if the patient eſcaped with life.

The third ſtage of this diſeaſe I ſuppoſe is irremediable; when a lover has previouſly been much encouraged, and at length meets with neglect or diſdain; the maniacal idea is ſo painful as not to be for a moment relievable by the exertions of reverie, but is inſtantly followed by furious or melancholy inſanity; and ſuicide, or revenge, have frequently been the conſequence. As was lately exemplified in Mr. Hackman, who ſhot Miſs Ray in the lobby of the playhouſe. So the poet deſcribes the paſſion of Dido,

———Moriamur inultæ?—
At moriamur, ait,—ſic, ſic, juvat ire ſub umbras!

The ſtory of Medæa ſeems to have been contrived by Ovid, who was a good judge of the ſubject, to repreſent the ſavage madneſs occaſioned by ill-requited love. Thus the poet,

Earth has no rage like love to hatred turn'd,
Nor hell a fury like a woman ſcorn'd.

DRYDEN.

5. Amor

5. *Amor ſui.* Vanity conſiſts of an agreeable reverie, and is well ridiculed in the ſtory of Narciſſus, who ſo long contemplated his own beautiful image in the water, that he died from neglect of taking ſuſtenance. I once ſaw a handſome young man, who had been ſo much flattered by his parents, that his vanity roſe ſo near to inſanity, that one might diſcern by his perpetual attention to himſelf, and the difficulty with which he arranged his converſation, that the idea of himſelf intruded itſelf at every comma or pauſe of his diſcourſe. In this degree vanity muſt afford great pleaſure to the poſſeſſor; and when it exiſts within moderate bounds, may contribute much to the happineſs of ſocial life.

My friend Mr. —— once complained to me, that he was much troubled with baſhfulneſs in company, and believed that it aroſe from his want of perſonal vanity; on this account he determined on a journey to Paris, when Paris was the center of politeneſs; he there learnt to dreſs, to dance, and to move his hands gracefully in converſation; and returned a moſt conſummate coxcomb. But after a very few years he relapſed into ruſticity of dreſs and manners.

M. M. The cure of vanity may be attempted by exceſs of flattery, which will at length appear ridiculous, or by its familiarity will ceaſe to be deſired. I remember to have heard a ſtory of a nobleman in the court of France, when France had a court, who was ſo diſagreeably vain in converſation, that the king was pleaſed to direct his cure, which was thus performed. Two gentlemen were directed always to attend him, one was to ſtand behind his chair, and the other at a reſpectful diſtance before him; whenever his lordſhip began to ſpeak, one of them always pronounced, "Lord Gallimaufre is going to ſay the beſt thing in the world." And, as ſoon as his lordſhip had done ſpeaking, the other attendant pronounced, "Lord Gallimaufre has ſpoken the beſt thing in the world." Till in a few weeks this noble lord was ſo diſguſted with praiſe that he ceaſed to be vain; and his majeſty diſmiſſed his keepers.

6. *Noſtalgia.*

6. *Noſtalgia.* Maladie de Pais. Calenture. An unconquerable deſire of returning to one's native country, frequent in long voyages, in which the patients become ſo inſane as to throw themſelves into the ſea, miſtaking it for green fields or meadows. The Swiſs are ſaid to be particularly liable to this diſeaſe, and when taken into foreign ſervice frequently to deſert from this cauſe, and eſpecially after hearing or ſinging a particular tune, which was uſed in their village dances, in their native country, on which account the playing or ſinging this tune was forbid by the puniſhment of death. Zwingerus.

> Dear is that ſhed, to which his ſoul conforms,
> And dear that hill, which lifts him to the ſtorms.
>
> GOLDSMITH.

7. *Spes religioſa.* Superſtitious hope. This maniacal hallucination in its milder ſtate produces, like ſentimental love, an agreeable reverie; but when joined with works of ſupererogation, it has occaſioned many enormities. In India devotees conſign themſelves by vows to moſt painful and unceaſing tortures, ſuch as holding up their hands, till they cannot retract them; hanging up by hooks put into the thick ſkin over their ſhoulders, ſitting upon ſharp points, and other ſelf torments. While in our part of the globe faſting and mortification, as flagellation, has been believed to pleaſe a merciful deity! The ſerenity, with which many have ſuffered cruel martyrdoms, is to be aſcribed to this powerful reverie.

Mr. ——, a clergyman, formerly of this neighbourhood, began to bruiſe and wound himſelf for the ſake of religious mortification, and paſſed much time in prayer, and continued whole nights alone in the church. As he had a wife and family of ſmall children, I believed the caſe to be incurable; as otherwiſe the affection and employment in his family connections would have oppoſed the beginning of this inſanity. He was taken to a madhouſe without effect, and after he he returned home, continued to beat and bruiſe himſelf, and by this kind of mortification, and by ſometimes long faſting, he at length became emaciated and died. I once told him in converſation, that "God was a merciful being, and could not delight in cruelty, but that I ſuppoſed he worſhipped the devil." He was ſtruck with this idea, and promiſed me not to beat himſelf for three days, and I believe kept his word for one day. If this idea had been frequently forced on his mind, it might probably have been of ſervice.

When theſe works of ſupererogation have been of a public nature, what cruelties, murders, maſſacres, has not this inſanity introduced into the world!—A commander, who had been very active in leading and encouraging the bloody deeds of St. Bartholomew's day at Paris, on confeſſing his ſins to a worthy eccleſiaſtic on his death-bed, was aſked, "Have you nothing to ſay about St. Bartholomew?" "On that day," he replied, "God Almighty was obliged to me!"—The fear of hell is another inſanity, which will be ſpoken of below.

8. *Superbia ſtemmatis.* Pride of family has frequently formed a maniacal hallucination, which in its mild ſtate has conſiſted in agreeable reverie, but when it has been ſo painful as to demand homage from others, it has frequently induced inſane exertions. This inſanity ſeems to have exiſted in the flouriſhing ſtate of Rome, as now all over Germany, and is attacked by Juvenal with great ſeverity, a ſmall part of which I ſhall here give as a method of cure. Sat. 8.

Say, what avails the pedigree, that brings
Thy boaſted line from heroes or from kings;
Though many a mighty lord, in parchment roll'd,
Name after name, thy coxcomb hands unfold;
Though wreathed patriots crowd thy marble halls,
Or ſteel-clad warriors frown along the walls;
While on broad canvas in the gilded frame
All virtues flouriſh, and all glories flame?—

Say,

Say,—if ere noon with idiot laugh you lie
Wallowing in wine, or cog the dubious die,
Or act unshamed, by each indignant bust,
The midnight orgies of promiscuous lust!——
Go, lead mankind to Virtue's holy shrine,
With morals mend them, and with arts refine,
Or lift, with golden characters unfurl'd,
The flag of peace, and still a warring world!—
—So shall with pious hands immortal Fame
Wreathe all her laurels round thy honour'd name,
High o'er thy tomb with chissel bold engrave,
"THE TRULY NOBLE ARE THE GOOD AND BRAVE."

9. *Ambitio.* Inordinate desire of fame. A carelessness about the opinions of others is said by Xenophon to be the source of impudence; certainly a proper regard for what others think of us frequently incites us to virtuous actions, and deters us from vicious ones; and increases our happiness by enlarging our sphere of sympathy, and by flattering our vanity.

Abstract what others feel, what others think.
All pleasures sicken, and all glories sink.

POPE.

When this reverie of ambition excites to conquer nations, or to enslave them, it has been the source of innumerable wars, and the occasion of a great devastation of mankind. Cæsar is reported to have boasted, that he had destroyed three millions of his enemies, and one million of his friends.

The works of Homer are supposed to have done great injury to mankind by inspiring the love of military glory. Alexander was said to sleep with them always on his pillow. How like a mad butcher amid a flock of sheep appears the hero of the Iliad, in the following fine lines of Mr. Pope, which conclude the twentieth book.

His fiery courſers, as the chariot rolls,
Tread down whole ranks, and cruſh out heroes' ſouls;
Daſh'd from their hoofs, as o'er the dead they fly,
Black bloody drops the ſmoaking chariot dye;—
The ſpiky wheels through heaps of carnage tore,
And thick the groaning axles dropp'd with gore;
High o'er the ſcene of death ACHILLES ſtood,
All grim with duſt, all horrible with blood;
Yet ſtill inſatiate, ſtill with rage on flame,
Such is the luſt of never dying fame!

The cure muſt be taken from moral writers. Woolaſton ſays, Cæſar conquered Pompey; that is, a man whoſe name conſiſted of the letters C. æ. ſ. a. r. conquered a long time ago a man, whoſe name conſiſted of the letters P. o. m. p. e. y. and that this is all that remains of either of them. Juvenal alſo attacks this mode of inſanity, Sat. X. 166.

—I, demens, et ſævas curre per alpes,
Ut pueris placeas, et declamatio fias!

Which is thus tranſlated by Dr. Johnſon,

And left a name, at which the world grew pale,
To point a moral, or adorn a tale!

10. *Mæror.* Grief. A perpetual voluntary contemplation of all the circumſtances of ſome great loſs, as of a favourite child. In general the painful ideas gradually decreaſe in energy, and at length the recollection becomes more tender and leſs painful. The letter of Sulpicius to Cicero on the loſs of his daughter is ingenious. The example of David on the loſs of his child is heroic.

A widow lady was left in narrow circumſtances with a boy and a girl, two beautiful and lively children, the one ſix and the other ſeven years of age; as her circumſtances allowed her to keep but one maid-

maid-ſervant, theſe two children were the ſole attention, employment, and conſolation of her life; ſhe fed them, dreſſed them, ſlept with them, and taught them herſelf; they were both ſnatched from her by the gangrenous ſore throat in one week: ſo that ſhe loſt at once all that employed her, as well as all that was dear to her. For the firſt three or four days after their death, when any friend viſited her, ſhe ſat upright, with her eyes wide open, without ſhedding tears, and affected to ſpeak of indifferent things. Afterwards ſhe began to weep much, and for ſome weeks talked to her friends of nothing elſe but her dear children. But did not for many years, even to her dying hour, get quite over a gloom, which was left upon her countenance.

In violent grief, when tears flow, it is eſteemed a good ſymptom; becauſe then the actions cauſed by ſenſitive aſſociation take the place of thoſe cauſed by volition; that is, they prevent the voluntary exertions of ideas, or muſcular actions, which conſtitute inſanity.

The ſobbing and ſighing attendant upon grief are not convulſive movements, they are occaſioned by the ſenſorial power being ſo expended on the painful ideas, and their connections, that the perſon neglects to breathe for a time, and then a violent ſigh or ſob is neceſſary to carry on the blood, which oppreſſes the pulmonary veſſels, which is then performed by deep or quick inſpirations, and laborious expirations. Sometimes nevertheleſs the breath is probably for a while voluntarily held, as an effort to relieve pain. The paleneſs and ill health occaſioned by long grief is ſpoken of in Claſs IV. 2. 1. 9.

The melioration of grief by time, and its being at length even attended with pleaſure, depends on our retaining a diſtinct idea of the loſt object, and forgetting for a time the idea of the loſs of it. This pleaſure of grief is beautifully deſcribed by Akenſide. Pleaſures of Imagination, Book II. l. 680.

———Aſk the faithful youth,
Why the cold urn of her, whom long he loved,
So often fills his arms; ſo often draws
His lonely footſteps at the ſilent hour
To pay the mournful tribute of his tears?
Oh! he will tell thee, that the wealth of worlds
Should ne'er ſeduce his boſom to forego
That ſacred hour; when, ſtealing from the noiſe
Of care and envy, ſweet remembrance ſoothes
With Virtue's kindeſt looks his aching breaſt,
And turns his tears to rapture.

M. M. Conſolation is beſt ſupplied by the Chriſtian doctrine of a happy immortality. In the pagan religion the power of dying was the great conſolation in irremediable diſtreſs. Seneca ſays, "no one need be unhappy unleſs by his own fault." And the author of Telemachus begins his work by ſaying, that Calypſo could not conſole herſelf for the loſs of Ulyſſes, and found herſelf unhappy in being immortal. In the firſt hours of grief the methods of conſolation uſed by uncle Toby, in Triſtram Shandy, is probably the beſt; "he ſat down in an arm chair by the bed of his diſtreſſed friend, and ſaid nothing."

11. *Tædium vitæ.* The inanity of ſublunary things has afforded a theme to philoſophers, moraliſts, and divines, from the earlieſt records of antiquity; "Vanity of vanities!" ſays the preacher, "all is vanity!" Nor is there any one, I ſuppoſe, who has paſſed the meridian of life, who has not at ſome moments felt the nihility of all things.

Wearineſs of life in its moderate degree has been eſteemed a motive to action by ſome philoſophers. See Sect. XXXIV. 2. 3. But in thoſe men, who have run through the uſual amuſements of life early in reſpect to their age; and who have not induſtry or ability to cultivate

cultivate thoſe ſciences, which afford a perpetual fund of novelty, and of conſequent entertainment, are liable to become tired of life, as they ſuppoſe there is nothing new to be found in it, that can afford them pleaſure; like Alexander, who is ſaid to have ſhed tears, becauſe he had not another world to conquer.

Mr. ———, a gentleman about fifty, of poliſhed manners, who in a few months afterwards deſtroyed himſelf, ſaid to me one day, "a ride out in the morning, and a warm parlour and a pack of cards in the afternoon, is all that life affords." He was perſuaded to have an iſſue on the top of his head, as he complained of a dull head-ach, which being unſkilfully managed, deſtroyed the pericranium to the ſize of an inch in diameter; during the time this took in healing, he was indignant about it, and endured life, but ſoon afterwards ſhot himſelf.

Mr. ——, a gentleman of Gray's Inn, ſome years ago was prevailed upon by his friends to diſmiſs a miſtreſs, by whom he had a child, but who was ſo great a termagant and ſcold, that ſhe was believed to uſe him very ill, and even to beat him. He became melancholy in two days from the want of his uſual ſtimulus to action, and cut his throat on the third ſo completely, that he died immediately.

Mr. Anſon, the brother to the late Lord Anſon, related to me the following anecdote of the death of Lord Sc———. His Lordſhip ſent to ſee Mr. Anſon on the Monday preceding his death, and ſaid, "You are the only friend I value in the world, I determined therefore to acquaint you, that I am tired of the inſipidity of life, and intend to-morrow to leave it." Mr. Anſon ſaid, after much converſation, that he was obliged to leave town till Friday, and added, "As you profeſs a friendſhip for me, do me this laſt favour, I entreat you, live till I return." Lord Sc——— believed this to be a pious artifice to gain time, but neverthеleſs agreed, if he ſhould return by four o'clock on that day. Mr. Anſon did not return till five, and found,

by

by the countenances of the domeſtics, that the deed was done. He went into his chamber and found the corpſe of his friend leaning over the arm of a great chair, with the piſtol on the ground by him, the ball of which had been diſcharged into the roof of his mouth, and paſſed into his brain.

Mr. —— and Mr. ——, two young men, heirs to conſiderable fortunes, ſhot themſelves at the age of four or five and twenty, without their friends being able to conjecture any cauſe for thoſe raſh actions. One of them I had long known to expreſs himſelf with diſſatisfaction of the world; at eighteen years of age he complained, that he could not entertain himſelf; he tried to ſtudy the law at Cambridge, and afterwards went abroad for a year or two by my advice; but returned diſſatisfied with all things. As he had had an eruption for ſome years on a part of his face, which he probably endeavoured to remove by external applications; I was induced to aſcribe his perpetual ennui to the pain or diſagreeable ſenſation of a diſeaſed liver. The other young gentleman ſhot himſelf in his bed-room, and I was informed that there was found written on a ſcrap of paper on his table, "I am impotent, and therefore not fit to live." From whence there was reaſon to conclude, that this was the hallucinatio maniacalis, the delirious idea, which cauſed him to deſtroy himſelf. The caſe therefore belongs to mania mutabilis, and not to tædium vitæ.

M. M. Some reſtraint in exhauſting the uſual pleaſures of the world early in life. The agreeable cares of a matrimonial life. The cultivation of ſcience, as of chemiſtry, natural philoſophy, natural hiſtory, which ſupplies an inexhauſtible ſource of pleaſurable novelty, and relieves ennui by the exertions it occaſions.

In many of theſe caſes, whence irkſomeneſs of life has been the oſtenſible cauſe of ſuicide, there has probably exiſted a maniacal hallucination, a painful idea, which the patient has concealed even to his dying hour; except where the mania has evidently ariſen from hereditary

ditary or acquired diſeaſe of the membranous or glandular parts of the ſyſtem.

12. *Pulchitudinis deſiderium.* The loſs of beauty, either by diſeaſe, as by the ſmall-pox, or by age, as life advances, is ſometimes painfully felt by ladies, who have been much flattered on account of it. There is a curious caſe of this kind related in Le Sage's Bachelor of Salamanca, which is too nicely deſcribed to be totally imaginary.

In this ſituation ſome ladies apply to what are termed coſmetics under various names, which crowd the newſpapers. Of theſe the white has deſtroyed the health of thouſands; a calx, or magiſtery, of biſmuth is ſuppoſed to be ſold in the ſhops for this purpoſe; but it is either, I am informed, in part or entirely white lead or ceruſſa. The pernicious effects of the external uſe of thoſe ſaturnine applications are ſpoken of in gutta roſea, Claſs II. 1. 4. 6. The real calx of biſmuth would probably have the ſame ill effect. As the red paint is prepared from cochineal, which is an animal body, leſs if any injury ariſes from its uſe, as it only lies on the ſkin like other filth.

The tan of the ſkin occaſioned by the ſun may be removed by lemon juice evaporated by the fire to half its original quantity, or by diluted marine acid; which cleans the cuticle, by eroding its ſurface, but requires much caution in the application; the marine acid muſt be diluted with water, and when put upon the hand or face, after a ſecond of time, as ſoon as the tan diſappears, the part muſt be waſhed with a wet towel and much warm water. Freckles lie too deep for this operation, nor are they in general removeable by a bliſter, as I once experienced. See Claſs I. 2. 2. 9.

It is probable, that thoſe materials which ſtain ſilk, or ivory, might be uſed to ſtain the cuticle, or hair, permanently; as they are all animal ſubſtances. But I do not know, that any trials of this kind have been made on the ſkin. I endeavoured in vain to whiten the

back

back of my hand by marine acid oxygenated by manganeſe, which ſo inſtantly whitens cotton.

The cure therefore muſt be ſought from moral writers, and the cultivation of the graces of the mind, which are frequently a more valuable poſſeſſion than celebrated beauty.

13. *Paupertatis timor.* The fear of poverty is one kind of avarice; it is liable to affect people who have left off a profitable and active buſineſs; as they are thus deprived of their uſual exertions, and are liable to obſerve the daily expenditure of money, without calculating the ſource from whence it flows. It is alſo liable to occur with a ſudden and unexpected increaſe of fortune. Mr. ——, a ſurgeon, about fifty years of age, who was always rather of a parſimonious diſpoſition, had a large houſe, with a fortune of forty thouſand pounds, left him by a diſtant relation; and in a few weeks became inſane from the fear of poverty, lamenting that he ſhould die in a jail or workhouſe. He had left off a laborious country buſineſs, and the daily perception of profit in his books; he alſo now ſaw greater expences going forwards in his new houſe, than he had been accuſtomed to obſerve, and did not ſo diſtinctly ſee the ſource of ſupply; which ſeems to have occaſioned the maniacal hallucination.—This idea of approaching poverty is a very frequent and very painful diſeaſe; ſo as to have induced many to become ſuicides, who were in good circumſtances; more perhaps than any other maniacal hallucination, except the fear of hell.

The covetouſneſs of age is more liable to affect ſingle men, than thoſe who have families; though an accumulation of wealth would ſeem to be more deſirable to the latter. But an old man in the former ſituation, has no perſonal connections to induce him to open his purſe; and having loſt the friends of his youth, and not eaſily acquiring new ones, feels himſelf alone in the world; feels himſelf unprotected, as his ſtrength declines, and is thus led to depend for aſſiſtance on money, and on that account wiſhes to accumulate it. Whereas the father of a family has not only thoſe connections, which demand the frequent expenditure of money, but feels a conſolation in the friendſhip of his children, when age may render their good offices neceſſary to him.

M. M. I have been well informed of a medical perſon in good cir-circumſtances in London, who always carries an account of his affairs, as debtor and creditor, in his pocket-book; and looks over it frequently in a day, when this diſeaſe returns upon him; and thus, by counteracting the maniacal hallucination, wiſely prevents the increaſe of his inſanity. Another medical perſon, in London, is ſaid to have cured himſelf of this diſeaſe by ſtudying mathematics with great attention; which exertions of the mind relieved the pain of the maniacal hallucination.

Many moral writers have ſtigmatiſed this inſanity; the covetous, they ſay, commit crimes and mortify themſelves without hopes of reward; and thus become miſerable both in this world and the next. Thus Juvenal:

> Cum furor haud dubius, cum ſit manifeſta phrenitis,
> Ut locuples moriaris, egenti vivere fato!

The covetous man thought he gave good advice to the ſpendthrift, when he ſaid, "Live like me," who well anſwered him,

> ———"Like you, Sir John?
> "That I can do, when all I have is gone!"
>
> POPE.

14. *Lethi timor.* The fear of death perpetually employs the thoughts of theſe patients; hence they are deviſing new medicines, and applying to phyſicians and quacks without number. It is confounded with hypochondriaſis, Claſs I. 2. 3. 9. in popular converſation, but is in reality an inſanity.

A young gentleman, whom I advifed to go abroad as a cure for this difeafe, affured me, that during the three years he was in Italy and France he never paffed a quarter of an hour without fearing he fhould die. But has now for above twenty years experienced the contrary.

The fufferers under this malady are generally at once difcoverable by their telling you, amidft an unconnected defcription of their complaints, that they are neverthelefs not afraid of dying. They are alfo eafily led to complain of pains in almoft any part of the body, and are thus foon difcovered.

M. M. As the maniacal hallucination has generally arifen in early infancy from fome dreadful account of the ftruggles and pain of dying, I have fometimes obferved, that thefe patients have received great confolation from the inftances I have related to them of people dying without pain. Some of thefe, which I think curious, I fhall concifely relate, as a part of the method of cure.

Mr. ——, an elderly gentleman, had fent for me one whole day before I could attend him; on my arrival he faid he was glad to fee me, but that he was now quite well, except that he was weak, but had had a pain in his bowels the day before. He then lay in bed with his legs cold up to the knees, his hands and arms cold, and his pulfe fcarcely difcernible, and died in about fix hours. Mr. ——, another gentleman about fixty, lay in the act of dying, with difficult refpiration like groaning, but in a kind of ftupor or coma vigil, and every ten or twelve minutes, while I fat by him, he waked, looked up, and faid, "who is it groans fo, I am fure there is fomebody dying in the room," and then funk again into a kind of fleep. From thefe two cafes there appeared to be no pain in the act of dying, which may afford confolation to all, but particularly to thofe who are afflicted with the fear of death.

15. *Orci*

15. *Orci timor*. The fear of hell. Many theatric preachers among the Methodiſts ſucceſsfully inſpire this terror, and live comfortably upon the folly of their hearers. In this kind of madneſs the poor patients frequently commit ſuicide; although they believe they run headlong into the hell, which they dread! Such is the power of oratory, and ſuch the debility of the human underſtanding!

Thoſe, who ſuffer under this inſanity, are generally the moſt innocent and harmleſs people; who are then liable to accuſe themſelves of the greateſt imaginary crimes, and have ſo much intellectual cowardice, that they dare not reaſon about thoſe things, which they are directed by their prieſts to believe, however contradictory to human apprehenſion, or derogatory to the great Creator of all things. The maniacal hallucination at length becomes ſo painful, that the poor inſane flies from life to become free from it.

M. M. Where the intellectual cowardice is great, the voice of reaſon is ineffectual; but that of ridicule may ſave many from thoſe mad-making doctors; though it is too weak to cure thoſe, who are already hallucinated. Foot's Farces are recommended for this purpoſe.

16. *Satyriaſis*. An ungovernable deſire of venereal indulgence. The remote cauſe is probably the ſtimulus of the ſemen; whence the phallus becomes diſtended with blood by the arterial propulſion of it being more ſtrongly excited than the correſpondent venous abſorption. At the ſame time a new ſenſe is produced in the other termination of the urethra; which, like itching, requires ſome exterior friction to facilitate the removal of the cauſe of the maniacal actions, which may probably be increaſed in thoſe caſes by ſome aſſociated hallucinations of ideas. It differs from priapiſmus chronicus in the deſire of its appropriated object, which is not experienced in the latter, Claſs I. 1. 4. 6. and from the priapiſmus amatorius, Claſs II.

 1. 7.

1. 7. 9. in the maniacal actions in consequence of desire. The furor uterius, or nymphomania, is a similar disease.

M. M. Venesection. Cathartics. Torpentia. Marriage.

17. *Ira.* Anger is caused by the pain of offended pride. We are not angry at breaking a bone, but become quite insane from the smallest stroke of a whip from an inferior. Ira furor brevis. Anger is not only itself a temporary madness, but is a frequent attendant on other insanities, and as, whenever it appears, it distinguishes insanity from delirium, it is generally a good sign in fevers with debility.

An injury voluntarily inflicted on us by others excites our exertions of self-defence or of revenge against the perpetrator of it; but anger does not succeed in any great degree unless our pride is offended; this idea is the maniacal hallucination, the pain of which sometimes produces such violent and general exertions of our muscles and ideas, as to disappoint the revenge we meditate, and vainly to exhaust our sensorial power. Hence angry people, if not further excited by disagreeable language, are liable in an hour or two to become humble, and sorry for their violence, and willing to make greater concessions than required.

M. M. Be silent, when you feel yourself angry. Never use loud oaths, violent upbraidings, or strong expressions of countenance, or gesticulations of the arms, or clenched fists; as these by their former associations with anger will contribute to increase it. I have been told of a sergeant or corporal, who began moderately to cane his soldiers, when they were awkward in their exercise, but being addicted to swearing and coarse language, he used soon to enrage himself by his own expressions of anger, till toward the end he was liable to beat the delinquents unmercifully.

18. *Rabies.* Rage. A desire of biting others, most frequently attendant on canine madness. Animals in great pain, as in the colica saturnina, are said to bite the ground they lie upon, and even their own flesh. I have seen patients bite the attendants, and even their own arms, in the epilepsia dolorifica. It seems to be an exertion to relieve pain, as explained in Sect. XXXIV. 1. 3. The dread of water in hydrophobia is occasioned by the repeated painful attempts to swallow it, and is therefore not an essential or original part of the disease called canine madness. See Class III. 1. 1. 15.

There is a mania reported to exist in some parts of the east, in which a man is said to run a muck; and these furious maniacs are believed to have induced their calamity by unlucky gaming, and afterwards by taking large quantities of opium; whence the pain of despair is joined with the energy of drunkenness; they are then said to sally forth into the most populous streets, and to wound and slay all they meet, till they receive their own death, which they desire to procure without the greater guilt, as they suppose, of suicide.

M. M. When there appears a tendency to bite in the painful epilepsy, the end of a rolled-up towel, or a wedge of soft wood, should be put into the mouth of the patient. As a bullet is said sometimes to be given to a soldier, who is to be severely flogged, that he may by biting it better bear his punishment.

19. *Citta.* A desire to swallow indigestible substances. I once saw a young lady, about ten years of age, who filled her stomach with the earth out of a flower-pot, and vomited it up with small stones, bits of wood, and wings of insects amongst it. She had the bombycinous complexion, and looked like a chlorotic patient, though so young; this generally proceeds from an acid in the stomach.

M. M. A vomit. Magnesia alba. Armenian bole. Rhubarb. Bark. Steel. A blister. See Class I. 2. 4. 5.

20. *Cacositia.*

20. *Cacoſitia.* Averſion to food. This may ariſe, without diſeaſe of the ſtomach, from connecting nauſeous ideas to our uſual food, as by calling a ham a hog's a——. This madneſs is much inculcated by the ſtoic philoſophy. See Antoninus' Meditations. See two caſes of patients who refuſed to take nouriſhment, Claſs III. 1. 2. 1.

Averſions to peculiar kinds of food are thus formed early in life by aſſociation of ſome maniacal hallucination with them. I remember a child, who on taſting the griſtle of ſturgeon, aſked what griſtle was? And being told it was like the diviſion of a man's noſe, received an ideal hallucination; and for twenty years afterwards could not be perſuaded to taſte ſturgeon.

The great fear or averſion, which ſome people experience at the ſight of ſpiders, toads, crickets, and the like, have generally had a ſimilar origin.

M. M. Aſſociate agreeable ideas with thoſe which diſguſt; as call a ſpider ingenious, a frog clean and innocent; and repreſs all expreſſions of diſguſt by the countenance, as ſuch expreſſions contribute to preſerve, or even to increaſe, the energy of the ideas aſſociated with them; as mentioned above in Species 17. Ira.

21. *Syphilis imaginaria.* The fear that they are infected with the venereal diſeaſe, when they have only deſerved it, is a very common inſanity amongſt modeſt young men; and is not to be cured without applying artfully to the mind; a little mercury muſt be given, and hopes of a cure added weekly and gradually by interview or correſpondence for ſix or eight weeks. Many of theſe patients have been repeatedly ſalivated without curing the mind!

22. *Pſora imaginaria.* I have twice ſeen an imaginary itch, and twice an imaginary diabætes, where there was not the leaſt veſtige of either of thoſe diſeaſes, and once an imaginary deafneſs, where the patient heard perfectly well. In all theſe caſes the hallucinated idea

is ſo powerfully excited, that it is not to be changed ſuddenly by occular ſenſation, or reaſon. Yet great perſeverance in the frequently preſenting contrary ideas will ſometimes ſlowly remove this hallucination, or in great length of time oblivion, or forgetfulneſs, performs a cure, by other means in vain attempted.

23. *Tabes imaginaria.* This imaginary diſeaſe, or hallucination, is cauſed by the ſuppoſed too great frequency of parting with the ſemen, and had long impoſed upon the phyſician as well as the patient, till Mr. John Hunter firſt endeavoured to ſhew, that in general the morbid effects of this pollution was in the imagination; and that thoſe were only liable to thoſe effects in general, who had been terrified by the villainous books, which pretend to prevent or to cure it, but which were purpoſely written to vend ſome quack medicine. Moſt of thoſe unhappy patients, whom I have ſeen, had evidently great impreſſion of fear and ſelf-condemnation on their minds, and might be led to make contradictory complaints in almoſt any part of the body, and if their confeſſions could be depended on, had not uſed this pollution to any great exceſs.

M. M. 1. Aſſure them if the loſs of the ſemen happens but twice a week, it will not injure them. 2. Marry them. The laſt is a certain cure; whether the diſeaſe be real or imaginary. Cold partial bath, and aſtringent medicines frequently taken, only recal the mind to the diſeaſe, or to the delinquency; and thence increaſe the imaginary effects and the real cauſe, if ſuch exiſts. Mr. —— deſtroyed himſelf to get free from the pain of fear of the ſuppoſed ill conſequences of ſelf-pollution, without any other apparent diſeaſe; whoſe parents I had in vain adviſed to marry him, if poſſible.

24. *Sympathia aliena.* Pity. Our ſympathy with the pleaſures and pains of others diſtinguiſhes men from other animals; and is probably the foundation of what is termed our moral ſenſe; and the

ſource

ſource of all our virtues. See Sect. XXII. 3. 3. When our ſympathy with thoſe miſeries of mankind, which we cannot alleviate, riſes to exceſs, the mind becomes its own tormentor; and we add to the aggregate ſum of human miſery, which we ought to labour to diminiſh; as in the following eloquent lamentation from Akenſide's Pleaſures of Imagination, Book II. l. 200.

——————————Dark,
As midnight ſtorms, the ſcene of human things
Appear'd before me; deſerts, burning ſands,
Where the parch'd adder dies; the frozen ſouth;
And deſolation blaſting all the weſt
With rapine and with murder. Tyrant power
Here ſits enthroned in blood; the baleful charms
Of ſuperſtition there infect the ſkies,
And turn the ſun to horror. Gracious Heaven!
What is the life of man? Or cannot theſe,
Not theſe portents thy awful will ſuffice?
That, propagated thus beyond their ſcope,
They riſe to act their cruelties anew
In my afflicted boſom, thus decreed
The univerſal ſenſitive of pain,
The wretched heir of evils not its own!

A poet of antiquity, whoſe name I do not recollect, is ſaid to have written a book deſcribing the miſeries of the world, and to have deſtroyed himſelf at the concluſion of his taſk. This ſympathy, with all ſenſitive beings, has been carried ſo far by ſome individuals, and even by whole tribes, as the Gentoos, as not only to reſtrain them from killing animals for their ſupport, but even to induce them to permit inſects to prey upon their bodies. Such is however the condition of mortality, that the firſt law of nature is, "Eat or be eaten." We cannot long exiſt without the deſtruction of other animal or vegetable beings, either in their mature or their embryon ſtate. Unleſs the fruits, which ſurround the ſeeds of ſome vegetables, or the honey ſtolen

ſtolen from them by the bee, may be ſaid to be an exception to this aſſertion. See Botannic Garden, P. I. Cant. I. l. 278. Note. Hence, from the neceſſity of our nature, we may be ſuppoſed to have a right to kill thoſe creatures, which we want to eat, or which want to eat us. But to deſtroy even inſects wantonly ſhews an unreflecting mind or a depraved heart.

Nevertheleſs mankind may be well divided into the ſelfiſh and the ſocial; that is, into thoſe whoſe pleaſures ariſe from gratifying their appetites, and thoſe whoſe pleaſures ariſe from their ſympathizing with others. And according to the prevalence of theſe oppoſing propenſities we value or diſlike the poſſeſſor of them.

In conducting the education of young people, it is a nice matter to inſpire them with ſo much benevolent ſympathy, or compaſſion, as may render them good and amiable; and yet not ſo much as to make them unhappy at the ſight of incurable diſtreſs. We ſhould endeavour to make them alive to ſympathize with all remediable evils, and at the ſame time to arm them with fortitude to bear the ſight of ſuch irremediable evils, as the accidents of life muſt frequently preſent before their eyes. About this I have treated more at large in a plan for the conduct of a boarding ſchool for ladies, which I intend to publiſh in the courſe of the next year.

25. *Educatio heroica.* From the kinds and degrees of inſanities already enumerated, the reader will probably recollect many more from his own obſervation; he will perceive that all extraordinary exertions of voluntary action in conſequence of ſome falſe idea or hallucination, which ſtrongly affects us, may philoſophically, though not popularly, be termed an inſanity; he will then be liable to divide theſe voluntary exertions into diſagreeable, pernicious, deteſtable, or into meritorious, delectable, and even amiable, inſanities. And will laſtly be induced to conceive, that a good education conſiſts in the art of producing ſuch happy hallucinations of ideas, as may be followed by ſuch vo-

luntary exertions, as may be termed meritorious or amiable insanities.

The old man of the mountain in Syria, who governed a small nation of people called Assassines, is recorded thus to have educated those of his army who were designed to assassinate the princes with whom he was at war. A young man of natural activity was chosen for the purpose, and thrown into a deep sleep by opium mixed with his food; he was then carried into a garden made to represent the paradise of Mahomet, with flowers of great beauty and fragrance, fruits of delicious flavor, and beautiful houries beckoning him into the shades. After a while, on being a second time stupified with opium, the young enthusiast was reconveyed to his apartment; and on the next day was assured by a priest, that he was designed for some great exploit, and that by obeying the commands of their prince, immortal happiness awaited him.

Hence it is easy to collect how the first impressions made on us by accidental circumstances in our infancy continue through life to bias our affections, or mislead our judgments. One of my acquaintance can trace the origin of his own energies of action from some such remote sources; which justifies the observation of M. Rousseau, that the seeds of future virtues or vices are oftener sown by the mother, than the tutor.

## ORDO II.

*Decreased Volition.*

## GENUS I.

*With decreased Actions of the Muscles.*

OUR muscles become fatigued by long contraction, and cease for a time to be excitable by the will; owing to exhaustion of the sensorial power, which resides in them. After a short interval of relaxation the muscle regains its power of voluntary contraction; which is probably occasioned by a new supply of the spirit of animation. In weaker people these contractions cease sooner, and therefore recur more frequently, and are attended with shorter intervals of relaxation, as exemplified in the quickness of the pulse in fevers with debility, and in the tremors of the hands of aged or feeble people.

After a common degree of exhaustion of the sensorial power in a muscle, it becomes again gradually restored by the rest of the muscle; and even accumulated in those muscles, which are most frequently used; as in those which constitute the capillaries of the skin after having been rendered torpid by cold. But in those muscles, which are generally obedient to volition, as those of locomotion, though their usual quantity of sensorial power is restored by their quiescence, or in sleep (for sleep affects these parts of the system only), yet but little accumulation of it succeeds. And this want of accumulation of the sensorial power in these muscles, which are chiefly subservient to volition, explains to us one cause of their greater tendency to paralytic affection.

It muſt be obſerved, that thoſe parts of the ſyſtem, which have been for a time quieſcent from want of ſtimulus, as the veſſels of the ſkin, when expoſed to cold, acquire an accumulation of ſenſorial power during their inactivity; but this does not happen at all, or in much leſs quantity, from their quieſcence after great expenditure of ſenſorial power by a previous exceſſive ſtimulus, as after intoxication. In this caſe the muſcles or organs of ſenſe gradually acquire their natural quantity of ſenſorial power, as after ſleep; but not an accumulation or ſuperabundance of it. And by frequent repetitions of exhauſtion by great ſtimulus, theſe veſſels ceaſe to acquire their whole natural quantity of ſenſorial power; as in the ſchirrous ſtomach, and ſchirrous liver, occaſioned by the great and frequent ſtimulus of vinous ſpirit; which may properly be termed irritative paralyſis of thoſe parts of the ſyſtem.

In the ſame manner in common palſies the inaction of the paralytic muſcle ſeems not to be owing to defect of the ſtimulus of the will, but to exhauſtion of ſenſorial power. Whence it frequently follows great exertion, as in Sect. XXXIV. 1. 7. Thus ſome parts of the ſyſtem may ceaſe to obey the will, as in common paralyſis; others may ceaſe to be obedient to ſenſation, as in the impotency of age; others to irritation, as in ſchirrous viſcera; and others to aſſociation, as in impediment of ſpeech; yet though all theſe may become inexcitable, or dead, in reſpect to that kind of ſtimulus, which has previouſly exhauſted them, whether of volition, or ſenſation, or irritation, or aſſociation, they may ſtill in many caſes be excited by the others.

SPECIES.

## SPECIES.

1. *Laſſitudo.* Fatigue or wearineſs after much voluntary exertion. From the too great expenditure of ſenſorial power the muſcles are with difficulty brought again into voluntary contraction; and ſeem to require a greater quantity or energy of volition for this purpoſe. At the ſame time they ſtill remain obedient to the ſtimulus of agreeable ſenſation, as appears in tired dancers finding a renovation of their aptitude to motion on the acquiſition of an agreeable partner; or from a tired child riding on a gold-headed cane, as in Sect. XXXIV. 2. 6. Theſe muſcles are likewiſe ſtill obedient to the ſenſorial power of aſſociation, becauſe the motions, when thus excited, are performed in their deſigned directions, and are not broken into variety of geſticulation, as in St. Vitus's dance.

A laſſitude likewiſe frequently occurs with yawning at the beginning of ague-fits; where the production of ſenſorial power in the brain is leſs than its expenditure. For in this caſe the torpor may either originate in the brain, or the torpor of ſome diſtant parts of the ſyſtem may by ſympathy affect the brain, though in a leſs proportionate degree than the parts primarily affected.

2. *Vacillatio ſenilis.* Some elderly people acquire a ſee-ſaw motion of their bodies from one ſide to the other, as they ſit, like the oſcillation of a pendulum. By theſe motions the muſcles, which preſerve the perpendicularity of the body, are alternately quieſcent, and exerted; and are thus leſs liable to fatigue or exhauſtion. This therefore reſembles the tremors of old people above mentioned, and not thoſe ſpaſmodic movements of the face or limbs, which are called tricks, deſcribed in Claſs IV. 3. 2. 2. which originate from exceſs of ſenſorial

ſenſorial power, or from efforts to relieve diſagreeable ſenſation, and are afterwards continued by habit.

3. *Tremor ſenilis.* Tremor of old age conſiſts of a perpetual trembling of the hands, or of the head, or of other muſcles, when they are exerted; and is erroneouſly called paralytic; and ſeems owing to the ſmall quantity of animal power reſiding in the muſcular fibres. Theſe tremors only exiſt when the affected muſcles are excited into action, as in lifting a glaſs to the mouth, or in writing, or in keeping the body upright; and ceaſe again, when no voluntary exertion is attempted, as in lying down. Hence theſe tremors evidently originate from the too quick exhauſtion of the leſſened quantity of the ſpirit of animation. So many people tremble from fear or anger, when too great a part of the ſenſorial power is exerted on the organs of ſenſe, ſo as to deprive the muſcles, which ſupport the body erect, of their due quantity.

4. *Brachiorum paralyſis.* A numbneſs of the arms is a frequent ſymptom in hydrops thoracis, as explained in Claſs I. 2. 3. 14. and in Sect. XXIX. 5. 2.; it alſo accompanies the aſthma dolorificum, Claſs III. 1. 1. 11. and is owing probably to the ſame cauſe in both. In the colica ſaturnina a paralyſis affects the wriſts, as appears on the patient extending his arm horizontally with the palm downwards, and is often attended with a tumor on the carpal or metacarpal bones. See Claſs IV. 2. 2. 10.

Mr. M——, a miner and well-ſinker, about three years ago, loſt the power of contracting both his thumbs; the balls or muſcles of the thumbs are much emaciated, and remain paralytic. He aſcribes his diſeaſe to immerſing his hands too long in cold water in the execution of his buſineſs. He ſays his hands had frequently been much benumbed before, ſo that he could not without difficulty clench them;

but

but that they recovered their motion, as ſoon as they began to glow, after he had dried and covered them.

In this caſe there exiſted two injurious circumſtances of different kinds; one the violent and continued action of the muſcles, which deſtroys by exhauſting the ſenſorial power; and the other, the application of cold, which deſtroys by defect of ſtimulus. The cold ſeems to have contributed to the paralyſis by its long application, as well as the continued exertion; but as during the torpor occaſioned by the expoſure to cold, if the degree of it be not ſo great as to extinguiſh life, the ſenſorial power becomes accumulated; there is reaſon to believe, that the expoſing a paralytic limb to the cold for a certain time, as by covering it with ſnow or iced water for a few minutes, and then covering it with warm flannel, and this frequently repeated, might, by accumulation of ſenſorial power, contribute to reſtore it to a ſtate of voluntary excitability. As this accumulation of ſenſorial power, and conſequent glow, ſeems, in the preſent caſe, ſeveral times to have contributed to reſtore the numbneſs or inability of thoſe muſcles, which at length became paralytic. See Claſs I. 2. 3. 21.

M. M. Ether externally. Friction. Saline warm bath. Electricity.

5. *Raucedo paralytica.* Paralytic hoarſeneſs conſiſts in the almoſt total loſs of voice, which ſometimes continues for months, or even years, and is occaſioned by inability or paralyſis of the recurrent nerves, which ſerve the muſcles of vocality, by opening or cloſing the larynx. The voice generally returns ſuddenly, even ſo as to alarm the patient. A young lady, who had many months been affected with almoſt a total loſs of voice, and had in vain tried variety of advice, recovered her voice in an inſtant, on ſome alarm as ſhe was dancing at an aſſembly. Was this owing to a greater exertion of voli-

 tion

tion than usual? like the dumb young man, the son of Crœsus, who is related to have cried out, when he saw his father's life endangered by the sword of his enemy, and to have continued to speak ever afterwards. Two young ladies in this complaint seemed to be cured by electric shocks passed through the larynx every day for a fortnight. See Raucedo catarrhalis, Class II. 1. 3. 5.

M. M. An emetic. Electric shocks. Mustard-seed, a large spoonful swallowed whole, or a little bruised, every morning. Valerian. Burnt sponge. Blisters on each side of the larynx. Sea-bathing. A gargle of decoction of seneca. Friction. Frequent endeavours to shout and sing.

6. *Vesicæ urinariæ paralysis.* Paralysis of the bladder is frequently a symptom in inirritative fever; in this case the patient makes no water for a day or two; and the tumor of the bladder distended with urine may be seen by the shape of the abdomen, as if girt by a cord below the navel, or distinguished by the hand. Many patients in this situation make no complaint, and suffer great injury by the inattention of their attendants; the water must be drawn off once or twice a day by means of a catheter, and the region of the bladder gently pressed by the hand, whilst the patient be kept in a sitting or erect posture.

M. M. Bark. Wine. Opium, a quarter of a grain every six hours. Balsam of copaiva or of Peru. Tincture of cantharides 20 drops twice a day, or repeated small blisters.

7. *Recti paralysis.* Palsy of the rectum. The rectum intestinum, like the urinary bladder in the preceding article, possesses voluntary power of motion; though these volitions are at times uncontrollable by the will, when the acrimony of the contained feces, or their bulk, stimulate it to a greater degree. Hence it happens, that this part is liable

liable to loſe its voluntary power by paralyſis, but is ſtill liable to be ſtimulated into action by the contained feces. This frequently occurs in fevers, and is a bad ſign as a ſymptom of general debility; and it is the ſenſibility of the muſcular fibres of this and of the urinary bladder remaining, after the voluntarity has ceaſed, which occaſions theſe two reſervoirs ſo ſoon to regain, as the fever ceaſes, their obedience to volition; becauſe the paralyſis is thus ſhewn to be leſs complete in thoſe caſes than in common hemiplegia; as in the latter the ſenſe of touch, though perhaps not the ſenſe of pain, is generally deſtroyed in the paralytic limb.

M. M. A ſponge introduced within the ſphincter ani to prevent the conſtant diſcharge, which ſhould have a ſtring put through it, by which it may be retracted.

8. *Pareſis voluntaria.* Indolence; or inaptitude to voluntary action. This debility of the exertion of voluntary efforts prevents the accompliſhment of all great events in life. It often originates from a miſtaken education, in which pleaſure or flattery is made the immediate motive of action, and not future advantage; or what is termed duty. This obſervation is of great value to thoſe, who attend to the education of their own children. I have ſeen one or two young married ladies of fortune, who perpetually became uneaſy, and believed themſelves ill, a week after their arrival in the country, and continued ſo uniformly during their ſtay; yet on their return to London or Bath immediately loſt all their complaints, and this repeatedly; which I was led to aſcribe to their being in their infancy ſurrounded with menial attendants, who had flattered them into the exertions they then uſed. And that in their riper years, they became torpid for want of this ſtimulus, and could not amuſe themſelves by any voluntary employment; but required ever after, either to be amuſed by other people, or to be flattered into activity. This I ſuppoſe, in the other ſex, to have ſupplied one ſource of ennui and ſuicide.

9. *Catalepſis* is ſometimes uſed for fixed ſpaſmodic contractions or tetanus, as deſcribed in Sect. XXXIV. 1. 5. and in Claſs III. 1. 1. 13. but is properly ſimply an inaptitude to muſcular motion, the limbs remaining in any attitude in which they are placed. One patient, whom I ſaw in this ſituation, had taken much mercury, and appeared univerſally torpid. He ſat in a chair in any poſture he was put, and held a glaſs to his mouth for many minutes without attempting to drink, or withdrawing his hand. He never ſpoke, and it was at firſt neceſſary to compel him to drink broth; he recovered in a few weeks without relapſe.

10. *Hemiplegia.* Palſy of one ſide conſiſts in the total diſobedience of the affected muſcles to the power of volition. As the voluntary motions are not perpetually exerted, there is little ſenſorial power accumulated during their quieſcence, whence they are leſs liable to recover from torpor, and are thus more frequently left paralytic, or diſobedient to the power of volition, though they are ſometimes ſtill alive to painful ſenſation, as to the prick of a pin, and to heat; alſo to irritation, as in ſtretching and yawning; or to electric ſhocks. Where the paralyſis is complete the patient ſeems gradually to learn to uſe his limbs over again by repeated efforts, as in infancy; and, as time is required for this purpoſe, it becomes difficult to know, whether the cure is owing to the effect of medicines, or to the repeated efforts of the voluntary power.

The diſpute, whether the nerves decuſſate or croſs each other before they leave the cavities of the ſkull or ſpine, ſeems to be decided in the affirmative by comparative anatomy; as the optic nerves of ſome fiſh have been ſhewn evidently to croſs each other; as ſeen by Haller, Elem. Phyſiol. t. v. p. 349. Hence the application of bliſters, or of ether, or of warm fomentations, ſhould be on the ſide of the head oppoſite to that of the affected muſcles. This ſubject ſhould neverthele ſs be nicely determined, before any one ſhould trepan for the

the hydrocephalus internus, when the diseafe is fhewn to exift only on one fide of the brain, by a fquinting affecting but one eye; as propofed in Clafs I. 2. 5. 4. Dr. Sommering has fhewn, that a true decuffation of the optic nerves in the human fubject actually exifts, Elem. of Phyfiology by Blumenbach, tranflated by C. Caldwell, Philadelphia. This further appears probable from the oblique direction and infertion of each optic nerve, into the fide of the eye next to the nofe, in a direct line from the oppofite fide of the brain.

The vomiting, which generally attends the attack of hemiplegia, is mentioned in Sect. XX. 8. and is fimilar to that attending vertigo in fea-ficknefs, and at the commencement of fome fevers. Black ftools fometimes attend the commencement of hemiplegia, which is probably an effufion of blood from the biliary duct, where the liver is previoufly affected; or fome blood may be derived to the inteftines by its efcaping from the vena cava into the receptacle of chyle during the diftrefs of the paralytic attack; and may be conveyed from thence into the inteftines by the retrograde motions of the lacteals; as probably fometimes happens in diabætes. See Sect. XXVII. 2. Palfy of one fide of the face is mentioned in Clafs II. 1. 4. 6. Paralyfis of the lacteals, of the liver, and of the veins, which are defcribed in Sect. XXVIII. XXX. and XXVII. do not belong to this clafs, as they are not difeafes of voluntary motions.

M. M. The electric fparks and fhocks, if ufed early in the difeafe, are frequently of fervice. A purge of aloes, or calomel. A vomit. Blifter. Saline draughts. Then the bark. Mercurial ointment or fublimate, where the liver is evidently difeafed; or where the gutta rofea has previoufly exifted. Sudden alarm. Frequent voluntary efforts. Externally ether. Volatile alcali. Fomentation on the head. Friction. When children, who have fuffered an hemiplegia, begin to ufe the affected arm, the other hand fhould be tied up for half an hour three or four times a day; which obliges them at their play to uſe more frequent voluntary efforts with the diſeaſed limb, and thus ſooner to reſtore the diſſevered aſſociations of motion.

Dr. J. Alderſon has lately much recommended the leaves of rhus toxicodendon (ſumach), from one gr. to iv. of the dried powder to be taken three or four times a day. Eſſay on Rhus Toxic. Johnſon, London, 1793. But it is difficult to know what medicine is of ſervice, as the movements of the muſcles muſt be learned, as in infancy, by frequent efforts.

11. *Paraplegia.* A palſy of the lower half of the body divided horizontally. Animals may be conceived to have double bodies, one half in general reſembling ſo exactly the other, and being ſupplied with ſeparate ſets of nerves; this gives riſe to hemiplegia, or palſy of one half of the body divided vertically; but the paraplegia, or palſy of the lower parts of the ſyſtem, depends on an injury of the ſpinal marrow, or that part of the brain which is contained in the vertebræ of the back; by which all the nerves ſituated below the injured part are deprived of their nutriment, or precluded from doing their proper offices; and the muſcles, to which they are derived, are in conſequence diſobedient to the power of volition.

This ſometimes occurs from an external injury, as a fall from an eminence; of which I ſaw a deplorable inſtance, where the bladder and rectum, as well as the lower limbs, were deprived of ſo much of their powers of motion, as depended on volition or ſenſation; but I ſuppoſe not of that part of it, which depends on irritation. In the ſame manner as the voluntary muſcles in hemiplegia are ſometimes brought into action by irritation, as in ſtretching or pendiculation, deſcribed in Sect. VII. 1. 3.

But the moſt frequent cauſe of paraplegia is from a protuberance of one of the ſpinal vertebræ; which is owing to the innutrition or ſoftneſs of bones, deſcribed in Claſs I. 2. 2. 17. The cure of this deplorable diſeaſe is frequently effected by the ſtimulus of an iſſue placed

on

on each ſide of the prominent ſpine, as firſt publiſhed by Mr. Pott. The other means recommended in ſoftneſs of bones ſhould alſo be attended to; both in reſpect to the internal medicines, and to the mechanical methods of ſupporting, or extending the ſpine; which laſt, however, in this caſe requires particular caution.

12. *Somnus.* In ſleep all voluntary power is ſuſpended, ſee Sect. XVIII. An unuſual quantity of ſleep is often produced by weakneſs. In this caſe ſmall doſes of opium, wine, and bark, may be given with advantage. For the periods of ſleep, ſee Claſs IV. 2. 4. 1.

The ſubſequent ingenious obſervations on the frequency of the pulſe, which ſometimes occurs in ſleep, are copied from a letter of Dr. Currie of Liverpool to the author.

" Though reſt in general perhaps renders the healthy pulſe ſlower, yet under certain circumſtances the contrary is the truth. A full meal without wine or other ſtrong liquor does not increaſe the frequency of my pulſe, while I ſit upright, and have my attention engaged. But if I take a recumbent poſture after eating, my pulſe becomes more frequent, eſpecially if my mind be vacant, and I become drowſy; and, if I ſlumber, this increaſed frequency is more conſiderable with heat and fluſhing.

" This I apprehend to be a general truth. The obſervation may be frequently made upon children; and the reſtleſs and feveriſh nights experienced by many people after a full ſupper are, I believe, owing to this cauſe. The ſupper occaſions no inconvenience, whilſt the perſon is upright and awake; but, when he lies down and begins to ſleep, eſpecially if he does not perſpire, the ſymptoms above mentioned occur. Which may be thus explained in part from your principles. When the power of volition is aboliſhed, the other ſenſorial actions are increaſed. In ordinary ſleep this does not occaſion increaſed frequency of the pulſe; but where ſleep takes place during the proceſs of digeſtion, the digeſtion itſelf goes on with increaſed rapidity.

pidity. Heat is excited in the ſyſtem faſter than it is expended; and operating on the ſenſitive actions, it carries them beyond the limitation of pleaſure, producing, as is common in ſuch caſes, increaſed frequency of pulſe.

"It is to be obſerved, that in ſpeaking of the heat generated under theſe circumſtances, I do not allude to any chemical evolution of heat from the food in the proceſs of digeſtion. I doubt if this takes place to any conſiderable degree, for I do not obſerve that the parts incumbent on the ſtomach are increaſed in heat during the moſt hurried digeſtion. It is on ſome parts of the ſurface, but more particularly on the extremities of the body, that the increaſed heat excited by digeſtion appears, and the heat thus produced ariſes, as it ſhould ſeem, from the ſympathy between the ſtomach and the veſſels of the ſkin. The parts moſt affected are the palms of the hands and the ſoles of the feet. Even there the thermometer ſeldom riſes above 97 or 98 degrees, a temperature not higher than that of the trunk of the body; but three or four degrees higher than the common temperature of theſe parts, and therefore producing an uneaſy ſenſation of heat, a ſenſation increaſed by the great ſenſibility of the parts affected.

"That the increaſed heat excited by digeſtion in ſleep is the cauſe of the accompanying fever, ſeems to be confirmed by obſerving, that if an increaſed expenditure of heat accompanies the increaſed generation of it (as when perſpiration on the extremities or ſurface attends this kind of ſleep) the frequent pulſe and fluſhed countenance do not occur, as I know by experiment. If, during the feveriſh ſleep already mentioned, I am awakened, and my attention engaged powerfully, my pulſe becomes almoſt immediately ſlower, and the fever gradually ſubſides."

From theſe obſervations of Dr. Currie it appears, that, while in common ſleep the actions of the heart, arteries, and capillaries, are ſtrengthened by the accumulation of ſenſorial power during the ſuſpenſion of voluntary action, and the pulſe in conſequence becomes fuller and ſlower ; in the feveriſh ſleep above deſcribed the actions of the heart, arteries, and capillaries, are quickened as well as ſtrengthened by their conſent with the increaſed actions of the ſtomach, as well as by the ſtimulus of the new chyle introduced into the circulation. For the ſtomach, and all other parts of the ſyſtem, being more ſenſible and more irritable during ſleep, Sect. XVIII. 15. and probably more ready to act from aſſociation, are now exerted with greater velocity as well as ſtrength, conſtituting a temporary fever of the ſenſitive irritated kind, reſembling the fever excited by wine in the beginning of intoxication; or in ſome people by a full meal in their waking hours. Sect. XXXV. 1.

On waking, this increaſed ſenſibility and irritability of the ſyſtem ceaſes by the renewed exertions of volition ; in the ſame manner as more violent exertions of volition deſtroy greater pains ; and the pulſe in conſequence ſubſides along with the increaſe of heat ; if more violent efforts of volition are exerted, the ſyſtem becomes ſtill leſs affected by ſenſation or irritation. Hence the fever and vertigo of intoxication are leſſened by intenſe thinking, Sect. XXI. 8 ; and inſane people are known to bear the pain of cold and hunger better than others, Sect. XXXIV. 2. 5 ; and laſtly, if greater voluntary efforts exiſt, as in violent anger or violent exerciſe, the whole ſyſtem is thrown into more energetic action, and a voluntary fever is induced, as appears by the red ſkin, quickened pulſe, and increaſe of heat ; whence dropſies and fevers with debility are not unfrequently removed by inſanity.

Hence the exertion of the voluntary power in its natural degree diminiſhes the increaſed ſenſibility, and irritability, and probably the increaſed aſſociability, which occurs during ſleep ; and thus reduces the frequency of the pulſe in the feveriſh ſleep after a full meal. In its more powerful ſtate of exertion, it diminiſhes or deſtroys ſenſations and irritations, which are ſtronger than natural, as in intoxica-

tion,

tion, or which precede convulſions, or inſanity. In its ſtill more powerful degree, the ſuperabundance of this ſenſorial power actuates and invigorates the whole moving ſyſtem, giving ſtrength and frequency to the pulſe, and an univerſal glow both of colour and of heat, as in violent anger, or outrageous inſanities.

If, in the feveriſh ſleep above deſcribed, the ſkin becomes cooled by the evaporation of much perſpirable matter, or by the application of cooler air, or thinner clothes, the actions of the cutaneous capillaries are leſſened by defect of the ſtimulus of heat, which counteracts the increaſe of ſenſibility during ſleep, and the pulſations of the heart and arteries become ſlower from the leſſened ſtimulus of the particles of blood thus cooled in the cutaneous and pulmonary veſſels. Hence the admiſſion of cold air, or ablution with ſubtepid or with cold water, in fevers with hot ſkin, whether they be attended with arterial ſtrength, or arterial debility, renders the pulſe ſlower; in the former caſe by diminiſhing the ſtimulus of the blood, and in the latter by leſſening the expenditure of ſenſorial power. See Suppl. I. 8. and 15.

13. *Incubus.* The night-mare is an imperfect ſleep, where the deſire of locomotion is vehement, but the muſcles do not obey the will; it is attended with great uneaſineſs, a ſenſe of ſuffocation, and frequently with fear. It is cauſed by violent fatigue, or drunkenneſs, or indigeſtible food, or lying on the back, or perhaps from many other kinds of uneaſineſs in our ſleep, which may originate either from the body or mind.

Now as the action of reſpiration is partly voluntary, this complaint may be owing to the irritability of the ſyſtem being too ſmall to carry on the circulation of the blood through the lungs during ſleep, when the voluntary power is ſuſpended. Whence the blood may accumulate in them, and a painful oppreſſion ſupervene; as in ſome hæmorrhages

rhages of the lungs, which occur during ſleep; and in patients much debilitated by fevers. See Somnus interruptus, Claſs I. 2. 1. 3. and I. 2. 1. 9.

Great fatigue with a full ſupper and much wine, I have been well informed by one patient, always produced this diſeaſe in himſelf to a great degree. Now the general irritability of the ſyſtem is much decreaſed by fatigue, as it exhauſts the ſenſorial power; and ſecondly, too much wine and ſtimulating food will again diminiſh the irritability of ſome parts of the ſyſtem, by employing a part of the ſenſorial power, which is already too ſmall, in digeſting a great quantity of aliment; and in increaſing the motions of the organs of ſenſe in conſequence of ſome degree of intoxication, whence difficulty of breathing may occur from the inirritability of the lungs, as in Claſs I. 2. 1. 3.

M. M. To ſleep on a hard bed with the head raiſed. Moderate ſupper. The bark. By ſleeping on a harder bed the patient will turn himſelf more frequently, and not be liable to ſleep too profoundly, or lie too long in one poſture. To be awakened frequently by an alarm clock.

14. *Lethargus.* The lethargy is a ſlighter apoplexy. It is ſuppoſed to originate from univerſal preſſure on the brain, and is ſaid to be produced by compreſſing the ſpinal marrow, where there is a deficiency of the bone in the ſpina bifida. See Sect. XVIII. 20. Whereas in the hydrocephalus there is only a partial preſſure of the brain; and probably in nervous fevers with ſtupor the preſſure on the brain may affect only the nerves of the ſenſes, which lie within the ſkull, and not thoſe nerves of the medulla oblongata, which principally contribute to move the heart and arteries; whence in the lethargic or apoplectic ſtupor the pulſe is ſlow as in ſleep, whereas in nervous fever the pulſe is very quick and feeble, and generally ſo in hydrocephalus.

In cafes of obftructed kidneys, whether owing to the tubuli uriniferi being totally obftructed by calculous matter, or by their paralyfis, a kind of drowfinefs or lethargy comes on about the eighth or ninth day, and the patient gradually finks. See Clafs I. 1. 3. 9.

15. *Syncope epileptica*, is a temporary apoplexy, the pulfe continuing in its natural ftate, and the voluntary power fufpended. This terminates the paroxyfms of epilepfy.

When the animal power is much exhaufted by the preceding convulfions, fo that the motions from fenfation as well as thofe from volition are fufpended; in a quarter or half an hour the fenforial power becomes reftored, and if no pain, or irritation producing pain, recurs, the fit of epilepfy ceafes; if the pain recurs, or the irritation, which ufed to produce it, a new fit of convulfion takes place, and is fucceeded again by a fyncope. See Epilepfy, Clafs III. 1. 1. 7.

16. *Apoplexia*. Apoplexy may be termed an univerfal palfy, or a permanent fleep. In which, where the pulfe is weak, copious bleeding muft be injurious; as is well obferved by Dr. Heberden, Tranf. of the College.

Mr. ———, about 70 years of age, had an apoplectic feizure. His pulfe was ftrong and full. One of the temporal arteries was opened, and about ten ounces of blood fuddenly taken from it. He feemed to receive no benefit from this operation; but gradually funk, and lived but a day or two.

If apoplexy arifes from the preffure of blood extravafated on the brain, one moderate venefection may be of fervice to prevent the further effufion of blood; but copious venefection muft be injurious by weakening the patient; fince the effufed blood muft have time, as in common vibices or bruifes, to undergo a chemico-animal procefs, fo to change its nature as to fit it for abforption; which may take

 two

two or three weeks, which time a patient weakened by repeated venesection or arteriotomy may not survive.

Mrs. ———, about 40 years old, had an apoplectic seizure after great exertion from fear; she had lain about 24 hours without speech, or having swallowed any liquid. She was then forcibly raised in bed, and a spoonful of solution of aloes in wine put into her mouth, and the end of the spoon withdrawn, that she might more easily swallow the liquid.—This was done every hour, with broth, and wine and water intervening, till evacuations were procured; which with other means had good effect, and she recovered, except that a considerable degree of hemiplegia remained, and some imperfection of her speech.

Many people, who have taken so much vinous spirit as to acquire the temporary apoplexy of intoxication, and are not improperly said to be dead-drunk, have died after copious venesection, I suppose in consequence of it. I once saw at a public meeting two gentlemen in the drunken apoplexy; they were totally insensible with low pulse, on this account they were directed not to lose blood, but to be laid on a bed with their heads high, and to be turned every half hour; as soon as they could swallow, warm tea was given them, which evacuated their stomachs, and they gradually recovered, as people do from less degrees of intoxication.

M. M. Cupping on the occiput. Venesection once in moderate quantity. Warm fomentations long continued and frequently repeated on the shaved head. Solution of aloes. Clysters with solution of aloe and oil of amber. A blister on the spine. An emetic. Afterwards the bark, and small doses of chalybeates. Small electric shocks through the head. Errhines. If small doses of opium?

17. *Mors a frigore.* Death from cold. The unfortunate travellers, who almost every winter perish in the snow, are much exhausted by their efforts to proceed on their journey, as well as benumbed by cold.

 And

And as much greater exercise can be borne without fatigue in cold weather than in warm; because the excessive motions of the cutaneous vessels are thus prevented, and the consequent waste of sensorial power; it may be inferred, that the fatigued traveller becomes paralytic from violent exertion as well as by the application of cold.

Great degrees of cold affect the motions of those vessels most, which have been generally excited into action by irritation; for when the feet are much benumbed by cold, and painful, and at the same time almost insensible to the touch of external objects, the voluntary muscles retain their motions, and we continue to walk on; the same happens to the fingers of children in throwing snow-balls, the voluntary motions of the muscles continue, though those of the cutaneous vessels are benumbed into inactivity.

Mr. Thompson, an elderly gentleman of Shrewsbury, was seized with hemiplegia in the cold bath; which I suppose might be owing to some great energy of exertion, as much as to the coldness of the water. As in the instance given of Mr. Nairn, who, by the exertion to save his relation, perished himself. See Sect. XXXIV. 1. 7.

Whence I conclude, that though heat is a fluid necessary to muscular motion, both perhaps by its stimulus, and by its keeping the minute component parts of the ultimate fibrils of the muscles or organs of sense at a proper distance from each other; yet that paralysis, properly so called, is the consequence of exhaustion of sensorial power by exertion. And that the accumulations of it during the torpor of the cutaneous vessels by exposure to cold, or of some internal viscus in the cold fits of agues, are frequently instrumental in recovering the use of paralytic limbs, or of the motions of other paralytic parts of the system. See Spec. 4. of this genus.

Animal bodies resist the power of cold probably by their exertions in consequence of the pain of cold, see Botan. Gard. V. 1. additional note xii. But if these increased exertions be too violent, so as to exhaust the sensorial power in producing unnecessary motions, the animal

mal will probably ſooner periſh. Thus a moderate quantity of wine or ſpirit repeated at proper intervals of time might be of ſervice to thoſe, who are long expoſed to exceſſive cold, both by increaſing the action of the capillary veſſels, and thus producing heat, and perhaps by increaſing in ſome degree the ſecretion of ſenſorial power in the brain. But the contrary muſt happen when taken immoderately, and not at due intervals. A well atteſted hiſtory was once related to me of two men, who ſet out on foot to travel in the ſnow, one of whom drank two or three glaſſes of brandy before they began their journey, the other contented himſelf with his uſual diet and potation; the former of whom periſhed in ſpite of any aſſiſtance his companion could afford him; and the other performed his journey with ſafety. In this caſe the ſenſorial power was exhauſted by the unneceſſary motions of incipient intoxication by the ſtimulus of the brandy, as well as by the exertions of walking; which ſo weakened the dram-drinker, that the cold ſooner deſtroyed him; that is, he had not power to produce ſufficient muſcular or arterial action, and in conſequence ſufficient heat, to ſupply the great expenditure of it. Hence the capillaries of the ſkin firſt ceaſe to act, and become pale and empty; next thoſe which are immediately aſſociated with them, as the extremities of the pulmonary artery, as happens on going into the cold bath. By the continued inaction of theſe parts of the vaſcular ſyſtem the blood becomes accumulated in the internal arteries, and the brain is ſuppoſed to be affected by its compreſſion; becauſe theſe patients are ſaid to ſleep, or to become apoplectic, before they die. I overtook a fiſhman aſleep on his panniers on a very cold froſty night, but on waking him he did not appear to be in any degree of ſtupor. See Claſs I. 2. 2. 1.

When travellers are benighted in deep ſnow, they might frequently be ſaved by covering themſelves in it, except a ſmall aperture for air; in which ſituation the lives of hares, ſheep, and other animals, are ſo often preſerved. The ſnow, both in reſpect to its component parts, and to the air contained in its pores, is a bad conductor of heat,

and

and will therefore well keep out the external cold; and as the water, when part of it diſſolves, is attracted into the pores of the remainder of it, the ſituation of an animal beneath it is perfectly dry; and, if he is in contact with the earth, he is in a degree of heat between 48, the medium heat of the earth, and 32, the freezing point; that is, in 40 degrees of heat, in which a man thus covered will be as warm as in bed. See Botan. Garden, V. II. notes on Anemone, Barometz, and Muſchus. If theſe facts were more generally underſtood, it might annually ſave the lives of many.

After any part of the vaſcular ſyſtem of the body has been long expoſed to cold, the ſenſorial power is ſo much accumulated in it, that on coming into a warm room the pain of hotach is produced, and inflammation, and conſequent mortification, owing to the great exertion of thoſe veſſels, when again expoſed to a moderate degree of warmth. See Sect. XII. 5. Whence the propriety of applying but very low degrees of heat to limbs benumbed with cold at firſt, as of ſnow in its ſtate of diſſolving, which is at 32 degrees of heat, or of very cold water. A French writer has obſerved, that if frozen apples be thawed gradually by covering them with thawing ſnow, or immerſing them in very cold water, that they do not loſe their taſte; if this fact was well aſcertained, it might teach us how to preſerve other ripe fruits in ice-houſes for winter conſumption.

ORDO

## ORDO II.

*Decreaſed Volition.*

## GENUS II.

*With decreaſed Actions of the Organs of Senſe.*

### SPECIES.

1. *Recollectionis jactura.* Loſs of recollection. This is the defect of memory in old people, who forget the actions of yeſterday, being incapable of voluntary recollection, and yet remember thoſe of their youth, which by frequent repetition are introduced by aſſociation or ſuggeſtion. This is properly the paralyſis of the mind; the organs of ſenſe do not obey the voluntary power; that is, our ideas cannot be recollected, or acted over again by the will.

After an apoplectic attack the patients, on beginning to recover, find themſelves moſt at a loſs in recollecting proper names of perſons or places; as thoſe words have not been ſo frequently aſſociated with the ideas they ſtand for, as the common words of a language. Mr. ———, a man of ſtrong mind, of a ſhort necked family; many of whom had ſuffered by apoplexy, after an apoplectic fit on his recovering the uſe of ſpeech, after repeated trials to remember the name of a perſon or place, applauded himſelf, when he ſucceeded, with ſuch a childiſh ſmile on the partial return of his ſagacity, as very much affected me.—Not long, alas! to return; for another attack in a few weeks deſtroyed the whole.

I ſaw a child after the ſmall-pox, which was left in this ſituation; it was lively, active, and even vigorous; but ſhewed that kind of ſurpriſe,

ſurpriſe, which novelty excites, at every object it viewed; and that as often as it viewed it. I never heard the termination of the caſe.

2. *Stultitia voluntaria.* Voluntary folly. The abſence of voluntary power and conſequent incapacity to compare the ideas of preſent and future good. Brute animals may be ſaid to be in this ſituation, as they are in general excited into action only by their preſent painful or pleaſurable ſenſations. Hence though they are liable to ſurpriſe, when their paſſing trains of ideas are diſſevered by violent ſtimuli; yet are they not affected with wonder or aſtoniſhment at the novelty of objects; as they poſſeſs but in a very inferior degree, that voluntary power of comparing the preſent ideas with thoſe previouſly acquired, which diſtinguiſhes mankind; and is termed analogical reaſoning, when deliberatively exerted; and intuitive analogy, when uſed without our attention to it, and which always preſerves our hourly trains of ideas conſiſtent with truth and nature. See Sect. XVII. 3. 7.

3. *Credulitas.* Credulity. Life is ſhort, opportunities of knowledge rare; our ſenſes are fallacious, our reaſonings uncertain, mankind therefore ſtruggles with perpetual error from the cradle to the coffin. He is neceſſitated to correct experiment by analogy, and analogy by experiment; and not always to reſt ſatisfied in the belief of facts even with this two-fold teſtimony, till future opportunities, or the obſervations of others, concur in their ſupport.

Ignorance and credulity have ever been companions, and have miſled and enſlaved mankind; philoſophy has in all ages endeavoured to oppoſe their progreſs, and to looſen the ſhackles they had impoſed; philoſophers have on this account been called unbelievers: unbelievers of what? of the fictions of fancy, of witchcraft, hobgobblins, apparitions, vampires, fairies; of the influence of ſtars on human actions, miracles wrought by the bones of ſaints, the flights of ominous birds,

 the

the predictions from the bowels of dying animals, expounders of dreams, fortune-tellers, conjurors, modern prophets, necromancy, cheiromancy, animal magnetism, with endless variety of folly? These they have disbelieved and despised, but have ever bowed their hoary heads to Truth and Nature.

Mankind may be divided in respect to the facility of their belief or conviction into two classes; those, who are ready to assent to single facts from the evidence of their senses, or from the serious assertions of others; and those, who require analogy to corroborate or authenticate them.

Our first knowledge is acquired by our senses; but these are liable to deceive us, and we learn to detect these deceptions by comparing the ideas presented to us by one sense with those presented by another. Thus when we first view a cylinder, it appears to the eye as a flat surface with different shades on it, till we correct this idea by the sense of touch, and find its surface to be circular; that is, having some parts gradually receding further from the eye than others. So when a child, or a cat, or a bird, first sees its own image in a looking-glass, it believes that another animal exists before it, and detects this fallacy by going behind the glass to examine, if another tangible animal really exists there.

Another exuberant source of error consists in the false notions, which we receive in our early years from the design or ignorance of our instructors, which affect all our future reasoning by their perpetual intrusions; as those habits of muscular actions of the face or limbs, which are called tricks, when contracted in infancy continue to the end of our lives.

A third great source of error is the vivacity of our ideas of imagination, which perpetually intrude themselves by various associations, and compose the farrago of our dreams; in which, by the suspension of volition, we are precluded from comparing the ideas of one sense with those of another, or the incongruity of their successions with the usual course of nature, and thus to detect their fallacy. Which we do in our waking hours by a perpetual voluntary exertion, a process of the mind above mentioned, which we have termed intuitive analogy. Sect. XVII. 3. 7.

This analogy presupposes an acquired knowledge of things, hence children and ignorant people are the most credulous, as not possessing much knowledge of the usual course of nature; and secondly, those are most credulous, whose faculty of comparing ideas, or the voluntary exertion of it, is slow or imperfect. Thus if the power of the magnetic needle of turning towards the north, or the shock given by touching both sides of an electrized coated jar, was related for the first time to a philosopher, and to an ignorant person; the former would be less ready to believe them, than the latter; as he would find nothing similar in nature to compare them to, he would again and again repeat the experiment, before he would give it his entire credence; till by these repetitions it would cease to be a single fact, and would therefore gain the evidence of analogy. But the latter, as having less knowledge of nature, and less facility of voluntary exertion, would more readily believe the assertions of others, or a single fact, as presented to his own observation. Of this kind are the bulk of mankind; they continue throughout their lives in a state of childhood, and have thus been the dupes of priests and politicians in all countries and in all ages of the world.

In regard to religious matters, there is an intellectual cowardice instilled into the minds of the people from their infancy; which prevents their inquiry: credulity is made an indispensable virtue; to inquire or exert their reason in religious matters is denounced as sinful; and in the catholic church is punished with more severe penances than moral crimes. But in respect to our belief of the supposed medical facts, which are published by variety of authors; many of whom are ignorant, and therefore credulous; the golden rule of David Hume may be applied with great advantage. "When two miraculous assertions

ſertions oppoſe each other, believe the leſs miraculous." Thus if a perſon is ſaid to have received the ſmall-pox a ſecond time, and to have gone through all the ſtages of it, one may thus reaſon: twenty thouſand people have been expoſed to the variolous contagion a ſecond time without receiving the variolous fever, to every one who has been ſaid to have thus received it; it appears therefore leſs miraculous, that the aſſertor of this ſuppoſed fact has been deceived, or wiſhes to deceive, than that it has ſo happened contrary to the long experienced order of nature.

M. M. The method of cure is to increaſe our knowledge of the laws of nature, and our habit of comparing whatever ideas are preſented to us with thoſe known laws, and thus to counteract the fallacies of our ſenſes, to emancipate ourſelves from the falſe impreſſions which we have imbibed in our infancy, and to ſet the faculty of reaſon above that of imagination.

## *The Orders and Genera of the Fourth Claſs of Diſeaſes.*

# CLASS IV.

### DISEASES OF ASSOCIATION.

## ORDO I.

*Increaſed Aſſociate Motions.*

### GENERA.

1. Catenated with irritative motions.
2. Catenated with ſenſitive motions.
3. Catenated with voluntary motions.
4. Catenated with external influences.

## ORDO II.

*Decreaſed Aſſociate Motions.*

### GENERA.

1. Catenated with irritative motions.
2. Catenated with ſenſitive motions.
3. Catenated with voluntary motions.
4. Catenated with external influences.

## ORDO III.

*Retrograde Aſſociate Motions.*

### GENERA.

1. Catenated with irritative motions.
2. Catenated with ſenſitive motions.
3. Catenated with voluntary motions.
4. Catenated with external influences.

The

## *The Orders, Genera, and Species, of the Fourth Claſs of Diſeaſes.*

---

# CLASS IV.

DISEASES OF ASSOCIATION.

## ORDO I.

*Increaſed Aſſociate Motions.*

### GENUS I.

*Catenated with Irritative Motions.*

SPECIES.

| | |
|---|---|
| 1. *Rubor vultûs pranſorum.* | Fluſhing of the face after dinner. |
| 2. *Sudor ſtragulis immerſorum.* | Sweat from covering the face in bed. |
| 3. *Ceſſatio ægritudinis cute excitata.* | Cure of ſickneſs by ſtimulating the ſkin. |
| 4. *Digeſtio aucta frigore cutaneo.* | Digeſtion increaſed by coldneſs of the ſkin. |
| 5. *Catarrhus a frigore cutaneo.* | Catarrh from cold ſkin. |
| 6. *Abſorptio cellularis aucta vomitu.* | Cellular abſorption increaſed by vomiting. |
| 7. *Syngultus nephriticus.* | Nephritic hiccough. |
| 8. *Febris irritativa.* | Irritative fever. |

GENUS

## GENUS II.

*Catenated with Senſitive Motions.*

### SPECIES.

| | |
|---|---|
| 1. *Lacrymarum fluxus ſympatheticus.* | Sympathetic tears. |
| 2. *Sternutatio a lumine.* | Sneezing from light. |
| 3. *Dolor dentium a Stridore.* | Tooth-edge from grating ſounds. |
| 4. *Riſus ſardonicus.* | Sardonic ſmile. |
| 5. *Salivæ fluxus cibo viſo.* | Flux of ſaliva at ſight of food. |
| 6. *Tenſio mamularum viſo puerulo.* | Tenſion of the nipples of lacteſcent women at ſight of the child. |
| 7. *Tenſio penis in hydrophobia.* | Tenſion of the penis in hydrophobia. |
| 8. *Teneſmus calculoſus.* | Teneſmus from ſtone. |
| 9. *Polypus narium ex aſcaride.* | Polypus of the noſe from aſcarides. |
| 10. *Crampus ſurarum in diarrhœa.* | Cramp from diarrhœa. |
| 11. *Zona ignea nephritica.* | Nephritic ſhingles. |
| 12. *Eruptio variolarum.* | Eruption of ſmall-pox. |
| 13. *Gutta roſea ſtomatica.* | Stomatic roſy drop. |
| 14. ——— *hepatica.* | Hepatic roſy drop. |
| 15. *Podagra.* | Gout. |
| 16. *Rheumatiſmus.* | Rheumatiſm. |
| 17. *Eryſipelas.* | Eryſipelas. |
| 18. *Teſtium tumor in gonorrhœa.* | Swelled teſtis in gonorrhœa. |
| 19. ——— *in parotitide.* | ——— in mumps. |

GENUS

## GENUS III.

*Catenated with Voluntary Motions.*

### SPECIES.

| | |
|---|---|
| 1. *Deglutitio invita.* | Involuntary deglutition. |
| 2. *Nictitatio invita.* | ———— nictitation. |
| 3. *Risus invitus.* | ———— laughter. |
| 4. *Lusus digitorum invitus.* | ———— actions with the fingers. |
| 5. *Unguium morsiuncula invita.* | ———— biting the nails. |
| 6. *Vigilia invita.* | ———— watchfulness. |

## GENUS IV.

*Catenated with External Influences.*

### SPECIES.

| | |
|---|---|
| 1. *Vita ovi.* | Life of an egg. |
| 2. *Vita hiemi-dormientium.* | Life of winter-sleepers. |
| 3. *Pullulatio arborum.* | Budding of trees. |
| 4. *Orgasmatis venerei periodus.* | Periods of venereal desire. |
| 5. *Brachii concussio electrica.* | Electric shock through the arm. |
| 6. *Oxygenatio sanguinis.* | Oxygenation of the blood. |
| 7. *Humectatio corporis.* | Humectation of the body. |

ORDO

## ORDO II.

*Decreased Associate Motions.*

## GENUS I.

*Catenated with Irritative Motions.*

### SPECIES.

| | |
|---|---|
| 1. *Cutis frigida pransorum.* | Chillness after dinner. |
| 2. *Pallor urinæ pransorum.* | Pale urine after dinner. |
| 3. ——— *a frigore cutaneo.* | ——— from cold skin. |
| 4. *Pallor ex ægritudine.* | Paleness from sickness. |
| 5. *Dyspnœa a balneo frigido.* | Shortness of breath from cold bathing. |
| 6. *Dyspepsia a pedibus frigidis.* | Indigestion from cold feet. |
| 7. *Tussis a pedibus frigidis.* | Cough from cold feet. |
| 8. ——— *hepatica.* | Liver-cough. |
| 9. ——— *arthritica.* | Gout-cough. |
| 10. *Vertigo rotatoria.* | Vertigo rotatory. |
| 11. ——— *visualis.* | ——— visual. |
| 12. ——— *ebriosa.* | ——— inebriate. |
| 13. ——— *febriculosa.* | ——— feverish. |
| 14. ——— *cerebrosa.* | ——— from the brain. |
| 15. *Murmur aurium vertiginosum.* | Noise in the ears. |
| 16. *Tactus, gustus, olfactus vertiginosi.* | Vertiginous touch, taste, smell. |
| 17. *Pulsus mollis a vomitione.* | Soft pulse in vomiting. |
| 18. ——— *intermittens a ventriculo.* | Intermittent pulse from the stomach. |
| 19. *Febris inirritativa.* | Inirritative fever. |

## GENUS II.

*Catenated with Senſitive Motions.*

### SPECIES.

| | |
|---|---|
| 1. *Torpor genæ a dolore dentis.* | Coldneſs of the cheek from tooth-ach. |
| 2. *Stranguria a dolore veſicæ.* | Strangury from pain of the bladder. |
| 3. ——— *convulſiva.* | Convulſive ſtrangury. |
| 4. *Dolor termini ductûs choledochi.* | Pain of the end of the bile-duct. |
| 5. *Dolor pharyngis abacido gaſtrico.* | Pain of the throat from gaſtric acid. |
| 6. *Pruritus narium a vermibus.* | Itching of the noſe from worms. |
| 7. *Cephalæa.* | Head-ach. |
| 8. *Hemicrania et otalgia.* | Partial head-ach, and ear-ach. |
| 9. *Dolor humeri in hepatidide.* | Pain of ſhoulder in hepatitis. |
| 10. *Torpor pedum variolâ erumpente.* | Cold feet in eruption of ſmall-pox. |
| 11. *Teſtium dolor nephriticus.* | Nephritic pain of teſtis. |
| 12. *Dolor digiti minimi ſympatheticus.* | Pain of little finger from ſympathy. |
| 13. *Dolor brachii in hydrope pectoris.* | Pain of the arm in dropſy of the cheſt. |
| 14. *Diarrhœa a dentitione.* | Diarrhœa from toothing. |

## GENUS III.

*Catenated with Voluntary Motions.*

### SPECIES.

| | |
|---|---|
| 1. *Titubatio linguæ.* | Impediment of ſpeech. |
| 2. *Chorea ſancti viti.* | St. Vitus' dance. |
| 3. *Riſus.* | Laughter. |
| 4. *Tremor ex irâ.* | Trembling from anger. |
| 5. *Rubor ex irâ.* | Redneſs from anger. |
| 6. ——— *criminati.* | Bluſh of guilt. |
| 7. *Tarditas paralytica.* | Slowneſs from palſy. |
| 8. ——— *ſenilis.* | ——— of age. |

## GENUS IV.

*Catenated with External Influences.*

### SPECIES.

| | |
|---|---|
| 1. *Somni periodus.* | Periods of ſleep. |
| 2. *Studii inanis periodus.* | ——— of reverie. |
| 3. *Hemicraniæ periodus.* | ——— of head-ach. |
| 4. *Epilepſiæ dolorificæ periodus.* | ——— of painful epilepſy. |
| 5. *Convulſionis dolorificæ periodus.* | ——— of painful convulſion. |
| 6. *Tuſſis periodicæ periodus.* | ——— of periodic cough. |
| 7. *Catameniæ periodus.* | ——— of catamenia. |
| 8. *Hæmorrhoidis periodus.* | ——— of the piles. |
| 9. *Podagræ periodus.* | ——— of the gout. |

10. *Eryſipelatis periodus.* Periods of eryſipelas.
11. *Febrium periodus.* ——— of fevers.

## ORDO III.

### *Retrograde Aſſociate Motions.*

### GENUS I.

*Catenated with Irritative Motions.*

#### SPECIES.

1. *Diabætes irritata.* Diabetes from irritation.
2. *Sudor frigidus in aſthmate.* Cold ſweat in aſthma.
3. *Diabætes a timore.* Diabetes from fear.
4. *Diarrhœa a timore.* Diarrhœa from fear.
5. *Pallor et tremor a timore.* Paleneſs and trembling from fear.
6. *Palpitatio cordis a timore.* Palpitation of the heart from fear.
7. *Abortio a timore.* Abortion from fear.
8. *Hyſteria a timore.* Hyſterics from fear.

### GENUS II.

*Catenated with Senſitive Motions.*

#### SPECIES.

1. *Nauſea idealis.* Nauſea from ideas.
2. ——— *a conceptu.* Nauſea from conception.
3. *Vomitio vertiginoſa.* Vomiting from vertigo.
4. ——— *a calculo in uretere.* ——— from ſtone in the ureter.
5. ——— *ab inſultu paralytico.* ——— from ſtroke of palſy.

6. *Vomitio*

6. *Vomitio a titilatione faucium.* Vomiting from tickling the throat.
7. ——— *cute ſympathetica.* ——— from ſympathy with the ſkin.

## GENUS III.

*Catenated with Voluntary Motions.*

### SPECIES.

1. *Ruminatio.* Rumination.
2. *Vomitio voluntaria.* Voluntary vomiting.
3. *Eructatio voluntaria.* ——— eructation.

## GENUS. IV.

*Catenated with External Influences.*

### SPECIES.

1. *Catarrhus periodicus.* Periodical catarrh.
2. *Tuſſis periodica.* Periodic cough.
3. *Hiſteria a frigore.* Hyſterics from cold.
4. *Nauſea pluvialis.* Sickneſs againſt rain.

# CLASS IV.

## DISEASES OF ASSOCIATION.

## ORDO I.

*Increaſed Aſſociate Motions.*

## GENUS I.

*Catenated with Irritative Motion.*

THE importance of the ſubſequent claſs not only conſiſts in its elucidating all the ſympathetic diſeaſes, but in its opening *a road to the knowledge of fever.* The difficulty and novelty of the ſubject muſt plead in excuſe for the preſent imperfect ſtate of it. The reader is entreated previouſly to attend to the following circumſtances for the greater facility of inveſtigating their intricate connections; which I ſhall enumerate under the following heads.

A. Aſſociate motions diſtinguiſhed from catenations.
B. Aſſociate motions of three kinds.
C. Aſſociations affected by external influences.
D. Aſſociations affected by other ſenſorial motions.
E. Aſſociations catenated with ſenſation.
F. Direct and reverſe ſympathy.
G. Aſſociations affected four ways.
H. Origin of aſſociations.
I. Of the action of vomiting.
K. Tertian aſſociations.

A. *Associate Motions distinguished from Catenations.*

Associate motions properly mean only those, which are caused by the sensorial power of association. Whence it appears, that those fibrous motions, which constitute the introductory link of an associate train of motions, are excluded from this definition, as not being themselves caused by the sensorial power of association, but by irritation, or sensation, or volition. I shall give for example the flushing of the face after dinner; the capillary vessels of the face increase their actions in consequence of their catenation, not their association, with those of the stomach; which latter are caused to act with greater energy by the irritation excited by the stimulus of food. These capillaries of the face are associated with each other reciprocally, as being all of them excited by the sensorial power of association; but they are only catenated with those of the stomach, which are not in this case associate motions but irritative ones. The common use of the word association for almost every kind of connection has rendered this subject difficult; from which inaccuracy I fear some parts of this work are not exempt.

B. *Associate Motions of three Kinds.*

Those trains or tribes of associate motions, whose introductory link consists of an irritative motion, are termed irritative associations; as when the muscles of the eyelids close the eye in common nictitation. Those, whose introductory link consists of a sensitive motion, are termed sensitive associations; as when the pectoral and intercostal muscles act in sneezing. And lastly, those, whose introductory link consists of a voluntary motion, are termed voluntary associations;

6 as

as when the muſcles of the lower limbs act in concert with thoſe of the arm in fencing.

### C. *Aſſociations affected by external Influences.*

Circles of aſſociate motions, as well as trains and tribes of them, are liable to be affected by external influences, which conſiſt of etherial fluids, and which, by penetrating the ſyſtem, act upon it perhaps rather as a cauſa ſine quâ non of its movements, than directly as a ſtimulus; except when they are accumulated in unuſual quantity. We have a ſenſe adapted to the perception of the exceſs or defect of one of theſe fluids; I mean that of elementary heat; in which all things are immerſed. See Claſs IV. 1. 4. 1. But there are others of them, which as we have no power to evade their influence, ſo we have no ſenſe to perceive it; theſe are the ſolar, and lunar, and terreſtrial gravitation, in which alſo all things are immerſed; the electric aura, which pervades us, and is perpetually varying, See Claſs IV. 1. 4. 5; the magnetic fluid, Claſs IV. 1. 4. 6; and laſtly, the great life-preſerver oxygen gas, and the aqueous vapour of the atmoſphere, ſee Claſs IV. 1. 4. 6. and 7. and 2.

Of theſe external influences thoſe of heat, and of gravity, have diurnal periods of increaſe and decreaſe; beſides their greater periods of monthly or annual variation. The manner in which they act by periodical increments on the ſyſtem, till ſome effect is produced, is ſpoken of in Sect. XXXII. 3. and 6.

### D. *Aſſociations affected by other Senſorial Motions.*

Circles and trains of aſſociate motions are alſo liable to be affected by their catenations with other ſenſorial powers, as of irritation, or ſenſation, or volition; which other ſenſorial powers either thus ſimply form ſome of the links of the catenation, or add to the energy of the aſſociated motions. Thus when vomiting is cauſed by the ſtimulus of a ſtone in the ureter, the ſenſation of pain ſeems to be a link of the catenation rather than an efficient cauſe of the vomiting. But when the capillary veſſels of the ſkin increaſe their action from the influence of external heat, they are excited both by the ſtimulus of unuſual heat, as well as by the ſtimulus of the blood, and by their accuſtomed aſſociation with the actions of the heart and arteries. And laſtly, in the bluſh of anger the ſenſorial power of volition is added to that of aſſociation, and irritation, to excite the capillaries of the face with increaſed action. See Claſs IV. 2. 3. 5.

E. *Aſſociations catenated with Senſation.*

Pain frequently accompanies aſſociate trains or circles of motion without its being a cauſe, or a link, of them, but ſimply an attendant ſymptom; though it frequently gives name to the diſeaſe, as head-ach. Thus in the cramp of the calves of the legs in diarrhœa, the increaſed ſenſorial power of aſſociation is the proximate cauſe; the preceding increaſed action of the bowels is the remote cauſe; and the proximate effect is the violent contractions of the muſculi gaſtrocnemii; but the pain of theſe muſcles is only an attendant ſymptom, or a remote effect. See Sect. XVIII. 15. Other ſenſitive aſſociations are mentioned in Claſs IV. 1. 2. and IV. 1. 2. 15.

Thus, if the fluſhing of the face above mentioned after dinner be called a diſeaſe, the immediate or proximate cauſe is the increaſed power of aſſociation, the remote cauſe is the increaſed irritative motions of the ſtomach in conſequence of the ſtimulus of food and wine. The diſeaſe or proximate effect conſiſts in the increaſed actions of the cutaneous veſſels of the face; and the ſenſation of heat, the exiſtence

of

of heat, and the red colour, are attendants or ſymptoms, or remote effects, of the increaſed actions of theſe cutaneous veſſels.

F. *Direct and reverſe Sympathy.*

The increaſed actions of the primary part of the trains of aſſociated motions are ſometimes ſucceeded by increaſed actions of the ſecondary part of the train; and ſometimes by decreaſed actions of it. So likewiſe the decreaſed actions of the primary part of a train of aſſociate motions are ſometimes ſucceeded by decreaſed actions of the ſecondary part, and ſometimes by increaſed actions of it. The former of theſe ſituations is called direct ſympathy, and the latter reverſe ſympathy. In general I believe, where the primary part of the train of aſſociated motions is exerted more than natural, it produces direct ſympathy in ſtrong people, and reverſe ſympathy in weak ones, as a full meal makes ſome people hot, and others chill. And where the primary part of the train is exerted leſs than natural, it produces direct ſympathy in weak people, and reverſe ſympathy in ſtrong ones, as on being expoſed for a certain length of time on horſeback in a cold day gives indigeſtion and conſequent heart-burn to weak people, and ſtrengthens the digeſtion, and induces conſequent hunger in ſtrong ones. See Sect. XXXV. 1.

This may perhaps be more eaſily underſtood, by conſidering ſtrength and weakneſs, when applied to animal bodies, as conſiſting in the quantity of ſenſorial power reſiding in the contracting fibres, and the quantity of ſtimulus applied, as ſhewn in Sect. XII. 2. 1. Now when defective ſtimulus, within certain limits, is partially applied to parts ſubject to perpetual motion, the expenditure of ſenſorial power is for a while leſſened, but not its general production in the brain, nor its derivation into the weakly-ſtimulated part. Hence in ſtrong people, or ſuch whoſe fibres abound with ſenſorial power, if

 the

the firſt tribe of an aſſociate train of motions be deprived in part of its accuſtomed ſtimulus, its action becomes diminiſhed; and the ſenſorial power becomes accumulated, and by its ſuperabundance, or overflowing as it were, increaſes the action of the ſecond tribe of the aſſociate actions by reverſe ſympathy. As expoſing the warm ſkin for a moderate time to cold air increaſes the action of the ſtomach, and thus ſtrengthens the power of digeſtion.

On the reverſe, when additional ſtimulus within certain limits is partially applied to parts, which are deficient in reſpect to the natural quantity of ſenſorial power, the expenditure of ſenſorial power is increaſed, but in a leſs degree than the increaſed production of it in the brain, or its increaſed derivation into the ſtrongly-ſtimulated organ. Hence in weak people, or ſuch whoſe fibres are deficient of ſenſorial power, if the firſt tribe of an aſſociate train of motions be ſubjected for a while to greater ſtimulus than uſual, a greater production of ſenſorial power, or a greater derivation of it into the ſtimulated parts occurs; which by its exceſs, or overflowing as it were, increaſes the actions of the ſecond tribe of the aſſociate motions by direct ſympathy. Thus when vomiting occurs with cold extremities, a bliſter on the back in a few hours occaſions univerſal warmth of the ſkin, and ſtops the vomiting. And when a diarrhœa occurs with pale ſkin and cold extremities, the pricking of the points of a flannel ſhirt, worn next the ſkin, occaſions univerſal warmth of it, and checks or cures the diarrhœa.

In ſome aſſociate trains of action nevertheleſs reverſe ſympathies more frequently occur than direct ones, and in others direct ones more frequently than reverſe ones. Thus in continued fever with debility there appears to be a reverſe ſympathy between the capillary veſſels of the ſtomach and thoſe of the ſkin; becauſe there exiſts a total averſion to ſolid food, and conſtant heat on the ſurface of the body. Yet theſe two ſyſtems of veſſels are at other times actuated by direct ſympathy, as when paleneſs attends ſickneſs, or cold feet induces

duces indigeſtion. This ſubject requires to be further inveſtigated, as it probably depends not only on the preſent or previous plus or minus of the ſenſorial power of aſſociation, but alſo on the introduction of other kinds of ſenſorial power, as in Claſs IV. 1. 1. D; or the increaſed production of it in the brain, or the greater mobility of one part of a train of actions than another.

Thus when much food or wine is taken into the ſtomach, if there be no ſuperfluity of ſenſorial power in the ſyſtem, that is, none to be ſpared from the continual actions of it, a paleneſs and chillneſs ſucceeds for a time; becauſe now the expenditure of it by the increaſed actions of the ſtomach is greater than the preſent production of it. In a little time however the ſtimulus of the food and wine increaſes the production of ſenſorial power in the brain, and this produces a ſuperfluity of it in the ſyſtem; in conſequence of which the ſkin now becomes warm and florid, which was at firſt cold and pale; and thus the reverſe ſympathy is ſhortly converted into a direct one; which is probably owing to the introduction of a ſecond ſenſorial power, that of pleaſurable ſenſation.

On the contrary, when an emetic drug produces ſickneſs, the ſkin is at firſt pale for a time by direct ſympathy with the capillaries of the ſtomach; but in a few minutes, by the accumulation of ſenſorial power in the ſtomach during its leſs active ſtate in ſickneſs, the capillaries of the ſkin, which are aſſociated with thoſe of the ſtomach, act with greater energy by reverſe ſympathy, and a florid colour returns. Where the quantity of action is diminiſhed in the firſt part of a train of motions, whether by previous diminution of ſenſorial power, or preſent diminution of ſtimulus, the ſecond part of the train becomes torpid by direct ſympathy. And when the quantity of action of the firſt part becomes increaſed by the accumulation of ſenſorial power during its previous torpor, or by increaſe of ſtimulus, the actions of the ſecond part of it likewiſe become increaſed by direct ſympathy.

In

In moderate hunger the ſkin is pale, as before dinner, and in moderate ſickneſs, as no great accumulation of ſenſorial power has commenced; but in violent hunger, and in greater torpor of the ſtomach, as from contagious matter, the accumulation of ſenſorial power becomes ſo great as to affect the arterial and capillary ſyſtem, and fever is produced in both caſes.

In contagious fevers with arterial debilities commencing with torpor of the ſtomach, why is the action of the heart weakened, and that of the capillaries increaſed? Is it becauſe the mobility of the heart is leſs than that of the ſtomach, and the mobility of the capillaries greater? Or is it becauſe the aſſociation between the muſcular fibres of the ſtomach and thoſe of the heart have been uniformly aſſociated by direct ſympathy; and the capillaries of the ſtomach and thoſe of the ſkin have been more frequently aſſociated by reverſe ſympathy?

Where the actions of the ſtomach have been previouſly exhauſted by long ſtimulus, as on the day after intoxication, little or no accumulation of ſenſorial power occurs, during the torpor of the organ, beyond what is required to replace the deficiency of it, and hence fever ſeldom follows intoxication. And a repetition of the ſtimulus ſometimes becomes neceſſary even to induce its natural action, as in dram-drinkers.

Where there has been no previous exhauſtion of ſenſorial power, and the primary link of aſſociate motions is violently actuated by the ſenſorial power of ſenſation, the ſecondary link is alſo violently actuated by direct ſympathy, as in inflammatory fevers. Where however the ſenſorial power of the ſyſtem is leſs than natural, the ſecondary link of aſſociated motions becomes torpid by reverſe ſympathy, as in the inoculated ſmall-pox during the eruption on the face the feet are frequently cold.

G. *Aſſociations*

### G. *Aſſociations affected four Ways.*

Hence aſſociated trains or circles of motions may be affected four different ways. 1. By the greater or leſs energy of action of the firſt link with which they are catenated, and from which they take their names; as irritative, ſenſitive, or voluntary aſſociations. 2. By being excited by two or more ſenſorial powers at the ſame time, as by irritation and aſſociation, as in the inſtance of the application of the ſtimulus of increaſed external heat to the cutaneous capillaries. 3. By catenation with other ſenſorial powers, as with pain or pleaſure, which are in this caſe not the proximate cauſe of motion, but which, by becoming a link of catenation, excites the ſenſorial power of aſſociation into action; as the pain at the neck of the gall-bladder occaſioned by a gall-ſtone is transferred to the other end of that canal, and becomes a link of catenation between the action of the two extremities of it. 4. The influence of ethereal fluids, as of heat and gravitation. To which laſt perhaps might be added moiſture and oxygen gas as conſtituting neceſſary parts of the ſyſtem, rather than ſtimuli to excite it into action.

### H. *The Origin of Aſſociations.*

Some trains or circles of aſſociate motions muſt have been formed before our nativity, as thoſe of the heart, arteries, and capillaries; others have been aſſociated, as occaſion required them, as the muſcles of the diaphragm and abdomen in vomiting; and others by perpetual habit, as thoſe of the ſtomach with the heart and arteries directly, as in weak pulſe during ſickneſs; with the capillaries directly, as in the fluſhed ſkin after dinner; and laſtly, with the cellular abſorbents reverſely, as in the increaſed abſorption in anaſarca during ſickneſs; and with the irritative motions of the organs of ſenſe reverſely, as in vertigo, or ſea-ſickneſs. Some of theſe aſſociati ns ſhall be here ſhortly deſcribed to facilitate the inveſtigation of others.

Firſt, other congeries of glands occupy but a particular part of the ſyſtem, or conſtitute a particular organ, as the liver, or kidneys; but thoſe glands, which ſecrete the mucus, and perſpirable matter, which are called capillaries, are of very great extent; they receive the blood from the arteries, ſeparate from it the mucus, which lines every cell, and covers every cavity of body; and the perſpirable matter, which ſoftens and lubricates the whole ſurface of the ſkin, and the more extenſive ſurface of the air-veſſels, which compoſe the lungs. Theſe are ſupplied with blood by the perpetual action of the heart and arteries, and have therefore their motions aſſociated with the former, and with each other, by ſympathy, which is ſometimes direct, and ſometimes reverſe.

One branch of this aſſociation, the capillaries of the ſkin, are very irritable by the increaſed quantities of cold and heat, another branch, that of the lungs, has not the perception of cold and heat, but is liable by direct ſympathy to act in concert with the former, as in going into the cold bath. And it is probable the capillaries of the internal membranes are likewiſe directly affected by their ſympathy with thoſe of the ſkin, as appears from the defect of ſecretion in ulcers during the cold fits of agues.

The motions of this extenſive ſyſtem of capillaries, thus aſſociated by direct ſympathy, are alſo aſſociated with thoſe of the heart and arteries, ſometimes by reverſe and ſometimes by direct ſympathy; and thus conſtitute ſimple fever. The cold paroxyſm of which conſiſts in their torpor, and the hot one in their orgaſm, or increaſed activity.

I. *Of*

## I. *Of the Action of Vomiting.*

The manner, in which the ſtomach and the diaphragm and abdominal muſcles acquire their aſſociate action in vomiting, requires ſome attention. It is not probable, that this action of vomiting occurs before nativity; as the uniform application of the nutritive liquor amnii to the mouth of the fœtus, and the uniform expenditure of its nouriſhment, would not ſeem to give occaſion to too great temporary repletion of the ſtomach; and would preclude the deglutition of any improper material. After nativity the ſtomach of the child may be occaſionally too much diſtended with milk; as previous hunger may induce it to overgorge itſelf; and by repeated efforts the act of vomiting is learned, as a means of getting free from a diſagreeable ſenſation. Thus when any diſguſtful material, as a bitter drug, is taken into the mouth; certain retrograde motions of the tongue and lips are produced, for the purpoſe of putting the diſagreeable material out of the mouth again.

When the ſtomach is diſagreeably ſtimulated by the diſtention or acrimony of the aliment, a ſimilar effort to regurgitate it muſt occur; and by repeated trials the action of the diaphragm and abdominal muſcles by ſqueezing the ſtomach aſſiſts its retrograde exertion to diſgorge its contents. In the ſame manner when a piece of gravel is puſhed into the urethra, or a piece of indurated bile into the neck of the gall-bladder, after they have been in vain preſſed forward by the uſual motions of thoſe ducts, they return into the bladders of gall and urine by the retrograde motions of them.

That this is one mode, in which vomiting is induced, appears from the inſtantaneous rejection from the ſtomach occaſioned by ſome nauſeous drug, or from ſome nauſeous idea; and laſtly, from the voluntary power, which ſome people have been ſaid to have acquired, of emptying their ſtomachs, much in the ſame manner as ruminating animals bring up the graſs from their firſt ſtomach.

There are nevertheleſs many modes by which theſe inverted motions of the ſtomach and œſophagus are induced, and which it is of conſequence to diſtinguiſh from each other. The firſt is the mode above deſcribed, where an effort is made to diſlodge ſomething, which ſtimulates the ſtomach into diſagreeable ſenſation; and which is returned by repeated exertions; as when a nauſeous drug is taken into the mouth, or a bit of ſand falls into the eye, or a drop of water into the wind-pipe. In this the periſtaltic motions of the ſtomach are firſt ſtopped, and then reverted by painful ſenſation; and the abdominal muſcles and diaphragm by repeated efforts become aſſociated with them. Now as leſs ſenſorial power is expended on the retrograde actions of the ſtomach, and of the lymphatics, which open their mouths on its ſurface, than by their natural motions, an accumulation of ſenſorial power in the fibres of the ſtomach follows the exhibition of an emetic, and on that account an emetic will ſometimes ſtop a ſpontaneous vomiting which was owing to ſenſorial deficiency. See Sect. XXXV. 1. 3. and Art. V. 2. 1.

As bitters and metallic ſalts, exhibited in ſmall doſes, ſtimulate the ſtomach into greater action, as appears by their increaſing the power of digeſtion, and yet become emetic, when given in larger doſes; one might ſuſpect, that they became emetic by inducing debility, and conſequent retrograde actions of the ſtomach, by their previouſly exhauſting the ſenſorial power by their great ſtimulus; which might be effected in a moment without producing pain, and in conſequence without our perceiving it. But on the contrary, there does not in general appear on the exhibition of emetics to be any previous exhauſtion of ſenſorial power; becauſe there is evidently an accumulation of it during the ſickneſs, as appears from the digeſtion being ſtronger afterwards; and from the increaſed action of the cellular and cutaneous abſorbents during its operation. See Art. V. 2. 1.

Another mode, by which vomiting is induced, is owing to debility or deficiency of ſenſorial power, from the previous exhauſtion of it; as on the day after intoxication, or which occurs in people enfeebled with the gout, and in dropſy, and in ſome fevers with debility. In theſe, when the vomiting ceaſes, there is no appearance of accumulation of ſenſorial power, as the digeſtion ſtill remains weak and imperfect.

Another mode by which ſickneſs or vomiting is induced, is by defect of ſtimulus, as in great hunger; and in thoſe, who have been habituated to ſpice and ſpirit with their meals, who are liable to be ſick after taking food without theſe additional ſtimuli. Other means of inducing ſickneſs by vertigo, or by nauſeous ideas, will be mentioned below.

We ſhall only add, that the motions of the muſcular fibres of the ſtomach are aſſociated with thoſe of the heart and arteries by direct ſympathy, as appears by the weakneſs of the pulſe during the exhibition of an emetic; and that the abſorbents of the ſtomach are aſſociated with the cellular and cutaneous abſorbents by reverſe ſympathy, as is ſhewn by the great abſorption of the mucus of the cells in anaſarca during ſickneſs; at the ſame time that the abſorbents of the ſtomach invert their actions, and pour the mucus and water thus abſorbed into that viſcus.

In cold paroxyſms of fever the ſtomach partakes of the general torpor, and vomiting is induced by its debility, either by its aſſociation with the torpid capillaries, or other torpid parts, or by its own torpor commencing firſt, and cauſing the cold fit. The diſordered motions of the ſtomach frequently ſeem to be the cauſe or primary ſeat of fever, as where contagious miaſmata are ſwallowed with the ſaliva, and where fever is produced by ſea-ſickneſs, which I once ſaw. Nevertheleſs a diſorder of the ſtomach does not always induce fever, as in that caſe it ſhould conſtantly attend indigeſtion, and vertigo, and ſea-ſickneſs; but is itſelf frequently induced by aſſociation with the diſordered movements of other parts of the ſyſtem, as when it ariſes from gravel in the ureter, or from a percuſſion on the head.

The connexion of the motions of the ſtomach with irritative ideas, or motions of the organs of ſenſe, in vertigo, is ſhewn in Sect. XX. and thus it appears, that many circles of aſſociation are either directly or reverſely aſſociated, or catenated, with this viſcus; which will much contribute to unfold ſome of the ſymptoms of fever.

K. *Tertian Aſſociations.*

The third link of aſſociate trains of motion is ſometimes actuated by reverſe ſympathy, with the ſecond link, and that by reverſe ſympathy with the firſt link; ſo that the firſt and third link may act by direct ſympathy, and the intermediate one by reverſe ſympathy. Of this inſtances are given in the ſyngultus nephriticus, Claſs IV. 1. 1. 7. and IV. 2. 1. At other times the tertian or quartan links of aſſociate motions are actuated by direct ſympathy; and that ſometimes forwards and ſometimes backwards in reſpect to the uſual order of thoſe trains of aſſociate motions, as in Claſs IV. 1. 2. 1.

SPECIES.

1. *Rubor vultûs pranſorum.* Fluſhing of the face after dinner is explained in Sect. XXXV. 1. In the beginning of intoxication the whole ſkin becomes florid from the aſſociation of the actions of the cutaneous arteries with thoſe of the ſtomach, becauſe vinous ſpirit excites the fibres of the ſtomach into more violent action than the ſtimulus of common food; and the cutaneous capillaries of the face, from their more frequent expoſure to the viciſſitudes of cold and heat, poſſeſs

possess more mobility or irritability than those of other parts of the skin, as further explained in Sect. XXXIII. 2. 10. Vinegar is liable to produce this flushing of the face, which probably is owing to the quantity of vinous spirit it contains, as I believe the unfermented vegetable acids do not produce this effect. In every kind of blush the arterial blood is propelled into the capillaries faster than the venous absorption can carry it forwards into the veins, in this respect resembling the tensio phalli.

Can the beginning vinous or acetous fermentation of the aliment in weak stomachs contribute to this effect? or is it to be ascribed to the greater power of association between the arteries of the face and the fibres of the stomach in some people than in others?

M. M. Eat and drink less at a time, and more frequently. Put 20 drops of weak acid of vitriol into water to be drank at meals. Let the dress over the stomach and bowels be loose. Use no fermented liquors, or vinegar, or spice.

2. *Sudor stragulis immersorum.* Sweat from being covered in bed. In the commencement of an epidemic fever, in which the perpetual efforts to vomit was a distressing symptom, Dr. Sydenham discovered, that if the patient's head was for a short time covered over with the bed clothes, warmth was produced, and a sweat broke out upon the skin, and the tendency to vomit ceased. In this curious fact two trains of associated motions are excited into increased action. First, the vessels of the lungs are known to have their motion associated with those of the skin by the difficulty of breathing on going into the cold bath, as described in Sect. XXXII. 3. 2. Hence, when the vessels of the lungs become excited into stronger action, by the bad air under the bed clothes, warmed and adulterated by frequent breathing, those of the external skin soon become excited by their association into more energetic action, and generate more heat along with a greater secretion of perspirable matter. Secondly, the sympathy

pathy between the ſtomach and ſkin is evident in variety of circumſtances; thus the cold air of froſty days applied to the ſkin for a ſhort time increaſes the action of the ſtomach by reverſe ſympathy, but decreaſes it if continued too long by direct ſympathy; ſo in the circumſtance above mentioned the action of the ſtomach is increaſed by direct ſympathy with that of the ſkin; and the tendency to vomit, which was owing to its diminiſhed action, ceaſes.

3. *Ceſſatio ægritudinis cute excitatâ.* The cure of ſickneſs by ſtimulating the ſkin. This is explained in the preceding article; and further noticed in IV. 2. 2. 4. and in IV. 1. 1. f.

Similar to theſe is the effect of a bliſter on the back in relieving ſickneſs, indigeſtion, and heart-burn; and, on the contrary, by theſe ſymptoms being frequently induced by coldneſs of the extremities. The bliſter ſtimulates the cutaneous veſſels into greater action; whence warmth and pain are produced at the ſame time, and the fibres of the ſtomach are excited into greater action by their aſſociation with thoſe of the ſkin. It does not appear, that the concomitant pain of the bliſter cauſes the increaſed energy of the ſtomach, becauſe the motions of it are not greater than natural; though it is ſometimes difficult to determine, whether the primary part of ſome aſſociated trains be connected with irritative or ſenſitive motions.

In the ſame manner a flannel ſhirt, to one who has not been in the habit of wearing one, ſtimulates the ſkin by its points, and thus ſtops vomiting in ſome caſes; and is particularly efficacious in checking ſome chronical diarrhœas, which are not attended with fever; for the abſorbents of the ſkin are thus ſtimulated into greater action, with which thoſe of the inteſtines conſent by direct ſympathy.

This effect cannot be aſcribed to the warmth alone of the flannel ſhirt, as being a covering of looſe texture, and confining air in its pores, like a ſponge, which air is known to be a bad conductor of heat,

heat, ſince in that caſe its uſe ſhould be equally efficacious, if it were worn over a linen ſhirt; and an increaſed warmth of the room of the patient would be equally ſerviceable.

4. *Digeſtio aucta frigore cutaneo.* Digeſtion increaſed by coldneſs of the ſkin. Every one has experienced the increaſe of his appetite after walking in the cool air in froſty days; for there is at this time not only a ſaving of ſenſorial power by the leſs exertion of the cutaneous veſſels; but, as theſe conſent with thoſe of the ſtomach and bowels, this ſaving of ſenſorial power is transferred by reverſe ſympathy from the cutaneous capillaries and abſorbents to thoſe of the ſtomach and inteſtines.

Hence weak people ſhould uſe the cold air of winter as a cold bath; that is, they ſhould ſtay in it but a ſhort time at once, but ſhould immerſe themſelves in it many times a day.

5. *Catarrhus a frigore cutaneo.* Catarrh from cold ſkin. This has been already explained in Claſs I. 1. 2. 7. and is further deſcribed in Sect. XXXV. 1. 3. In this diſeaſe the veſſels of the membrane, which lines the noſtrils, are excited into greater action; when thoſe of the ſkin, with which they are aſſociated, are excited into leſs action by the deficiency of external heat, by reverſe ſympathy; and though the pain of cold attends the torpor of the primary link of this aſſociation, yet the increaſed motions of the membrane of the noſtrils are aſſociated with thoſe of the cutaneous veſſels, and not with the pain of them, becauſe no inflammation follows.

6. *Abſorptio cellularis aucta vomitu.* In the act of vomiting the irritative motions of the ſtomach are inverted, and of the abſorbents, which open their mouths into it; while the cutaneous, cellular, and pulmonary abſorbents are induced, by reverſe ſympathy with them,

to act with greater energy. This is seen in cases of anasarca, when long sickness and vomiting are caused by squills, or antimonial salts, or most of all by the decoction of digitalis purpurea, foxglove; and Mr. J. Hunter mentions a case, in which a large bubo, which was just ready to break, was absorbed in a few days by sickness at sea. Treatise on the Blood, p. 501, which is thus accounted for; less sensorial power is expended during sickness by the decreased action of the fibres of the stomach, and of its absorbents; as shewn in Sect. XXXV. 1. 3. whence an accumulation of it is produced, and there is in consequence a greater quantity of sensorial power for the exertion of those motions, which are associated with the absorbents of the stomach by reverse sympathy.

The reverse sympathy between the lacteal and lymphatic branches of the absorbent system have been produced by the one branch being less excited to act, when the other supplies sufficient fluid or nutriment to the sanguiferous vessels. Thus when the stomach is full, and the supply of chyle and mucus and water is in sufficient quantity; the pulmonary, cellular, and cutaneous lymphatics are not excited into action; whence the urine is pale, and the skin moist, from the defect of absorption on those surfaces.

7. *Syngultus nephriticus.* When a stone irritates the ureter, and that even without its being attended with pain or fever, sometimes a chronical hiccough occurs, and continues for days and weeks, instead of sickness or vomiting; which are the common symptoms. In this case the motions of the stomach are decreased by their sympathy with those of the ureter, which are increased by the stimulus of the stone in it; and the increased motions of the diaphragm seem to exist in consequence of their association with the stomach by a second reverse sympathy. This hiccough may nevertheless admit of another explanation, and be supposed to be a convulsive exertion of the diaphragm

phragm to relieve the disagreeable sensation of the stomach in consequence of its disordered irritative associations; and in that case it would belong to Class III. 1. 1. See Class IV. 2. 1. for another example of tertiary association.

M. M. Venesection. Emetic. Calomel. Cathartic, opium, oil of cinnamon from two to ten drops. Aerated alcaline water. Peruvian bark.

8. *Febris irritativa.* Irritative fever, described in Class I. 1. 1. 1. The diseases above explained in this genus are chiefly concerning the sympathies of the absorbent system, or the alimentary canal, which are not so much associated with the arterial system, as to throw it into disorder, when they are slightly deranged; but when any great congeries of conglomerate glands, which may be considered as the extremities of the arterial system, are affected with torpor, the whole arterial system and the heart sympathize with the torpid glands, and act with less energy; which constitutes the cold fit of fever; which is therefore at first a decreased action of the associate organ; but as this decrease of action is only a temporary effect, and an increase of exertion both of the torpid glands, and of the whole arterial system, soon follows; the hot fit of irritative fever, or fever with strong pulse, properly belongs to this class and genus of diseases.

## ORDO I.

*Increaſed Aſſociate Motions.*

## GENUS II.

*Catenated with Senſitive Motions.*

The primary links of the aſſociated actions of this genus are either produced or attended by painful or pleaſurable ſenſation. The ſecondary links of the firſt ten ſpecies are attended with increaſed motions without inflammation, thoſe of the remainder are attended with inflammation. All inflammations, which do not ariſe in the part which was previouſly torpid, belong to this genus; as the gout, rheumatiſm, eryſipelas. It is probable many other inflammations may, by future obſervation, require to be tranſplanted into this claſs.

The circles of ſenſitive aſſociate motions conſiſt chiefly of the excretory ducts of the capillaries and of the mouths of the abſorbent veſſels, which conſtitute the membranes; and which have been induced into action at the ſame time; or they conſiſt of the terminations of canals; or of parts which are endued with greater ſenſibility than thoſe which form the firſt link of the aſſociation. An inſtance of the firſt of thoſe is the ſympathy between the membranes of the alveolar proceſſes of the jaws, and the membranes above or beneath the muſcles about the temples in hemicrania. An inſtance of the ſecond is in the ſympathy between the excretory duct of the lacrymal gland, and the naſal duct of the lacrymal ſack. And an inſtance of the third is the ſympathy between the membranes of the liver, and the ſkin of the face in the gutta roſea of inebriates.

SPECIES.

## SPECIES.

1. *Lacrymarum fluxus ſympatheticus.* A flow of tears from grief or joy. When the termination of the duct of the lacrymal ſac in the noſtrils becomes affected either by painful or pleaſurable ſenſations, in conſequence of external ſtimulus, or by its aſſociation with agreeable or diſagreeable ideas, the motions of the lacrymal gland are at the ſame time exerted with greater energy, and a profuſion of tears ſucceeds by ſenſitive aſſociation, as explained in Sect. XVI. 8. 2.

In this caſe there exiſts a chain of aſſociated actions, the ſecretion of the lacrymal gland is increaſed by whatever ſtimulates the ſurface of the eye, at the ſame time the increaſed abundance of tears ſtimulates the puncta lacrymalia into greater action; and the fluid thus abſorbed ſtimulates the lacrymal ſac, and its naſal duct in the noſe into greater action. In a contrary direction of this chain of aſſociation the preſent increaſe of action is induced. Firſt, the naſal duct of the lacrymal ſac is excited into increaſed action by ſome pleaſurable or painful idea, as deſcribed in Sect. XVI. 8. 2. 2d. The puncta lacrymalia or other extremity of the lacrymal ſac ſympathizes with it (as the two ends of all other canals ſympathize with each other). 3d. With theſe increaſed motions of the puncta lacrymalia thoſe of the excretory duct of the lacrymal gland are aſſociated from their having ſo perpetually acted together. And, laſtly, with the increaſed actions of the excretory duct of this gland are aſſociated thoſe of the other end of it by their frequently acting together; in the ſame manner as the extremities of other canals are aſſociated; and thus a greater flow of tears is poured into the eye.

When a flow of tears is produced in grief, it is believed to relieve the violence of it, which is worthy a further inquiry. Painful ſenſations, when great, excite the faculty of volition; and the perſon continues voluntarily to call up or perform thoſe ideas, which occaſion

 the

the painful ſenſation; that is, the afflicted perſon becomes ſo far inſane or melancholy; but tears are produced by the ſenſorial faculty of aſſociation, and ſhew that the pain is ſo far relieved as not to excite the exceſſive power of volition, or inſanity, and are therefore a ſign of the abatement of the painful ſtate of grief, rather than a cauſe of that abatement. See Claſs III. 1. 2. 10.

2. *Sternutatio a lumine.* Some perſons ſneeze from looking up at the light ſky in a morning after coming out of a dark bedroom. The olfactory nerves are brought into too great action by their ſympathy with the optic nerves, or by their reſpective ſympathies with ſome intervening parts, as probably with the two extremities of the lacrymal ſac; that is, with the puncta lacrymalia and the naſal duct. See Claſs II. 1. 1. 3.

3. *Dolor dentium Stridore.* Tooth-edge from grating ſounds, and from the touch of certain ſubſtances, and even from imagination alone, is deſcribed and explained in Sect. XVI. 10. The increaſed actions of the alveolar veſſels or membranes are aſſociated with the ideas, or ſenſual motions of the auditory nerves in the firſt caſe; and of thoſe of the ſenſe of touch, in the ſecond caſe; and by imagination, or ideas exerted of painful ſenſation alone, in the laſt.

4. *Riſus ſardonicus.* A diſagreeable ſmile attends inflammations of the diaphragm ariſing from the aſſociations of the reiterated exertions of that muſcle with thoſe of the lips and cheeks in laughing. See Diaphragmitis, Claſs II. 1. 2. 6.

5. *Salivæ fluxus cibo viſo.* The flow of ſaliva into the mouths of hungry animals at the ſight or ſmell of food is ſeen in dogs ſtanding round a dinner-table. The increaſed actions of the ſalivary glands have been uſually produced by the ſtimulus of agreeable food on their excretory

excretory ducts during the mastication of it; and with this increased action of their excretory ducts the other terminations of those glands in the capillary arteries have been excited into increased action by the mutual association of the ends of canals; and at the same time the pleasurable ideas, or sensual motions, of the sense of smell and of sight have accompanied this increased secretion of saliva. Hence this chain of motions becomes associated with those visual or olfactory ideas, or with the pleasure, which produces or attends them.

6. *Tensio mamularum viso puerulo.* The nipples of lactescent women are liable to become turgid at the sight of their young offspring. The nipple has generally been rendered turgid by the titillation of the lips or gums of the child in giving suck; the visible idea of the child has thus frequently accompanied this pleasurable sensation of parting with the milk, and turgescence of the tubes, which constitute the nipple. Hence the visual idea of the child, and the pleasure which attends it, become associated with those increased arterial actions, which swell the cells of the mamula, and extend its tubes; which is very similar to the tensio phalli visâ muliere nudâ etiam in insomnio.

7. *Tensio penis in hydrophobia.* An erection of the penis occurs in the hydrophobia, and is a troublesome symptom, as observed by Cœlius Aurelianus, Fothergill, and Vaughn, and would seem to be produced by an unexplained sympathy between the sensations about the fauces and the penis. In men the hair grows about both these parts, the voice changes, and the neck thickens at puberty. In the mumps, when the swellings about the throat subsides, the testicles are liable to swell. Venereal infection received by the penis is very liable to affect the throat with ulcers. Violent coughs, with soreness or rawness about the fauces are often attended with erection of the penis; which is also said to happen to male animals, that are hanged; which

which laſt circumſtance has generally been aſcribed to the obſtruction of the circulation of the blood, but is more probably occaſioned by the ſtimulus of the cord in compreſſing the throat; ſince if it was owing to impeded circulation it ought equally to occur in drowning animals.

In men the throat becomes ſo thickened at the time of puberty, that a meaſure of this is uſed to aſcertain the payment of a poll-tax on males in ſome of the iſlands of the Mediterranean, which commences at puberty; a ſtring is wrapped twice round the thinneſt part of the neck, the ends of it are then put into each corner of the mouth; and if, when thus held in the teeth, it paſſes readily over the head, the ſubject is taxable.

It is difficult to point out by what circumſtance the ſenſitive motions of the penis and of the throat and noſe become aſſociated; I can only obſerve, that theſe parts are ſubjected to greater pleaſurable ſenſations than any other parts of the body; one being deſigned to preſerve ourſelves by the pleaſure attending the ſmell and deglutition of food, and the other to enſure the propagation of our ſpecies; and may thus gain an aſſociation of their ſenſitive motion by their being eminently ſenſible to pleaſure. See Claſs I. 3. 1. 11. and III. 1. 1. 15. and Sect. XVI. 5.

In the female ſex this aſſociation between the face, throat, noſe, and pubis does not exiſt; whence no hair grows on their chins at the time of puberty, nor does their voices change, or their necks thicken. This happens probably from there being in them a more exquiſite ſenſitive ſympathy between the pubis and the breaſts. Hence their breaſts ſwell at the time of puberty, and ſecrete milk at the time of parturition. And in the parotitis, or mumps, the breaſts of women ſwell, when the tumor of the parotitis ſubſides. See Claſs I. 1. 2. 15. Whence it would appear, that their breaſts poſſeſs an intermediate ſympathy between the pubis and the throat; as they are the ſeat of a paſſion, which men do not poſſeſs, that of ſuckling children.

8. *Teneſmus*

8. *Tenesmus calculosus.* The sphincter of the rectum becomes painful or inflamed from the association of its sensitive motions with those of the sphincter of the bladder, when the latter is stimulated into violent pain or inflammation by a stone.

9. *Polypus narium ex ascaridibus?* The stimulation of ascarides in the rectum produces by sensitive sympathy an itching of the nose, as explained in IV. 2. 2. 6; and in three children I have seen a polypus in the nose, who were all affected with ascarides; to the perpetual stimulation of which, and the consequent sensitive association, I was led to ascribe the inflammation and thickening of the membrane of the nostrils.

10. *Crampus surarum in cholera.* A cramp of the muscles of the legs occurs in violent diarrhœa, or cholera, and from the use of too much acid diet in gouty habits. This seems to sympathize with uneasy sensation in the bowels. See Class III. 1. 1. 14. This association is not easily accounted for, but is analogous in some degree to the paralysis of the muscles of the arms in colica saturnina. It would seem, that the muscles of the legs in walking get a sympathy with the lower parts of the intestines, and those of the arms in variety of employment obtain a sympathy with the higher parts of them. See Cholera and Ileus.

11. *Zona ignea nephritica.* Nephritic shingles. The external skin about the loins and sides of the belly I suppose to have greater mobility in respect to sensitive association, than the external membrane of the kidney; and that their motions are by some unknown means thus associated. When the torpor or beginning inflammation of this membrane ceases, the external skin becomes inflamed in its stead, and a kind of herpes, called the shingles, covers the loins and sides of the belly. See Class II. 1. 5. 9.

12. *Eruptio variolarum.* After the inflammation of the inoculated arm has ſpread for a quarter of a lunation, it affects the ſtomach by reverſe ſympathy; that is, the actions of the ſtomach are aſſociated with thoſe of the ſkin; and as much ſenſorial power is now exerted on the inflamed ſkin, the other part of this ſenſitive aſſociation is deprived of its natural ſhare, and becomes torpid, or inverts its motions. After this torpor of the ſtomach has continued a time, and much ſenſorial power is thus accumulated; other parts of the ſkin, which are alſo aſſociated with it, as that of the face firſt, are thrown into partial inflammation; that is, the eruptions of the ſmall-pox appear on the face.

For that the variolous matter affects the ſtomach previous to its eruption on the ſkin appears from the ſickneſs at the commencement of the fever; and becauſe, when the morbid motions affect the ſkin, thoſe of the ſtomach ceaſe; as in the gout and eryſipelas, mentioned below. The conſent between the ſtomach and the ſkin appears in variety of other diſeaſes; and as they both conſiſt of ſurfaces, which abſorb and ſecrete a quantity of moiſture, their motions muſt frequently be produced together or in ſucceſſion; which is the foundation of all the ſympathies of animal motions, whether of the irritative, ſenſitive, or voluntary kinds.

Now as the ſkin, which covers the face, is expoſed to greater variations of heat and cold than any other part of the body; it probably poſſeſſes more mobility to ſenſitive aſſociations, not only than the ſtomach, but than any other part of the ſkin; and is thence affected at the eruption of the ſmall-pox with violent action and conſequent inflammation, by the aſſociation of its motions with thoſe of the ſtomach, a day before the other parts of the ſkin; and becomes fuller of puſtules, than any other part of the body. See Claſs II. 1. 3. 9.

It might be ſuppoſed, that the ſucceſſive ſwelling of the hands, when the face ſubſides, at the height of the ſmall-pox, and of the feet, when the hands ſubſide, were governed by ſome unknown associations of those parts of the system; but these successions of tumor and subsidence more evidently depend on the times of the eruption of the pustules on those parts, as they appear a day sooner on the face than on the hands, and a day sooner on the hands than on the feet, owing to the greater comparative mobility of those parts of the skin.

13. *Gutta rosea stomatica.* Stomatic red face. On drinking cold water, or cold milk, when heated with exercise, or on eating cold vegetables, as raw turnips, many people in harvest-time have been afflicted with what has been called a surfeit. The stomach becomes painful, with indigestion and flatulency, and after a few days an eruption of the face appears, and continues with some relief, but not with entire relief; as both the pimpled face and indigestion are liable to continue even to old age.

M. M. Venesection. A cathartic with calomel. Then half a grain of opium twice a day for many weeks. If saturated solution of arsenic three or five drops twice or thrice a day for a week?

14. *Gutta rosea hepatica.* The rosy drop of the face of some drinking people is produced like the gout described below, in consequence of an inflamed liver. In these constitutions the skin of the face being exposed to greater variation of heat and cold than the membranes of the liver, possesses more mobility than those hepatic membranes; and hence by whatever means these membranes are induced to sympathize, when this sensitive association occurs, the cutaneous vessels of the face run into greater degrees of those motions, which constitute inflammation, than previously existed in the membranes of the liver; and then those motions of the liver cease. See Class II. 1. 4. 6.

An inflammation of the liver so frequently attends the great potation of vinous spirit, there is reason to suspect, that this viscus itself becomes inflamed by sensitive association with the stomach; or that, when one

 termination

termination of the bile-duct, which enters the duodenum is stimulated violently, the other end may become inflamed by sensitive association.

15. *Podagra.* The gout, except when it affects the liver or stomach, seems always to be a secondary disease, and, like the rheumatism and erysipelas mentioned below, begins with the torpor of some distant part of the system.

The most frequent primary seat of the gout I suppose to be the liver, which is probably affected with torpor not only previous to the annual paroxysms of the gout; but to every change of its situation from one limb to another. The reasons, which induce me to suspect the liver to be first affected, are not only because the jaundice sometimes attends the commencement of gout, as described in Sect. XXIV. 2. 8. but a pain also over the pit of the stomach, which I suppose to be of the termination of the bile-duct in the duodenum, and which is erroneously supposed to be the gout of the stomach, with indigestion and flatulency, generally attends the commencement of the inflammation of each limb. See Arthritis ventriculi, Class I. 2. 4. 6. In the two cases, which I saw, of the gout in the limbs being preceded by jaundice, there was a cold shivering fit attended the inflammation of the foot, and a pain at the pit of the stomach; which ceased along with the jaundice, as soon as the foot became inflamed. This led me to suspect, that there was a torpor of the liver, and perhaps of the foot also, but nevertheless the liver might also in this case be previously inflamed, as observed in Sect. XXIV. 2. 8.

Now as the membranes of the joints of the feet suffer greater variations of heat and cold than the membranes of the liver, and are more habituated to extension and contraction than other parts of the skin in their vicinity; I suppose them to be more mobile, that is, more liable to run into extremes of exertion or quiescence; and are thence more susceptible of inflammation, than such parts as are less exposed to great variations of heat and cold, or of extension and contraction.

When a stone presses into the sphincter of the bladder, the glans penis is affected with greater pain by sympathy, owing to its greater sensibility, than the sphincter of the bladder; and when this pain commences, that of the sphincter ceases, when the stone is not too large, or pushed too far into the urethra. Thus when the membrane, which covers the ball of the great toe, sympathizes with some membranous part of a torpid or inflamed liver; this membrane of the toe falls into that kind of action, whether of torpor or inflammation, with greater energy, than those actions excited in the diseased liver; and when this new torpor or inflammation commences, that with which it sympathises ceases; which I believe to be a general law of associated inflammations.

The paroxysms of the gout would seem to be catenated with solar influence, both in respect to their larger annual periods, and to their diurnal periods—See Sect. XXXVI. 3. 6—as the former occur about the same season of the year, and the latter commence about an hour before sun-rise; nevertheless the annual periods may depend on the succession of great vicissitudes of cold and heat, and the diurnal ones on our increased sensibility to internal sensations during sleep, as in the fits of asthma, and of some epilepsies. See Sect. XVIII. 15.

In respect to the pre-remote cause or disposition to the gout, there can be no doubt of its individually arising from the potation of fermented or spirituous liquors in this country; whether opium produces the same effect in the countries, where it is in daily use, I have never been well informed. See Sect. XXI. 10, where this subject is treated of; to which I have to add, that I have seen some, and heard of others, who have moderated their paroxysms of gout, by diminishing the quantity of fermented liquors, which they had been accustomed to; and others who, by a total abstinence from fermented

 liquors,

liquors, have entirely freed themſelves from this excruciating malady; which otherwiſe grows with our years, and curtails or renders miſerable the latter half, or third, of the lives of thoſe, who are ſubject to it. The remote cauſe is whatever induces temporary torpor or weakneſs of the ſyſtem; and the proximate cauſe is the inirritability, or defective irritation, of ſome part of the ſyſtem; whence torpor and conſequent inflammation. The great Sydenham ſaw the beneficial effects of the abſtinence from fermented liquors in preventing the gout, and adds, "if an empiric could give ſmall-beer only to "gouty patients as a noſtrum, and perſuade them not to drink any "other ſpirituous fluids, that he might reſcue thouſands from this "diſeaſe, and acquire a fortune for his ingenuity." Yet it is to be lamented, that this accurate obſerver of diſeaſes had not reſolution to practiſe his own preſcription, and thus to have ſet an example to the world of the truth of his doctrine; but, on the contrary, recommends Madeira, the ſtrongeſt wine in common uſe, to be taken in the fits of the gout, to the detriment of thouſands; and is ſaid himſelf to have periſhed a martyr to the diſeaſe, which he knew how to ſubdue!

As example has more forcible effect than ſimple aſſertion, I ſhall now conciſely relate my own caſe, and that of one of my moſt reſpected friends. E. D. was about forty years of age, when he was firſt ſeized with a fit of the gout. The ball of his right great toe was very painful, and much ſwelled and inflamed, which continued five or ſix days in ſpite of veneſection, a briſk cathartic with ten grains of calomel, and the application of cold air and cold water to his foot. He then ceaſed to drink ale or wine alone; confining himſelf to ſmall beer, or wine diluted with about thrice its quantity of water. In about a year he ſuffered two other fits of the gout, in leſs violent degree. He then totally abſtained from all fermented liquors, not even taſting ſmall-beer, or a drop of any kind of wine; but eat plentifully of fleſh-meat, and all kinds of vegetables, and fruit, uſing for his drink at meals chiefly water alone, or lemonade, or cream and water; with tea and coffee between them as usual.

By this abstinence from fermented liquors he kept quite free from the gout for fifteen or sixteen years; and then began to take small-beer mixed with water occasionally, or wine and water, or perry and water, or cyder and water; by which indulgence after a few months he had again a paroxysm of gout, which continued about three days in the ball of his toe; which occasioned him to return to his habit of drinking water, and has now for above twenty years kept in perpetual health, except accidental colds from the changes of the seasons. Before he abstained from fermented or spirituous liquors, he was frequently subject to the piles, and to the gravel, neither of which he has since experienced.

In the following case the gout was established by longer habit and greater violence, and therefore required more cautious treatment. The Rev. R. W. was seized with the gout about the age of thirty-two, which increased so rapidly that at the age of forty-one he was confined to his room seven months in that year; he had some degree of lameness during the intervals, with chalky swellings of his heels and elbows. As the disease had continued so long and so violently, and the powers of his digestion were somewhat weakened, he was advised not entirely to leave off all fermented liquors; and as small-beer is of such various strength, he was advised to drink exactly two wine glasses, about four ounces, of wine mixed with three or four times its quantity of water, with or without lemon and sugar, for his daily potation at dinner, and no other fermented liquor of any kind; and was advised to eat flesh-meat with any kind of boiled vegetables, and fruit, with or without spice. He has now scrupulously continued this regimen for above five years, and has had an annual moderate gouty paroxysm of a few weeks, instead of the confinement of so many months, with great health and good spirits during the intervals.

The

The following is a more particular account of the hiſtory of this caſe; being part of a letter which Mr. Wilmot wrote on that ſubject at my entreaty.

"I entered into the army with an excellent conſtitution at the age of fifteen. The corps I ſerved in was diſtinguiſhed by its regularity, that is, the regular allowance of the meſs was only one pint of wine per man each day; unleſs we had company to dine with us; then, as was the general cuſtom of the time, the bottle circulated without limit. This mode of living, though by no means conſidered as exceſs for men, was certainly too great for a youth of my age. This ſtyle of living I continued, when with the regiment, till the latter end of the year 1769, when I had the misfortune to ſleep in a damp bed at Sheffield on a journey to York, but arrived there before I felt the ill effects of it. I was then ſeized with a violent inflammatory rheumatiſm with great inflammation of my eyes, and was attended by Dr. Dealtry; ſo violent was the diſorder, that I was bled for it eight times in leſs than a fortnight; and was three months, before I could conſider my health perfectly re-eſtabliſhed. Dr. Dealtry told me, that I ſhould be ſubject to ſimilar attacks for many years; and that he had no doubt, from the tendency he found in my habit to inflammation, that, when I was farther advanced in life, I ſhould change that complaint for the gout. He predicted truly; for the three ſucceeding winters I had the ſame complaint, but not ſo violently; the fourth winter I eſcaped, and imputed my eſcape to the continuance of cold bathing during the whole of that winter; after that I never eſcaped it, till I had a regular and ſevere fit of the gout: after the firſt attack of rheumatic fever I was more abſtemious in my manner of living, though when in company I never ſubjected myſelf to any great reſtraint. In the year 1774 I had quitted the army, and being in a more retired ſituation, was ſeldom led into any exceſs; in 1776 and 1777 I was in the habit of drinking a good deal of wine very frequently, though not conſtantly. After that period till the year

year 1781, I drank a larger quantity of wine regularly, but very ſeldom to any degree of intoxication. I lived much at that time in the ſociety of ſome gentlemen, who uſually drank nearly a bottle of wine daily after dinner. I muſt here however obſerve, that at no part of my life was I accuſtomed to drink wine in an evening, and very ſeldom drank any thing more than a ſingle half-pint glaſs of ſome ſort of ſpirits diluted with much water. Till the year 1781 I had always been accuſtomed to uſe very violent and continued exerciſe on horſeback; in the winter months I purſued all field diverſions, and in the ſummer months I rode frequent and long journeys; and with this exerciſe was liable to perſpire to great exceſs; beſides which I was ſubject to very profuſe night-ſweats, and had frequently boils break out all over me, eſpecially in the ſpring and autumn; for which I took no medicine, except a little flour of ſulphur with cream of tartar in honey.

" You will obſerve I bring every thing down to the date of 1781. In the month of October in that year, when I was juſt entered into the thirty-ſecond year of my age, I had the firſt attack of gout; that fit was very ſevere, and of many weeks continuance. I now determined upon a more abſtemious method of living, in reſpect to wine; and indeed the ſociety, in which I had before been accuſtomed to live, being conſiderably changed, I had leſs frequent temptations to exceſs. From this time I enjoyed the moſt perfect good ſtate of health till Auguſt 1784, when I had my ſecond attack of gout. I never perfectly recovered from this attack through the ſucceeding winter, and in March 1785 was adviſed to try the Bath waters, and drank them under the direction of one of the faculty of that place. I was there ſoon ſeized with a fever, and a ſlight attack of gout in one knee. I ſhould obſerve, that when I ſet out from home, I was in a weak and low ſtate, and unequal to much fatigue; as appeared by my having a fainting fit one day on the road, after having travelled only about fifty miles; in the courſe of the ſummer I had two or three more ſlight attacks of gout of leſs conſequence, till the month of October; when I was afflicted with it all over me in ſuch a manner, as to be without the poſſibility of the leaſt degree of removal for ſome days; and was about two months without being able to get into the air. This was the ſevereſt attack I had then experienced; though I have ſince had ſeveral equally ſevere. In the courſe of this ſummer I had a fall with my horſe; and ſoon after it, having diſcovered an enlargement on one elbow, I concluded I had hurt it at that time; but in the courſe of this laſt attack having a ſimilar enlargement on the other elbow, I found my miſtake, and that they were collections of gouty matter; theſe increaſed to the ſize of pullet's eggs, and continue in that ſtate. I had ſoon after ſimilar enlargements on my heels; the right heel being ſeverely bruiſed, I was under the neceſſity of having it lanced, and a large quantity of chalky matter was diſcharged from it; and have ſince that time frequently had chalky matter taken from it, and ſometimes ſmall bits of apparently perfect chalk. My right hand ſoon was afflicted in the ſame way, and I have ſcarcely a joint on thoſe fingers now in a natural ſtate. My left hand has eſcaped tolerably well. After this laſt attack (viz. October 1785), I had two or three ſlight attacks before the month of June 1787, when I had a very ſevere intermittent fever; from that time I continued very well till the latter end of the year, when I began to feel the gout about me very much, but was not confined by it. I was in this ſtate adviſed to try what is called the American Recipe (gum guaiacum and nitre diſſolved in ſpirits); it had apparently been of eſſential ſervice to a friend of mine, who from the inability to walk a mile for ſome years, was believed to be reſtored by the uſe of this medicine to a good ſtate of health, ſo as to walk ten miles a day. In addition to this medicine I drank, as my common beverage with my meals, ſpruce beer. I had ſo high an opinion of this medicine in the gout, and of ſpruce beer as an antiſcorbutic, that I contemplated with much ſatisfaction, and with

with very little doubt, the perfect restoration of my health and strength; but I was miserably deceived; for in September 1788 I was seized with the gout in a degree that none but arthritics, and indeed but few of those, can easily conceive. From this time till August 1789 I scarcely ever passed a comfortable day; seven months of this time I had been confined, my health seemed much impaired, my strength was diminished, and my appetite almost gone. In this state my friends pressed me to consult you. I was unwilling for some time to do it, as I had lost all hope of relief; however, when I had determined to apply to you, I likewise determined to give up every prejudice of my own respecting my case, and to adhere most strictly to your advice. On the 20th of August 1789 I consulted you, on the 25th I entered upon the regimen, which you prescribed, and which was as follows.

"Drink no malt liquor on any account. Let your beverage at "dinner consist of two glasses of wine diluted with three half-pints "of water. On no account drink any more wine or spirituous "liquors in the course of the day; but, if you want more liquid, "take cream and water, or milk and water, or lemonade, with tea, "coffee, chocolate. Use the warm bath twice a week for half an "hour before going to bed, at the degree of heat which is most "grateful to your sensations. Eat meat constantly at dinner, and "with it any kind of tender vegetables you please. Keep the body "open by two evacuations daily, if possible without medicine, if not "take the size of a nutmeg of lenitive electuary occasionally, or five "grains of rhubarb every night. Use no violent exercise, which "may subject yourself to sudden changes from heat to cold; but as "much moderate exercise as may be, without being much fatigued "or starved with cold. Take some supper every night; a small "quantity of animal food is preferred; but if your palate refuses "this, take vegetable food, as fruit pie, or milk; something should "be eaten, as it might be injurious to you to fast too long." To

 the

the whole of this I adhered moſt ſcrupulouſly, and ſoon found my appetite improve, and with it my ſtrength and ſpirits. I had in December a ſevere attack, and two or three ſlight ones in the courſe of twelve months; but the improvement in the general ſtate of my health induced me to perſevere. On the 18th of Auguſt 1790 I had another ſevere attack, but it went off eaſier than before, and I ſoon recovered ſufficiently to go to Buxton, which you adviſed me to, and from which I reaped great benefit; nevertheleſs on the 29th of December I had a ſlight attack in compariſon of ſome that I had before experienced, and from that time I was free from gout, and enjoyed my health perfectly well till the fourth week in October 1791; from that till the third week in October 1792; from that till the third week in October 1793; and from that till June 1794. From what happened for the laſt three years I dreaded the month of October; but I eſcaped then, and have enjoyed my health moſt perfectly ever ſince till within the laſt week, that I have had a ſlight attack in one knee, which is nearly gone, without any ſymptom to lead me to ſuppoſe that it will go further.

"I adhered to your advice moſt ſcrupulouſly for the firſt year; and in regard to the not drinking malt liquor, and taking only the two glaſſes of wine with water, I have never deviated but two days; and then the firſt day I only drank one glaſs of ale and one glaſs of Champaigne; on the ſecond only one glaſs of Champaigne. With regard to the warm bath, I only uſe it now when I have gouty ſymptoms upon me, and in ſuch ſituations I find it of infinite ſervice; and in other reſpects I continue to live according to your direction.

"Many perſons have laughed at the idea of my perſeverance in a ſyſtem, which has not been able to *cure* the gout after five years trial; but ſuch perſons are either ignorant of what I before ſuffered, or totally unacquainted with the nature of the diſorder. Under the bleſſing of Providence, by an adherence to your advice, I am reaping all

all the benefit you flattered me I might expect from it, viz. my attacks less frequent, my sufferings less acute, and an improvement in the general state of my health.

"I have been particular in this account of myself at your request, and am, Sir, &c.

MORLEY, near DERBY,
February 10th, 1795.

ROBERT WILMOT."

There are situations nevertheless in which a paroxysm of gout has been believed to be desirable, as relieving the patient from other disagreeable diseases, or debilities, or sensations. Thus when the liver is torpid, a perpetual uneasiness and depression of spirits occur; which a fit of gout is supposed to cure by a metastasis of the disease. Others have acquired epileptic fits, probably from the disagreeable sensation of a chronically inflamed liver; which they suppose the pain and inflammation of gout would relieve. When gouty patients become much debilitated by the progress of the disease, they are liable to dropsy of the chest, which they suppose a fit of the gout would relieve. But in all these cases the attempt to procure a paroxysm of gout by wine, or aromatics, or volatiles, or blisters, or mineral waters, seldom succeeds; and the patients are obliged to apply to other methods of relief adapted to their particular cases. In the two former situations small repeated doses of calomel, or mercurial unction on the region of the liver may succeed, by giving new activity to the vessels of the liver, either to secrete or to absorb their adapted fluids, and thus to remove the cause of the gout, rather than to promote a fit of it. In the last case the tincture of digitalis, and afterwards the class of sorbentia, must be applied to.

M. M. In young strong patients the gout should be cured by venesection and cathartics and diluents, with poultices externally. But it has a natural crisis by producing calcareous matter on the inflamed membrane, and therefore in old enfeebled people it is safest to



wait

wait for this crisis, attending to the natural evacuations and the degree of fever; and in young ones, where it is not attended with much fever, it is customary and popular not to bleed, but only to keep the body open with aloes, to use gentle sudorifics, as neutral salts, and to give the bark at the decline of the fit; which is particularly useful where the patient is much debilitated. See Arthritis ventriculi, Class I. 2. 4. 6. and Sect. XXV. 17.

When there is not much fever, and the patient is debilitated with age, or the continuance of the disease, a moderate opiate, as twenty drops of tincture of opium, or one grain of solid opium, may be taken every night with advantage. Externally a paste made with double the quantity of yeast is a good poultice; and booterkins made with oiled silk, as they confine the perspirable matter, keep the part moist and supple, and thence relieve the pain like poultices.

The only safe way of moderating the disease is by an uniform and equal diminution, or a total abstinence from fermented liquors, with the cautions directed in Sect. XII. 7. 8. The continued use of strong bitters, as of Portland's powder, or bark, has been frequently injurious, as spoken of in the Materia Medica, Art. IV. 2. 11.

One of my acquaintance, who was much afflicted with the gout, abstained for about half a year from beer and wine; and not having resolution to persist, returned to his former habits of potation in less quantity; and observed that he was then for one winter stronger and freer from the gout than usual. This however did not long continue, as the disease afterwards returned with its usual or increased violence. This I think is a circumstance not unlikely to occur, as opium has a greater effect after its use has been a while intermitted; and the debility or torpor, which is the cause of gout, is thus for a few months prevented by the greater irritability of the system, acquired during the lessened use of fermented liquor.

For the same reason an ounce of spirituous tincture of guaiacum, or of bark, is said to have for some time prevented returns of the gout;

 which

which has afterwards, like all other great ſtimuli when long continued, been ſucceeded by greater debility, and deſtroyed the patient. This ſeems to have been exemplified in the caſe of the ingenious Dr. Bown, ſee Preface to his Elementa Medicinæ; he found temporary relief from the ſtimulus of wine, regardleſs of its future effects.

16. *Rheumatiſmus.* Acute rheumatiſm. There is reaſon to ſuſpect, that rheumatic inflammations, like the gouty ones, are not a primary diſeaſe; but that they are the conſequence of a tranſlation of morbid action from one part of the ſyſtem to another. This idea is countenanced by the frequent change of place of rheumatic-like gouty inflammations, and from their attacking two ſimilar parts at the ſame time, as both ankles and both wriſts, and theſe attacks being in ſucceſſion to each other. Whereas it is not probable that both feet or both hands ſhould at the ſame time be equally expoſed to any external cauſe of the diſeaſe, as to cold or moiſture; and leſs ſo that theſe ſhould occur in ſucceſſion. Laſtly, from the inflammatory diatheſis in this diſeaſe being more difficult to ſubdue, and more dangerous in event, than other common inflammations, eſpecially to pregnant women, and in weak conſtitutions.

From this idea of the rheumatiſm being not a primary diſeaſe, like the gout, but a transferred morbid action owing to the previous torpor of ſome other part of the ſyſtem, we perceive why it attacks weak people with greater pertinacity than ſtrong ones; reſiſting or recurring again and again after frequent evacuations, in a manner very different from primary inflammations; becauſe the cauſe is not removed, which is at a diſtance from the ſeat of the inflammation.

This alſo accounts for rheumatic inflammations ſo very rarely terminating in ſuppuration, becauſe like the gout the original cauſe is not in the inflamed part, and therefore does not continue to act after the

the inflammation commences. Inſtead of ſuppuration in this diſeaſe, as well as in the gout, a quantity of mucus or coagulable lymph is formed on the inflamed membrane; which in the gout changes into chalkſtones, and in the rheumatiſm is either reabſorbed, or lies on the membrane, producing pains on motion long after the termination of the inflammation, which pains are called chronic rheumatiſm. The membranes, which have thus been once or repeatedly inflamed, become leſs mobile, or leſs liable to be affected by ſympathy, as appears by the gout affecting new parts, when the joints of the foot have been frequently inflamed by it; hence as the cauſe of the inflammation does not exiſt in the inflamed part, and as this part becomes leſs liable to future attacks, it ſeldom ſuppurates.

Secondly, when rheumatiſm affects the muſcles of the cheſt, it produces ſymptoms ſimilar to pleuriſy, but are diſtinguiſhed from that by the patient having previouſly ſuffered rheumatic affections in other parts, and by the pertinacity or continuance of the inflammatory ſtate of the patient, this ſhould be termed pleurodyne rheumatica.

Thirdly, when rheumatic inflammation affects the bowels, it produces a diſeaſe very different from enteritis, or common inflammation of the bowels, and ſhould be termed enteralgia rheumatica. The pain is leſs than in enteritis, and the diſeaſe of longer continuance, with harder pulſe, and the blood equally ſizy. It is attended with frequent dejections, with much mucus, and previous griping pains, but without vomiting; and differs perhaps from dyſentery from its not being attended with bloody ſtools, and not being infectious.

Fourthly, there is another kind of rheumatiſm attended with debility, which ſuppurates, and ſhould be termed rheumatiſmus ſuppurans. It is generally believed to be the gout, till ſuppuration takes place on the ſwelled joint; and, as the patient ſinks, there are ſloughs formed over the whole mouth; and he ſeems to be deſtroyed by inflammation or gangrene of the mucous membranes. I have

twice

twice ſeen this diſeaſe in patients about ſixty. Some other diſeaſes are erroneouſly called rheumatic, as hemicrania, and odontalgia. See Sect. XXVI. 3.

M. M. In the three former kinds veneſection repeatedly. Cathartics. Antimonials. Diluents. Neutral ſalts. Oil. Warm bath. Afterwards the bark. Opium with or without ipecacuanha; but not till the patient is conſiderably weakened. Sweats forced early in the diſeaſe do injury. Opium given early in the diſeaſe prolongs it. In the laſt kind, gentle ſtimulants, as wine and water, mucilage, ſorbentia.

The following is a caſe of ſuppurative rheumatiſm. Mr. F——, about ſixty, was ſuppoſed to have the gout in his hand, which however ſuppurated, and it was then called the ſuppurative rheumatiſm. He had lived rather intemperately in reſpect to wine, and was now afflicted with a tendency to inflammation of the mucous membranes. As he lay on the bed half reſupine, propped up with pillows, and alſo ſlept in that poſture, his lower jaw dropped by its own weight, when the voluntary power of the muſcles was ſuſpended. The mucus of his mouth and throat became quite dry, and at length was ſucceeded with ſloughs; this was a moſt diſtreſſing circumſtance to him, and was in vain endeavoured to be relieved by ſupporting his jaw by ſlender ſteel ſprings fixed to his night-cap, and by ſprings of elaſtic gum. The ſloughs ſpread and ſeemed to accelerate his death. See Claſs I. 1. 3. 2.

17. *Eryſipelas.* The eryſipelas differs from the zona ignea, and other ſpecies of herpes, in its being attended with fever, which is ſometimes of the ſenſitive irritated or inflammatory kind, with ſtrong and full pulſe; and at other times with weak pulſe and great inirritability, as when it precedes or attends mortifications. See Claſs II. 1. 3. 2.

Like the zona ignea above deſcribed, it ſeems to be a ſecondary diſeaſe, having for its primary part the torpor or inflammation of ſome

internal

internal or diſtant membrane, as appears from its ſo frequently attending wounds; ſometimes ſpreading from iſſues over the whole limb, or back, by ſympathy with a tendon or membrane, which is ſtimulated by the peaſe in them. In its more violent degree I ſuppoſe that it ſympathizes with ſome extenſive internal membranes, as of the liver, ſtomach, or brain. Another reaſon, which countenances this idea, is, that the inflammation gradually changes its ſituation, one part healing as another inflames; as happens in reſpect to more diſtant parts in gout and rheumatiſm; and which ſeems to ſhew, that the cauſe of the diſeaſe is not in the ſame place with the inflammation. And thirdly, becauſe the eryſipelas of the face and head is liable to affect the membranes of the brain; which were probably in theſe caſes the original or primary ſeat of the diſeaſe; and laſtly, becauſe the fits of eryſipelas, like thoſe of the gout, are liable to return at certain annual or monthly periods, as further treated of in Claſs II. 1. 3. 2.

Many caſes of eryſipelas from wounds or bruiſes are related in Deſault's Surgical Journal, Vol. II. in which poultices are ſaid to do great injury, as well as oily or fatty applications. Saturnine ſolutions were ſometimes uſed with advantage. A grain of emetic tartar given to clear the ſtomach and bowels, is ſaid to be of great ſervice.

18. *Teſtium tumor in gonorrhœa.* Mr. Hunter in his Treatiſe on the Venereal Diſeaſe obſerves, that the tumor of the teſtes in gonorrhœa ariſes from their ſympathy with the inflammation of the urethra; and that they are not ſimilar to the actions ariſing from the application of venereal matter, whether by abſorption or otherwiſe; as they ſeldom or never ſuppurate; and when ſuppuration happens, the matter produced is not venereal. Treatiſe on Venereal Diſeaſe, p. 53.

19. *Teſtium tumor in parotidite.* The ſympathy between ſome parts about the throat and the genitals has been treated of in Claſs IV. 1. 2. 7. The ſwelling of the teſtes, when that of the parotis ſubſides, ſeems

to

to arife from the affociation of fucceffive action; as the tenfion of the penis in hydrophobia appears to arife from the previous fynchronous affociations of the fenfitive motions of thefe parts; but the manner of the production of both thefe affociations is yet very obfcure. In women a fwelling of the breafts often fucceeds the decline of the mumps by another wonderful fympathy. See Clafs IV. 1. 2. 7. and I. 1. 2. 15. In many perfons a delirium fucceeds the fwelling of the parotis, or the fubfequent ones of the teftes or breafts; which is fometimes fatal, and feems to arife from a fympathy of fucceffive action, and not of fynchronous action, of the membranes of the brain with thofe of the parotide glands. Sometimes a ftupor comes on inftead of this delirium, which is relieved by fomenting the fhaved head for an hour or two. See Clafs II. 1. 3. 4.

## ORDO I.

*Increaſed Aſſociate Motions.*

## GENUS III.

*Catenated with Voluntary Motions*

### SPECIES.

1. *Deglutitio invita.* When any one is told not to ſwallow his ſaliva, and that eſpecially if his throat be a little ſore, he finds a neceſſity of immediately ſwallowing it; and this the more certainly, the more he voluntarily endeavours not to do ſo.

In this caſe the voluntary power exerted by our attention to the pharinx renders it more ſenſible to irritation, and therefore occaſions it to be more frequently induced to ſwallow the ſaliva. Here the irritation induces a volition to ſwallow it, which is more powerful than the deſire not to ſwallow it. See XXIV. 1. 7. So in reverie, when the voluntary power was exerted on any of the ſenſes, as of ſight or taſte, the objects of thoſe ſenſes became perceived; but not otherwiſe. Sect. XIX. 6. This is a troubleſome ſymptom in ſome ſore throats.

M. M. Mucilage, as ſugar and gum arabic. Warm water held in the mouth frequently, as a fomentation to the inflamed throat.

2. *Nictitatio invita.* Involuntary winking with the eye-lids, and twitchings of the face, are originally induced by an endeavour to relieve ſome diſagreeable ſenſations about inflamed eyes, as the dazzling of light; and afterwards theſe motions become catenated with other motions

motions or ſenſations, ſo as not to be governed by the will. Here the irritation firſt produces a volition to wink, which by habit becomes ſtronger than the anti-volition not to wink.

This ſubject is rendered difficult from the common acceptation of the word, volition, including previous deliberation, as well as the voluntary exertion, which ſucceeds it. In the volitions here ſpoken of there is no time for deliberation or choice of objects, but the voluntary act immediately ſucceeds the ſenſation which excites it.

M. M. Cover the affected parts with a ſticking plaſter or a bliſter. Paſs a fine needle and thread through a part of the ſkin over the muſcle, which moves, and attach the other end of the thread by a ſticking plaſter to a diſtant part. An iſſue behind the ear. To practiſe daily by a looking-glaſs to ſtop the motions with the hand. See the cure of a caſe of the leaping of a muſcle of the arm, Sect. XVII. 1. 8. See Convulſio debilis, Claſs III. 1. 1. 5.

3. *Riſus invitus.* Involuntary laughter. When the pleaſure ariſing from new combinations of words and ideas, as in puns; or of other circumſtances, which are ſo trivial, as to induce no voluntary exertion to compare or conſider their preſent importance or their future conſequence; the pleaſure is liable to riſe into pain; that is, the ideas or ſenſual motions become exerted too violently for want of ſome antitheſiſtic ideas; in the ſame manner as thoſe muſcles, which have weak antagoniſts, as thoſe of the calf of the leg, are liable to fall into cramp or painful contraction. In this ſituation a ſcream is begun to relieve this pain of ideas too violently exerted, which is ſtopped again ſoon, as explained in Sect. XXXIV. 1. 4. and Claſs III. 1. 1. 4. and IV. 2. 3. 3.

The pain, into which this pleaſure riſes, which would excite the ſcream of laughter, has been felt forcibly by every one; when they have been under ſuch circumſtances, as have induced them to reſtrain it by a counter-volition; till at length the increaſed aſſociate motions

 produce

produce ſo much pain as to overcome the counter-volition, and the patient burſts out into indecent laughter, contrary to his will in the common acceptation of that word.

4. *Luſus digitorum invitus.* An awkward playing with the fingers in ſpeaking in public. Theſe habits are began through baſhfulneſs, and ſeem rather at firſt deſigned to engage the attention in part, and thus prevent the diſagreeable ideas of mauvaiſe hont; as timorous boys whiſtle, when they are obliged to walk in the dark; and as it is ſometimes neceſſary to employ raw ſoldiers in perpetual manœuvres, as they advance to the firſt charge.

5. *Unguium morſiuncula invita.* Biting the nails is a depraved habit ariſing from ſimilar cauſes as thoſe of the laſt article.

M. M. Dip the fingers in ſolution of aloes.

6. *Vigilia invita.* Watchfulneſs, where the perſon wiſhes, and endeavours to fall aſleep, properly belongs to this place, as the wiſh or volition to ſleep prevents the deſired effect; becauſe ſleep conſiſts in an abolition of volition. See Claſs III. 1. 2. 3.

ORDO

## ORDO I.

### *Increaſed Aſſociate Motions.*

## GENUS IV.

### *Catenated with External Influences.*

## SPECIES.

1. *Vita ovi.* Life of an egg. The eggs of fowls were ſhewn by Mr. J. Hunter to reſiſt the freezing proceſs in their living ſtate more powerfully, than when they were killed by having the yolk and white ſhook together. Philoſ. Tranſ. It may be aſked, does the heat during the incubation of eggs act as a ſtimulus exciting the living principle into activity? Or does it act ſimply as a cauſa ſine quâ non, as an influence, which penetrating the maſs, removes the particles of it to a greater diſtance from each other, ſo as to allow their movement over each other, in the ſame manner as heat is conceived to produce the fluidity of water; not by ſtimulus, but by its penetrating influence? Or may elementary heat in its uncombined ſtate be ſuppoſed to act only as an influence neceſſary to life in its natural quantity; whence torpor and death follows the eduction of it from the body; but in its increaſed ſtate above what is natural, or uſual, that it acts as a ſtimulus; which we have a ſenſe to perceive; and which excites many parts of the ſyſtem into unnatural action? See Claſs IV. 1. 1. C.

2. *Vita hiemi-dormientium.* The torpor of inſects, and birds, and quadrupeds, during the cold ſeaſon, has been called ſleep; but I ſup-

pose it must differ very much from that state of animal life, since not only all voluntary power is suspended, but sensation and vascular motion has ceased, and can only be restored by the influence of heat. There have been related instances of snails, which have recovered life and motion on being put into water after having experienced many years of torpidity, or apparent death, in the cabinets of the curious. Here the water as well as the heat are required not only as a stimulus, but as a causa sine quâ non of fluidity and motion, and consequent life.

3. *Pullulatio arborum.* The annual revivescence of the buds of trees seems not only to be owing to the influence of the returning warmth of the spring, but also to be catenated with solar gravitation; because seeds and roots and buds, which are analogous to the eggs of animals, put forth their shoots by a less quantity of heat in spring, than they had undergone in the latter part of autumn, which may however be ascribed to their previous torpid state, and consequent accumulation of sensorial power, or irritability; as explained in Botanic Garden, Part II. Cant. I. l. 322. note. Other circumstances, which countenance the idea, that vegetation is affected by solar gravitation, as well as by heat, may be observed in the ripening of the seeds of plants both in those countries where the summers are short, and in those where they are long. And by some flowers closing their bells at noon, or soon after; and hence seem to sleep rather at solar diurnal periods, than from the influence of cold, or the deficiency of light.

4. *Orgasmatis venerei periodus.* The venereal orgasm of birds and quadrupeds commences or returns about the vernal or autumnal equinoxes, and thence seems in respect to their great periods to be governed by solar influence. But if this orgasm be disappointed of its object, it is said to recur at about monthly periods, as observed in mares

mares and bitches in this refpect refembling the female catamenia. See Sect. XXXVI. 2. 3. and Sect. XVI. 13.

5. *Brachii concuffio electrica.* The movement of the arm, even of a paralytic patient, when an electric fhock is paffed through it, is owing to the ftimulus of the excefs of electricity. When a piece of zinc and filver, each about the fize of a crown-piece, are placed one under the upper lip, and the other on the tongue, fo as the outer edges may be brought into contact, there is an appearance of light in the eyes, as often as the outer edges of thefe metals are brought into contact or feparated; which is another inftance of the ftimulus of the paffage of electric fhocks through the fibres of the organs of fenfe, as well as through the mufcular fibres. See Sect. XII. 1. 1. and firft addit. note to Vol. I. of this work. But in its natural ftate electricity feems only to act as an influence on animal and vegetable bodies; of the falutary or injurious effects of which we have yet no precife knowledge.

Yet if regular journals were kept of the variations of atmofpheric electricity, it is probable fome difcoveries of its influence on our fyftem might in time be difcovered. For this purpofe a machine on the principle of Mr. Bennet's electric doubler might be applied to the pendulum of a clock, fo as to manifeft, and even to record the daily or hourly variations of aerial electricity. Which has already been executed, and applied to the pendulum of a Dutch wooden clock, by Mr. Bennet, curate of Wirkfworth in Derbyfhire.

Befides the variations of the degree or kind of atmofpheric electricity, fome animals, and fome men, feem to poffefs a greater power of accumulating this fluid in themfelves than others. Of which a famous hiftory of a Ruffian prince was lately publifhed; who, during the clear and fevere frofts of that country, could not move himfelf in bed without luminous corrufcations. Such may have been the cafe of thofe people, who have been related to have taken fire fpontaneoufly, and

and to have been reduced to aſhes. The electric concuſſion from the gymnotus electricus, and torpedo, are other inſtances of the power of the animal ſyſtem to accumulate electricity, as in theſe it is uſed as a weapon of defence, or for the purpoſe of taking their prey.

Some have believed that the accumulation or paſſage of the magnetic fluid might affect the animal ſyſtem, and have aſſerted that the application of a large magnet to an aching tooth has quickly effected a cure. If this experiment is again tried in odontalgia, or hemicrania, the painful membrane of the tooth or head ſhould be included between the ſouth and north poles of a horſe-ſhoe magnet, or between the contrary poles of two different magnets, that the magnetiſm may be accumulated on the torpid part.

6. *Oxygenatio ſanguinis.* The variation of the quantity of oxygen gas exiſting in the atmoſphere muſt affect all breathing animals; in its exceſs this too muſt be eſteemed a ſtimulus; but in its natural quantity would ſeem to act as an influence, or cauſe, without which animal life cannot exiſt even a minute. It is hoped that Dr. Beddoes's plan for a pneumatic infirmary, for the purpoſe of putting this and various other airs to the teſt of experiment, will meet with public encouragement, and render conſumption, aſthma, cancer, and many diſeaſes conquerable, which at preſent prey with unremitted devaſtation on all orders and ages of mankind.

7. *Humectatio corporis.* Water, and probably the vapour of water diſſolved or diffuſed in the atmoſphere, unites by mechanical attraction with the unorganized cuticle, and ſoftens and enlarges it; as may be ſeen in the looſe and wrinkled ſkin of the hands of waſherwomen; the ſame probably occurs to the mucous membrane of the lungs in moiſt weather; and by thickening it increaſes the difficulty of reſpiration of ſome people, who are ſaid to be aſthmatical. So far water may be ſaid to act as an influx or influence, but when it is taken up by the mouths

mouths of the abſorbent ſyſtem, it muſt excite thoſe mouths into action, and then acts as a ſtimulus.

There appears from hence to be four methods by which animal bodies are penetrated by external things. 1. By their ſtimulus, which induces the abſorbent veſſels to imbibe them. 2. By mechanical attraction, as when water ſoftens the cuticle. 3. By chemical attraction, as when oxygen paſſes through the membranes of the air-veſſels of the lungs, and combines with the blood. And laſtly, by influx without mechanical attraction, chemical combination, or animal abſorption, as the univerſal fluids of heat, gravitation, electricity, magnetiſm, and perhaps of other ethereal fluids yet unknown.

## ORDO II.

*Decreased Associate Motions.*

## GENUS I.

*Catenated with Irritative Motions.*

As irritative muscular motions are attended with pain, when they are exerted too weakly, as well as when they are exerted too strongly; so irritative ideas become attended with sensation, when they are exerted too weakly, as well as when they are exerted too strongly. Which accounts for these ideas being attended with sensation in the various kinds of vertigo described below.

There is great difficulty in tracing the immediate cause of the deficiences of action of some links of the associations of irritative motions; first, because the trains and tribes of motions, which compose these links, are so widely extended as to embrace almost the whole animal system; and secondly, because when the first link of an associated train of actions is exerted with too great energy, the second link by reverse sympathy may be affected with torpor. And then this second link may transmit, as it were, this torpor to a third link, and at the same time regain its own energy of action; and it is possible this third link may in like manner transmit its torpor to a fourth, and thus regain its own natural quantity of motion.

I shall endeavour to explain this by an example taken from sensitive associated motions, as the origin of their disturbed actions is more easily detected. This morning I saw an elderly person, who had gradually lost all the teeth in his upper jaw, and all of the under except three of the molares; the last of these was now loose, and occasionally

ſionally painful; the fangs of which were almoſt naked, the gums being much waſted both within and without the jaw. He is a man of attentive obſervation, and aſſured me, that he had again and again noticed, that, when a pain commenced in the membranes of the alveolar proceſs of the upper jaw oppoſite to the looſe tooth in the under one (which had frequently occurred for ſeveral days paſt), the pain of the looſe tooth ceaſed. And that, when the pain afterwards extended to the ear and temple on that ſide, the pain in the membranes of the upper jaw ceaſed. In this caſe the membranes of the alveolar proceſs of the upper jaw became torpid, and conſequently painful, by their reverſe ſympathy with the too violent actions of the inflamed membranes of the looſe tooth; and then by a ſecondary ſympathy the membranes about the ear and temple became torpid, and painful; and thoſe of the alveolar proceſs of the upper jaw regained their natural quantity of action, and ceaſed to be painful. A great many more nice and attentive obſervations are wanted to elucidate theſe curious circumſtances of aſſociation, which will be found to be of the greateſt importance in the cure of many diſeaſes, and lead us to the knowledge of fever.

## SPECIES.

1. *Cutis frigida pranſorum.* Chillneſs after dinner frequently attends weak people, or thoſe who have been exhauſted by exerciſe; it ariſes from the great expenditure of the ſenſorial power on the organs of digeſtion, which are ſtimulated into violent action by the aliment; and the veſſels of the ſkin, which are aſſociated with them, become in ſome meaſure torpid by reverſe ſympathy; and a conſequent chillneſs ſucceeds with leſs abſorption of atmoſpheric moiſture. See the ſubſequent article.

2. *Pallor urinæ pranſorum.* The paleneſs of urine after a full meal is an inſtance of reverſe aſſociation; where the ſecondary part of a train of aſſociate motions acts with leſs energy in conſequence of the greater exertions of the primary part. After dinner the abſorbent veſſels of the ſtomach and inteſtines are ſtimulated into greater action, and drink up the newly taken aliment; while thoſe, which are ſpread in great number on the neck of the bladder, abſorb leſs of the aqueous part of the urine than uſual, which is therefore diſcharged in a more dilute ſtate; and has been termed crude by ſome medical writers, but it only indicates, that ſo great a proportion of the ſenſorial power is expended on digeſtion and abſorption of the aliment, that other parts of the ſyſtem act for a time with leſs energy. See Claſs IV. 1. 1. 6.

3. *Pallor urinæ a frigore cutaneo.* There is a temporary diſcharge of pale water, and a diarrhœa, induced by expoſing the ſkin to the cold air; as is experienced by boys, who ſtrip themſelves before bathing. In this caſe the mouths of the cutaneous lymphatics become torpid by the ſubduction of their accuſtomed degree of heat, and thoſe of the bladder and inteſtines become torpid by direct ſympathy; whence leſs of the thinner part of the urinary ſecretion, and of the mucus of the inteſtines, is reabſorbed. See Sect. XXIX. 4. 6. This effect of ſuddenly cooling the ſkin by the aſperſion of cold water has been uſed with ſucceſs in coſtiveneſs, and has produced evacuations, when other means have failed. When young infants are afflicted with griping joined with coſtiveneſs, I have ſometimes directed them to be taken out of a warm bed, and carried about for a few minutes in a cool room, with almoſt inſtant relief.

4. *Pallor ex ægritudine.* When ſickneſs of ſtomach firſt occurs, a paleneſs of the ſkin attends it; which is owing to the aſſociation or catenation between the capillaries of the ſtomach and the cataneous ones;

ones; which at firſt act by direct ſympathy. But in a ſhort time there commences an accumulation of the ſenſorial power of aſſociation in the cutaneous capillaries during their ſtate of inactivity, and then the ſkin begins to glow, and ſweats break out, from the increaſed action of the cutaneous glands or capillaries, which is now in reverſe ſympathy with thoſe of the ſtomach. So in continued fevers, when the ſtomach is totally torpid, which is known by the total averſion to ſolid food, the cutaneous capillaries are by reverſe ſympathy in a perpetual ſtate of increaſed activity, as appears from the heat of the ſkin.

5. *Dyſpnœa a balneo frigido.* The difficulty of breathing on going up to the middle in cold water is owing to the irritative aſſociation or catenation of the action of the extreme veſſels of the lungs with thoſe of the ſkin. So that when the latter are rendered torpid or inactive by the application of ſudden cold, the former become inactive at the ſame time, and retard the circulation of the blood through the lungs, for this difficulty of breathing cannot be owing to the preſſure of the water impeding the circulation downwards, as it happens equally by a cold ſhower-bath, and is ſoon conquered by habitual immerſions. The capillaries of the ſkin are rendered torpid by the ſubduction of the ſtimulus of heat, and by the conſequent diminution of the ſenſorial power of irritation. The capillaries of the lungs are rendered torpid by the diminution of the ſenſorial power of aſſociation, which is now excited in leſs quantity by the leſſened actions of the capillaries of the ſkin, with which they are catenated. So that at this time both the cutaneous and pulmonary capillaries are principally actuated, as far as they have any action, by the ſtimulus of the blood. But in a ſhort time the ſenſorial powers of irritation, and of aſſociation, become accumulated, and very energetic action of both theſe membranes ſucceed. Which thus reſemble the cold and hot fit of an intermittent fever.

6. *Dyspepsia a pedibus frigidis.* When the feet are long cold, as in riding in cold and wet weather, some people are very liable to indigestion and consequent heart-burn. The irritative motions of the stomach become torpid, and do their office of digestion imperfectly, in consequence of their association with the torpid motions of the vessels of the extremities. Fear, as it produces paleness and torpidity of the skin, frequently occasions temporary indigestion in consequence of this association of the vessels of the skin with those of the stomach; as riding in very bad roads will give flatulency and indigestion to timorous people.

A short exposure to cold air increases digestion, which is then owing to the reverse sympathy between the capillary vessels of the skin, and of the stomach. Hence when the body is exposed to cold air, within certain limits of time and quantity of cold, a reverse sympathy of the stomach and the skin first occurs, and afterwards a direct sympathy. In the former case the expenditure of sensorial power by the skin being lessened, but not its production in the brain; the second link of the association, viz. the stomach, acquires a greater share of it. In the latter case, by the continuation of the deficient stimulus of heat, the torpor becomes extended to the brain itself, or to the trunks of the nerves; and universal inactivity follows.

7. *Tussis a pedibus frigidis.* On standing with the feet in thawing snow, many people are liable to incessant coughing. From the torpidity of the absorbent vessels of the lungs, in consequence of their irritative associations with those of the skin, they cease to absorb the saline part of the secreted mucus; and a cough is thus induced by the irritation of this saline secretion; which is similar to that from the nostrils in frosty weather, but differs in respect to its immediate cause; the former being from association with a distant part, and the latter from defect of the stimulus of heat on the nostrils themselves. See Catarrhus frigidus, Class I. 2. 3. 3.

8. *Tussis*

8. *Tussis hepatica.* The cough of inebriates, which attends the enlargement of the liver, or a chronical inflammation of its upper membrane, is supposed to be produced by the inconvenience the diaphragm suffers from the compression or heat of the liver. It differs however essentially from that attending hepatitis, from its not being accompanied with fever. And is perhaps rather owing to irritative association, or reverse sympathy, between the lungs and the liver. As occurs in sheep, which are liable to a perpetual dry cough, when the fleuk-worm is preying on the substance of their livers. See Class II. 1. 1. 5.

M. M. From half a grain to a grain of opium twice a day. A drachm of mercurial ointment rubbed on the region of the liver every night for eight or ten times.

9. *Tussis arthritica.* Gout-cough. I have seen a cough, which twice recurred at a few years distance in the same person, during his fits of the gout, with such pertinacity and violence as to resist venesection, opiates, bark, blisters, mucilages, and all the usual methods employed in coughs. It was for a time supposed to be the hooping-cough, from the violence of the action of coughing; it continued two or three weeks, the patient never being able to sleep more than a few minutes at once during the whole time, and being propped up in bed with pillows night and day.

As no fever attended this violent cough, and but little expectoration, and that of a thin and frothy kind, I suspected the membrane of the lungs to be rather torpid than inflamed, and that the saline part of the mucus not being absorbed stimulated them into perpetual exertion. And lastly, that though the lungs are not sensible to cold and heat, and probably therefore less mobile; yet, as they are nevertheless liable to consent with the torpor of cold feet, as described in Species 6 of this Genus, I suspected this torpor of the lungs to succeed the gout in the feet, or to act a vicarious part for them.

10. *Vertigo*

10. *Vertigo rotatoria.* In the vertigo from circumgyration the irritative motions of viſion are increaſed; which is evinced from the pleaſure that children receive on being rocked in a cradle, or by ſwinging on a rope. For whenever ſenſation ariſes from the production of irritative motion with leſs energy than natural, it is of the diſagreeable kind, as from cold or hunger; but when it ariſes from their production with greater energy than natural, if it be confined within certain limits, it is of the pleaſurable kind, as by warmth or wine. With theſe increaſed irritative motions of viſion, I ſuppoſe thoſe of the ſtomach are performed with greater energy by direct ſympathy; but when the rotatory motions, which produce this agreeable vertigo, are continued too long, or are too violent, ſickneſs of the ſtomach follows; which is owing to the decreaſed action of that organ from its reverſe ſympathy with the increaſed actions of the organ of viſion. For the expenditure of ſenſorial power by the organ of viſion is always very great, as appears by the ſize of the optic nerves; and is now ſo much increaſed as to deprive the next link of aſſociation of its due ſhare. As mentioned in Article 6 of this Genus.

In the ſame manner the undulations of water, or the motions of a ſhip, at firſt give pleaſure by increaſing the irritative motions belonging to the ſenſe of viſion; but produce ſickneſs at length by expending on one part of the aſſociated train of irritative actions too much of that ſenſorial power, which uſually ſerved the whole of it; whence ſome other parts of the train acquire too little of it, and perform their actions in conſequence too feebly, and thence become attended with diſagreeable ſenſation.

It muſt alſo be obſerved, that when the irritative motions are ſtimulated into unuſual action, as in inebriation, they become ſucceeded by ſenſation, either of the pleaſurable or painful kind; and thus a new link is introduced between the irritative motions thus excited, and thoſe which uſed to ſucceed them; whence the aſſociation is either diſſevered or much weakened, and thus the vomiting in ſea-

ſickneſs

ſickneſs occurs from the defect of the power of aſſociation, rather than from the general deficiency of ſenſorial power.

When a blind man turns round, or when one, who is not blind, revolves in the dark, a vertigo is produced belonging to the ſenſe of touch. A blind man balances himſelf by the ſenſe of touch, which being a leſs perfect means of determining ſmall quantities of deviation from the perpendicular, occaſions him to walk more carefully upright than thoſe, who balance themſelves by viſion. When he revolves, the irritative aſſociations of the muſcular motions, which were uſed to preſerve his perpendicularity, become diſordered by their new modes of ſucceſſive exertion; and he begins to fall. For his feet now touch the floor in manners or directions different from thoſe they have been accuſtomed to; and in conſequence he judges leſs perfectly of the ſituation of the parts of the floor in reſpect to that of his own body, and thus loſes his perpendicular attitude. This may be illuſtrated by the curious experiment of croſſing one finger over the next to it, and feeling of a nut or bullet with the ends of them. When, if the eyes be cloſed, the nut or bullet appears to be two, from the deception of the ſenſe of touch.

In this vertigo from gyration, both of the ſenſe of ſight, and of the ſenſe of touch, the primary link of the aſſociated irritative motions is increaſed in energy, and the ſecondary ones are increaſed at firſt by direct ſympathy; but after a time they become decreaſed by reverſe ſympathy with the primary link, owing to the exhauſtion of ſenſorial power in general, or to the power of aſſociation in particular; becauſe in the laſt caſe, either pleaſurable or painful ſenſation has been introduced between the links of a train of irritative motions, and has diſſevered, or much enfeebled them.

Dr. Smyth, in his Eſſay on Swinging in Pulmonary Conſumption, has obſerved, that ſwinging makes the pulſe ſlower. Dr. Ewart of Bath confirmed this obſervation both on himſelf and on Col. Cathcart, who was then hectic, and that even on ſhipboard, where ſome degree of vertigo might be ſuppoſed previouſly to exiſt. Dr. Currie of Liverpool not only confirmed this obſervation frequently on himſelf, when he was alſo phthiſical, but found that equitation had a ſimilar effect on him, uniformly retarding his pulſe. This curious circumſtance cannot ariſe from the general effect of exerciſe, or fatigue, as in thoſe caſes the pulſe becomes weaker and quicker; it muſt therefore be aſcribed to a degree of vertigo, which attends all thoſe modes of motion, which we are not perpetually accuſtomed to.

Dr. Currie has further obſerved, that " in caſes of great debility the voluntary muſcular exertion requiſite in a ſwing produces wearineſs, that is, increaſes debility; and that in ſuch inſtances he had frequently noticed, that the diminution of the frequency of the pulſe did not take place, but the contrary." Theſe circumſtances may thus be accounted for.

The links of aſſociation, which are effected in the vertigo occaſioned by unuſual motion, are the irritative motions of the ſenſe of viſion, thoſe of the ſtomach, and thoſe of the heart and arteries. When the irritative ideas of viſion are exerted with greater energy at the beginning of vertigo, a degree of ſenſation is excited, which is of the pleaſurable kind, as above mentioned; whence the aſſociated trains of irritative motions of the ſtomach, and heart, and arteries, act at firſt with greater energy, both by direct ſympathy, and by the additional ſenſorial power of ſenſation. Whence the pulſe of a conſumptive patient becomes ſtronger and conſequently ſlower.

But if this vertigo becomes much greater in degree or duration, the firſt link of this train of aſſociated irritative motions expends too much of the ſenſorial power, which was uſually employed on the whole train; and the motions of the ſtomach become in conſequence exerted with leſs energy. This appears, becauſe in this degree of vertigo ſickneſs ſupervenes, as in ſea-ſickneſs, which has been ſhewn to be owing to leſs energetic action of the ſtomach. And the motions of the heart and arteries then become weaker, and in conſequence

quence more frequent, by their direct ſympathy with the leſſened actions of the ſtomach. See Supplement, I. 12. and Claſs II. 1. 6. 7. The general weakneſs from fatigue is owing to a ſimilar cauſe, that is, to the too great expenditure of ſenſorial power in the increaſed actions of one part of the ſyſtem, and the conſequent deficiency of it in other parts, or in the whole.

The abatement of the heat of the ſkin in hectic fever by ſwinging, is not only owing to the increaſed ventilation of cool air, but to the reverſe ſympathy of the motions of the cutaneous capillaries with thoſe of the heart and arteries; which occurs in all fevers with arterial debility, and a hot or dry ſkin. Hence during moderate ſwinging the action of the heart and arteries becomes ſtronger and ſlower, and the action of the capillaries, which was before too great, as appeared by the heat of the ſkin, now is leſſened by their reverſe ſympathy with that of the heart and arteries. See Supplement, I. 8.

11. *Vertigo viſualis.* Viſual vertigo. The vertigo rotatoria deſcribed above, was induced by the rotation or undulation of external objects, and was attended with increaſed action of the primary link of the aſſociated motions belonging to viſion, and with conſequent pleaſure. The vertigo viſualis is owing to leſs perfect viſion, and is not accompanied with pleaſurable ſenſation. This frequently occurs in ſtrokes of the palſy, and is then ſucceeded by vomiting; it ſometimes precedes epileptic fits, and often attends thoſe, whoſe ſight begins to be impaired by age.

In this vertigo the irritative ideas of the apparent motions of objects are leſs diſtinct, and on that account are not ſucceeded by their uſual irritative aſſociations of motion; but excite our attention. Whence the objects appear to librate or circulate according to the motions of our heads, which is called dizzineſs; and we loſe the means of balancing ourſelves, or preſerving our perpendicularity, by viſion. So that in this vertigo the motions of the aſſociated organs are decreaſed

 by

by direct ſympathy with their primary link of irritation; as in the preceding caſe of ſea-ſickneſs they are decreaſed by reverſe ſympathy.

When vertigo affects people about fifty years of age, their ſight has generally been ſuddenly impaired; and from their leſs accurate viſion they do not ſoon enough perceive the apparent motions of objects; like a perſon in a room, the walls of which are ſtained with the uniform figures of lozenges, explained in Sect. XX. 1. This is generally aſcribed to indigeſtion; but it ceaſes ſpontaneouſly, as the patient acquires the habit of balancing himſelf by leſs diſtinct objects.

A gentleman about 50 was ſeized with an uncommon degree of vertigo, ſo as to fall on the ground, and not to be able to turn his head, as he ſat up either in his chair or in his bed, and this continued eight or ten weeks. As he had many decayed teeth in his mouth, and the vertigo was preceded and ſometimes accompanied by pains on one ſide of his head, the diſeaſe of a tooth was ſuſpected to be the cauſe. And as his timidity was too great to admit the extraction of thoſe which were decayed; after the trial of cupping repeatedly, fomentations on his head, repeated bliſters, with valerian, Peruvian bark, muſk, opium, and variety of other medicines; mercurials were uſed, both externally and internally, with deſign to inflame the membranes of the teeth, and by that means to prevent the torpor of the action of the membranes about the temple, and parietal bone; which are catenated with the membranes of the teeth by irritative aſſociation, but not by ſenſitive aſſociation. The event was, that as ſoon as the gums became ſore with a ſlight ptyaliſm, the pains about the head and vertigo gradually diminiſhed, and during the ſoreneſs of his gums entirely ceaſed; but I believe recurred afterwards, though in leſs degree.

The idea of inflaming the membranes of the teeth to produce increaſed ſenſation in them, and thus to prevent their irritative connection with thoſe of the cranium, was taken from the treatment of triſmus,

mus, or locked jaw, by endeavouring to inflame the injured tendon; which is ſaid to prevent or to remove the ſpaſm of the muſcles of the jaw. See Claſs III. 1. 1. 13. and 15.

M. M. Emetics. Bliſters. Iſſues about the head. Extraction of decayed teeth. Slight ſalivation. Sorbentia. Incitantia.

12. *Vertigo ebrioſa.* Vertigo from intoxication is owing to the aſſociation of the irritative ideas of viſion with the irritative motions of the ſtomach. Whence when theſe latter become much increaſed by the immoderate ſtimulus of wine, the irritative motions of the retina are produced with leſs energy by reverſe ſympathy, and become at the ſame time ſucceeded by ſenſation in conſequence of their decreaſed action. See Sect. XXI. 3. and XXXV. 1. 2. So converſely when the irritative motions of viſion are increaſed by turning round, or by our unaccuſtomed agitation at ſea, thoſe of the ſtomach become inverted by reverſe ſympathy, and are attended in conſequence with diſagreeable ſenſation. Which decreaſed action of the ſtomach is in conſequence of the increaſed expenditure of the ſenſorial power on the irritative ideas of viſion, as explained in Vertigo rotatoria.

Whence though a certain quantity of vinous ſpirit ſtimulates the whole ſyſtem into increaſed action, and perhaps even increaſes the ſecretion of ſenſorial power in the brain; yet as ſoon as any degree of vertigo is produced, it is a proof, that by the too great expenditure of ſenſorial power by the ſtomach, and its neareſt aſſociated motions, the more diſtant ones, as thoſe of viſion, become imperfectly exerted. From hence may be deduced the neceſſity of exhibiting wine in fevers with weak pulſe in only appropriated quantity; becauſe if the leaſt intoxication be induced, ſome part of the ſyſtem muſt act more feebly from the unneceſſary expenditure of ſenſorial power.

13. *Vertigo febriculoſa.* Vertigo in fevers either proceeds from the general deficiency of ſenſorial power belonging to the irritative aſſociations,

ations, or to a greater expenditure of it on ſome links of the trains and tribes of aſſociated irritative motions. There is however a ſlighter vertigo attending all people, who have been long confined in bed, on their firſt riſing; owing to their having been ſo long unuſed to the apparent motions of objects in their erect poſture, or as they paſs by them, that they have loſt in part the habit of balancing themſelves by them.

14. *Vertigo cerebroſa.* Vertigo from injuries of the brain, either from external violence, or which attend paralytic attacks, are owing to the general deficiency of ſenſorial power. In theſe diſtreſsful ſituations the vital motions, or thoſe immediately neceſſary to life, claim their ſhare of ſenſorial power in the firſt place, otherwiſe the patient muſt die; and thoſe motions, which are leſs neceſſary, feel a deficiency of it, as theſe of the organs of ſenſe and muſcles; which conſtitute vertigo; and laſtly the voluntary motions, which are ſtill leſs immediately neceſſary to life, are frequently partially deſtroyed, as in palſy; or totally, as in apoplexy.

15. *Murmur aurium vertiginoſum.* The vertiginous murmur in the ears, or noiſe in the head, is compared to the undulations of the ſound of bells, or to the humming of bees. It frequently attends people about 60 years of age; and like the viſual vertigo deſcribed above is owing to our hearing leſs perfectly from the gradual inirritability of the organ on the approach of age; and the diſagreeable ſenſation of noiſe attending it is owing to the leſs energetic action of theſe irritative motions; which not being ſufficiently diſtinct to excite their uſual aſſociations become ſucceeded by our attention, like the indiſtinct view of the apparent motions of objects mentioned in vertigo viſualis. This may be better underſtood from conſidering the uſe, which blind men make of theſe irritative ſounds, which they have taught themſelves to attend to, but which eſcape the notice of others. The late

late blind Juſtice Fielding walked for the firſt time into my room, when he once viſited me, and after ſpeaking a few words ſaid, "this room is about 22 feet long, 18 wide, and 12 high;" all which he gueſſed by the ear with great accuracy. Now if theſe irritative ſounds from the partial loſs of hearing do not correſpond with the ſize or uſual echoes of the places, where we are; their catenation with other irritative ideas, as thoſe of viſion, becomes diſſevered or diſturbed; and we attend to them in conſequence, which I think unravels this intricate circumſtance of noiſes being always heard in the head, when the ſenſe of hearing begins to be impaired, from whatever cauſe it occurs.

This ringing in the ears alſo attends the vertigo from intoxication; for the irritative ideas of ſound are then more weakly excited in conſequence of the deficiency of the ſenſorial power of aſſociation. As is known by this alſo being attended with diſagreeable ſenſation, and by its accompanying other diſeaſes of debility, as ſtrokes on the head, fainting fits, and paralytic ſeizures. For in this vertigo from intoxication ſo much ſenſorial power in general is expended on the increaſed actions of the ſtomach, and its neareſt connections, as the capillaries of the ſkin; that there is a deficiency for the purpoſes of the other irritative aſſociations of motions uſually connected with it. This auditory vertigo attends both the rotatory and the viſual vertigo above mentioned; in the former it is introduced by reverſe ſympathy, that is, by the diminution of ſenſorial power; too great a quantity of it being expended on the increaſed irritative motions of viſion; in the latter it is produced either by the ſame cauſes which produce the viſual vertigo, or by direct ſympathy with it. See Sect. XX. 7.

M. M. Stimulate the internal ear by ether, or with eſſential oil diluted with expreſſed oil, or with a ſolution of opium in wine, or in water. Or with ſalt and water.

 16. *Tactus*,

16. *Tactus, gustus, olfactus vertiginosi.* Vertiginous touch, taste, and smell. In the vertigo of intoxication, when the patient lies down in bed, it sometimes happens even in the dark, that the bed seems to librate under him, and he is afraid of falling out of it. The same occurs to people, who are sea-sick, even when they lie down in the dark. In these the irritative motions of the nerves of touch, or irritative tangible ideas, are performed with less energy, in one case by reverse sympathy with the stomach, in the other by reverse sympathy with the nerves of vision, and in consequence become attended with sensation, and produce the fear of falling by other associations.

A vertigo of the sense of touch may be produced, if any one turns round for a time with his eyes shut, and suddenly stops without opening them; for he will for a time seem to be still going forwards; which is difficult to explain. See the notes at the end of the first and second volume belonging to Sect. XX. 6.

In the beginning of some fevers, along with incessant vomiting, the patients complain of disagreeable tastes in their mouth, and disagreeable odours; which are to be ascribed to the general debility of the great trains and tribes of associated irritative motions, and to be explained from their direct sympathy with the decreased action of a sick stomach; or from the less secretion of sensorial power in the brain. These organs of sense are constantly stimulated into action by the saliva or by the air; hence, like the sense of hunger, when they are torpid from want of stimulus, or from want of sensorial power; pain or disagreeable sensation ensues, as of hunger, or faintness, or sickness in one case; and the ideas of bad tastes or odours in the other. This accords with the laws of causation, Sect. IV. 5.

17. *Pulsus mollis in vomitione.* The softness of the pulse in the act of vomiting is caused by direct association between the heart and the stomach; as explained in Sect. XXV. 17. A great slowness of the pulsation

pulſation of the heart ſometimes attends ſickneſs, and even with intermiſſions of it, as in the exhibition of too great a doſe of digitalis.

18. *Pulſus intermittens a ventriculo.* When the pulſe firſt begins to intermit, it is common for the patient to bring up a little air from his ſtomach; which if he accompliſhes before the intermiſſion occurs, always prevents it; whence that this debility of the heart is owing to the direct aſſociation of its motions with thoſe of the ſtomach is well evinced. See Sect. XXV. 17.

I this morning ſaw Mr. ——, who has long had at times an unequal pulſe, with indigeſtion and flatulency, and occaſional aſthma; he was ſeized two days ago with diarrhœa, and this morning with ſickneſs, and his pulſe was every way unequal. After an emetic his pulſe ſtill continued very intermittent and unequal. He then took ſome breakfaſt of toaſt and butter, and tea, and to my great ſurpriſe his pulſe became immediately perfectly regular, about 100 in a minute, and not weak, by this ſtimulus on his ſtomach.

A perſon, who for many years had had a frequent intermiſſion of his pulſe, and occaſional palpitation of his heart, was relieved from them both for a time by taking about four drops of a ſaturated ſolution of arſenic three or four times a day for three or four days. As this intermiſſion of the pulſe is occaſioned by the direct aſſociation of the motions of the heart with thoſe of the ſtomach, the indication of cure muſt be to ſtrengthen the action of the ſtomach by the bark. Spice. Moderate quantities of wine. A bliſter. Half a grain of opium twice a day. Solution of arſenic?

19. *Febris inirritativa.* Inirritative fever deſcribed in Claſs I. 2. 1. 1. belongs to this place, as it conſiſts of diſordered trains and tribes of aſſociated irritative motions, with leſſened actions of the aſſociated organs. In this fever the pulſations of the heart and arteries are weakened or leſſened, not only in the cold paroxyſm, as in the irrita-

tive fever, but alſo in the hot paroxyſm. The capillary arteries or glands have their actions nevertheleſs increaſed after the firſt cold fit, as appears by the greater production of heat, and the glow of arterial blood, in the cutaneous veſſels; and laſtly, the action of the ſtomach is much impaired or deſtroyed, as appears by the total want of appetite to ſolid food. Whence it would ſeem, that the torpid motions of the ſtomach, whatever may occaſion them, are a very frequent cauſe of continued fever with weak pulſe; and that theſe torpid motions of the ſtomach do not ſufficiently excite the ſenſorial power of aſſociation, which contributes in health to actuate the heart and arteries along with the irritation produced by the ſtimulus of the blood; and hence the actions of theſe organs are weaker. And laſtly, that the accumulation of the ſenſorial power of aſſociation, which ought to be expended on the motions of the heart and arteries, becomes now exerted on the cutaneous and pulmonary capillaries. See Supplement I. 8. and Sect. XXXV. 1. 1. and XXXIII. 2. 10.

I have dwelt longer on the vertiginous diſeaſes in this genus, both becauſe of their great intricacy, and becauſe they ſeem to open a road to the knowledge of fever, which conſiſts of aſſociated trains and tribes of irritative or ſenſitive motions, which are ſometimes mixed with the vertiginous ones, and ſometimes ſeparate from them.

## ORDO II.

### *Decreafed Affociate Motions.*

## GENUS II.

### *Catenated with Senfitive Motions.*

IN this genus the fenforial power of affociation is exerted with lefs energy, and thence the actions produced by it are lefs than natural; and pain is produced in confequence, according to the fifth law of animal caufation, Sect. IV. This pain is generally attended with coldnefs of the affected part, and is feldom fucceeded by inflammation of it. This decreafed action of the fecondary link of the affociated motions, belonging to this genus, is owing to the previous exhauftion of fenforial power either in the increafed actions of the primary link of the affociated motions, or by the pain which attends them; both which are frequently the confequence of the ftimulus of fomething external to the affected fibres.

As pain is produced either by excefs or defect of the natural exertions of the fibres, it is not, confidered feparately, a criterion of the prefence of either. In the affociations belonging to this genus the fenfation of pain or pleafure produces or attends the primary link of the affociated motions, and very often gives name to the difeafe.

When great pain exifts without caufing any fibrous motions, I conjecture that it contributes to exhauft or expend the general quantity of fenforial power; becaufe people are fatigued by enduring pain, till at length they fleep. Which is contrary to what I had perhaps erroneoufly fuppofed in Sect. XXXV. 2. 3. If it caufes fibrous motions, it then takes the name of fenfation, according to the definition of ſenſation in Sect. II. 2. 9.; and increaſed fibrous action or inflammation is the conſequence. This circumſtance of the general exhauſtion of ſenſorial power by the exiſtence of pain will aſſiſt in explaining many of the diſeaſes of this genus.

Many of the canals of the body, as the urethra, the bile-duct, the throat, have the motions of their two extremities aſſociated by having been accuſtomed to feel pleaſurable or painful ſenſations at the ſame time or in ſucceſſion. This is termed ſenſitive aſſociation, though thoſe painful or pleaſurable ſenſations do not cauſe the motions, but only attend them; and are thus perhaps, ſtrictly ſpeaking, only catenated with them.

## SPECIES.

1. *Torpor genæ a dolore dentis.* In tooth-ach there is generally a coldneſs of the cheek, which is ſenſible to the hand, and is attended in ſome degree with the pain of cold. The cheek and tooth have frequently been engaged in pleaſurable action at the ſame time during the maſticating of our food; whence they have acquired ſenſitive aſſociations. The torpor of the cheek may have for its cauſe the too great expenditure of ſenſorial power by the painful ſenſation of the membranes of the diſeaſed tooth; whence the membranes of the cheek aſſociated with thoſe of the alveolar proceſs are deprived of their natural ſhare of it, and become torpid; thus they produce leſs ſecretions, and leſs heat, and the pain of cold is the conſequence. This torpor of the veſſels of the cheek cannot be produced by the activity of the ſenſorial power of ſenſation; for then they would act more violently than natural, or become inflamed. And though the pain by exhauſting ſo much ſenſorial power may be a remote cauſe, it is the defect of the power of aſſociation, which is the immediate cauſe of the torpor of the cheek.

After ſome hours this pain occaſioned by the torpor of the veſſels of the

the cheek either gradually ceases along with the pain of the diseased tooth; or, by the accumulation of sensorial power during their state of torpor, the capillaries of the cheek act with greater violence, and produce more secretions, and heat, and consequent tumour, and inflammation. In this state the pain of the diseased tooth ceases; as the sensorial power of sensation is now expended on the inflamed vessels of the cheek. It is probable that most other internal membranous inflammations begin in a similar manner; whence there may seem to be a double kind of sensitive association; first, with decreased action of the associated organ, and then with increased action of it; but the latter is in this case simply the consequence of the former; that is, the tumor or inflammation of the cheek is in consequence of its previous quiescence or torpor.

2. *Stranguria a dolore vesicæ.* The strangury, which has its origin from pain at the neck of the bladder, consists of a pain in the external extremity of the urethra or of the glans penis of men, and probably in the external termination of the urethra or of the clitoris of women; and is owing to the sympathy of these with some distant parts, generally with the other end of the urethra; and endeavour and difficulty of making water attends this pain.

Its remote cause is from the internal or external use of cantharides, which stimulate the neck of the bladder; or from a stone, which whenever it is pushed into the neck of the bladder, gives this pain of strangury, but not at other times; and hence it is felt most severely in this case after having made water.

The sensations or sensitive motions of the glans penis, and of the sphincter of the bladder, have been accustomed to exist together during the discharge of the urine; and hence the two ends of the urethra sympathize by association. When there is a stone at the neck of the bladder, which is not so large or rough as to inflame the part, the sphincter of the bladder becomes stimulated into pain; but as the glans

glans penis is for the purpoſes of copulation more ſenſitive than the ſphincter of the bladder, as ſoon as it becomes affected with pain by the aſſociation above mentioned, the ſenſation at the neck of the bladder ceaſes; and then the pain of the glans penis would ſeem to be aſſociated with the irritative motions only of the ſphincter of the bladder, and not with the ſenſitive ones of it. But a circumſtance ſimilar to this occurs in epileptic fits, which at firſt are induced by diſagreeable ſenſation, and afterwards ſeem to occur without previous pain, from the ſuddenneſs in which they follow and relieve the pain, which occaſioned them. From this analogy I imagine the pain of the glans penis is aſſociated with the pain of the ſphincter of the bladder; but that *as ſoon as the greater pain in a more ſenſible part is produced; the leſſer one, which occaſioned it, ceaſes*; and that this is one of the laws of ſenſitive aſſociation. See Sect. XXXV. 2. 1.

A young man had by an accident ſwallowed a large ſpoonful or more of tincture of cantharides; as ſoon as he began to feel the pain of ſtrangury, he was adviſed to drink large quantities of warmiſh water; to which, as ſoon as it could be got, ſome gum arabic was added. In an hour or two he drank by intervals of a few minutes about two gallons of water, and diſcharged his urine every four or five minutes. A little blood was voided towards the end, but he ſuffered no ill conſequence.

M. M. Warm water internally. Clyſters of warm water. Fomentation. Opium. Solution of fixed alkali ſuperſaturated with carbonic acid. A bougie may be uſed to puſh back a ſtone into the bladder. See Claſs I. 1. 3. 10.

3. *Stranguria convulſiva.* The convulſive ſtrangury, like that before deſcribed, is probably occaſioned by the torpor or defective action of the painful part in conſequence of the too great expenditure of ſenſorial power on the primary link of the aſſociated motions, as no heat or inflammation attends this violent pain. This kind of ſtrangury recurs

recurs by ſtated periods, and ſometimes ariſes to ſo great a degree, that convulſion or temporary madneſs terminates each period of it. It affects women oftener than men, is attended with cold extremities without fever, and is diſtinguiſhed from the ſtone of the bladder by the regularity of its periods, and by the pain being not increaſed after making water.

On introducing the catheter ſometimes part of the urine will come away and not the whole, which is difficult to explain; but may ariſe from the weakneſs of the muſcular fibres of the bladder; which are not liable ſuddenly to contract themſelves ſo far as to exclude the whole of the urine. In ſome old people, who have experienced a long retention of urine, the bladder never regains the power of completely emptying itſelf; and many who are beginning to be weak from age can make water a ſecond time, a few minutes after they ſuppoſed they had emptied the bladder.

I have believed this pain to originate from ſympathy with ſome diſtant part, as from aſcarides in the rectum, or from piles in women; or from caruncles in the urethra about the caput gallinaginis in men; and that the pain has been in the glans or clitoris by reverſe ſympathy of theſe more ſenſible parts with thoſe above mentioned.

M. M. Veneſection. Opium in large quantities. Warm bath. Balſams. Bark. Tincture of cantharides. Bougie, and the treatment for hæmorrhoids. Leeches applied to the ſphincter ani. Aerated alcaline water. Soap and ſal ſoda. Opium in clyſters given an hour before the expected return. Smoke of tobacco in clyſters. Arſenic?

4. *Dolor termini inteſtinalis ductûs choledochi.* Pain at the inteſtinal end of the gall-duct. When a gall-ſtone is protruded from the gall-bladder a little way into the end of the gall-duct, the pain is felt at the other end of the gall-duct, which terminates in the duodenum. For the actions of the two terminations of this canal are aſſociated together

from

from the ſame ſtreams of bile paſſing through them in ſucceſſion, exactly as the two terminations of the urethra have their actions aſſociated, as deſcribed in Species 2 and 3 of this genus. But as the inteſtinal termination of the bile-duct is made more ſenſible for the purpoſe of bringing down more bile, when it is ſtimulated by new ſupplies of food from the ſtomach, it falls into violent pain from aſſociation; and then the pain on the region of the gall-bladder ceaſes, exactly as above explained in the account of the pain of the glans penis from a ſtone in the ſphincter of the bladder.

The common bile-duct opens into the inteſtine exactly at what is called the pit of the ſtomach; and hence it has ſometimes happened, that this pain from aſſociation with the ſenſation of a gall-ſtone at the other end of the bile-duct has been miſtaken for a pain of the ſtomach.

For the method of cure ſee Claſs I. 1. 3. 8. to which ſhould be added the uſe of ſtrong electric ſhocks paſſed through the bile-duct from the pit of the ſtomach to the back, and from one ſide to the other. A caſe of the good effect of electricity in the jaundice is related in Sect. XXX. 2. And another caſe, where it promoted the paſſage of a painful gall-ſtone, is deſcribed by Dr. Hall, experienced on himſelf. Tranſ. of the College at Philadelphia, Vol. I. p. 192.

Half a pint of warm water two or three times a day is much recommended to dilute the inſpiſſated bile.

5. *Dolor pharyngis ab acido gaſtrico.* The two ends of the throat ſympathize by ſenſitive aſſociation in the ſame manner as the other canals above mentioned, namely, the urethra and the bile-duct; hence when too great acidity of undigeſted aliment, or the carbonic acid air, which eſcapes in fermentation, ſtimulates the cardia ventriculi, or lower end of the gula, into pain; the pharinx, or upper end of it, is affected with greater pain, or a diſagreeable ſenſation of heat.

6. *Pruritus narium a vermibus.* The itching of the noſe from worms in the inteſtines is another curious inſtance of the ſenſitive aſſociations of the motions of membranes; eſpecially of thoſe which conſtitute the canals of the body. Previous to the deglutition of agreeable food, as milk in our earlieſt infancy, an agreeable odour affects the membrane, which lines the noſtrils; and hence an aſſociation ſeems to take place between the agreeable ſenſations produced by food in the ſtomach and bowels, and the agreeable ſenſations of the noſtrils. The exiſtence of aſcarides in the rectum I believe produces this itching of the noſtrils more than the worms in other parts of the inteſtines; as we have already ſeen, that the terminations of canals ſympathize more than their other parts, as in the urethra and gall-ducts. See Claſs I. 1. 5. 9. IV. 1. 2. 9.

7. *Cephalæa.* Head-ach. In cold fits of the ague, the head-ach ariſes from conſent with ſome torpid viſcus, like the pain of the loins. After drunkenneſs the head-ach is very common, owing to direct ſympathy of the membranes of the head with thoſe of the ſtomach; which is become torpid after the too violent ſtimulus of the preceding intoxication; and is hence removeable by ſpirit of wine, or opium, exhibited in ſmaller quantities. In ſome conſtitutions theſe head-achs are induced, when the feet are expoſed to much external cold; in this caſe the feet ſhould be covered with oiled ſilk, which prevents the evaporation of the perſpirable matter, and thence diminiſhes one cauſe of external cold.

M. M. Valerian in powder two drams three or four times a day is recommended. The bark. Chalybeates. A grain of opium twice a day for a long time. From five to ten drops of the ſaturated ſolution of arſenic two or three times a day. See Claſs I. 2. 4. 11. A lady once aſſured me, that when her head-ach was coming on, ſhe drank three pints (pounds) of hot water, as haſtily as ſhe could; which prevented the progreſs of the diſeaſe. A ſolution of arſenic is

 recommended

recommended by Dr. Fowler of York. Very ſtrong errhines are ſaid ſometimes to cure head-achs taken at the times the pain recurs, till a few drops of blood iſſue from the noſtrils. As one grain of turpeth mineral (vitriolic calx of mercury) mixed with ten grains of fine ſugar. Euphorbium or cayan pepper mixed with ſugar, and uſed with caution as an errhine. See the M. M. of the next Species.

8. *Hemicrania.* Pain on one ſide of the head. This diſeaſe is attended with cold ſkin, and hence whatever may be the remote cauſe, the immediate one ſeems to be want of ſtimulus, either of heat or diſtention, or of ſome other unknown ſtimulus in the painful part; or in thoſe, with which it is aſſociated. The membranes in their natural ſtate are only irritable by diſtention; in their diſeaſed ſtate, they are ſenſible like muſcular fibres. Hence a diſeaſed tooth may render the neighbouring membranes ſenſible, and is frequently the cauſe of this diſeaſe.

Sometimes the ſtomach is torpid along with the pained membrane of the head; and then ſickneſs and inappetency attends either as a cauſe or conſequence. The natural cure of hemicrania is the accumulation of ſenſorial power during the reſt or ſickneſs of the patient. Mrs. —— is frequently liable to hemicrania with ſickneſs, which is probably owing to a diſeaſed tooth; the paroxyſm occurs irregularly, but always after ſome previous fatigue, or other cauſe of debility. She lies in bed, ſick, and without taking any ſolid food, and very little of fluids, and thoſe of the aqueous kind, and, after about 48 or 50 hours, riſes free from complaint. Similar to this is the recovery from cold paroxyſms of fever, from the torpor occaſioned by fear, and from ſyncope; which are all owing to the accumulation of ſenſorial power during the inactivity of the ſyſtem. Hence it appears, that, though when the ſenſorial power of volition is much exhauſted by fatigue, it can be reſtored by eight or ten hours of ſleep; yet, when the ſenſorial power of irritation is exhauſted by fatigue, that it requires

requires two whole ſolar or lunar days of reſt, before it can be reſtored.

The late Dr. Monro aſſerted in his lectures, that he cured the hemicrania, or megrim, by a ſtrong vomit, and a briſk purge immediately after it. This method ſucceeds beſt if opium and the bark are given in due quantity after the operation of the cathartic; and with ſtill more certainty, if bleeding in ſmall quantity is premiſed, where the pulſe will admit of it. See Sect. XXXV. 2. 1.

The pain generally affects one eye, and ſpreads a little way on that ſide of the noſe, and may ſometimes be relieved by preſſing or cutting the nerve, where it paſſes into the bone of the orbit above the eye. When it affects a ſmall defined part on the parietal bone on one ſide, it is generally termed Clavus hyſtericus, and is always I believe owing to a diſeaſed dens molaris. The tendons of the muſcles, which ſerve the office of maſtication, have been extended into pain at the ſame time, that the membranous coverings of the roots of the teeth have been compreſſed into pain, during the biting or maſtication of hard bodies. Hence when the membranes, which cover the roots of the teeth, become affected with pain by a beginning decay, or perhaps by the torpor or coldneſs of the dying part of the tooth, the tendons and membranous faſcia of the muſcles about the ſame ſide of the head become affected with violent pain by their ſenſitive aſſociations: and as ſoon as this aſſociated pain takes place, the pain of the tooth entirely ceaſes, as explained in the ſecond ſpecies of this genus.

A remarkable circumſtance attends this kind of hemicrania, viz. that it recurs by periods like thoſe of intermittent fevers, as explained in the Section on Catenation of Motions; theſe periods ſometimes correſpond with alternate lunar or ſolar days like tertian agues, and that even when a decaying tooth is evidently the cauſe; which has been evinced by the cure of the diſeaſe by extracting the tooth. At

 other

other times they obſerve the monthly lunations, and ſeem to be induced by the debility, which attends menſtruation.

The dens ſapientiæ, or laſt tooth of the upper jaw, frequently decays firſt, and gives hemicrania over the eye on the ſame ſide. The firſt or ſecond grinder in the under-jaw is liable to give violent pain about the middle of the parietal bone, or ſide of the head, on the ſame ſide, which is generally called the Clavus hyſtericus, of which an inſtructive caſe is related in Sect. XXXV. 2. 1.

M. M. Detect and extract the diſeaſed tooth. Cut the affected nerve, or ſtimulate the diſeaſed membrane by acu-puncture. Veneſection to ſix ounces by the lancet or by leeches. A ſtrong emetic and a ſubſequent cathartic; and then an opiate and the bark. Paſs ſmall electric ſhocks through the pained membrane, and through the teeth on the ſame ſide. Apply vitriolic ether externally, and a grain of opium with camphor internally, to the cheek on the affected ſide, where a diſeaſed tooth may be ſuſpected. Foment the head with warm vinegar. Drink two large ſpoonfuls of vinegar. Stimulate the gums of the ſuſpected teeth by oil of cloves, by opium. See Claſs I. 1. 4. 4. Snuff volatile ſpirit of vinegar up the noſtrils. Laſtly, in permanent head-achs, as in permanent vertigo, I have ſeen good effect by the uſe of mercurial ointment rubbed on the ſhaved head or about the throat, till a mild ſalivation commences, which by inflaming the membranes of the teeth may prevent their irritative ſympathy with thoſe of the cranium. Thus by inflaming the tendon, which is the cauſe of locked jaw, and probably by inflaming the wound, which is the cauſe of hydrophobia, thoſe diſeaſes may be cured, by diſuniting the irritative ſympathy between thoſe parts, which may not poſſeſs any ſenſitive ſympathy. This idea is well worth our attention.

*Otalgia.* Ear-ach is another diſeaſe occaſioned by the ſympathy of the membranes of the ear with thoſe which inveſt or ſurround a decaying

caying tooth, as I have had frequent reaſon to believe; and is frequently relieved by filling the ear with tincture of opium. See Claſs I. 2. 4.

9. *Dolor humeri in hepatidide.* In the efforts of excluding the fæces and urine the muſcles of the ſhoulders are exerted to compreſs the air in the lungs, that the diaphragm may be preſſed down. Hence the diſtention of the tendons or fibres of theſe muſcles is aſſociated with the diſtention of the tendons or fibres of the diaphragm; and when the latter are pained by the enlargement or heat of the inflamed liver, the former ſympathize with them. Sometimes but one ſhoulder is affected, ſometimes both; it is probable that many other pains, which are termed rheumatic, have a ſimilar origin, viz. from ſenſitive aſſociations.

As no inflammation is produced in conſequence of this pain of the ſhoulder, it ſeems to be owing to inaction of the membranous part from defect of the ſenſorial power of aſſociation, of which the primary link is the inflamed membrane of the liver; which now expends ſo much of the ſenſorial power in general by its increaſed action, that the membranes about the ſhoulder, which are links of aſſociation with it, become deprived of their uſual ſhare, and conſequently fall into torpor.

10. *Torpor pedum in eruptione variolarum.* At the commencement of the eruption of the ſmall-pox, when the face and breaſt of children are very hot, their extremities are frequently cold. This I aſcribe to ſenſitive aſſociation between the different parts of the ſkin; whence when a part acts too violently, the other part is liable to act too weakly; and the ſkin of the face being affected firſt in the eruption of the ſmall-pox, the ſkin of the feet becomes cold in conſequence by reverſe ſympathy.

M. M. Cover the feet with flannel, and expoſe the face and boſom to cool air, which in a very ſhort time both warms the feet and cools the face; and hence what is erroneouſly called a raſh, but which is probably a too haſty eruption of the ſmall-pox, diſappears; and afterwards fewer and more diſtinct eruptions of the ſmall-pox supervene.

11. *Teſtium dolor nephriticus.* The pain and retraction of the teſticle on the ſame ſide, when there is a ſtone in the ureter, is to be aſcribed to ſenſitive aſſociation; whether the connecting cauſe be a branch of the ſame nerve, or from membranes, which have been frequently affected at the ſame time.

12. *Dolor digiti minimi ſympatheticus.* When any one accidentally ſtrikes his elbow againſt any hard body, a tingling pain runs down to the little finger end. This is owing to ſenſitive aſſociation of motions by means of the ſame branch of a nerve, as in hemicrania from a decaying tooth the pain is owing to the ſenſitive aſſociation of tendons or membranes.

13. *Dolor brachii in hydrope pectoris.* The pain in the left arm which attends ſome dropſies of the cheſt, is explained in Sect. XXIX. 5. 2. 10. which reſembles the pain of the little finger from a percuſſion of the nerve at the elbow in the preceding article. A numbneſs of this kind is produced over the whole leg, when the crural nerve is much compreſſed by ſitting for a time with one leg croſſed over the other.

Mr. ——, about ſixty, had for two years been affected with difficulty of reſpiration on any exertion, with pain about the ſternum, and of his left arm; which laſt was more conſiderable than is uſual

 in

in dropſy of the cheſt; ſome months ago the pain of his arm, after walking a mile or two, became exceſſive, with coldneſs and numbneſs; and on the next day the back of the hand, and a part of the arm ſwelled, and became inflamed, which relieved the pain; and was taken for the gout, and continued ſeveral days. He after ſome months became dropſical both in reſpect to his cheſt and limbs, and was ſix or ſeven times perfectly relieved by one dram of ſaturated tincture of digitalis, taken two or three times a day for a few days in a glaſs of peppermint water. He afterwards breathed oxygen gas undiluted, in the quantity of ſix or eight gallons a day for three or four weeks without any effect, and ſunk at length from general debility.

In this inſtructive caſe I imagine the preſſure or ſtimulus of one part of the nerve within the cheſt cauſed the other part, which ſerves the arm, to become torpid, and conſequently cold by ſympathy; and that the inflammation was the conſequence of the previous torpor and coldneſs of the arm, in the ſame manner as the ſwelling and inflammation of the cheek in tooth-ach, in the firſt ſpecies of this genus; and that many rheumatic inflammations are thus produced by ſympathy with ſome diſtant part.

14. *Diarrhœa a dentitione.* The diarrhœa, which frequently attends dentition, is the conſequence of indigeſtion; the aliment acquires chemical changes, and by its acidity acts as a cathartic; and changes the yellow bile into green, which is evacuated along with indigeſted parts of the coagulum of milk. The indigeſtion is owing to the torpor of the ſtomach and inteſtines cauſed by their aſſociation with the membranes of the gums, which are now ſtimulated into great exertion with pain; both which contribute to expend the general quantity of ſenſorial power, which belongs to this membranous aſſociation; and thus the

ſtomach

ſtomach and inteſtines act with leſs than their natural energy. This is generally eſteemed a favourable ſymptom in difficult dentition, as the pain of the alveolar membranes exhauſts the ſenſorial power without producing convulſions for its relief. See Claſs I. 1. 4. 5. And the diarrhœa ceaſes, as the tooth advances.

ORDO

## ORDO II.

### *Decreased Associate Motions.*

## GENUS III.

### *Catenated with Voluntary Motions.*

## SPECIES.

1. *Titubatio linguæ.* Impediment of ſpeech is owing to the aſſociations of the motions of the organs of ſpeech being interrupted or diſſevered by ill-employed ſenſation or ſenſitive motions, as by awe, baſhfulneſs, ambition of ſhining, or fear of not ſucceeding, and the perſon uſes voluntary efforts in vain to regain the broken aſſociations, as explained in Sect. XVII. 1. 10. and XVII. 2. 10.

The broken aſſociation is generally between the firſt conſonant and the ſucceeding vowel; as in endeavouring to pronounce the word parable, the p is voluntarily repeated again and again, but the remainder of the word does not follow, becauſe the aſſociation between it and the next vowel is diſſevered.

M. M. The art of curing this defect is to cauſe the ſtammerer to repeat the word, which he finds difficult to ſpeak, eight or ten times without the initial letter, in a ſtrong voice, or with an aſpirate before it, as arable, or harable; and at length to ſpeak it very ſoftly with the initial letter p, parable. This ſhould be practiſed for weeks or months upon every word, which the ſtammerer heſitates in pronouncing. To this ſhould be added much commerce with mankind, in order to acquire a careleſſneſs about the opinions of others.

2. *Chorea St. Viti.* In the St. Vitus's dance the patient can at any time lie ſtill in bed, which ſhews the motions not to be convulſive; and he can at different times voluntarily exert every muſcle of his body; which evinces, that they are not paralytic. In this diſeaſe the principal muſcle in any deſigned motion obeys the will; but thoſe muſcles, whoſe motions were aſſociated with the principal one, do not act; as their aſſociation is diſſevered, and thus the arm or leg is drawn outward, or inward, or backward, inſtead of upward or forward, with various geſticulations exactly reſembling the impediment of ſpeech.

This diſeaſe is frequently left after the itch has been too haſtily cured. See Convulſio dolorifica, Claſs III. 1. 1. 6. A girl about eighteen, after wearing a mercurial girdle to cure the itch, acquired the Chorea St. Viti in ſo univerſal a manner, that her ſpeech became affected as well as her limbs; and there was evidently a diſunion of the common trains of ideas; as the itch was ſtill among the younger children of the family, ſhe was adviſed to take her ſiſter as a bedfellow, and thus received the itch again; and the dance of St. Vitus gradually ceaſed. See Claſs II. 1. 5. 6.

M. M. Give the patient the itch again. Calomel a grain every night, or ſublimate a quarter of a grain twice a day for a fortnight. Steel. Bark. Warm-bath. Cold-bath. Opium. Veneſection once at the beginning of the diſeaſe. Electricity. Perpetual ſlow and repeated efforts to move each limb in the deſigned direction, as in the titubatio linguæ above deſcribed.

3. *Riſus.* Laughter is a perpetual interruption of voluntary exertion by the interpoſition of pleaſurable ſenſation; which not being checked by any important conſequences riſes into pain, and requires to be relieved or moderated by the frequent repetition of voluntary exertion. See Sect. XXXIV. 1. 4. and Claſs III. 1. 1. 4. and IV. 1. 3. 3.

4. *Tremor*

4. *Tremor ex irâ.* The trembling of the limbs from anger. The interruption of the voluntary aſſociations of motions by anger, originates from too great a part of the ſenſorial power being exerted on the organs of ſenſe; whence the muſcles, which ought to ſupport the body upright, are deprived of their due quantity, and tremble from debility. See Claſs III. 2. 1. 1.

5. *Rubor ex irâ.* Redneſs from anger. Anger is an exceſs of averſion, that is of voluntarity not yet employed. It is excited by the pain of offended pride; when it is employed it becomes outrage, cruelty, inſanity. The cutaneous capillaries, eſpecially thoſe of the face, are more mobile, that is, more eaſily excited into increaſed action, or more eaſily become torpid, from leſs variation of ſenſorial power, than any other parts of the ſyſtem, which is owing to their being perpetually ſubject to the viciſſitudes of heat and cold, and of extenſion and corrugation. Hence, when an exceſs of voluntarity exiſts without being immediately expended in the actions of the large muſcles, the capillary arteries and glands acquire more energetic action, and a fluſhed ſkin is produced, with increaſed ſecretion of perſpirable matter, and conſequent heat, owing to the pauſe or interruption of voluntary action; and thus the actions of theſe cutaneous veſſels become aſſociated between the iraſcent ideas and iraſcent muſcular actions, which are thus for a time interrupted.

6. *Rubor criminati.* The bluſhing of accuſed people, whether guilty or not, appears to be owing to circumſtances ſimilar to that of anger; for in theſe ſituations there is always a ſudden voluntarity, or wiſh, of clearing their characters ariſes in the mind of the accuſed perſon; which, before an opportunity is given for it to be expended on the large muſcles, influences the capillary arteries and glands, as in the preceding article. Whence the increaſed actions of the capillaries, and the conſequent redneſs and heat, become exerted between

the voluntary ideas of ſelf-defence, and the muſcular actions neceſſary for that purpoſe; which laſt are thus for a time interrupted or delayed.

Even in the bluſh of modeſty or baſhfulneſs there is a ſelf-condemnation for ſome ſuppoſed defect, or indecorum, and a ſudden voluntarity, or wiſh, of ſelf-defence; which not being expended in actions of the larger muſcles excites the capillaries into action; which in theſe ſubjects are more mobile than in others.

The bluſh of young girls on coming into an aſſembly room, where they expect their dreſs, and ſteps, and manner to be examined, as in dancing a minuet, may have another origin; and may be conſidered as a hot fit of returning confidence, after a previous cold fit of fear.

7. *Tarditas paralytica.* By a ſtroke of the palſy or apoplexy it frequently happens, that thoſe ideas, which were aſſociated in trains, whoſe firſt link was a voluntary idea, have their connection diſſevered; and the patient is under the neceſſity by repeated efforts ſlowly to renew their aſſociations. In this ſituation thoſe words, which have the feweſt other words aſſociated with them, as the proper names of perſons or places, are the moſt difficult to recollect. And in thoſe efforts of recollection the word oppoſite to the word required is often produced, as hot for cold, winter for ſummer, which is owing to our aſſociating our ideas of things by their oppoſites as well as by their ſimilitudes, and in ſome inſtances perhaps more frequently, or more forcibly. Other paralytic patients are liable to give wrong names to external objects, as uſing the word pigs for ſheep, or cows for horſes; in this caſe the aſſociation between the idea of the animal and the name of it is diſſevered; but the idea of the claſs or genus of the thing remains; and he takes a name from the firſt of the ſpecies, which preſents itſelf, and ſometimes can correct himſelf, till he finds the true one.

8. *Tarditas ſenilis.* Slowneſs of age. The difficulty of aſſociating ideas

ideas increases with our age; as may be observed from old people forgetting the business of the last hour, unless they impress it strongly, or by frequent repetition, though they can well recollect the transactions of their youth. I saw an elderly man, who could reason with great clearness and precision and in accurate language on subjects, which he had been accustomed to think upon; and yet did not know, that he had rang the bell by his fire-side in one minute afterwards; nor could then recollect the object he had wanted, when his servant came.

Similar to this is the difficulty which old people experience in learning new bodily movements, that is, in associating new muscular actions, as in learning a new trade or manufactury. The trains of movements, which obey volition, are the last which we acquire; and the first, which are disassociated.

ORDO

## ORDO II.

*Decreased Associate Motions.*

## GENUS IV.

*Catenated with External Influences.*

As the diseases, which obey solar or lunar periods, commence with torpor or inactivity, such as the cold paroxysms of fevers, the torpor and consequent pain of hemicrania, and the pains which precede the fits of epilepsy and convulsion, it would seem, that these diseases are more generally owing to the diminution than to the excess of solar or lunar gravitation; as the diseases, which originate from the influence of the matter of heat, are much more generally in this country produced by the defect than by the excess of that fluid.

The periodic returns of so many diseases coincide with the diurnal, monthly, and annual rounds of time; that any one, who would deny the influence of the sun and moon on the periods of quotidian, tertian, and quartan fevers, must deny their effect on the tides, and on the seasons. It has generally been believed, that solar and lunar effect was exerted on the blood; which was thus rendered more or less stimulant to the system, as described in Sect. XXXII. 6. But as the fluid matter of gravitation permeates and covers all things, like the fluid matter of heat; I am induced to believe, that gravitation acts in its medium state rather as a causa sine quâ non of animal motion, like heat; which may disorder the system chemically or mechanically, when it is diminished; but may nevertheless stimulate it, when increased, into animal exertion.

Without heat and motion, which ſome philoſophers ſtill believe to be the ſame thing, as they ſo perpetually appear together, the particles of matter would attract and move towards each other, and the whole univerſe freeze or coaleſce into one ſolid maſs. Theſe therefore counteract the gravitation of bodies to one center; and not only prevent the planets from falling into the ſun, but become either the efficient cauſes of vegetable and animal life, or the cauſes without which life cannot exiſt; as by their means the component particles of matter are enabled to ſlide over each other with all the various degrees of fluidity and repulſion.

As the attraction of the moon countervails or diminiſhes the terrene gravitation of bodies on the ſurface of the earth; a tide riſes on that ſide of the earth, which is turned towards the moon; and follows it, as the earth revolves. Another tide is raiſed at the ſame time on the oppoſite ſide of the revolving earth; which is owing to the greater centrifugal motion of that ſide of the earth, which counteracts the gravitation of bodies near it's ſurface. For the earth and moon may be conſidered as two cannon balls of different ſizes held together by a chain, and revolving once a month round a common center of gravity between them, near the earth's ſurface; at the ſame time that they perform their annual orbits round the ſun. Whence the centrifugal force of that ſide of the earth, which is fartheſt from this center of motion, round which the earth and moon monthly revolve, is conſiderably greater, than the centrifugal force of that ſide of the earth, which is neareſt it; to which ſhould be added, that this centrifugal force not only contributes to diminiſh the terrene gravitation of bodies on the earth's ſurface on that ſide furtheſt from this center of motion, but alſo to increaſe it on that ſide, which is neareſt it.

Another circumſtance, which tends to raiſe the tide on the part of the earth's ſurface, which is moſt diſtant from the moon, is, that the attraction of the moon is leſs on that part of the ocean, than it is on the

the other parts of the earth. Thus the moon may be ſuppoſed to attract the water on the ſide of the earth neareſt it with a power equal to three; and to attract the central parts of the earth with a power equal to two; and the water on the part of the earth moſt diſtant from the moon with a power only equal to one. Hence on the ſide of the earth moſt diſtant from the moon, the moon's attraction is leſs, and the centrifugal force round their common center of motion is greater; both which contribute to raiſe the tides on that ſide of the earth. On the ſide of the earth neareſt the moon, the moon's attraction is ſo much greater as to raiſe the tides; though the centrifugal force of the ſurface of the earth round their common center of motion in ſome degree oppoſes this effect.

On theſe accounts, when the moon is in the zenith or nadir, the gravitation of bodies on the earth's ſurface will be greateſt at the two oppoſite quadratures; that is, the greateſt gravitation of bodies on the earth's ſurface towards her center during the lunar day is about ſix hours and an half after the ſouthing, or after the northing of the moon.

Circumſtances ſimilar to theſe, but in a leſs degree, muſt occur in reſpect to the ſolar influence on terreſtrial bodies; that is, there muſt be a diminution of the gravity of bodies near the earth's ſurface at noon, when the ſun is over them; and alſo at midnight from the greater centrifugal force of that ſide of the earth, which is moſt diſtant from the center, round which the earth moves in her annual orbit, than on the ſide neareſt that center. Whence it likewiſe follows, that the gravitation of bodies towards the earth is greateſt about ſix hours after noon, and after midnight.

Now when the ſun and moon have their united gravitation on the ſame ſide of the earth, as at the new moon; or when the ſolar attraction coincides with the greater centrifugal motion of that ſide of the earth, which is furtheſt diſtant from the moon, as at the full moon; and when this happens about noon or midnight, the gravitation of

terrene

terrene bodies towards the earth will be greater about ſix hours after noon, and after midnight, than at any other part of the lunar period; becauſe the attraction of both theſe luminaries is then exerted on thoſe ſides of the earth over which they hang, which at other times of the month are more or leſs exerted on other parts of it.

Laſtly, as heat and motion counteract the gravitation of the particles of bodies to each other, and hence become either the efficient cauſes of vegetable and animal life, or the cauſes without which life cannot exiſt, it ſeems to follow, that when our gravitation towards the earth's center is greateſt, the powers of life ſhould be the leaſt; and hence that thoſe diſeaſes, which begin with torpor, ſhould occur about ſix hours after the ſolar or lunar noon, or about ſix hours after the ſolar or lunar midnight; and this moſt frequently about ſix hours after or before the new or full moon; and eſpecially when theſe happen at noon or at midnight; or laſtly, according to the combination of theſe powers in diminiſhing or increaſing the earth's attraction to bodies on its ſurface.

The returns or exacerbations of many fevers, both irritative and inflammatory, about ſix in the evening, and of the periodic cough deſcribed in Sect. XXXVI. 3. 9. countenance this theory. Tables might be made out to ſhew the combined powers of the ſun and moon in diminiſhing the gravitation of bodies on the earth's ſurface, at every part of their diurnal, monthly, and annual periods; and which might facilitate the elucidation of this ſubject. But I am well aware of the difficulty of its application to diſeaſes, and hope theſe conjectures may induce others to publiſh more numerous obſervations, and more concluſive reaſonings.

## SPECIES.

1. *Somni periodus.* The periods of ſleeping and of waking are ſhortened or prolonged by ſo many other circumſtances in animal life, beſides the minute difference between diurnal and nocturnal ſolar gravitation, that it can ſcarcely be aſcribed to this influence. At the ſame time it is curious to obſerve, that vegetables in reſpect to their times of ſleeping more regularly obſerve the hour of the day, than the preſence or abſence of light, or of heat, as may be ſeen by conſulting the calendar of Flora. Botanic Garden, Part II. Canto 2. l. 165. note.

Some diſeaſes, which at firſt ſight might be ſuppoſed to be influenced by ſolar periods, ſeem to be induced by the increaſing ſenſibility of the ſyſtem to pain during our ſleeping hours; as explained in Sect. XVIII. 15. Of theſe are the fits of aſthma, of ſome epilepſies, and of ſome hæmoptoes; all which diſturb the patient after ſome hours ſleep, and are therefore to be aſcribed to the increaſe of our dormant ſenſibility. There may likewiſe be ſome doubt, whether the commencement of the pain of gout in the foot, as it generally makes its attack after ſleep, ſhould be aſcribed to the increaſed ſenſibility in ſleep, or to ſolar influence?

M. M. When aſthmatic or epileptic fits or hæmoptoe occur after a certain number of hours of ſleep, the patient ſhould be forcibly awakened before the expected time by an alarm clock, and drink a cup of chocolate or lemonade.—Or a grain of opium ſhould be given at going to bed.—In one caſe to prevent the too great increaſe of ſenſibility by ſhortening the time of ſleep; and in the other by increaſing the irritative motions, and expending by that means a part of the ſenſorial power.

2. *Studii*

2. *Studii inanis periodus.* Claſs III. 1. 2. 2. The cataleptic ſpaſm which preceded the reverie and ſomnambulation in the patient, whoſe caſe is related in Sect. XIX. 2. occurred at exactly the ſame hour, which was about eleven in the morning for many weeks; till thoſe periods were diſturbed by large doſes of opium; and muſt therefore be referred to ſome effect of ſolar gravitation. In the caſe of Maſter A. Sect. XXXIV. 3. as the reverie began early in the morning during ſleep, there may be a doubt, whether this commenced with torpor of ſome organ catenated with ſolar gravitation; or was cauſed by the exiſtence of a previous torpid part, which only became ſo painful as to excite the exertions of reverie by the perpetual increaſe of ſenſibility during the continuance of ſleep, as in ſome fits of epilepſy, aſthma, and hæmoptoe mentioned in the preceding article.

3. *Hemicraniæ periodus.* Periods of hemicrania. Claſs IV. 2. 2. 8. The torpor and conſequent pain of ſome membranes on one ſide of the head, as over one eye, is frequently occaſioned by a decaying tooth, and is liable to return every day, or on alternate days at ſolar or lunar periods. In this caſe large quantities of the bark will frequently cure the diſeaſe, and eſpecially if preceded by veneſection and a briſk cathartic; but if the offending tooth can be detected, the moſt certain cure is its extraction. Theſe partial head-achs are alſo liable to return at the greater lunar periods, as about once a month. Five drops from a two-ounce phial of a ſaturated ſolution of arſenic twice a day for a week or two have been ſaid to prevent the returns of this diſeaſe. See a Treatiſe on Arſenic by Dr. Fowler, of York. Strong errhines have alſo been recommended.

4. *Epilepſiæ dolorificæ periodus.* Claſs III. 1. 1. 8. The pain which induces after about an hour the violent convulſions or inſanity, which conſtitute the painful epilepſy, generally obſerve ſolar diurnal periods for four or five weeks, and are probably governed by ſolar and

lunar times in respect to their greater periods; for I have observed that the daily paroxysms, unless disturbed by large doses of opium, recur at very nearly the same hour, and after a few weeks the patients have recovered to relapse again at the interval of a few months. But more observations are wanted upon this subject, which might be of great advantage in preventing the attacks of this disease; as much less opium given an hour before its expected daily return will prevent the paroxysm, than is necessary to cure it, after it has commenced.

5. *Convulsionis dolorificæ periodus.* Class III. 1. 1. 6. The pains, which produce these convulsions, are generally left after rheumatism, and come on when the patients are become warm in bed, or have been for a short time asleep, and are therefore perhaps rather to be ascribed to the increasing sensibility of the system during sleep, than to solar diurnal periods, as in Species first and second of this Genus.

6. *Tussis periodicæ periodus.* Periodic cough, Class IV. 2. 1. 9. returns at exact solar periods; that described in Sect. XXXVI. 3. 9. recurred about seven in the afternoon for several weeks, till its periods were disturbed by opium, and then it recurred at eleven at night for about a week, and was then totally destroyed by opium given in very large quantities, after having been previously for a few days omitted.

7. *Catameniæ periodus.* Periods of menstruation. The correspondence of the periods of the catamenia with those of the moon was treated of in Sect. XXXII. 6. and can admit of no more doubt, than that the returns of the tides are governed by lunar influence. But the manner in which this is produced, is less evident; it has commonly been ascribed to some effect of the lunar gravitation on the circulating blood, as mentioned in Sect. XXXII. 6. But it is more

analogous

analogous to other animal phenomena to suppose that the lunar gravitation immediately affects the solids by its influx or stimulus. Which we believe of the fluid element of heat, in which we are equally immersed; and of the electric fluid, which also surrounds and pervades us. See Sect. XXXVI. 2. 3.

If the torpor of the uterine veins, which induces the monthly periods of the catamenia, be governed by the increase of terrene gravitation; that is, by the deficiency of the counter-influence of solar and lunar gravitation; why does not it occur most frequently when the terrene gravitation is the greatest, as about six hours after the new moon, and next to that at about six hours after the full moon? This question has its difficulty; first, if the terrene gravitation be greatest about six hours after the new moon, it must become less and less about the same time every lunar day, till the end of the first quarter, when it will be the least; it must then increase daily till the full. After the full the terrene gravitation must again decrease till the end of the third quarter, when it will again be the least, and must increase again till the new moon; that is, the solar and lunar counter-gravitation is greatest, when those luminaries are vertical, at the new moon, and full moon, and least about six hours afterwards. If it was known, whether more menstruations occur about six hours after the moon is in the zenith or nadir; and in the second and fourth quarters of the moon, than in the first and third; some light would be thrown on this subject; which must in that respect wait for future observations.

Secondly, if the lunar influence produces a very small degree of quiescence, suppose of the uterine veins, at first; and if that recurs at certain periods, as of lunar days, or about 25 hours, even with less power to produce quiescence than at first; yet the quiescence will daily increase by the acquired habit acting at the same time, as explained in Sect. XII. 3. 3. till at length so great a degree of quiescence will be induced as to cause the inaction of the veins of the uterus, and consequent

consequent venous hæmorrhage. See Sect. XXXII. 6. Class I. 2. 1. 11. IV. 1. 4. 4. See the introduction to this Genus.

8. *Hæmorrhoidis periodus.* The periods of the piles depend on the torpor of the veins of the rectum, and are believed to recur nearly at monthly intervals. See Sect. XXVII. 2. and Class I. 2. 1. 6.

9. *Podagræ periodus.* The periods of gout in some patients recur at annual intervals, as in the case related above in Class IV. 1. 2. 15. in which the gouty paroxysm returned for three successive years on nearly the same day of the month. The commencement of the pain of each paroxysm is generally a few hours after midnight, and may thence either be induced by diurnal solar periods, or by the increasing sensibility during sleep, as mentioned in the first species of this genus.

10. *Erysipelatis periodus.* Some kinds of erysipelas which probably originate from the association of the cutaneous vessels with a diseased liver, occur at monthly periods, like the hæmorrhois or piles; and others at annual periods like the gout; as a torpor of some part I suppose always precedes the erysipelatous inflammation, the periods should accord with the increasing influence of terrene gravitation, as described in the introduction to this Genus, and in Species the seventh of it. Other periods of diseases referable to solar and lunar influence are mentioned in Sect. XXXVI. and many others will probably be discovered by future observation.

11. *Febrium periodus.* Periods of fevers. The commencement of the cold fits of intermittent fevers, and the daily exacerbations of other fevers, so regularly recur at diurnal solar or lunar periods, that it is impossible to deny their connection with gravitation; as explained in Sect. XXXVI. 3. Not only these exacerbations of fever, and their remissions, obey the diurnal solar and lunar periods; but the prepara-

tory circumſtances, which introduce fevers, or which determine their criſiſes, appear to be governed by the parts of monthly lunar periods, and of ſolar annual ones. Thus the variolous fever in the natural ſmall-pox commences on the 14th day, and in the inoculated ſmall-pox on the ſeventh day. The fever and eruption in the diſtinct kind take up another quarter of a lunation, and the maturation another quarter.

The fever, which is termed canine madneſs, or hydrophobia, is believed to commence near the new or full moon; and, if the cauſe is not then great enough to bring on the diſeaſe, it ſeems to acquire ſome ſtrength, or to lie dormant, till another, or perhaps more powerful lunation calls it into action. In the ſpring, about three or four years ago, a mad dog very much worried one ſwine confined in a ſty, and bit another in the ſame ſty in a leſs degree; the former became mad, refuſed his meat, was much convulſed, and died in about four days; this diſeaſe commenced about a month after the bite. The other ſwine began to be ill about a month after the firſt, and died in the ſame manner.

ORDO

## ORDO III.

*Retrograde Aſſociate Motions.*

## GENUS I.

*Catenated with Irritative Motions.*

THOSE retrograde aſſociate motions, the firſt links of which are catenated with irritative motions, belong to this genus. All the retrograde motions are conſequent to debility, or inactivity, of the organ; and therefore properly belong to the genera of decreaſed actions both in this and the former claſſes.

### SPECIES.

1. *Diabœtes irritata.* When the abſorbents of the inteſtines are ſtimulated too ſtrongly by ſpirit of wine, as in the beginning of drunkenneſs, the urinary abſorbents invert their motions. The ſame happens from worms in the inteſtines. In other kinds of diabetes may not the remote cauſe be the too ſtrong action of the cutaneous abſorbents, or of the pulmonary ones? May not in ſuch caſes oil externally or internally be of ſervice? or warm bathing for an hour at a time? In hyſteric inverſions of motion is ſome other part too much ſtimulated? or pained from the want of ſtimulus?

2. *Sudor frigidus in aſthmate.* The cauſe of the paroxyſms of humoral aſthma is not well underſtood; I ſuppoſe it to be owing to a torpidity or inaction of the abſorbents belonging to the pulmonary veſſels,

vessels, as happens probably to other viscera at the commencement of intermittent fevers, and to a consequent accumulation of fluids in them; which at length producing great irritation or uneasy sensation causes the violent efforts to produce the absorption of it. The motions of the cutaneous absorbent vessels by their association with those of the pulmonary ones become retrograde, and effuse upon the skin a fluid, which is said to be viscid, and which adheres in drops.

A few days ago I saw a young man of delicate constitution in what was called a fit of the asthma; he had about two months before had a peripneumony, and had been ever since subject to difficult respiration on exertion, with occasional palpitation of his heart. He was now seized about eight at night after some exertion of mind in his business with cold extremities, and difficulty of breathing. He gradually became worse, and in about half an hour, the palpitation of his heart and difficult respiration were very alarming; his whole skin was cold and pale, yet he did not shudder as in cold paroxysm of fever; his tongue from the point to the middle became as cold as his other extremities, with cold breath. He seemed to be in the act of dying, except that his pulse continued equal in time, though very quick. He lost three ounces of blood, and took ten drops of laudanum with musk and salt of hartshorn, and recovered in an hour or two without any cold sweat.

There being no cold sweat seems to indicate, that there was no accumulation of serous fluid in the lungs; and that their inactivity, and the coldness of the breath, was owing to the sympathy of the air-cells with some distant part. There was no shuddering produced, because the lungs are not sensible to heat and cold; as any one may observe by going from a warm room into a frosty air, and the contrary. So the steam of hot tea, which scalds the mouth, does not affect the lungs with the sensation of heat. I was induced to believe,

 that

that the whole cold fit might be owing to ſuppuration in ſome part of the cheſt; as the general difficulty of breathing ſeemed to be increaſed after a few days with pulſe of 120, and other ſigns of empyema. Does the cold ſweat, and the occurrence of the fits of aſthma after ſleep, diſtinguiſh the humoral aſthma from the cold paroxyſm of intermittents, or which attends ſuppuration, or which precedes inflammation?—I heard a few weeks afterwards, that he ſpit up much matter at the time he died.

3. *Diabætes a timore.* The motions of the abſorbent veſſels of the neck of the bladder become inverted by their conſent with thoſe of the ſkin; which are become torpid by their reverſe ſympathy with the painful ideas of fear, as in Sect. XVI. 8. 1. whence there is a great diſcharge of pale urine, as in hyſteric diſeaſes.

The ſame happens from anxiety, where the painful ſuſpenſe is continued, even when the degree of fear is ſmall; as in young men about to be examined for a degree at the univerſities the frequency of making water is very obſervable. When this anxiety is attended with a ſleepleſs night, the quantity of pale urine is amazingly great in ſome people, and the micturition very frequent.

M. M. Opium. Joy. Conſolations of friendſhip.

4. *Diarrhœa a timore.* The abſorbent veſſels of the inteſtines invert their motions by direct conſent with the ſkin; hence many liquid ſtools as well as much pale urine are liable to accompany continued fear, along with coldneſs of the ſkin. The immediate cauſe of this is the decreaſed ſenſorial power of aſſociation, which intervenes between the actions of the abſorbents of the cold ſkin, and thoſe of the inteſtinal abſorbents; the motions of the latter become on that account weakened and at length retrograde. The remote cauſe is the

torpor

torpor of the veſſels of the ſkin catenated with the pain of fear, as explained in Sect. XVI. 8. 1.

The capillaries of the ſkin conſent more generally by direct ſympathy with thoſe of the lower inteſtines, and of the bladder; but by reverſe ſympathy more generally with thoſe of the ſtomach and upper inteſtines. As appears in fevers, where the hot ſkin accompanies indigeſtion of the ſtomach; and in diarrhœas attended with cold extremities.

The remote cauſe is the torpor of the ſkin owing to its reverſe ſympathy with the painful ſenſual motions, or ideas, of fear; which are now actuated with great energy, ſo as to deprive the ſecond link of aſſociated motions of their due ſhare of ſenſorial power. It is alſo probable, that the pain of fear itſelf may contribute to exhauſt the ſenſorial power, even when it produces no muſcular action. See Claſs IV. 2. 2.

5. *Pallor et tremor a timore.* A retrograde action of the capillaries of the ſkin producing paleneſs, and a torpor of the muſcular fibres of the limbs occaſioning trembling, are cauſed by their reverſe aſſociations with the ideas or imaginations of fear; which are now actuated with violent energy, and accompanied with great pain. The cauſe of theſe aſſociations are explained in Sect. XVI. 8. 1.

Theſe torpid actions of the capillaries and muſcles of the limbs are not cauſed immediately by the painful ſenſation of fear; as in that caſe they would have been increaſed and not decreaſed actions, as occurs in anger; where the painful volition increaſes the actions of the capillaries, exciting a bluſh and heat of the ſkin. Whence we may gain ſome knowledge of what is meant by depreſſing and exciting paſſions; the former conſiſting of ideas attended with pain, which pain occaſions no muſcular actions, like the pain of cold head-ach;

the latter being attended with volitions, and consequent muscular exertions.

That is, the pain of fear, and the pain of anger, are produced by the exertion of certain ideas, or motions of certain nerves of sense; in the former case, the painful sensation of fear produces no muscular actions, yet it exhausts or employs so much sensorial power, that the whole system acts more feebly, or becomes retrograde; but some parts of it more so than others, according to their early associations described in Sect. XVI. 8. 1. hence the tremor of the limbs, palpitation of heart, and even syncope. In anger the painful volition produces violent muscular actions; but if previous to these any deliberation occurs, a flushed countenance sometimes, and a red skin, are produced by this superabundance of volition exerted on the arterial system; but at other times the skin becomes pale, and the legs tremble, from the exhaustion or expenditure of the sensorial power by the painfu volitions of anger on the organs of sense, as by the painful sensations of fear above mentioned.

Where the passion of fear exists in a great degree, it exhausts or expends so much sensorial power, either simply by the pain which attends it, or by the violent and perpetual excitement of the terrific imaginations or ideas, that not only a cold and pale skin, but a retrograde motion of the cutaneous absorbents occurs, and a cold sweat appears upon the whole surface of the body, which probably sometimes increases pulmonary absorption; as in Class II. 1. 6. 4. and as in the cold sweats, which attend the paroxysms of humoral asthma. Hence anxiety, which is a continued pain of fear, so universally debilitates the constitution as to occasion a lingering death; which happens much more frequently than is usually supposed; and these victims of continued anxiety are said to die of a broken heart. Other kinds of paleness are described in Class I. 2. 2. 2.

M. M. Opium. Wine. Food. Joy.

6. *Palpitatio*

6. *Palpitatio cordis a timore.* The palpitation of the heart from fear is owing to the weak action of it, and perhaps ſometimes to the retrograde exertion of the ventricules and auricles; becauſe it ſeems to be affected by its aſſociation with the capillaries, the actions of which, with thoſe of the arteries and veins, conſtitute one great circle of aſſociate motions. Now when the capillaries of the ſkin become torpid, coldneſs and paleneſs ſucceed; and with theſe are aſſociated the capillaries of the lungs, whence difficult reſpiration; and with theſe the weak and retrograde actions of the heart. At the ſame time the abſorbents of the ſkin, and of the bladder, and of the inteſtines, ſometimes become retrograde, and regurgitate their contents; as appears by the pale urine in large quantities, which attends hyſteric complaints along with this palpitation of the heart; and from the cold ſweats, and diarrhœa; all which, as well as the hyſteric complaints, are liable to be induced or attended by fear.

When fear has ſtill more violently affected the ſyſtem, there have been inſtances where ſyncope, and ſudden death, or a total ſtoppage of the circulation, have ſucceeded: in theſe laſt caſes, the pain of fear has employed or exhauſted the whole of the ſenſorial power, ſo that not only thoſe muſcular fibres generally exerted by volition ceaſe to act, whence the patient falls down; and thoſe, which conſtitute the organs of ſenſe, whence ſyncope; but laſtly thoſe, which perform the vital motions, become deprived of ſenſorial power, and death enſues. See Claſs I. 2. 1. 4. and I. 2. 1. 10. Similar to this in ſome epileptic fits the patient firſt ſuddenly falls down, without even endeavouring to ſave himſelf by his hands before the convulſive motions come on. In this caſe the great exertion of ſome ſmall part in conſequence of great irritation or ſenſation exhauſts the whole ſenſorial power, which was lodged in the extremities of the locomotive nerves, for a ſhort time, as in ſyncope; and as ſoon as

7 theſe

these muscles are again supplied, convulsions supervene to relieve the painful sensation. See Class III. 1. 1. 7.

7. *Abortio a timore.* Women miscarry much more frequently from a fright, than from bodily injury. A torpor or retrograde motion of the capillary arteries of the internal uterus is probably the immediate cause of these miscarriages, owing to the association of the actions of those vessels with the capillaries of the skin, which are rendered torpid or retrograde by fear. By this contraction of the uterine arteries, the fine vessels of the placenta, which are inserted into them, are detruded, or otherwise so affected, that the placenta separates at this time from the uterus, and the fetus dies from want of oxygenation. A strong young woman, in the fifth or sixth month of her pregnancy, who has since borne many children, went into her cellar to draw beer; one of the servant boys was hid behind a barrel, and started out to surprise her, believing her to be the maid-servant; she began to flood immediately, and miscarried in a few hours. See Sect. XXXIX. 6. 5. and Class I. 2. 1. 14.

8. *Hysteria a timore.* Some delicate ladies are liable to fall into hysteric fits from sudden fright. The peristaltic motions of the bowels and stomach, and those of the œsophagus, make a part of the great circle of irritative motions with those of the skin, and many other membranes. Hence when the cutaneous vessels become torpid from their reverse sympathy with the painful ideas of fear; these of the bowels, and stomach, and œsophagus, become first torpid by direct sympathy with those of the skin, and then feebly and ineffectually invert the order of their motions, which constitutes a paroxysm of the hysteric disease. See Class I. 3. 1. 10. These hysteric paroxysms are sometimes followed by convulsions, which

which belong to Claſs III. as they are exertions to relieve pain; and ſometimes by death. See Species 9 of this Genus, and Claſs I. 2. 1. 4.

Indigeſtion from fear is to be aſcribed in the ſame manner to the torpor of the ſtomach, owing to its aſſociation with the ſkin. As in Claſs IV. 1. 2. 5. IV. 2. 1.

## ORDO III.

*Retrograde Associate Motions.*

## GENUS II.

*Catenated with Sensitive Motions.*

### SPECIES.

1. *Nausea idealis.* Nausea from disgustful ideas, as from nauseous stories, or disgustful sights, or smells, or tastes, as well as vomiting from the same causes, consists in the retrograde actions of the lymphatics of the throat, and of the œsophagus, and stomach; which are associated with the disgustful ideas, or sensual motions of sight, or hearing, or smell, or taste; for as these are decreased motions of the lymphatics, or of the œsophagus, or stomach, they cannot immediately be excited by the sensorial power of painful sensation, as in that case they ought to be increased motions. So much sensorial power is employed for a time on the disgustful idea, or expended in the production of inactive pain, which attends it, that the other parts of the associated chain of action, of which this disgustful idea is now become a link, is deprived of their accustomed share; and therefore first stop, and then invert their motions. Owing to deficiency of sensorial power, as explained more at large in Sect. XXXV. 1. 3.

2. *Nausea a conceptu.* The nausea, which pregnant women are so subject to during the first part of gestation, is owing to the reverse sympathy between the uterus and stomach, so that the increased action of the former, excited by the stimulus of the growing embryon, which

which I believe is ſometimes attended with ſenſation, produces decreaſed actions of the latter with the diſagreeable ſenſation of ſickneſs with indigeſtion and conſequent acidity. When the fetus acquires ſo much muſcular power as to move its limbs, or to turn itſelf, which is called quickening, this ſickneſs of pregnancy generally ceaſes.

M. M. Calcined magneſia. Rhubarb. Half a grain of opium twice a day. Recumbent poſture on a ſofa.

3. *Vomitio vertiginoſa.* Sea-ſickneſs, the irritative motions of viſion, by which we balance ourſelves, and preſerve our perpendicularity, are diſturbed by the indiſtinctneſs of their objects; which is either owing to the ſimilarity of them, or to their diſtance, or to their apparent or unuſual motions. Hence theſe irritative motions of viſion are exerted with greater energy, and are in conſequence attended with ſenſation; which at firſt is agreeable, as when children ſwing on a rope; afterwards the irritative motions of the ſtomach, and of the abſorbent veſſels, which open their mouths into it, become inverted by their aſſociations with them by reverſe ſympathy.

For the action of vomiting, as well as the diſagreeable ſenſation of ſickneſs, are ſhewn to be occaſioned by defect of the ſenſorial power; which in this caſe is owing to the greater expenditure of it by the ſenſe of viſion. On the ſame account the vomiting, which attends the paſſage of a ſtone through the ureter, or from an inflammation of the bowels, or in the commencement of ſome fevers, is cauſed by the increaſed expenditure of the ſenſorial power by the too great action of ſome links of the aſſociations of irritative motions; and there being in conſequence a deficiency of the quantity required for other links of this great catenation.

It muſt be obſerved, that the expenditure of ſenſorial power by the

the retinas of the eyes is very great; which may be eſtimated by the perpetual uſe of thoſe organs during our waking hours, and during moſt of our ſleeping ones; and by the large diameters of the two optic nerves, which are nearly the ſize of a quill, or equal to ſome of the principal nerves, which ſerve the limbs.

4. *Vomitio a calculo in uretere.* The action of vomiting in conſequence of the increaſed or decreaſed actions of the ureter, when a ſtone lodges in it. The natural actions of the ſtomach, which conſiſt of motions ſubject to intermitted irritations from the fluids, which paſs through it, are aſſociated with thoſe of the ureter; and become torpid, and conſequently retrograde, by intervals, when the actions of the ureter becomes torpid owing to previous great ſtimulus from the ſtone it contains; as appears from the vomiting exiſting when the pain is leaſt. When the motions of the ureter are thus leſſened, the ſenſorial power of aſſociation, which ought to actuate the ſtomach along with the ſenſorial power of irritation, ceaſes to be excited into action; and in conſequence the actions of the ſtomach become leſs energetic, and in conſequence retrograde.

For as vomiting is a decreaſed action of the ſtomach, as explained in Sect. XXXV. 1. 3. it cannot be ſuppoſed to be produced by the pain of gravel in the ureter alone, as it ſhould then be an increaſed action, not a decreaſed one.

The perpetual vomiting in ileus is cauſed in like manner by the defective excitement of the ſenſorial power of aſſociation by the bowel, which is torpid during the intervals of pain; and the ſtomach ſympathizes with it. See Enteritis, Claſs II. 1. 2. 11. Does this ſymptom of vomiting indicate, whether the diſeaſe be above or below the valve of the colon? Does not the ſofter pulſe in ſome kinds of enteritis depend on the ſympathy of the heart and arteries with the ſickneſs of the ſtomach? See Ileus and Cholera.

Hence this ficknefs, as well as the ficknefs in fome fevers, cannot be efteemed an effort of nature to diflodge any offenfive material; but like the fea-ficknefs defcribed above, and in Sect. XX. 4. is the confequence of the affociations of irritative or fenfitive motions. See Clafs I. 1. 3. 9.

5. *Vomitio ab infultu paralytico.* Paralytic affections generally commence with vomiting, the fame frequently happens from a violent blow with a ftick on the head; this curious connection of the brain and ftomach has not been explained; as it refembles the ficknefs in confequence of vertigo at fea, it would feem to arife from a fimilar caufe, viz. from difturbed irritative or fenfitive affociations.

6. *Vomitio a titillatione faucium.* If the throat be flightly tickled with a feather, a naufea is produced, that is, an inverted action of the mouths of the lymphatics of the fauces, and by direct fympathy an inverted action of the ftomach enfues. As thefe parts have frequently been ftimulated at the fame time into pleafurable action by the deglutition of our daily aliment, their actions become ftrongly affociated. And as all the food, we fwallow, is either moift originally, or mixed with our moift faliva in the mouth; a feather, which is originally dry, and which in fome meafure repels the moift faliva, is difagreeable to the touch of the fauces; at the fame time this naufea and vomiting cannot be caufed by the difagreeable fenfation fimply, as then they ought to have been increafed exertions, and not decreafed ones, as fhewn in Section XXXV. 1. 3. But the mouths of the lymphatics of the fauces are ftimulated by the dry feather into too great action for a time, and become retrograde afterwards by the debility confequent to too great previous ftimulus.

7. *Vomitio cute ſympathetica.* Vomiting is ſucceſsfully ſtopped by the application of a bliſter on the back in ſome fevers, where the extremities are cold, and the ſkin pale. It was ſtopped by Sydenham by producing a ſweat on the ſkin by covering the head with the bed-clothes. See Claſs IV. 1. 1. 3. and Suppl. I. 1-1. 6.

ORDO

## ORDO III.

### *Retrograde Aſſociate Motions.*

## GENUS III.

### *Catenated with Voluntary Motions.*

### SPECIES.

1. *Ruminatio.* In the rumination of horned cattle the food is brought up from the firſt ſtomach by the retrograde motions of the ſtomach and œſophagus, which are catenated with the voluntary motions of the abdominal muſcles.

2. *Vomitio voluntaria.* Voluntary vomiting. Some human ſubjects have been ſaid to have obtained this power of voluntary action over the retrograde motions of the ſtomach and œſophagus, and thus to have been able to empty their ſtomach at pleaſure. See Sect. XXV. 6. This voluntary act of emptying the ſtomach is poſſeſſed by ſome birds, as the pigeon; who has an organ for ſecreting milk in its ſtomach, as Mr. Hunter obſerved; and ſoftens the food for its young by previouſly ſwallowing it; and afterwards putting its bill into theirs returns it into their mouths. See Sect. XXXIX. 4. 8. The pelicans uſe a ſtomach, or throat bag, for the purpoſe of bringing the fiſh, which they catch in the ſea to ſhore, and then eject them, and eat them at their leiſure. See Sect. XVI. 11. And I am well informed of a bitch, who having puppies in a ſtable at a diſtance from the houſe, ſwallowed the fleſh-meat, which was given her, in large pieces, and carrying it immediately to her whelps, brought it up out of her ſtomach, and laid it down before them.

3. *Eructatio voluntaria.* Voluntary eructation. Some, who have weak digestions, and thence have frequently been induced to eruct the quantity of air discharged from the fermenting aliment in their stomachs, have gradually obtained a power of voluntary eructation, and have been able thus to bring up hogsheads of air from their stomachs, whenever they pleased. This great quantity of air is to be ascribed to the increase of the fermentation of the aliment by drawing off the gas as soon as it is produced. See Sect. XXIII. 4.

## ORDO III.

### *Retrograde Associate Motions.*

## GENUS. IV.

### *Catenated with External Influences.*

## SPECIES.

1. *Catarrhus periodicus.* Periodical catarrh is not a very uncommon disease; there is a great discharge of a thin saline mucous material from the membranes of the nostrils, and probably from the maxillary and frontal sinuses, which recur once a day at exact solar periods; unless it be disturbed by the exhibition of opium; and resembles the periodic cough mentioned below. See Class I. 3. 2. 1. It is probably owing to the retrograde action of the lymphatics of the membranes affected, and produced immediately by solar influence.

2. *Tussis periodica.* Periodic cough, called nervous cough, and tussis ferina. It seems to arise from a periodic retrograde action of the lymphatics of the membrane, which lines the air-cells of the lungs. And the action of coughing, which is violently for an hour or longer, is probably excited by the stimulus of the thin fluid thus produced, as well as by the disagreeable sensation attending membranous inactivity; and resembles periodic catarrh not only in its situation on a mucous membrane, but in the discharge of a thin fluid. As it is partly restrainable, it does not come under the name of convulsion; and as it is not attended with difficult respiration, it cannot be called asthma; it is cured by very large doses of opium, see a case

and

and cure in Sect. XXXVI. 3. 9. see Clafs IV. 2. 4. 6. and feems immediately to be induced by folar influence.

3. *Hifteria a frigore.* Hyfteric paroxyfms are occafioned by whatever fuddenly debilitates the fyftem, as fear, or cold, and perhaps fometimes by external moifture of the air, as all delicate people have their days of greater or lefs debility, see Clafs IV. 3. 1. 8.

4. *Naufea pluvialis.* Sicknefs at the commencement of a rainy feafon is very common among dogs, who affift themfelves by eating the agroftris canina, or dog's grafs, and thus empty their ftomachs. The fame occurs with lefs frequency to cats, who make ufe of the fame expedient. See Sect. XVI. 11. I have known one perfon, who from his early years has always been fick at the beginning of wet weather, and ftill continues fo. Is this owing to a fympathy of the mucous membrane of the ftomach with the mechanical relaxation of the external cuticle by a moifter atmofphere, as is feen in the corrugated cuticle of the hands of wafhing-women? or does it fympathize with the mucous membrane of the lungs, which muft be affected along with the mucus on its furface by the refpiration of a moifter atmofphere?

SUPPLEMENT.

# SUPPLEMENT

## TO

# CLASS IV.

---

### *Sympathetic Theory of Fever.*

As fever consists in the increase or diminution of direct or reverse associated motions, whatever may have been the remote cause of them, it properly belongs to the fourth class of diseases; and is introduced at the end of the class, that its great difficulties might receive elucidation from the preceding parts of it. These I shall endeavour to enumerate under the following heads, trusting that the candid reader will discover in these rudiments of the theory of fever a nascent embryon, an infant Hercules, which Time may rear to maturity, and render serviceable to mankind.

I. Simple fever of two kinds.
II. Compound fever.
III. Termination of the cold fit.
IV. Return of the cold fit.
V. Sensation excited in fever.
VI. Circles of associated motions.
VII. Alternations of cold and hot fits.
VIII. Orgasm of the capillaries.

### I. *Simple Fever*.

1. When a ſmall part of the cutaneous capillaries with their mucous or perſpirative glands are for a ſhort time expoſed to a colder medium, as when the hands are immerſed in iced water for a minute, theſe capillary veſſels and their glands become torpid or quieſcent, owing to the eduction of the ſtimulus of heat. The ſkin then becomes pale, becauſe no blood paſſes through the external capillaries; and appears ſhrunk, becauſe their ſides are collapſed from inactivity, not contracted by ſpaſm; the roots of the hair are left prominent from the ſeceding or ſubſiding of the ſkin around them; and the pain of coldneſs is produced.

In this ſituation, if the uſual degree of warmth be applied, theſe veſſels regain their activity; and having now become more irritable from an accumulation of the ſenſorial power of irritation during their quieſcence, a greater exertion of them follows, with an increaſed glow of the ſkin, and another kind of pain, which is called the hot-ach; but no fever, properly ſo called, is yet produced; as this effect is not univerſal, nor permanent, nor recurrent.

2. If a greater part of the cutaneous capillaries with their mucous and perſpirative glands be expoſed for a longer time to cold, the torpor

por or quiescence becomes extended by direct sympathy to the heart and arteries; which is known by the weakness, and consequent frequency of the pulse in cold fits of fever.

This requires to be further explained. The movements of the heart and arteries, and the whole of the circulatory vessels, are in general excited into action by the two sensorial powers of irritation, and of association. The former is excited by stimulus, the latter by the previous actions of a part of the vital circle of motions. In the above situation the capillaries act weakly from defect of irritation, which is caused by deficient stimulus of heat; but the heart and arteries act weakly from defect of association, which is owing to the weak action of the capillaries; which does not now excite the sensorial power of association into action with sufficient energy.

After a time, either by the application of warmth, or by the increase of their irritability owing to the accumulation of the sensorial power of irritation during their previous quiescence, the capillary vessels and glands act with greater energy than natural; whence the red colour and heat of the skin. The heart and arteries acquire a greater strength of pulsation, and continue the frequency of it, owing to the accumulation of the sensorial power of association during their previous torpor, and their consequent greater associability; which is now also more strongly excited by the increased actions of the capillaries. And thus a fit of simple fever is produced, which is termed Febris irritativa; and consists of a torpor of the cutaneous capillaries with their mucous and perspirative glands, accompanied with a torpor of the heart and arteries; and afterwards of an increased action of all these vessels, by what is termed direct sympathy.

This fever, with strong pulse without inflammation, or febris irritativa, described in Class I. 1. 1. 1. is frequently seen in vernal intermittents, as the orgasm of the heart and arteries is then occasioned by their previous state of torpor; but more rarely I believe exists in the type of continued fever, except there be an evident remission, or

 approximation

approximation to a cold fit; at which time a new accumulation of the ſenſorial power of aſſociation is produced; which afterwards actuates the heart and arteries with unnatural vigour; or unleſs there be ſome ſtimulus perpetually acting on the ſyſtem, ſo as to induce an increaſed ſecretion of ſenſorial power in the brain, as occurs in ſlight degrees of intoxication. Since without one or other of theſe circumſtances in continued fevers without inflammation, that is, without the additional ſenſorial power of ſenſation being introduced, it ſeems difficult to account for the production of ſo great a quantity of ſenſorial power, as muſt be neceſſary to give perpetual increaſe of action to the whole ſanguiferous ſyſtem.

3. On the contrary, while the cutaneous capillaries with their mucous and perſpirative glands acquire an increaſed irritability, as above, by the accumulation of that ſenſorial power during their previous quieſcence, and thus conſtitute the hot fit of fever; if the heart and arteries do not acquire any increaſe of aſſociability, but continue in their ſtate of torpor, another kind of ſimple fever is produced; which is generally of the continued kind, and is termed Febris inirritativa; which conſiſts of a previous torpor of the capillaries of the ſkin, and of the heart and arteries by direct ſympathy with them; and afterwards of an orgaſm or increaſed action of the capillaries of the ſkin, with a decreaſed action, or continued torpor, of the heart and arteries by reverſe ſympathy with them. This orgaſm of the cutaneous capillaries, which appears by the bluſh and heat of the ſkin, is at firſt owing to the accumulation of the ſenſorial power of irritation during their pıevious torpid ſtate, as in the febris irritata above deſcribed; bnt which is afterwards ſupported or continued by the reverſe ſympathy of theſe capillaries with the torpid ſtate of the heart and arteries, as will be further explained in article 8 of this Supplement.

4. The

4. The renovated activity of the capillaries commences as ſoon or ſooner than that of the heart and arteries after the cold fit of irritative fever; and is not owing to their being forced open by the blood being impelled into them mechanically, by the renovated action of the heart and arteries; for theſe capillaries of the ſkin have greater mobility than the heart and arteries, as appears in the ſudden bluſh of ſhame; which may be owing to their being more liable to perpetual varieties of activity from their expoſure to the viciſſitudes of atmoſpheric heat. And becauſe in inirritative fevers, or thoſe with arterial debility, the capillaries acquire increaſed ſtrength, as is evinced by the heat of the ſkin, while the pulſations of the heart and arteries remain feeble.

5. It was ſaid above, that the cutaneous capillaries, when they were rendered torpid by expoſure to cold, either recovered their activity by the reapplication of external warmth; or by their increaſed irritability, which is cauſed by the accumulation of that ſenſorial power during their quieſcence. An example of the former of theſe may be ſeen on emerging from a very cold bath; which produces a fit of ſimple fever; the cold fit, and conſequent hot fit, of which may be prolonged by continuing in the bath; which has indeed proved fatal to ſome weak and delicate people, and to others after having been much exhauſted by heat and exerciſe. See Sect. XXXII. 3. 2. An example of the latter may be taken from going into a bath of about eighty degrees of heat, as into the bath at Buxton, where the bather firſt feels a chill, and after a minute becomes warm, though he remains in the ſame medium, owing to the increaſe of irritability from the accumulation of that ſenſorial power during the ſhort time, which the chilneſs continued.

6. Hence ſimple fevers are of two kinds; firſt, the febris irritativa, or fever with ſtrong pulſe; which conſiſts of a previous torpor of the heart, arteries, and capillaries, and a ſucceeding orgaſm of thoſe veſſels.

ſels. Secondly, the febris inirritativa, or fever with weak pulſe, which conſiſts of a previous torpor of the heart, arteries, and capillaries; and of a ſucceeding orgaſm of the capillaries, the torpor of the heart and arteries continuing. But as the frequency of the pulſe occurs both in the ſtate of torpor, and in that of orgaſm, of the heart and arteries; this conſtitutes a criterion to diſtinguiſh fever from other diſeaſes, which are owing to the torpor of ſome parts of the ſyſtem, as pareſis, and hemicrania.

7. The reader will pleaſe to obſerve, that where the cutaneous or pulmonary capillaries are mentioned, their mucous and perſpirative glands are to be underſtood as included; but that the abſorbents belonging to thoſe ſyſtems of veſſels, and the commencement of the veins, are not always included; as theſe are liable to torpor ſeparately, as in anaſarca, and petechiæ; or to orgaſm, or increaſed action, as in the exhibition of ſtrong emetics, or in the application of vinegar to the lips; yet he will alſo pleaſe to obſerve, that an increaſed or decreaſed action of theſe abſorbents and veins generally occurs along with that of the capillaries, as appears by the dry ſkin in hot fits of fever; and from there being generally at the ſame time no accumulation of venous blood in the cutaneous veſſels, which would appear by its purple colour.

## II. *Compound Fever.*

1. When other parts of the ſyſtem ſympathize with this torpor and orgaſm of the cutaneous capillaries, and of the heart and arteries; the fever-fit becomes more complicated and dangerous; and this in proportion to the number and conſequence of ſuch affected parts. Thus if the lungs become affected, as in going into very cold water, a ſhortneſs of breath occurs; which is owing to the collapſe or inac-

tivity (not to the active contraction, or spasm), of the pulmonary capillaries; which, as the lungs are not sensible to cold, are not subject to painful sensation, and consequent shuddering, like the skin. In this case after a time the pulmonary capillaries, like the cutaneous ones, act with increased energy; the breathing, which was before quick, and the air thrown out at each respiration in less quantity, and cool to the back of the hand opposed to it, now becomes larger in quantity, and warmer than natural; which however is not accompanied with the sensation of heat in the membrane, which lines the air-vessels of the lungs, as in the skin.

2. One consequence of this increased heat of the breath is the increased evaporation of the mucus on the tongue and nostrils. A viscid material is secreted by these membranes to preserve them moist and supple, for the purposes of the senses of taste and of smell, which are extended beneath their surfaces; this viscid mucus, when the aqueous part of it is evaporated by the increased heat of the respired air, or is absorbed by the too great action of the mucous absorbents, adheres closely on those membranes, and is not without difficulty to be separated from them. This dryness of the tongue and nostrils is a circumstance therefore worthy to be attended to; as it shews the increased action of the pulmonary capillaries, and the consequent increased heat of the expired air; and may thus indicate, when colder air should be admitted to the patient. See Class I. 1. 3. 1. The middle part of the tongue becomes dry sooner, and recovers its moisture later, than the edges of it; because the currents of respired air pass most over the middle part of it. This however is not the case, when the dryness of the tongue is owing only to the increased mucous absorption. When however a frequent cough attends pulmonary inflammation, the edges of the tongue are liable to be as much furred as the middle of it; as during the action of coughing the middle of the tongue is depressed, so as to form half a cylinder, to give a greater

aperture

aperture for the emiſſion of air from the larynx; and the edges of it become thus as much expoſed to the currents of air, as the middle parts of it.

3. When the internal capillaries or glands ſympathize with the cutaneous capillaries; or when any of them are previouſly affected with torpor, and the external or cutaneous capillaries are affected ſecondarily; other ſymptoms are produced, which render the paroxyſms of fever ſtill more complicate. Thus if the ſpleen or pancreas are primarily or ſecondarily affected, ſo as to be rendered torpid or quieſcent, they are liable to become enlarged, and to remain ſo even after the extinction of the fever-fit. Theſe in ſome intermittent fevers are perceptible to the hand, and are called ague-cakes; their tumour ſeems to be owing to the permanent torpor of the abſorbent ſyſtem, the ſecerning veſſels continuing to act ſome time afterwards. If the ſecretory veſſels of the liver are affected firſt with torpor, and afterwards with orgaſm, a greater ſecretion of bile is produced, which ſometimes cauſes a diarrhœa. If a torpor of the kidneys, and of the abſorbents of the bladder occurs, either primarily, or by ſympathy with the cutaneous capillaries, the urine is in ſmall quantity and pale, as explained in Claſs I. 2. 2. 5.; and if theſe ſecretory veſſels of the kidneys, and the abſorbents of the bladder act more ſtrongly than natural afterwards by their increaſed irritability or aſſociability, the urine becomes in larger quantity, and deeper coloured, or depoſits its earthy parts, as in Claſs I. 1. 2. 4. which has been eſteemed a favourable circumſtance. But if the urine be in ſmall quantity, and no ſediment appears in it, after the hot fit is over; it ſhews, that the ſecerning veſſels of the kidneys and the abſorbent veſſels of the bladder have not regained the whole of their activity, and thence indicates a greater tendency to a return of the cold fit.

4. When the ſtomach is affected with torpor either primarily; or ſecondarily

secondarily, by its sympathy with the cutaneous capillaries; or with some internal viscus; sickness occurs, with a total want of appetite to any thing solid; vomiting then supervenes, which may often be relieved by a blister on the skin, if the skin be cool and pale; but not if it be hot and flushed. The intestines cease to perform their office of absorption from a similar torpor; and a diarrhœa supervenes owing to the acrimony of their putrid, or of their acid contents. The loose undigested or fetid stools indicate the inability of the intestines to perform their proper office; as the mucus and gastric acid, which are vomited up, does that of the stomach; this torpor of the stomach is liable to continue after the cold paroxysm ceases, and to convert intermittent fevers into continued ones by its direct sympathy with the heart and arteries. See article 10 of this Supplement.

5. If the meninges of the brain sympathize with other torpid parts or are primarily affected, delirium, stupor, and perhaps hydrocephalus internus occur, see Class II. 1. 7. 1. and I. 2. 5. 10; and sometimes the pulse becomes slow, producing paresis instead of fever. But if the membranes, which cover the muscles about the head, or of the pericranium, become torpid by their sympathy with other torpid parts, or are primarily affected, a head-ach supervenes; which however generally ceases with the cold paroxysm of fever. For as when the sensorial power of volition is exhausted by labour, a few hours, or half a solar day, passed in sleep recruits the system by accumulation of this sensorial power; so when the sensorial power of irritation is exhausted, one or two solar or lunar days of rest or quiescence of the affected part will generally restore its action by accumulation of irritability, and consequent increase of association, as in hemicrania, Class IV. 2. 2. 8. But when the heart and arteries become torpid, either primarily, or by their sympathy with the stomach, this accumulation of the sensorial power of irritation can take place but slowly; *as to rest is death!* This explains the cause of the duration of fevers with

weak pulſe, which continue a quarter, or half, or three quarters, or a whole lunation, or ſtill longer, before ſufficient accumulation of irritability can be produced to reſtore their natural ſtrength of action.

6. If the abſorbent veſſels, which are ſpread around the neck of the bladder, become torpid by their direct ſympathy with the abſorbents of the ſkin in cold fits of fever; the urine, which is poured into the bladder in but ſmall quantity from the torpid kidneys, has nevertheleſs none of its aqueous ſaline part reabſorbed; and this ſaline part ſtimulates the bladder to empty itſelf frequently, though the urine is in ſmall quantity. Which is not therefore owing to any ſuppoſed ſpaſm of the bladder, for the action of it in excluding the urine is weak, and as much controlable by the will as in ordinary micturition.

7. If the beginnings or abſorbent mouths of the venous ſyſtem remain torpid, petechiæ or vibices are produced in fevers, ſimilar to thoſe which are ſeen in ſcurvy without fever. If the ſkin was frequently moiſtened for an hour, and at the ſame time expoſed to the common air, or to oxygen gas, it might contribute to turn the black colour of theſe points of extravaſated blood into ſcarlet, and thus by increaſing its ſtimulus facilitate its reabſorption? For oxygen gas penetrates moiſt animal membranes though not dry ones, as in the lungs during reſpiration.

8. When the ſenſorial power of ſenſation is introduced into the arterial ſyſtem, other kinds of compound fevers are produced, which will be ſpoken of in their place.

III. *Termination*

### III. *Termination of the cold Fit.*

1. If all the parts, which were affected with torpor, regain their irritability, and aſſociability, the cold paroxyſm of fever ceaſes; but as ſome of the parts affected were previouſly accuſtomed to inceſſant action, as the heart and arteries, and others only to intermitted action, as the ſtomach and inteſtines; and as thoſe, which are ſubjected during health to perpetual action, accumulate ſenſorial power faſter, when their motions are impeded, than thoſe which are ſubjected to intermitted action; it happens, that ſome of the parts, which were affected with torpor during the cold fit, recover their irritability or aſſociability ſooner than others, and more perfectly, or acquire a greater quantity of them than natural; as appears by the partial heat and fluſhings previous to the general hot fit.

Hence if all the parts, which were previouſly torpid, regain their due degree of irritability, or of aſſociability, the diſeaſe is removed, and health reſtored. If ſome or all of them acquire more than their natural degree of theſe ſenſorial powers; increaſed actions, and conſequent increaſed ſecretions, and greater heat occur, and conſtitute the hot fit of fever. If after this hot fit of fever all the parts, which had acquired too great irritability, or aſſociability, regain their natural degree of it; the diſeaſe is removed, and health reſtored. But if ſome of theſe parts do not regain their natural degree of theſe ſenſorial powers, the actions of thoſe parts remain imperfect, and are more or leſs injurious to the ſyſtem, according to the importance of their functions.

2. Thus if a torpor of the heart and arteries remains; the quick pulſe without ſtrength, which began in the cold fit, perſiſts; and a continued fever is produced. If the torpor of the ſtomach and inteſ-

tines remains, which are known by ſickneſs and undigeſted ſtools, the fever is liable to be of conſiderable length and danger ; the ſame if the kidnies and abſorbent ſyſtem retain ſome degree of torpor, as is ſhewn by the pale urine in not unuſual quantity. If part of the abſorbent ſyſtem remains torpid, as the abſorbent veſſels of the ſpleen, a tumour of that viſcus occurs, which may be felt by the hand; the ſame ſometimes happens to the liver; and theſe from their tendency to more complete torpor are afterwards liable to give occaſion to a return of the cold fit. If the cellular abſorbents do not completely recover their activity, a pale and bloated countenance with ſwelled legs mark their want of action.

3. As the termination of the cold fit is owing to the accumulation of the ſenſorial power of irritation and of aſſociation during the previous quieſcence of the ſyſtem ; and as thoſe parts, which are in perpetual action during health, are more ſubject to this accumulation during their torpor, or quieſcence ; one ſhould have imagined, that the heart and arteries would acquire this accumulation of ſenſorial power ſooner or in greater degree than other parts. This indeed ſo happens, where the pulſe is previouſly ſtrong, as in febris irritativa ; or where another ſenſorial power, as that of ſenſation, is exerted on the arterial ſyſtem, as in inflammations. The heart and arteries in theſe caſes ſoon recover from their torpor, and are exerted with great violence.

Many other parts of the ſyſtem ſubject to perpetual motion in health may reſt for a time without much inconvenience to the whole; as when the fingers of ſome people become cold and pale ; and during this complete reſt great accumulation of irritability may be produced. But where the heart and arteries are previouſly feeble, they cannot much diminiſh their actions, and certainly cannot reſt entirely, for that would be death ; and therefore in this caſe their accumulation of the ſenſorial power of irritation or of aſſociation is ſlowly produced,

 and

and a long fever supervenes in consequence; or sudden death, as frequently happens, terminates the cold fit.

Whence it appears, that in fevers with weak pulse, if the action of the heart, arteries, and capillaries could be diminished, or stopped for a short time without occasioning the death of the patient, as happens in cold bathing, or to persons apparently drowned, that a great accumulation of the sensorial powers of irritation or of association might soon be produced, and the pulse become stronger, and consequently slower, and the fever cease. Hence cold ablution may be of service in fevers with weak pulse, by preventing the expenditure and producing accumulation of the sensorial power of irritation or association. Stupor may be useful on the same account. Could a centrifugal swing be serviceable for this purpose, either by placing the head or the feet in the outward part of the circle, as described in Art. 15. 7. of this Supplement?

### IV. *Return of the cold Fit.*

1. If the increased action of the cutaneous and pulmonary capillaries, and of the heart and arteries, in febris irritativa continues long and with violence, a proportional expenditure or exhaustion of sensorial power occurs; which by its tendency to induce torpor of some part, or of the whole, brings on a return of the cold fit.

2. Another cause which contributes to induce torpor of the whole system by the sympathy of its parts with each other, is the remaining torpor of some viscus; which after the last cold paroxysm had not recovered itself, as of the spleen, liver, kidnies, or of the stomach and intestines, or absorbent vessels, as above mentioned.

3. Other

3. Other causes are the deficiency of the natural stimuli, as hunger, thirst, and want of fresh air. Other causes are great fatigue, want of rest, fear, grief, or anxiety of mind. And lastly, the influence of external ethereal fluids, as the defect of external heat, and of solar or lunar gravitation. Of the latter the return of the paroxysms of continued fevers about six o'clock in the evening, when the solar gravitation is the least, affords an example of the influence of it; and the usual periods of intermittents, whether quotidian, tertian, or quartan, which so regularly obey solar or lunar days, afford instances of the influence of those luminaries on these kinds of fevers.

4. If the tendency to torpor of some viscus is considerable, this will be increased at the time, when the terrene gravitation is greatest, as explained in the introduction to Class IV. 2. 4. and may either produce a cold paroxysm of quotidian fever; or it may not yet be sufficient in quantity for that purpose, but may nevertheless become greater, and continue so till the next period of the greatest terrene gravitation, and may then either produce a paroxysm of tertian fever; or may still become greater, and continue so till the next period of greatest terrene gravitation, and then produce a paroxysm of quartan ague. And lastly, the periodical times of these paroxysms may exceed, or fall short of, the time of greatest diurnal terrene gravitation according to the time of day, or period of the moon, in which the first fit began; that is, whether the diurnal terrene gravitation was then in an increasing or decreasing state.

V. *Sensation*

V. *Senſation excited in Fever.*

1. A curious obſervation is related by Dr. Fordyce in his Tract on Simple fever, page 168. He aſſerts, that thoſe people, who have been confined ſome time in a very warm atmoſphere, as of 120 or 130 degrees of heat, do not feel cold, nor are ſubject to paleneſs of their ſkins, on coming into a temperature of 30 or 40 degrees; which would produce great paleneſs and painful ſenſation of coldneſs in thoſe, who had been ſome time confined in an atmoſphere of only 86 or 90 degrees. Analogous to this, an obſerving friend of mine aſſured me, that once having ſat up to a very late hour with three or four very ingenious and humorous companions, and drank a conſiderable quantity of wine; both contrary to his uſual habits of life; and being obliged to riſe early, and to ride a long journey on the next day; he expected to have found himſelf weak and ſoon fatigued; but on the contrary he performed his journey with unuſual eaſe and alacrity; and frequently laughed, as he rode, at the wit of the preceding evening. In both theſe caſes a degree of pain or pleaſure actuated the ſyſtem; and thus a ſenſorial power, that of ſenſation, was ſuperadded to that of irritation, or volition. See Sect. XXXIV. 2. 6.

2. Similar to this, when the energetic exertions of ſome parts of the ſyſtem in the hot fit of fever ariſe to a certain exceſs, a degree of ſenſation is produced; as of heat, which particularly increaſes the actions of the cutaneous veſſels, which are more liable to be excited by this ſtimulus. When this additional ſenſorial power of ſenſation exiſts to a greater degree, the pulſe, which was before full, now becomes hard, owing to the inflammation of the vaſa vaſorum, or coats of the arteries. In theſe caſes, whether there is any topical inflammation or not, the fever ceaſes to intermit; but nevertheleſs there

are daily remiſſions and exacerbations of it; which recur for the moſt part about ſix in the evening, when the ſolar gravitation is the leaſt, as mentioned in Sect. XXXVI. 3. 7.

3. Thus the introduction of another ſenſorial power, that of ſenſation, converts an intermittent fever into a continued one. If it be attended with ſtrong pulſe, it is termed febris ſenſitiva irritata, or pyrexia, or inflammation; if with a weak pulſe, it is termed febris ſenſitiva inirritata, or typhus gravior, or malignant fever. The ſeat of the inflammation is in the glandular or capillary ſyſtem, as it conſiſts in the ſecretion of new fluids, or new fibres, which form new veſſels, as they harden, like the ſilk of the ſilk-worm. See Art. 15. of this Supplement.

## VI. *Circles of irritative Aſſociate Motions.*

1. There are ſome aſſociate motions, which are perpetually proceeding in our waking hours, and are catenated by their firſt link, or in ſome ſubſequent parts of the chain, with the ſtimuli or the influence of external things; which we ſhall here enumerate, as they contribute to the knowledge of fever. Of theſe are the irritative ideas, or ſenſual motions of the organs of ſenſe, and the muſcular motions aſſociated with them; which, when the chain is diſturbed or interrupted, excite the ſenſorial power of ſenſation, and proceed in confuſion. Thus if the irritative ideas of ſight are diſturbed, the paralactic motions of objects, which in general are unperceived, become ſenſible to us; and the locomotive muſcles aſſociated with them, which ought to preſerve the body erect, ſtagger from this decreaſe or interruption of the ſenſorial power of aſſociation; and vertigo is produced.

When

When the irritative ſenſual motions, or ideas, belonging to one ſenſe are increaſed or diminiſhed, the irritative ſenſual motions, or ideas, of the other ſenſes are liable to become diſturbed by their general catenations; whence occur noiſes in the ears, bad taſtes in the mouth, bad odours, and numbneſs or tingling of the limbs, as a greater or leſs number of ſenſes are affected. Theſe conſtitute concomitant circles of diſturbed irritative ideas; or make a part of the great circle of irritative ideas, or motions of the organs of ſenſe; and when thus diſturbed occaſion many kinds of hallucination of our other ſenſes, or attend on the vertigo of viſion.

2. Another great circle of irritative aſſociated motions conſiſts of thoſe of the alimentary canal; which are catenated with ſtimuli or with influences external to the ſyſtem, but continue to be exerted in our ſleeping as well as in our waking hours. When theſe aſſociations of motion are diſturbed by the too great or too ſmall ſtimulus of the food taken into the ſtomach, or by the too great exceſs or deprivation of heat, or by indigeſtible ſubſtances, or by torpor or orgaſm occaſioned by their aſſociation with other parts, various diſeaſes are induced under the names of apepſia, hypochondriaſis, hyſteria, diarrhœa, cholera, ileus, nephritis, fever.

3. A third circle of irritative aſſociate motions conſiſts of thoſe of the abſorbent ſyſtem; which may be divided into two, the lacteals, and the lymphatics. When the ſtomach and inteſtines are recently filled with food and fluid, the lacteal ſyſtem is ſtimulated into great action; at the ſame time the cellular, cutaneous, and pulmonary lymphatics act with leſs energy; becauſe leſs fluid is then wanted from thoſe branches, and becauſe more ſenſorial power is expended by the lacteal branch. On this account theſe two ſyſtems of abſorbents are liable to act by reverſe ſympathy; hence pale urine is made after a full dinner, as leſs of the aqueous part of it is imbibed by

 the

the urinary lymphatics; and hence the water in anaſarca of the lungs and limbs is ſpeedily abſorbed, when the actions of the lacteals of the ſtomach or inteſtines are weakened or inverted by the exhibition of thoſe drugs, which produce nauſea, or by violent vomiting, or violent cathartics.

Hence in diabetes the lacteal ſyſtem acts ſtrongly, at the ſame time that the urinary lymphatics invert their motions, and tranſmit the chyle into the bladder; and in diarrhœa from crapula, or too great a quantity of food and fluid taken at a time, the lacteals act ſtrongly, and abſorb chyle or fluids from the ſtomach and upper inteſtines; while the lymphatics of the lower inteſtines revert their motions, and tranſmit this over-repletion into the lower inteſtines, and thus produce diarrhœa; which accounts for the ſpeedy operation of ſome cathartic drugs, when much fluid is taken along with them.

4. Other circles of irritative aſſociate motions of great importance are thoſe of the ſecerning ſyſtem; of theſe are the motions of the larger congeries of glands, which form the liver, ſpleen, pancreas, gaſtric glands, kidneys, ſalivary glands, and many others; ſome of which act by direct and others by reverſe ſympathy with each other. Thus when the gaſtric glands act moſt powerfully, as when the ſtomach is filled with food, the kidneys act with leſs energy; as is ſhewn by the ſmall ſecretion of urine for the firſt hour or two after dinner; which reverſe ſympathy is occaſioned by the greater expenditure of ſenſorial power on the gaſtric glands, and to the newly abſorbed fluids not yet being ſufficiently animalized, or otherwiſe prepared, to ſtimulate the ſecretory veſſels of the kidneys.

But thoſe very extenſive glands, which ſecrete the perſpirable matter of the ſkin and lungs, with the mucus, which lubricates all the internal cells and cavities of the body, claim our particular attention. Theſe glands, as well as all the others, proceed from the capillary

pillary veſſels, which unite the arteries with the veins, and are not properly a part of them; the mucous and perſpirative glands, which ariſe from the cutaneous and pulmonary capillaries, are aſſociated by direct ſympathy; as appears from immerſion in the cold bath, which is therefore attended with a temporary difficult reſpiration; while thoſe from the capillaries of the ſtomach and heart and arteries are more generally aſſociated by reverſe ſympathy with thoſe of the cutaneous capillaries; as appears in fevers with weak pulſe and indigeſtion, and at the ſame time with a hot and dry ſkin.

The diſturbed actions of this circle of the aſſociate motions of the ſecerning ſyſtem, when the ſenſorial power of ſenſation is added to that of irritation, frequently produces inflammation, which conſiſts in the ſecretion of new fluids or new veſſels. Neverthelеſs, if theſe diſturbed actions be of the torpid kind, the pain, which attends them, is ſeldom productive of inflammation, as in hemicrania; but is liable to excite voluntary actions, and thus to expend much ſenſorial power, as in the ſhuddering in cold fits of fever, or in convulſions; or laſtly the pain itſelf, which attends torpid actions, is liable to expend or exhauſt much ſenſorial power without producing any increaſed actions; whence the low pulſe, and cold extremities, which uſually attend hemicrania; and hence when inert, or inactive ſenſation attends one link of aſſociated action, the ſucceeding link is generally rendered torpid, as a coldneſs of the cheek attends tooth-ach.

5. A fifth important circle of irritative motions is that of the ſanguiferous ſyſtem, in which the capillary veſſels are to be included, which unite the arterial and venous ſyſtems, both pulmonary and aortal. The diſturbed action of this ſyſtem of the heart and arteries, and capillaries, conſtitute ſimple fever; to which may be added, that the ſecerning and abſorbent veſſels appending to the capillaries, and the bibulous mouths of the veins, are in ſome meaſure at the ſame time generally affected.

6. Now, though the links of each of these circles of irritative motions are more strictly associated together, yet are they in greater or less degree associated or catenated with each other by direct or reverse sympathy. Thus the sickness, or inverted irritative motions of the stomach, are associated or catenated with the disturbed irritative ideas, or sensual motions, in vertigo; as in sea-sickness. This sickness of the stomach is also associated or catenated with the torpor of the heart and arteries by direct sympathy, and with the capillaries and absorbents by reverse sympathy; and are thus all of them liable occasionally to be disturbed, when one of them is diseased; and constitute the great variety of the kinds or symptoms of fevers.

## VII. *Alternation of the cold and hot Fits.*

1. When any cause occurs, which diminishes to a certain degree the supply of sensorial power in respect to the whole system; as suppose a temporary inexertion of the brain; what happens? First, those motions are exerted with less energy, which are not immediately necessary to life, as the locomotive muscles; and those ideas, which are generally excited by volition; at the same time this deficiency of voluntary motion is different from that which occurs in sleep; as in that the movements of the arterial system are increased in energy though not in frequency. Next, the motions of the alimentary canal become performed with less energy, or cease altogether; and a total want of appetite to solid food occurs, or sickness, or a diarrhoea occasioned by the indigested aliment. Then the absorbent vessels cease to act with their due energy; whence thirst, and pale urine, though in small quantities. Fourthly, the secerning vessels become affected by the general diminution of sensorial power; whence all the secreted fluids are produced in less quantity. And lastly, the sanguiferous canals feel the general torpor; the pulsations

of

of the heart and arteries become feeble, and consequently quick; and the capillaries of the skin become inactive, acquire less blood from the arteries, and are consequently paler and shrunk.

In this last circumstance of the torpor of the sanguiferous system consists inirritative fever; as all the others are rather accidental or concomitant symptoms, and not essential ones; as fewer or more of them may be present, or may exist with a greater or less degree of inactivity.

2. Now as the capillaries of the skin are exposed to greater varieties of heat and cold, than the heart and arteries, they are supposed to be more mobile, that is, more susceptible of torpor or exertion, or to inflammation, by external stimuli or influences, than the other parts of the sanguiferous system; and as the skin is more sensible to the presence of heat, than the internal parts of the body, the commencement of the cold paroxysms of fever generally either first exists in, or is first perceived by, the coldness and paleness of the skin; and the commencement of the hot fits by the heat and redness of it.

3. The accumulation of sensorial power occurs in these organs soonest, and in greatest quantity, during their quiescence, which were most perpetually in action during health; hence those parts of the system soonest recover from torpor in intermittent fever, and soonest fall into the contrary extreme of increased activity; as the sanguiferous system of the heart and arteries and capillaries. But of these the capillaries seem first to acquire a renovation of their action, as the heat of the skin becomes first renewed, as well as increased beyond its natural quantity, and this in some parts sooner than in others; which quantity of heat is however not to be estimated simply by the rise of the mercury in the thermometer, but also by the quantity carried away into the atmosphere, or diffused amongst other bodies

bodies in a given time; as more heat passes through water, which boils vehemently, than when it boils gently, though the rise of the thermometer in both cases continues the same. This fact may be known by boiling an egg in water, the white of which coagulates in much less time, if the water boils vehemently, than if it boils moderately, though the sensible heat of the water is the same in both cases.

Another cause, which induces the cutaneous capillaries to renew their actions sooner than the heart and arteries after immersion in the cold bath, is, that their torpor was occasioned by defect of irritation; whereas that of the heart and arteries was occasioned by defect of association; which defect of association was owing to the decreased actions of the capillaries, and is now again excited by their renewed action; which excitement must therefore be subsequent to that increased action of the capillaries; and in consequence the increased action of the heart and arteries at the commencement of the hot fit of some fevers is subsequent to the increased action of the cutaneous capillaries. There is, however, in this case an accumulation of the sensorial power of association in the heart and arteries, which must contribute to increase their orgasm in the hot fit, as well as the increased excitement of it by the increased action of the capillaries.

4. Now this increased action of the system, during the hot fit, by exhausting the sensorial powers of irritation and association, contributes to induce a renewal of the cold paroxysm; as the accumulation of those sensorial powers in the cold fit produces the increased actions of the hot fit; which two states of the system reciprocally induce each other by a kind of libration, or a plus and minus, of the sensorial powers of irritation and association.

If the exhaustion of sensorial power during the hot fit of fever only reduces the quantity of irritability and associability to its natural standard, the fever is cured, not being liable to return. If the quantity of theſe ſenſorial powers be reduced only ſo much, as not to produce a ſecond cold fit during the preſent quantity of external ſtimuli or influences; yet it may be ſo far reduced, that a very ſmall ſubtraction of ſtimulus, or of influence, may again induce a cold fit; ſuch as the coldneſs of the night-air, or the diminution of ſolar or lunar gravitation, as in intermittent fevers.

5. Another cauſe of the renovation of the cold fits of fever is from ſome parts of the ſyſtem not having completely recovered from the former cold paroxyſm; as happens to the ſpleen, liver, or other internal viſcus; which ſometimes remains tumid, and either occaſions a return of the cold fit by direct ſympathy with other parts of the body; or by its own want of action cauſes a diminution of the general quantity of heat, and thus facilitates the renovation of the torpor of the whole ſyſtem, and gives cauſe to intermittent fevers catenated with lunar or ſolar influence.

## VIII. *Orgaſm of the Capillaries.*

As the remaining torpor of ſome leſs eſſential part of the ſyſtem, as of the ſpleen, when the hot fit ceaſes, produces after one, two, or three days a return of cold fit by direct ſympathy with the cutaneous capillaries, when joined with ſome other cauſe of torpor, as the defect of ſolar or lunar influences, or the expoſure to cold or hunger, and thus gives origin to intermittent fever; ſo the remaining torpor of ſome more eſſential parts of the ſyſtem, as of the ſtomach and inteſtines, is probably the cauſe of the immediate recurrence of the cold paroxyſm, at the time the hot one ceaſes, by their direct ſympathy with the cutaneous capillaries, without the aſſiſtance of any other cauſe of torpor; and thus produces remittent fever. And laſtly the remaining torpor of ſome ſtill more eſſential parts of the

ſyſtem,

ſyſtem, as the heart and arteries, after the hot fit ought to ceaſe, is liable by reverſe ſympathy with the cutaneous capillaries to continue their orgaſm, and thus to render a fever continual, which would otherwiſe remit or intermit.

Many difficulties here occur, which we ſhall endeavour to throw ſome light upon, and leave to future inveſtigation; obſerving only hat difficulties were to be expected, otherwiſe fevers would long ſince have been underſtood, as they have employed the unremitted attention of the phyſicians of all ages of the world.

1. Why do the ſame parts of ſucceſſive trains of action ſometimes affect each other by direct, and ſometimes by reverſe ſympathy?—1ſt, When any irritative motion ceaſes, or becomes torpid, which was before in perpetual action; it is either deprived of its uſual ſtimulus, and thence the ſenſorial power of irritation is not excited; or it has been previouſly too much ſtimulated, and the ſenſorial power has been thus exhauſted.

In the former caſe an accumulation of ſenſorial power ſoon occurs, which is excitable by a renewal of the ſtimulus; as when the fingers, which have been immerſed ſome time in ſnow, are again expoſed to the uſual warmth of a room. Or, ſecondly, the ſenſorial power of irritation becomes ſo much accumulated, that the motions, which were torpid, are now performed by leſs ſtimulus than natural; as appears by the warmth, which ſoon occurs after the firſt chill in going into froſty air, or into the bath at Buxton, which is about eighty degrees of heat. Or, laſtly, this accumulation of the ſenſorial power of irritation ſo far abounds, that it increaſes the action of the next link of the aſſociated train or tribe of motions; thus on expoſing the ſkin to cold air, as in walking out in a froſty morning, the actions of the ſtomach are increaſed, and digeſtion ſtrengthened.

But where the torpor of ſome irritative motion is owing to the previous exhauſtion of the ſenſorial power of irritation by too great ſtimulus,

ſtimulus, the reſtoration of it occurs either not at all, or much more ſlowly than in the former inſtances; thus after intoxication the ſtomach is very ſlow in recovering its due quantity of the ſenſorial power of irritation, and never ſhews any accumulation of it.

2. When an aſſociate motion, as deſcribed in the introduction to Claſs IV. 1. 1. acts with leſs energy, the ſenſorial power of aſſociation is either not ſufficiently excited by the preceding fibrous motions; or it has been expended or exhauſted by the too violent actions of the preceding fibrous motions. In the former caſe there occurs an accumulation of the ſenſorial power of aſſociation; exactly as, where the uſual ſtimulus is withdrawn, there occurs an accumulation of the ſenſorial power of irritation. Thus when the actions of the capillaries of the ſkin are diminiſhed by immerſion in cold water, the capillaries of the lungs are rendered torpid by the want of the excitement of the ſenſorial power of aſſociation, owing to the leſſened actions of the previous fibrous motions, namely, of thoſe of the ſkin. Nevertheleſs as ſoon as the capillaries of the ſkin regain their increaſed activity by the accumulation of the ſenſorial power of irritation, theſe capillaries of the lungs act with greater energy alſo owing to their accumulated ſenſorial power of aſſociation. Theſe are inſtances of direct ſympathy, and conſtitute the cold and hot paroxyſms of intermittent fever; or the firſt paroxyſm of a continued one.

3. When the firſt link of a train of aſſociated motions, which is ſubject to perpetual action, becomes a conſiderable time torpid for want of being excited by the previous exertions of the irritative motions, with which it is catenated; the ſenſorial power of aſſociation becomes accumulated in ſo great a degree as to affect the ſecond link of the train of aſſociated motions, and to excite it into ſtronger action. Thus when the ſtomach is rendered torpid by contagious matter ſwallowed into it mixed with the ſaliva, the heart and arteries

act more feebly; becauſe the ſenſorial power of aſſociation, which uſed to be excited by the fibrous motions of the ſtomach, is not now excited; and in conſequence the motions of the heart and arteries act only by the ſenſorial power of irritation, which is excited by the ſtimulus of the blood.

But during this torpor of the ſtomach, and leſs action of the heart and arteries, ſo great an accumulation of the ſenſorial powers of irritation and of aſſociation occurs, that it adds to the action of the next link of this vital circle of actions, that is, to that of the cutaneous capillaries. Whence in this ſituation the torpor of the ſtomach occaſions a diminiſhed action of the heart and arteries by direct ſympathy, and may be ſaid to occaſion an increaſed one of the cutaneous capillaries by reverſe ſympathy; which conſtitute continued fever with weak pulſe.

Nor is this increaſed action of the capillaries in conſequence of the decreaſed action of the heart and arteries, as in fevers with weak pulſe, a ſingle fact in the animal economy; though it exiſts in this caſe in the greateſt degree or duration, becauſe the heart and arteries are perpetually in greater action than any other part of the ſyſtem. But a ſimilar circumſtance occurs, when the ſtomach is rendered inactive by defective excitement of the ſenſorial power of aſſociation, as in ſea-ſickneſs, or in nephritis. In theſe caſes the ſenſorial power of aſſociation becomes much accumulated in the ſtomach, and ſeems by its ſuperabundance to excite the abſorbent ſyſtem, which is ſo nearly connected with it, into great increaſe of action; as is known by the great quantity frequently in theſe ſituations rejected by vomit, which could not otherways be ſupplied. It is probable the increaſe of digeſtion by walking in froſty air, with many other animal facts, may by future obſervations be found to be dependent on this principle, as well as the increaſed action of the capillaries in continued fevers with weak pulſe.

Whereas in continued fever with ſtrong pulſe, which may perhaps occur

occur ſometimes on the firſt day even of the plague, the ſtomach with the heart and arteries and the capillaries act by direct ſympathy; that is, the ſtomach is excited into ſtronger action by increaſed irritation owing to the ſtimulus of contagious matter; theſe ſtronger irritative motions of the ſtomach excite a greater quantity of the ſenſorial power of aſſociation, which then actuates the heart and arteries with greater energy, as theſe are catenated with the ſtomach; and in the ſame manner the increaſed actions of the heart and arteries excite a greater quantity of the ſenſorial power of aſſociation, which actuates the cutaneous capillaries with increaſe of energy. See Claſs IV. 1. 1.

4. I ſhall dwell a little longer on this intricate ſubject. The commencement of fever-fits is known by the inactivity of the cutaneous capillaries, which inactivity is obſervable by the paleneſs and coldneſs of the ſkin, and alſo by the pain of coldneſs, which attends it. There is nevertheleſs in moſt caſes, except thoſe which are owing to expoſure to external cold, a torpor of the capillaries of ſome internal viſcus preceding this inactivity of the cutaneous capillaries; which is known by the tumour or hardneſs of the viſcus, or by an aching pain of it. The capillaries of the lungs are at the ſame time rendered inactive or torpid, as appears by the difficulty of breathing, and coldneſs of the breath in cold fits of fever, and in going into the cold bath; but the lungs are not affected with the pain either of coldneſs or of torpor.

One cauſe of this ſynchronous or ſucceſſive inactivity of the cutaneous capillaries, in conſequence of the previous torpor of ſome internal viſcus, may be owing to the deficiency of heat; which muſt occur, when any part becomes inactive; becauſe the ſecretions of that part ceaſe or are leſſened, and the quantity of heat of it in conſequence. But the principal cauſe of it I ſuppoſe to be owing to the defect of the ſenſorial power of aſſociation; which power of aſſociation is excited by ſome previous or concomitant motions of the parts of every great circle of actions. This appears on going into the cold bath, becauſe the ſhortneſs of breath inſtantly occurs, ſooner than one can conceive the diminution of the heat of the ſkin could affect the lungs by the want of its ſtimulus; but not ſooner than the defect of the ſenſorial power of aſſociation could affect them; becauſe this muſt ceaſe to be excited into action on the inſtant that the cutaneous capillaries ceaſe to act; whence in the firſt moment of contact of the cold water the cutaneous capillaries ceaſe to act from defect of irritation; which is cauſed by defect of the ſtimulus of heat; and in the ſecond moment the capillaries of the lungs ceaſe to act from the defect of aſſociation; which is cauſed by the defect of the motions of the cutaneous capillaries. Thus the univerſal torpor in the cold paroxyſm of fever is an example of direct ſympathy, though occaſioned in part by defect of irritation, and in part by defect of aſſociation.

5. Thus in walking out in a froſty morning the ſkin is cooled by the contact of the cold air, whence the actions of its capillaries are diminiſhed for want of their uſual ſtimulus of heat to excite a ſufficient quantity of the ſenſorial power of irritation. Hence there is at firſt a ſaving of ſenſorial power of irritation for the purpoſe of actuating the other parts of the ſyſtem with greater energy. Secondly the ſenſorial power of aſſociation, which uſed to be excited by the motions of the cutaneous capillaries, is now not ſo powerfully excited; and in conſequence the parts, which conſtitute the next links of the circles of aſſociated motions, are for a time actuated with leſs energy, and a temporary general chillneſs ſucceeds; which is ſo far ſimilar to the cold fit of intermittent fever.

In this ſituation there is a curious circumſtance occurs, which merits peculiar attention: after a ſhort time, though the external ſkin continues cool by its expoſure to the cold air, and the actions of its capillaries

capillaries are consequently diminiſhed, yet the capillaries of the ſtomach act with greater energy; as is known by increaſed digeſtion and conſequent hunger. This is to be aſcribed to the accumulation of the ſenſorial power of irritation, which now excites by its ſuperabundance, or overflowing, as it were, the ſtomach into increaſed action; though it is at the ſame time excited leſs powerfully than uſual by the ſenſorial power of aſſociation. Thus the accumulation of the ſenſorial power of irritation in the veſſels of the ſkin increaſes in this caſe the action of the ſtomach, in the ſame manner as an accumulation of the ſenſorial power of aſſociation in the heart and arteries in fevers with weak pulſe increaſes the action of the capillaries.

If nevertheleſs the coldneſs of the ſkin be too long continued, or exiſts in too great a degree, ſo as in ſome meaſure to impair the life of the part, no further accumulation of the ſenſorial power of irritation occurs; and in conſequence the actions of the ſtomach become leſs than natural by the defect of the ſenſorial power of aſſociation; which has ceaſed to be excited by the want of action of the cutaneous capillaries. Whence continued coldneſs of the feet is accompanied with indigeſtion and heartburn. See Claſs IV. 2. 1. 6.

6. Similar to this when the actions of the ſtomach are rendered torpid by the previous ſtimulus of a violent emetic, and its motions become retrograde in conſequence, a great quantity of ſenſorial power is exerted on the lymphatics of the lungs, and other parts of the body; which excites them into greater direct action, as is evinced by the exhibition of digitalis in anaſarca. In this ſituation I ſuppoſe the emetic drug ſtimulates the muſcular fibres of the ſtomach into too great action; and that in conſequence a great torpor ſoon ſucceeds; and that this inaction of the muſcular parts of the ſtomach is not followed by much accumulation of the ſenſorial power of irritation; becauſe that ſenſorial power is in great meaſure exhauſted by the

previous

previous exceſſive ſtimulus. But the lymphatics of the ſtomach have their actions leſſened by defect of the ſenſorial power of aſſociation, which is not now excited into action, owing to the leſſened motions of the muſcular parts of it, with which the lymphatics are aſſociated. The ſenſorial power of aſſociation becomes therefore accumulated in theſe lymphatics of the ſtomach, becauſe it is not excited into action; exactly as the power of irritation becomes accumulated in the hand, when immerſed in ſnow; and this accumulated ſenſorial power of aſſociation excites the lymphatic of the lungs and of other parts, which are moſt nearly aſſociated with thoſe of the ſtomach, into more energetic actions. Thus the muſcular fibres of the ſtomach act with the lymphatics of that organ in direct ſympathy; and the lymphatics of the ſtomach act in reverſe ſympathy with thoſe of the lungs and of other parts of the body; the former of which is cauſed by defect of the excitement of the ſenſorial power of aſſociation, and the latter by the accumulation of it.

Beſides the efficient cauſe, as above explained, the final cauſe, or convenience, of theſe organic actions are worthy our attention. In this caſe of an acrid drug ſwallowed into the ſtomach the reverted actions of the muſcular fibres of the ſtomach tend to eject its enemy; the reverted actions of its lymphatics pour a great quantity of fluids into the ſtomach for the purpoſe of diluting or waſhing off the noxious drug; and the increaſed actions of the other lymphatics ſupply theſe retrograde ones of the ſtomach with an inconceivable ſupply of fluids, as is ſeen in Ileus and Cholera.

7. The inquiſitive reader will excuſe my continuing this ſubject, though perhaps with ſome repetitions, as it envelopes the very eſſence of fever. When the firſt link of a train of actions is excited by exceſſive ſtimulus, or exceſſive irritability, and thus acts with unuſual energy by the increaſed quantity of irritation, theſe increaſed motions excite a greater quantity of the ſenſorial power of aſſociation,

 which

which causes increased motions in the second link, which is catenated with the first; and then the excessive action of this second link excites also a greater quantity of the sensorial power of association, which increases the motions of the third link of this chain of association, and thus the increase of the stimulus on the irritative motions, to which the chain of association is catenated, increases the action of the whole chain or circle of associated motions.

After a time the irritative motions become torpid by expenditure of the sensorial power of irritation, and then the power of association also becomes less exerted, both because it has been in part exhausted by too great action, and is now less excited by the lessened action of the irritative motions, which used to excite it. These are both instances of direct sympathy, and frequently constitute the cold and hot fit of intermittents.

But though the accumulation of the sensorial power of irritation during the quiescence of some motion owing to want of stimulus generally induces torpor in the first link of the train of associated motions catenated with it; as the capillaries of the lungs become torpid immediately on immersion of the skin into cold water; yet in some situations an orgasm or excess of action is produced in the first link of the associated motions thus catenated with irritative ones; as in the increased action of the stomach, when the skin is for a time exposed to cold air; which may in part be ascribed to the general increase of action of the whole system, owing to the diminished expenditure of sensorial power, but particularly of the parts, which have habitually acted together; as when one arm is paralytic the other is liable to more frequent or almost continual motion; and when one eye becomes blind the other frequently becomes stronger; which is well known to farriers, who are said sometimes to destroy the sight of one eye to strengthen that of the other in diseased horses.

Hence there is sometimes a direct sympathy, and sometimes a reverse

reverſe one ſucceeds the torpor occaſioned by defect of ſtimulus, the latter of which is perhaps owing to a certain time being required for the production of an accumulation of the ſenſorial power of irritation by the nervous branches of the torpid organ; which accumulation is now in part or entirely derived to the next link of the aſſociation, Thus in going into a coldiſh bath, as into a river in the ſummer months, we at firſt experience a difficulty of breathing from the torpid action of the pulmonary capillaries, owing to the deficient excitement of the ſenſorial power of aſſociation in conſequence of the torpor of the cutaneous capillaries. But in a very ſhort time, as in one minute, the ſenſorial power of irritation becomes accumulated by the inactivity of the cutaneous capillaries; and as its ſuperabundance becomes now expended on the pulmonary capillaries, the difficult reſpiration ceaſes; though the cutaneous capillaries continue torpid by their contact with the cold water, and conſequently the ſenſorial power of aſſociation, which uſed to contribute to actuate the pulmonary capillaries, is leſs excited.

8. In like manner when there exiſts an accumulation of the ſenſorial power of aſſociation, owing to defect of its excitement by ſome previous irritative or aſſociate motions, it is generally accompanied for a certain time by a torpor not only of the link firſt affected, but of the ſubſequent parts, or of the whole train of aſſociated motions, as in the cold fits of intermittent fevers. Yet after a time an increaſed action of the next links of aſſociated motions ſucceeds the torpor of the firſt, as the abſorbent veſſels of the lungs act more violently in conſequence of the deficient action of thoſe of the ſtomach; and the ſkin at the commencement of ſickneſs is pale and cold, but in a little time becomes fluſhed and warm.

Thus we ſee in aſſociate motions, which are rendered torpid by defect of excitement, that ſometimes a direct, and ſometimes a reverſe ſympathy ſucceeds in the ſubſequent links of the chain. But

I believe where a torpor of irritative or of the aſſociate motions is cauſed by a previous too great expenditure or exhauſtion of the ſenſorial powers of irritation or aſſociation, no increaſe of action in the ſubſequent link ever occurs, or not till after a very long time.

Thus when the ſtomach becomes torpid by previous violent exertion, and conſequent exhauſtion of the ſenſorial power of irritation, as after intoxication with wine or opium, or after the exhibition of ſome violent emetic drug, the torpor is communicated to the heart and arteries, as in continued fevers with weak pulſe. But where the torpor of the ſtomach is produced from defective aſſociation, as in ſea-ſickneſs; or in the ſickneſs which occurs, when a ſtone ſtimulates the ureter; no torpor is then communicated to the heart and arteries. For in the former caſe there is no accumulation of ſenſorial power in the ſtomach, which was previouſly exhauſted by too great ſtimulus; but in the latter caſe the accumulation of ſenſorial power in the ſtomach during its torpor is evinced by this circumſtance; that in ſea-ſickneſs the patients eat and drink voraciouſly at intervals; and the pulſe is generally not affected by the ſickneſs occaſioned by a ſtone in the ureter. For the action of the ſtomach is then leſſened, and in conſequence becomes retrograde, not owing to the exhauſtion of the ſenſorial power of irritation, but to the want of excitement of the ſenſorial power of aſſociation; which is cauſed by the defective action of the ureter, which becomes occaſionally torpid by the great ſtimulus of the ſtone it contains; or which is cauſed by the great exhauſtion of ſenſorial power by the pain; which affects the ureter without exciting inflammation, or increaſed action of it.

9. Thus though the ſtomach after the great ſtimulus of intoxication from exceſs of wine or opium will continue many hours without accumulation of ſenſorial power, as appears from the patient's experiencing no appetite at the intervals of ſickneſs; yet after long abſtinence from food, at length not only the exhauſted quantity of ſenſorial

power is renewed, but an accumulation of it at length occurs, and hunger returns. In this ſituation the ſtomach is generally about a whole day before it regains its uſual powers of digeſtion; but if it has been ſtill more violently ſtimulated, and its actions further impaired, a ſtill more permanent torpor along with a continued fever with weak pulſe is liable to occur; and a fourth part, or a half, or three fourths, or a whole lunar period paſſes, before it recovers its due irritability and conſequent action.

In ſimilar manner, after a perſon has been confined in a very warm room for ſome hours, the cutaneous capillaries, with their ſecretory and abſorbent veſſels, become exhauſted of their ſenſorial power of irritation by the too great violent exertions occaſioned by the unuſual ſtimulus of heat; and in coming into a colder atmoſphere an inactivity of the cutaneous veſſels exiſts at firſt for ſome time without accumulation of ſenſorial power; as is ſhewn by the continuance of the pain of cold and the paleneſs; but after a time both the pain of cold and paleneſs vaniſh, which now indicates an accumulation of the ſenſorial power of irritation, as leſs degrees of heat ſtimulate the ſyſtem into due action.

In the ſame manner, after any one has been ſome time in the ſummer ſunſhine, on coming into a dark cell he continues much longer before he can clearly diſtinguiſh objects, than if his eyes had only been previouſly expoſed to the light of a cloudy day in winter; becauſe the ſenſorial power of irritation, and conſequent ſenſation, had in the firſt caſe been previouſly much expended or exhauſted; and therefore required a much longer time before it could be produced in the brain, or derived to the optic nerves, in ſuch quantity as to reſtore the deficiency, and to cauſe an accumulation of it; whereas in the latter caſe no deficiency had occurred.

10. Thus the accumulation or deficiency of ſenſorial power in a torpid organ, which had previouſly been accuſtomed to perpetual action,

tion, depends on the manner in which it becomes torpid; that is, whether by great previous ſtimulus, or great previous excitement of the power of aſſociation; or by defect of its accuſtomed ſtimulus, or of its accuſtomed excitement of the power of aſſociation. In the former caſe the ſenſorial power is in an exhauſted ſtate, and therefore is not likely to become ſo ſoon accumulated, as after drunkenneſs, or expoſure to great heat, or to great light; in the latter a great accumulation of ſenſorial power occurs, as after expoſure to cold, or hunger, or darkneſs.

Hence when the ſtomach continues torpid by previous violent ſtimulus, as in the exhibition of digitalis, no accumulation of ſenſorial power of irritation ſupervenes; and in conſequence the motions of the heart and arteries, which are aſſociated with thoſe of the ſtomach, become weak, and ſlow, and intermittent, from the defect of the excitement of the ſenſorial power of aſſociation. But what follows? as the actions of the heart and arteries are leſſened by the deficient action of the ſenſorial power of aſſociation, and not by previous increaſed excitement of it; a great accumulation of the ſenſorial power of aſſociation occurs, which is exerted on the pulmonary and cutaneous abſorbents by reverſe ſympathy, and produces a great abſorption of the fluid effuſed into the cellular membrane in anaſarca, with dry ſkin; conſtituting one kind of atrophy.

But if at the ſame time the ſecerning veſſels of the ſtomach are ſtimulated into ſo violent activity as to induce great conſequent torpor, as probably happens when contagious matter is ſwallowed into the ſtomach with our ſaliva, thoſe of the heart and arteries act feebly from the deficient excitement of the power of aſſociation; and then the cutaneous and pulmonary ſecerning veſſels act with greater force than natural, owing to the accumulation of the ſenſorial power of aſſociation; and unnatural heat of the ſkin, and of the breath ſucceed; but without frequency of pulſe, conſtituting the pareſis irritativa of Claſs I. 2. 1. 2. And laſtly, if a paucity of blood attends this pareſis,

 or

or ſome other cauſe inducing a frequency of pulſe, the febris inirritativa, or fever with weak pulſe, is produced.

But on the contrary when the ſtomach has previouſly been rendered torpid by defect of ſtimulus, as by hunger, if food be too haſtily ſupplied, not only great exertion of the ſtomach itſelf ſucceeds, but fever with ſtrong pulſe is induced in conſequence; that is, the heart and arteries are excited into more energetic action by the exceſs of the power of aſſociation, which catenates their motions with thoſe of the ſtomach. For the redundancy of ſenſorial power of irritation, which was accumulated during the inactivity of the ſtomach, and is now called into action by ſtimulus, actuates that organ with increaſed energy, and excites by theſe increaſed motions the ſenſorial power of aſſociation; which has alſo been accumulated during the inactivity of the heart and arteries; and thus theſe organs alſo are now excited into greater action.

So after the ſkin has been expoſed ſome hours to greater heat than natural in the warm room, other parts, as the membranes of the noſtrils, or of the lungs, or of the ſtomach, are liable to become torpid from direct ſympathy with it, when we come into air of a moderate temperature; whence catarrhs, coughs, and fevers. But if this torpor be occaſioned by defect of ſtimulus, as after being expoſed to froſty air, the accumulation of ſenſorial power is exerted, and a glow of the ſkin follows, with increaſed digeſtion, full reſpiration, and more vigorous circulation.

11. It may be aſked, Why is there a great and conſtant accumulation of the ſenſorial power of aſſociation, owing to the torpor of the ſtomach and heart and arteries, in continued fever with weak pulſe; which is exerted on the cutaneous and pulmonary capillaries, ſo as to excite them into increaſed action for many weeks, and yet no ſuch exuberance of ſenſorial power produces fever in winter-ſleeping animals, or in chloroſis, or apepſia, or hyſteria?

In

In winter-ſleeping animals I ſuppoſe the whole nervous ſyſtem is torpid, or paralyſed, as in the ſleep of frozen people; and that the ſtomach is torpid in conſequence of the inactivity or quieſcence of the brain; and that all other parts of the body, and the cutaneous capillaries with the reſt, labour under a ſimilar torpor.

In chloroſis, I imagine, the actions of the heart and arteries, as well as thoſe of the cutaneous and pulmonary capillaries, ſuffer along with thoſe of the ſtomach from the deficient ſtimulus of the pale blood; and that though the liver is probably the ſeat of the original torpor in this diſeaſe, with which all other parts ſympathize from defect of the excitation of the ſenſorial power of aſſociation; yet as this torpor occurs in ſo ſmall a degree as not to excite a ſhüddering or cold fit, no obſervable conſequences are in general occaſioned by the conſequent accumulation of ſenſorial power. Sometimes indeed in chloroſis there does occur a frequent pulſe and hot ſkin; in which circumſtances I ſuppoſe the heart and arteries are become in ſome degree torpid by direct ſympathy with the torpid liver; and that hence not only the pulſe becomes frequent, but the capillaries of the ſkin act more violently by reverſe ſympathy with the heart and arteries, owing to the accumulation of the ſenſorial power of aſſociation in them during their torpid ſtate, as occurs in irritative fever. See Article 11 of this Supplement.

In apepſia chronica the actions of the ſtomach are not ſo far impaired or deſtroyed as totally to prevent the excitation of the ſenſorial power of aſſociation, which therefore contributes ſomething towards the actions of the heart and arteries, though leſs than natural, as a weak pulſe always I believe attends this diſeaſe.

There is a torpor of the ſtomach, and of the upper part of the alimentary canal in hyſteria, as is evident from the retrograde actions of the duodenum, ſtomach, and œſophagus, which conſtitute the globus hyſtericus, or ſenſation of a globe riſing into the throat. But as theſe retrograde actions are leſs than thoſe, which induce ſickneſs or vomiting,

ing, and are not occaſioned by previous exhauſtion of the ſenſorial power of irritation, they do not ſo totally prevent the excitement of the ſenſorial power of aſſociation, as to leſſen the motion of the heart and arteries ſo much as to induce fever; yet in this caſe, as in apepſia, and in chloroſis, the pulſations of the heart and arteries are weaker than natural, and are ſometimes attended with occaſionally increaſed action of the capillaries; as appears from the fluſhings of the face, and hot ſkin, which generally form an evening febricula in diſeaſes attended with weak digeſtion.

12. The increaſed action, or orgaſm, of the cutaneous, pulmonary, and cellular capillaries, with their ſecerning and abſorbent veſſels, in thoſe fevers which are attended with deficiency of vital action, exhauſts the patient both by the additional expenditure of ſenſorial power on thoſe organs of ſecretion, and by the too great abſorption of the mucus and fat of the body; whence great debility and great emaciation. Hence one great indication of cure of continued fever with arterial debility is to diminiſh the too great action of the capillaries; which is to be done by frequent ablutions, or bathing the whole ſkin in tepid or in cold water, as recommended by Dr. Currie of Liverpool (Philoſ. Tranſ. for 1792), for half an hour, twice a day, or at thoſe times when the ſkin feels dryeſt and hotteſt. Much cool air ſhould alſo be admitted, when the breath of the patient feels hot to one's hand; or when the tongue, eſpecially its middle part, is dry, and covered with a cruſt of indurated mucus; as theſe indicate the increaſed action of the pulmonary capillaries; in the ſame manner as the dry and hot ſkin indicates the orgaſm of the cutaneous capillaries; and the emaciation of the body that of the cellular ones.

For this purpoſe of abating the action of the capillaries by frequent ablution or fomentation, water of any degree of heat beneath that of the body will be of ſervice, and ought in accurate language to be called a cold bath; but the degree of coldneſs, where the patient is ſenſible,

ſenſible, ſhould in ſome meaſure be governed by his ſenſations ; as it is probable, that the degree of coldneſs, which is moſt grateful to him, will alſo be of the greateſt benefit to him. See Claſs III. 2. 1. 12. and Article 15 of this Supplement.

Another great uſe of frequent ablutions, or fomentations, or baths, in fevers, where the ſtomach is in ſome degree torpid, is to ſupply the ſyſtem with aqueous fluid by means of the cutaneous abſorbents; which is diſſipated faſter by the increaſed action of the ſecerning capillaries, than the ſtomach can furniſh, and occaſions great thirſt at the intervals of the ſickneſs.

## IX. *Torpor of the Lungs.*

1. The lungs in many caſes of contagion may firſt be affected with torpor, and the ſkin become cold by ſympathy; in the ſame manner as a cold ſkin on going into the cold bath induces difficulty of breathing. Or the ſtomach may become affected with torpor by its ſympathy with the lungs, as in the experiments of Mr. Watt with hydrocarbonate gas; a few reſpirations of which induced ſickneſs, and even ſyncope. When the ſtomach or ſkin is thus affected ſecondarily by aſſociation, an accumulation of ſenſorial power occurs much ſooner, than when theſe parts become torpid in conſequence of previous exceſs of ſtimulus; and hence they ſooner recover their accuſtomed action, and the fever ceaſes. The particles of contagious matter thus received by reſpiration ſomewhat reſemble in their effects the acid gaſes from burning ſulphur, or from charcoal; which, if they do not inſtantly deſtroy, induce a fever, and the patient ſlowly recovers.

2. I was ſome years ago ſtooping down to look, which way the water oozed from a moraſs, as a labourer opened it with a ſpade, to

 detect

detect the source of the spring, and inhaled a vapour, which occasioned an instant sense of suffocation. Immediately recoiling I believe I inhaled it but once; yet a few hours afterwards in the cool of the evening, when I returned home rather fatigued and hungry, a shivering and cold fit occurred, which was followed by a hot one; and the whole disease began and terminated in about twelve hours without return. In this case the power of fear, or of imagination, was not concerned; as I neither thought of the bad air of a morass before I perceived it; nor expected a fever-fit, till it occurred.

In this case the torpor commenced in the lungs, and after a few hours, by the addition of fatigue, and cold, and hunger, was propagated by direct sympathy to the rest of the system. An orgasm or increased action of the whole system was then induced by the accumulation of sensorial power of irritation in the lungs, and of association in the other organs; and when these subsided, the disease ceased. It may be asked, could a torpor of the capillaries of the air-vessels of the lungs be so suddenly produced by great stimulation?—It appears probable, that it might, because great exertion of irritative motions may be instantly produced without our perceiving them; that is, without their being attended by sensation, both in the lungs and stomach; and the organs may become torpid by the great expenditure of the sensorial power of irritation in an instant of time; as paralysis frequently instantly follows too great an exertion of voluntary power.

3. When the capillaries of the lungs act too violently, as in some continued fevers; which is known by the heat of the breath, and by the dryness of the tongue, especially of the middle part of it; not only cooler air might be admitted more freely into a sick room to counteract this orgasm of the pulmonary capillaries; but perhaps the patient might breathe with advantage a mixture of carbonic acid gas, or of hydrogene gas, or of azote with atmospheric air. And on the contrary, when there exists an evident torpor of the pulmonary capil-

6 laries,

laries, which may be known by the correſpondent chilneſs of the ſkin ; and by a tickling cough, which ſometimes attends cold paroxyſms of fever, and is then owing to the deficient abſorption of the pulmonary mucus, the ſaline parts of which ſtimulate the bronchiæ, or air-veſſels ; a mixture of one part of oxygen gas with 10 or 20 parts of atmoſpheric air might probably be breathed with great advantage.

## X. *Torpor of the Brain.*

As the inactivity or torpor of the abſorbent veſſels of the brain is the cauſe of hydrocephalus internus ; and as the deficiency of venous abſorption in the brain, or torpor of the extremities of its veins, is believed frequently to be the cauſe of apoplexies ; ſo there is reaſon to conclude, that the torpor of the ſecerning veſſels of the brain, which are ſuppoſed to produce the ſenſorial power, may conſtitute the immediate cauſe of ſome fevers with arterial debility. And alſo that the increaſed action of theſe ſecerning veſſels may ſomctimes conſtitute the immediate cauſe of fevers with arterial ſtrength.

It is neverthelеſs probable, that the torpor or orgaſm of the ſanguiferous, abſorbent, or ſecerning veſſels of the brain may frequently exiſt as a ſecondary effect, owing to their aſſociation with other organs, as the ſtomach or lungs ; and may thus be produced like the torpor of the heart and arteries in inirritative fevers, or like the orgaſm of thoſe organs in irritative fevers, or inflammatory ones.

Where there exiſts a torpor of the brain, might not very ſlight electric ſhocks paſſed frequently through it in all directions be uſed with advantage ? Might not fomentations of 94 or 96 degrees of heat on the head for an hour at a time, and frequently repeated, ſtimulate the brain into action ; as in the revival of winter-ſleeping animals by warmth ?

warmth? Ether externally might be frequently applied, and a blister on the shaved head.

Where the secerning vessels of the brain act with too great energy, as in some inflammatory fevers, might it not be diminished by laying the patient horizontally on a mill-stone, and whirling him, till sleep should be produced, as the brain becomes compressed by the centrifugal force? See Article 15 of this Supplement.

## XI. *Torpor of the Heart and Arteries.*

1. It was shewn in Class IV. 1. 1. 6. in IV. 2. 1. 2. and in Suppl. I. 6. 3. that a reverse sympathy generally exists between the lacteal and lymphatic branches of the absorbent system. Hence, when the motions of the absorbents of the stomach are rendered torpid or retrograde in fevers with arterial debility, those of the skin, lungs, and cellular membrane, act with increased energy. But the actions of the muscular fibres of the heart and arteries are at the same time associated with those of the muscular fibres of the stomach by direct sympathy. Both these actions occur during the operation of powerful emetics, as squill, or digitalis; while the motions of the stomach continue torpid or retrograde, the cellular and cutaneous absorbents act with greater energy, and the pulsations of the heart and arteries become weaker, and sometimes slower.

2. The increased action of the stomach after a meal, and of the heart and arteries at the same time from the stimulus of the new supply of chyle, seems originally to have produced, and to have established, this direct sympathy between them. As the increased action of the absorbents of the stomach after a meal has been usually attended with diminished action of the other branches of the absorbent system,

as mentioned in Clafs IV. 1. 1. 6. and has thus eftablifhed a reverfe fympathy between them.

2. Befides the reverfe fympathy of the abforbent veffels and the mufcles of the ftomach, and of the heart and arteries, with thofe of the fkin, lungs, and cellular membrane; there exifts a fimilar reverfe fympathy between the fecerning veffels or glands of the former of thefe organs with thofe of the latter; that is the mucous glands of the heart and arteries act generally by direct fympathy with thofe of the ftomach; and the mucous glands of the cellular membrane of the lungs, and of the fkin, act by reverfe fympathy with them both.

Hence when the ftomach is torpid, as in ficknefs, this torpor fometimes only affects the abforbent veffels of it; and then the abforbents of the cellular membrane and the fkin only act with increafed energy by reverfe fympathy. If the torpor affects the mufcular fibres of the ftomach, thofe of the heart and arteries act by direct fympathy with it, and a weak pulfe is produced, as in the exhibition of digitalis, but without increafe of heat. But if the torpor alfo affects the glands of the ftomach, the cutaneous and pulmonary glands act with greater energy by their reverfe fympathy with thofe of the ftomach, and of the heart and arteries; and great heat is produced along with increafed perfpiration both from the fkin and lungs.

3. There is fome difficulty in explaining, why the actions of the extenfive fyftem of capillary glands, which exift on every other membrane and cell in the body for the purpofe of fecreting mucus and perfpirable matter, fhould fo generally act by reverfe fympathy with thofe of the ftomach and upper part of the inteftines.. It was fhewn in Clafs IV. 1. 1. 6. that when the ftomach was filled with folid and fluid aliment, the abforbents of the cellular membrane, and of the bladder, and of the fkin acted with lefs energy; as the fluids, they were ufed to abforb and tranfmit into the circulation, were now

leſs wanted; and that hence by habit a reverſe ſympathy obtained between theſe branches of the abſorbents of the alimentary canal, and thoſe of the other parts of the body.

Now, as at this time leſs fluid was abſorbed by the cutaneous and cellular lymphatics, it would happen, that leſs would be ſecreted by their correſpondent ſecerning veſſels, or capillary glands; and that hence by habit, theſe ſecerning veſſels would acquire a reverſe ſympathy of action with the ſecerning veſſels of the alimentary canal.

Thus when the abſorption of the tears by the puncta lacrymalia is much increaſed by the ſtimulus of ſnuff; or of an affecting idea, on the naſal ducts, as explained in Sect. XVI. 8. 2. a great increaſe of the ſecretion of tears from the lacrymal glands is produced by the direct ſympathy of the action of theſe glands with thoſe of their correſpondent abſorbents; and that though in this caſe they are placed at ſo great a diſtance from each other.

4. A difficult queſtion here occurs; why does it happen, that in fevers with weak pulſe the contractions of the heart and arteries become at the ſame time more frequent; which alſo ſometimes occurs in chloroſis, and in ſome hyſteric and hypochondriac diſeaſes, and in ſome inſanities; yet at other times the weak pulſe becomes at the ſame time ſlow, as in the exhibition of digitalis, and in pareſis irritativa, deſcribed in Claſs I. 2. 1. 2. which may be termed a fever with ſlow pulſe? this frequency of pulſe can not depend on heat, becauſe it ſometimes exiſts without heat, as towards the end of ſome fevers with debility.

Now as apoplexies, which are ſometimes aſcribed to fulneſs of blood, are attended with ſlow pulſe; and as in animals dying in the ſlaughter-houſe from deficiency of blood the pulſe becomes frequent in extreme; may not the frequency of pulſe in fevers with arterial debility be in general owing to paucity of blood? as explained in Sect. XXXII. 2. 3. and its ſlowneſs in pareſis irritativa be cauſed by the

the debility being accompanied with due quantity of blood? or may not the former circumſtance ſometimes depend on a concomitant affection of the brain approaching to ſleep? or to the unuſual facility of the paſſage of the blood through the pulmonary and aortal capillaries? in which circumſtance the heart may completely empty itſelf at each pulſation, though its contractions may be weak. While the latter depends on the difficulty of the paſſage of the blood through the pulmonary or aortal capillaries, as in the cold fits of intermittents, and in ſome palpitations of the heart, and in ſome kinds of hæmoptoe? in theſe caſes the increaſed reſiſtance prevents the heart from emptying itſelf, and in conſequence a new diaſtole ſooner occurs, and thus the number of pulſations becomes greater in a given time.

5. In reſpect to the ſympathies of action, which produce or conſtitute fever with debility, the ſyſtem may be divided into certain provinces, which are aſſentient or oppoſite to each other First, the lacteals or abſorbent veſſels of the ſtomach, and upper part of the inteſtines; ſecondly, the lymphatics or all the other branches of the abſorbent veſſels, which ariſe from the ſkin, mucous membranes, cellular membranes, and the various glands. Theſe two diviſions act by reverſe ſympathy with each other in the hot fits of fever with debility, though by direct ſympathy in the cold ones. The third diviſion conſiſts of the ſecerning veſſels of the ſtomach and upper inteſtines; and the fourth of the ſecerning veſſels of all the other parts of the body, as the capillary glands of the ſkin, lungs, and cellular membrane, and the various other glands belonging to the ſanguiferous ſyſtem. Many of theſe frequently, but the capillaries always, act by reverſe ſympathy with thoſe of the third diviſion above mentioned in the hot fits of fever with debility, though by direct ſympathy with them in the cold fits. Fifthly, the muſcular fibres of the ſtomach, and upper inteſtines; and ſixthly, the muſcular fibres of the heart and arteries. The actions of theſe two laſt diviſions of moving

fibres

fibres act by direct ſympathy with each other, both in the cold and hot fits of fevers with debility.

The efficient cauſe of thoſe apparent ſympathies in fevers with weak pulſe may be thus underſtood. In the cold paroxyſm of fever with weak pulſe the part firſt affected I believe to be the ſtomach, and that it has become torpid by previous violent exertion, as by ſwallowing contagious matter mixed with ſaliva, and not by defect of ſtimulus, as from cold or hunger. The actions of this important organ, which ſympathizes with almoſt every part of the body, being thus much diminiſhed or nearly deſtroyed, the ſenſorial power of aſſociation is not excited; which in health contributes to move the heart and arteries, and all the reſt of the ſyſtem; whence an univerſal torpor occurs.

When the hot fit approaches, the ſtomach in fevers with ſtrong pulſe regains its activity by the accumulation of the ſenſorial power of either irritation, if it was the part firſt affected, or of aſſociation if it was affected in ſympathy with ſome other torpid part, as the ſpleen or liver; which accumulation is produced during its torpor. At the ſame time all the other parts of the ſyſtem acquire greater energy of action by the accumulation of the ſenſorial power of aſſociation, which was produced, during their inactivity in the cold fit.

But in fevers with weak pulſe the ſtomach, whoſe ſenſorial power of irritation had been previouſly exhauſted by violent action, acquires no ſuch quick accumulation of ſenſorial power, but remains in a ſtate of torpor after the hot fit commences. The heart and arteries remain alſo in a ſtate of torpor, becauſe there continues to be no excitement of their power of aſſociation owing to the torpid motions of the ſtomach; but hence it happens, that there exiſts at this time a great accumulation of the power of aſſociation in the leſs active fibres of the heart and arteries; which, as it is not excited and expended by them, increaſes the aſſociability of the next link of the aſſociated chain of motions, which conſiſts of the capillaries or other glands;

glands; and that in ſo great a degree as to actuate them with unnatural energy, and thus to produce a perpetual hot fit of fever. Becauſe the aſſociability of the capillaries is ſo much increaſed by the accumulation of this power, owing to the leſſened activity of the heart and arteries, as to over-balance the leſſened excitement of it by the weaker movements of the heart and arteries.

6. When the accumulation of the ſenſorial power of irritation cauſed by defect of ſtimulus is greater in the firſt link of a train of actions, to which aſſociated motions are catenated, than the deficiency of the excitement of the ſenſorial power of aſſociation in the next link, what happens?—the ſuperabundance of the unemployed ſenſorial power of the firſt link is derived to the ſecond; the aſſociability of which thus becomes ſo greatly increaſed, that it acts more violently than natural, though the excitement of its power of aſſociation by the leſſened action of the firſt link is leſs than natural. So that in this ſituation the withdrawing of an accuſtomed ſtimulus in ſome parts of the ſyſtem will decreaſe the irritative motions of that part, and at the ſame time occaſion an increaſe of the aſſociate motion of another part, which is catenated with it.

This circumſtance nevertheleſs can only occur in thoſe parts of the ſyſtem, whoſe natural actions are perpetual, and the accumulation of ſenſorial power on that account very great, when their activity is much leſſened by the deduction of their uſual ſtimulus; and are therefore only to be found in the ſanguiferous ſyſtem, or in the alimentary canal, or in the glands and capillaries. Of the firſt of which the following is an inſtance.

The reſpiration of a reduced atmoſphere, that is of air mixed with hydrogene or azote, quickens the pulſe, as obſerved in the caſe of Mrs. Eaton by Dr. Reynolds and Dr. Thornton; to which Dr. Beddoes adds in a note, that "he never ſaw an inſtance in which a lowered atmoſphere did not at the moment quicken the pulſe, while

it

it weakened the action of the heart and arteries." Considerations on Factitious Airs, by Thomas Beddoes and James Watt, Part III. p. 67. Johnson, London. By the assistance of this new fact the curious circumstance of the quick production of warmth of the skin on covering the head under the bed-clothes, which every one must at some time have experienced, receives a more satisfactory explanation, than that which is given in Class IV. 1. 1. 2. which was printed before this part of Dr. Beddoes's Considerations was published.

For if the blood be deprived of its accustomed quantity of oxygen, as in covering the head in bed, and thus breathing an air rendered impure by repeated respiration, or by breathing a factitious air with less proportion of oxygen, which in common respiration passes through the moist membranes of the lungs, and mixes with the blood, the pulsations of the heart and arteries become weaker, and consequently quicker, by the defect of the stimulus of oxygen. And as these vessels are subject to perpetual motion, the accumulation of the sensorial power of irritation becomes so great by their lessened activity, that it excites the vessels next connected, the cutaneous capillaries for instance, into more energetic actions, so as to produce increased heat of the skin, and greater perspiration.

How exactly this resembles a continued fever with weak and quick pulse!—in the latter the action of the heart and arteries are lessened by defect of the excitement of the sensorial power of association, owing to the torpor or lessened actions of the stomach; hence the accumulation of the sensorial power of association in this case, as the accumulation of that of irritation in the former, becomes so abundant as to excite into increased action the parts most nearly connected, as the cutaneous capillaries.

In respect to the circumstance mentioned by Sydenham, that covering the head in bed in a short time relieved the pertinacious sickness of the patient, it must be observed, that when the action of the heart and arteries become weakened by the want of the due stimulus

 of

of the proper quantity of oxygen in the blood, that an accumulation of the ſenſorial power of irritation occurs in the fibres of the heart and arteries, which then is expended on thoſe of the capillary glands, increaſing their actions and conſequent ſecretions and heat. And then the ſtomach is thrown into ſtronger action, both by the greater excitement of its natural quantity of the ſenſorial power of aſſociation by the increaſed actions of the capillaries, and alſo by ſome increaſe of aſſociability, as it had been previouſly a long time in a ſtate of torpor, or leſs activity than natural, as evinced by its perpetual ſickneſs.

In a manner ſomewhat ſimilar to this, is the redneſs of the ſkin produced in angry people by the ſuperabundance of the unemployed ſenſorial power of volition, as explained in Claſs IV. 2. 3. 5. Rubor ex irâ. From hence we learn how, when people in fevers with weak pulſe, or in dropſies, become inſane, the abundance of the unemployed ſenſorial power of volition increaſes the actions of the whole moving ſyſtem, and cures thoſe diſeaſes.

7. As the orgaſm of the capillaries in fevers with weak pulſe is immediately cauſed by the torpid actions of the heart and arteries, as above explained, this ſupplies us with another indication of cure in ſuch fevers, and that is to ſtimulate theſe organs. This may probably be done by ſome kind of medicines, which are known to paſs into the blood unchanged in ſome of their properties. It is poſſible that nitre, or its acid, may paſs into the blood and increaſe the colour of it, and thus increaſe its ſtimulus, and the ſame may be ſuppoſed of other ſalts, neutral or metallic? As rubia tinctoria, madder, colours the bones of young animals, it muſt paſs into the blood with its colouring matter at leaſt unchanged, and perhaps many other medicines may likewiſe affect the blood, and thus act by ſtimulating the heart and arteries, as well as by ſtimulating the ſtomach; which circumſtance deſerves further attention.

Another way of immediately ſtimulating the heart and arteries would be by transfuſing new blood into them. Is it poſſible that any other fluid beſides blood, as chyle, or milk, or water, could, if managed with great art, be introduced ſafely or advantageouſly into the vein of a living animal?

A third method of exciting the heart and arteries immediately is by increaſing the natural ſtimulus of the blood, and is well worthy experiment in all fevers with weak pulſe; and that conſiſts in ſupplying the blood with a greater proportion of oxygen; which may be done by reſpiration, if the patient was to breathe either oxygen gas pure, or diluted with atmoſpheric air, which might be given to many gallons frequently in a day, and by paſſing through the moiſt membranes of the lungs, according to the experiments of Dr. Prieſtley, and uniting with the blood, might render it more ſtimulant, and thus excite the heart and arteries into greater action! May not ſome eaſier method of exhibiting oxygen gas by reſpiration be diſcovered, as by uſing very ſmall quantities of hyper-oxygenated marine acid gas very much diluted with atmoſpheric air?

### XII. *Torpor of the Stomach and upper Inteſtines.*

1. The principal circumſtance, which ſupports the increaſed action of the capillaries in continued fever with weak pulſe, is their reverſe ſympathy with thoſe of the ſtomach and upper inteſtines, or with thoſe of the heart and arteries. The torpor of the ſtomach and upper inteſtines is apparent in continued fevers from the total want of appetite for ſolid food, beſides the ſickneſs with which fevers generally commence, and the frequent diarrhœa with indigeſted ſtools, at the ſame time the thirſt of the patient is ſometimes urgent at the intervals of the ſickneſs. Why the ſtomach can at this time take fluids by intervals, and not ſolids, is difficult to explain; except it be ſuppoſed, as ſome have affirmed, that the lacteal abſorbents are a dif-

 ferent

ferent branch from the lymphatic abſorbents, and that in this caſe the former only are in a ſtate of permanent torpor.

2. The torpor of the heart and arteries is known by the weakneſs of the pulſe. When the actions of the abſorbents of the ſtomach are diminiſhed by the exhibition of ſmall doſes of digitalis, or become retrograde by larger ones, the heart and arteries act more feebly by direct ſympathy; but the cellular, cutaneous, and pulmonary abſorbents are excited into greater action. Whence in anaſarca the fluids in the cellular membrane throughout the whole body are abſorbed during the ſickneſs, and frequently a great quantity of atmoſpheric moiſture at the ſame time; as appears by the very great diſcharge of urine, which ſometimes happens in theſe caſes; and in ileus the prodigious evacuations by vomiting, which are often a hundred fold greater than the quantity ſwallowed, evince the great action of all the other abſorbents during the ſickneſs of the ſtomach.

3. But when the ſtomach is rendered permanently ſick by an emetic drug, as by digitalis, it is not probable, that much accumulation of ſenſorial power is ſoon produced in this organ; becauſe its uſual quantity of ſenſorial power is previouſly exhauſted by the great ſtimulus of the foxglove; and hence it ſeems probable, that the great accumulation of ſenſorial power, which now cauſes the increaſed action of the abſorbents, is produced in conſequence of the inactivity of the heart and arteries; which inactivity is induced by deficient excitement of the ſenſorial power of aſſociation between thoſe organs and the ſtomach, and not by any previous exhauſtion of their natural quantity of ſenſorial power; whereas in ileus, where the torpor of the ſtomach, and conſequent ſickneſs, is induced by reverſe ſympathy with an inflamed inteſtine, that is, by diſſevered or defective aſſociation; the accumulation of ſenſorial power, which in that diſeaſe ſo violently actuates the cellular, pulmonary, and cutaneous abſorbents, is

 apparently

apparently produced by the torpor of the ſtomach and lacteals, and the conſequent accumulation of the ſenſorial power of aſſociation in them owing to their leſſened action in ſickneſs.

4. This accounts for the dry ſkin in fevers with weak pulſe, where the ſtomach and the heart and arteries are in a torpid ſtate, and for the ſudden emaciation of the body; becauſe the actions of the cellular and cutaneous abſorbents are increaſed by reverſe ſympathy with thoſe of the ſtomach, or with thoſe of the heart and arteries; that is by the expenditure of that ſenſorial power of aſſociation, which is accumulated in conſequence of the torpor of the ſtomach and heart and arteries, or of either of them; this alſo explains the ſudden abſorption of the milk in puerperal fevers; and contributes along with the heat of the reſpired air to the dryneſs of the mucous membrane of the tongue and noſtrils.

5. Beſides the reverſe ſympathy, with which the abſorbent veſſels of the ſtomach and upper inteſtines act in reſpect to all the other abſorbent veſſels, as in the exhibition of digitalis, and in ileus; there is another reverſe ſympathy exiſts between the capillaries, or ſecretory veſſels of the ſtomach, and thoſe of the ſkin. Which may nevertheleſs be occaſioned by the accumulation of ſenſorial power by the torpor of the heart and arteries, which is induced by direct ſympathy with the ſtomach; thus when the torpor of the ſtomach remains in a fever-fit, which might otherwiſe have intermitted, the torpor of the heart and arteries remains alſo by direct ſympathy, and the increaſed cutaneous capillary action, and conſequent heat, are produced by reverſe ſympathy; and the fever is thus rendered continual, owing primarily to the torpor of the ſtomach.

6. The reverſe ſympathy, which exiſts between the capillaries of the ſtomach and the cutaneous capillaries, appears by the chillneſs of ſome

ſome people after dinner; and contrary-wiſe by the digeſtion being ſtrengthened, when the ſkin is expoſed to cold air for a ſhort time; as mentioned in Claſs IV. 1. 1. 4. and IV. 2. 1. 1. and from the heat and glow on the ſkin, which attends the action of vomiting; for though when ſickneſs firſt commences, the ſkin is pale and cold; as it then partakes of the general torpor, which induces the ſickneſs; yet after the vomiting has continued ſome minutes, ſo that an accumulation of ſenſorial power exiſts in the capillaries of the ſtomach, and of the ſkin, owing to their diminiſhed action; a glow of the ſkin ſucceeds, with ſweat, as well as with increaſed abſorption.

7. Nevertheleſs in ſome circumſtances the ſtomach and the heart and arteries ſeem to act by direct ſympathy with the cutaneous capillaries, as in the fluſhing of the face and glow of the ſkin of ſome people after dinner; and as in fevers with ſtrong pulſe. In theſe caſes there appears to be an increaſed production of ſenſorial power, either of ſenſation, as in the bluſh of ſhame; or of volition, as in the bluſh of anger; or of irritation, as in the fluſhed face after dinner above mentioned.

This increaſed action of the capillaries of the ſkin along with the increaſed actions of the ſtomach and heart is perhaps to be eſteemed a ſynchronous increaſe of action, rather than a ſympathy between thoſe organs. Thus the fluſhing of the face after dinner may be owing to the ſecretion of ſenſorial power in the brain being increaſed by the aſſociation of that organ with the ſtomach, in a greater proportion than the increaſed expenditure of it, or may be owing alſo to the ſtimulus of new chyle received into the blood.

8. When the ſtomach and the heart and arteries are rendered torpid in fevers, not only the cutaneous, cellular, and pulmonary abſorbents are excited to act with greater energy; but alſo their correſpondent capillaries and ſecerning veſſels or glands, eſpecially perhaps thoſe

those of the skin, are induced into more energetic action. Whence greater heat, a greater secretion of perspirable matter, and of mucus; and a greater absorption of them both, and of aerial moisture. These reverse sympathies coincide with other animal facts, as in eruption of small pox on the face and neck the feet become cold, while the face and neck are much flushed; and in the hemiplagia, when one arm and leg become disobedient to volition, the patient is perpetually moving the other. Which are well accounted for by the accumulation of sensorial power in one part of an associated series of actions, when less of it is expended by another part of it; and by a deficiency of sensorial power in the second link of association, when too much of it is expended by the first.

9. This doctrine of reverse sympathy enables us to account for that difficult problem, why in continued fevers the increased action of the cutaneous, cellular, and pulmonary capillaries proceeds without interruption or return of cold fit; though perhaps with some exacerbations and remissions; and that during a quarter, or half, or three quarters, or a whole lunation; while at the same time the pulsations of the heart and arteries are weaker than natural.

To this should be added the direct sympathy, which exists between the peristaltic motions of the fibres of the stomach, and the pulsations of the heart. And that the stomach has become torpid by the too great stimulus of some poisonous or contagious matter; and this very intricate idea of continued fever with feeble pulse is reduced to curious simplicity.

The direct sympathy of the stomach and heart and arteries not only appears from the stronger and slower pulse of persons exhausted by fatigue, after they have drank a glass of wine, and eaten a few mouthfuls; but appears also from the exhibition of large doses of digitalis; when the patient labours under great and incessant efforts to vomit, at the same time that the actions of the absorbent system

are

are known to be much increaſed by the haſty abſorption of the ſerous fluid in anaſarca, the pulſations of the heart become ſlow and intermittent to an alarming degree. See Claſs IV. 2. 1. 17. and 18.

10. It would aſſiſt us much in the knowledge and cure of fevers, if we could always determine, which part of the ſyſtem was primarily affected; and whether the torpor of it was from previous exceſs or defect of ſtimulus; which the induſtry of future obſervers muſt diſcover. Thus if the ſtomach be affected primarily, and that by previous exceſs of ſtimulus, as when certain quantities of opium, or wine, or blue vitriol, or arſenic, are ſwallowed, it is ſome time in recovering the quantity of ſenſorial power previouſly exhauſted by exceſs of ſtimulus, before any accumulation of it can occur. But if it be affected with torpor ſecondarily, by ſympathy with ſome diſtant part; as with the torpid capillaries of the ſkin, that is by defective excitement of the ſenſorial power of aſſociation; or if it be affected by defect of ſtimulus of food or of heat; it ſooner acquires ſo much accumulation of ſenſorial power, as to be enabled to accommodate itſelf to its leſſened ſtimulus by increaſe of its irritability.

Thus in the hemicrania the torpor generally commences in a diſeaſed tooth, and the membranes about the temple, and alſo thoſe of the ſtomach become torpid by direct ſynchronous ſympathy; and pain of the head, and ſickneſs ſupervene; but no fever or quickneſs of pulſe. In this caſe the torpor of the ſtomach is owing to defect of the ſenſorial power of aſſociation, which is cauſed by the too feeble actions of the membranes ſurrounding the diſeaſed tooth, and thus the train of ſympathy ceaſes here without affecting the motions of the heart and arteries; but where contagious matter is ſwallowed into the ſtomach, the ſtomach after a time becomes torpid from exhauſtion of the ſenſorial power of irritation, and the heart and arteries act feebly from defect of the excitement of the power of aſſociation. In the former caſe the torpor of the ſtomach is conquered by accumulation of the

power

power of aſſociation in one or two whole days; in the latter it recovers by accumulation of the power of irritation in three or four weeks.

In intermittent fevers the ſtomach is generally I believe affected ſecondarily by ſympathy with the torpid cutaneous capillaries, or with ſome internal torpid viſcus, and on this account an accumulation of ſenſorial power ariſes in a few hours ſufficient to reſtore the natural irritability of this organ; and hence the hot fit ſucceeds, and the fever intermits. Or if this accumulation of ſenſorial power becomes exceſſive and permanent, the continued fever with ſtrong pulſe is produced, or febris irritativa.

In continued fevers the ſtomach is frequently I ſuppoſe affected with torpor by previous exceſs of ſtimulus, and conſequent exhauſtion of ſenſorial power, as when contagious matter is ſwallowed with the ſaliva, and it is then much ſlower in producing an accumulation of ſenſorial power ſufficient to reſtore its healthy irritability; which is a frequent cauſe of continued fever with weak pulſe or febris inirritativa. Which conſiſts, after the cold fit is over, in a more frequent and more feeble action of the heart and arteries, owing to their direct ſympathy with the muſcular fibres of the torpid ſtomach; together with an increaſed action of the capillaries, glands, and abſorbents of the ſkin, and cellular membrane, owing to their reverſe ſympathy with the torpid capillaries, glands, and abſorbents of the ſtomach, or with thoſe of the heart and arteries.

Or in more accurate language. 1. The febris inirritativa, or fever with weak pulſe, commences with torpor of the ſtomach, occaſioned by previous exhauſtion of ſenſorial power of irritation by the ſtimulus of contagious matter ſwallowed with the ſaliva. 2. The whole ſyſtem becomes torpid from defect of the excitement of the ſenſorial power of aſſociation owing to the too feeble actions of the ſtomach, this is the cold fit. 3. The whole ſyſtem, except the ſtomach with the upper inteſtines, and the heart and arteries, falls into increaſed action, or orgaſm, owing to accumulation of ſenſorial power of aſſo-

 ciation

ciation during their previous torpor, this is the hot fit. 4. The ſtomach and upper inteſtines have not acquired their natural quantity of ſenſorial power of irritation, which was previouſly exhauſted by violent action in conſequence of the ſtimulus of contagious matter, and the heart and arteries remain torpid from deficient excitement of the ſenſorial power of aſſociation owing to the too feeble actions of the ſtomach. 5. The accumulation of ſenſorial power of aſſociation in conſequence of the torpor of the heart and arteries occaſions a perpetual orgaſm, or increaſed action of the capillaries.

11. From hence it may be deducted firſt, that when the torpor of the ſtomach firſt occurs, either as a primary effect, or as a ſecondary link of ſome aſſociate train or circle of motions, a general torpor of the ſyſtem ſometimes accompanies it, which conſtitutes the cold fit of fever; at other times no ſuch general torpor occurs, as during the operation of a weak emetic, or during ſea-ſickneſs.

Secondly. After a time it generally happens, that a torpor of the ſtomach ceaſes, and its actions are renewed with increaſe of vigour by accumulation of ſenſorial power during its quieſcence; as after the operation of a weak emetic, or at the intervals of ſea-ſickneſs, or after the paroxyſm of an intermittent fever.

Thirdly. The ſtomach is ſometimes much ſlower in recovering from a previous torpor, and is then the remote cauſe of continued fever with weak pulſe; which is owing to a torpor of the heart and arteries, produced in conſequence of the deficient excitement of the power of aſſociation by the too weak actions of the ſtomach; and to an orgaſm of the capillaries of the other parts of the ſyſtem, in conſequence of the accumulation of ſenſorial power occaſioned by the inactivity of the heart and arteries.

Fourthly. The torpor of the ſtomach is ſometimes ſo complete, that probably the origin of its nerves is likewiſe affected, and then no accumulation of ſenſorial power occurs. In this caſe the patient dies for want of nouriſhment; either in three or four weeks, of the inirritative fever; or without quick pulſe, by what we have called pareſis irritativa. Or he continues many years in a ſtate of total debility. When this torpor ſuddenly commences, the patient generally ſuffers epileptic fits or temporary inſanity from the diſagreeable ſenſation of ſo great a torpor of the ſtomach; which alſo happens ſometimes at the eruption of the diſtinct ſmall pox; whence we have termed this diſeaſe anorexia epileptica. See Claſs II. 2. 2. 1. and III. 1. 1. 7. and Suppl. I. 14. 3.

Fifthly. When this torpor of the ſtomach is leſs in degree or extent, and yet without recovering its natural irritability by accumulation of ſenſorial power, as it does after the cold fit of intermittent fever, or after the operation of mild emetics, or during ſyncope; a permanent defect of its activity, and of that of the upper inteſtines, remains, which conſtitutes apepſia, cardialgia, hypochondriaſis, and hyſteria. See Claſs I. 3. 1. 3. and I. 2. 4. 5.

Sixthly. If the torpor of the ſtomach be induced by direct ſympathy, as in conſequence of a previous torpor of the liver, or ſpleen, or ſkin, an accumulation of ſenſorial power will ſooner be produced in the ſtomach; becauſe there has been no previous expenditure of it, the preſent torpor of the ſtomach ariſing from defect of aſſociation. Hence ſome fevers perfectly intermit, the ſtomach recovering its complete action after the torpor and conſequent orgaſm, which conſtitute the paroxyſm of fever, are terminated.

Seventhly. If the torpor of the ſtomach be owing to defect of irritation, as to the want of food, an accumulation of ſenſorial power ſoon occurs with an increaſe of digeſtion, if food be timely applied; or with violent inflammation, if food be given in too great quantity after very long abſtinence.

Eighthly. If the torpor of the ſtomach be induced by defect of pleaſurable ſenſation, as when ſickneſs is cauſed by the ſuggeſtion of nauſeous ideas; an accumulation of ſenſorial power ſoon occurs, and the ſickneſs ceaſes with the return of hunger; for in this caſe the inactivity of the ſtomach is occaſioned by the ſubduction of agreeable ſenſation, which acts as a ſubduction of ſtimulus, and not by exhauſting the natural quantity of ſenſorial power in the fibres or nerves of the ſtomach.

Ninthly. If the torpor of the ſtomach be induced by a twofold cauſe, as in ſea-ſickneſs. See Vertigo rotatoria. Claſs IV. 2. 1. 10. in which the firſt link of aſſociation acts too ſtrongly, and in conſequence expends more than uſual of the ſenſorial power of iritation; and ſecondly in which ſenſation is produced between the links of aſſociation, and diſſevers or enfeebles them; the accumulation of ſenſorial power ſoon occurs in the ſtomach; as no previous expenditure of it in that organ has occurred. Whence in ſea-ſickneſs the perſons take food with eagerneſs at times, when the vertigo caſes for a few minutes.

Tenthly. If the gaſtric torpor be induced by previous violent exertion, as after intoxication, or after contagious matter has been ſwallowed, or ſome poiſons, as digitalis, or arſenic; an accumulation of ſenſorial power very ſlowly ſucceeds; whence long ſickneſs, or continued fever, becauſe the quantity of ſenſorial power already waſted muſt firſt be renewed, before an accumulation of it can be produced.

12. This leads us to a ſecond indication of cure in continued fevers, which conſiſts in ſtrengthening the actions of the ſtomach; as the firſt indication conſiſted in decreaſing the actions of the cutaneous capillaries and abſorbents. The actions of the ſtomach may ſometimes be increaſed by exhibiting a mild emetic; as an accumulation of ſenſorial power in the fibres of the ſtomach is produced during their retrograde actions. Beſides the evacuation of any noxious material from the ſtomach and duodenum, and from the abſorbents,

 which

which open their mouths on their internal ſurfaces, by their retrograde motion.

It is probable, that when mild emetics are given, as ipecacuanha, or antimonium tartarizatum, or infuſion of chamomile, they are rejected by an inverted motion of the ſtomach and œſophagus in conſequence of diſagreeable ſenſation, as duſt is excluded from the eye; and theſe actions having by previous habit been found effectual, and that hence there is no exhauſtion of the ſenſorial power of irritation. But where ſtrong emetics are adminiſtered, as digitalis, or contagious matter, the previous exhauſtion of the ſenſorial power of irritation ſeems to be a cauſe of the continued retrograde actions and ſickneſs of the ſtomach. An emetic of the former kind may therefore ſtrengthen the power of the ſtomach immediately after its operation by the accumulation of ſenſorial power of irritation during its action. See Claſs IV. 1. 1.

Another method of decreaſing the action of the ſtomach for a time, and thence of increaſing it afterwards, is by the accumulation of the ſenſorial power of irritation during its torpor; is by giving ice, iced water, iced creams, or iced wine. This accounts for the pleaſure, which many people in fevers with weak pulſe expreſs on drinking cold beverage of any kind.

A ſecond method of exciting the ſtomach into action, and of decreaſing that of the capillaries in conſequence, is by the ſtimulus of wine, opium, bark, metallic ſalts of antimony, ſteel, copper, arſenic, given in ſmall repeated quantities; which ſo long as they render the pulſe ſlower are certainly of ſervice, and may be given warm or cold, as moſt agreeable to the patient. For it is poſſible, that the capillaries of the ſtomach may act too violently, and produce heat, at the ſame time that the large muſcles of it may be in a torpid ſtate; which curious circumſtance future obſervations muſt determine.

Thirdly. Hot fomentation on the region of the ſtomach might be of moſt eſſential ſervice by its ſtimulus, as heat penetrates the

ſyſtem

ſyſtem not by the abſorbent veſſels, but by external influence; whence the uſe of hot fomentation to the head in torpor of the brain; and the uſe of hot bath in caſes of general debility, which has been much too frequently neglected from a popular error occaſioned by the unmeaning application of the word relaxation to animal power. If the fluid of heat could be directed to paſs through particular parts of the body with as little diffuſion of its influence, as that of electricity in the ſhocks from the coated jar, it might be employed with ſtill greater advantage.

Fourthly. The uſe of repeated ſmall electric ſhocks through the region of the ſtomach might be of ſervice in fevers with weak pulſe, and well deſerves a trial; twenty or thirty ſmall ſhocks twice a day for a week or two would be a promiſing experiment.

Fifthly. A bliſter on the back, or ſides, or on the pit of the ſtomach, repeated in ſucceſſion, by ſtimulating the ſkin frequently ſtrengthens the action of the ſtomach by exciting the ſenſorial power of aſſociation; this eſpecially in thoſe fevers where the ſkin of the extremities, as of the hands or noſe or ears, ſooner becomes cold, when expoſed to the air, than uſual.

Sixthly. The action of the ſtomach may be increaſed by preventing too great expenditure of ſenſorial power in the link of previous motion with which it is catenated, eſpecially if the action of that link be greater than natural. Thus as the capillaries of the ſkin act too violently in fevers with weak pulſe, if theſe are expoſed to cold air or cold water, the ſenſorial power, which previouſly occaſioned their orgaſm, becomes accumulated, and tends to increaſe the action of the ſtomach; thus in thoſe fevers with weak pulſe and hot ſkin, if the ſtomach be ſtimulated by repeated ſmall doſes of bark and wine or opium, and be further excited at the ſame time by accumulation of ſenſorial power occaſioned by rendering the capillaries torpid by cold air or water, this twofold application is frequently attended with viſible good effect.

By

By thus ſtimulating the torpid ſtomach into greater action, the motions of the heart and arteries will likewiſe be increaſed by the greater excitement of the power of aſſociation. And the capillaries of the ſkin will ceaſe to act ſo violently, from their not poſſeſſing ſo great a ſuperfluity of ſenſorial power as during the greater quieſcence of the ſtomach and of the heart and arteries. Which is in ſome circumſtances ſimilar to the curious phenomenon mentioned in Claſs IV. 2. 2. 10; where, by covering the chill feet with flannel at the eruption of the ſmall-pox, the points of the flannel ſtimulate the ſkin of the feet into greater action, and the quantity of heat, which they poſſeſs, is alſo confined, or inſulated, and further increaſes by its ſtimulus the activity of the cutaneous veſſels of the feet; and by that circumſtance abates the too great action of the capillaries of the face, and the conſequent heat of it.

## XIII. *Caſe of continued fever.*

The following caſe of continued fever which I frequently ſaw during its progreſs, as it is leſs complicate than uſual, may illuſtrate this doctrine. Maſter S. D. an active boy about eight years of age, had been much in the ſnow for many days, and ſat in the claſſical ſchool with wet feet; he had alſo about a fortnight attended a writing ſchool, where many children of the lower order were inſtructed. He was ſeized on February the 8th, 1795, with great languor, and pain in his forehead, with vomiting and perpetual ſickneſs; his pulſe weak, but not very frequent. He took an emetic, and on the next day, had a bliſter, which checked the ſickneſs only for a few hours; his ſkin became perpetually hot, and dry; and his tongue white and furred; his pulſe when aſleep about 104 in a minute, and when awake about 112.

Fourth day of the diſeaſe. He has had another bliſter, the pain of of his head is gone, but the ſickneſs continues by intervals; he refuſes to take any ſolid food, and will drink nothing but milk, or milk and water, cold. He has two or three very liquid ſtools every day, which are ſomtimes green, but generally of a darkiſh yellow, with great flatulency both upwards and downwards at thoſe times. An antimonial powder was once given, but inſtantly rejected; a ſpoonful of decoction of bark was alſo exhibited with the ſame event. His legs are bathed, and his hands and face are moiſtened twice a day for half an hour in warmiſh water, which is nevertheleſs much colder than his ſkin.

Eighth day. His ſkin continues hot and dry without any obſervable remiſſions, with liquid ſtools and much flatulency and ſickneſs; his water when obſerved was of a ſtraw colour. He has aſked for cyder, and drinks nearly a bottle a day mixed with cold water, and takes three drops of laudanum twice a day.

Twelfth day. He continues much the ſame, takes no milk, drinks only cyder and water, ſkin hot and dry, tongue hot and furred, with liquid ſtools, and ſickneſs always at the ſame time; ſleeps much.

Sixteenth day. Was apparently more torpid, and once rather delirious; pulſe 112. Takes only capillaire and water; ſleeps much.

Twentieth day. Pulſe 100, ſkin dry but leſs hot, liquid ſtools not ſo frequent, he is emaciated to a great degree, he has eaten half a tea-cup full of cuſtard to day, drinks only capillaire and water, has thrice taken two large ſpoonfuls of decoction of bark with three drops of laudanum, refuſes to have his legs bathed, and will now take nothing but three drops of laudanum twice a day.

Twenty-fourth day. He has gradually taken more cuſtard every day, and began to attend to ſome new play things, and takes wine ſyllabub.

Twenty-eighth day. He daily grows ſtronger, eats eggs, and bread and butter, and ſleeps immediately after his food, can creep on his hands and knees, but cannot ſtand erect.

Thirty-ſecond day. He cannot yet ſtand alone ſafely, but ſeems hourly to improve in ſtrength of body, and activity of mind.

In this caſe the remote cauſe of his fever could not be well aſcertained, as it might be from having his feet cold for many ſucceſſive days, or from contagion; but the latter ſeems more probable, becauſe his younger brother became ill of a ſimilar fever about three weeks afterwards, and probably received the infection from him. The diſeaſe commenced with great torpor of the ſtomach, which was ſhewn by his total averſion to ſolid food, and perpetual ſickneſs; the watery ſtools, which were ſometimes green, or of a darkiſh yellow, were owing to the acrimony, or acidity, of the contents of the bowels; which as well as the flatulency were occaſioned by indigeſtion. This torpor of the ſtomach continued throughout the whole fever, and when it ceaſed, the fever ceaſed along with it.

The contagious material of this fever I ſuppoſe to have been mixed with the ſaliva, and ſwallowed into the ſtomach; that it excited the veſſels, which conſtitute the ſtomach, into the greateſt irritative motion like arſenic; *which might not be perceived, and yet might render that organ paralytic or inirritable in a moment of time*; as animals ſometimes die by one ſingle exertion, and conſequent paralyſis, without a ſecond ſtruggle; as by lightning, or being ſhot through the back part of the brain; of both which I have ſeen inſtances. I had once an opportunity of inſpecting two oxen, a few minutes after they were killed by lightning under a crab-tree on moiſt ground in long graſs; and obſerved, that they could not have ſtruggled, as the graſs was not preſſed or bent near them; I have alſo ſeen two horſes ſhot through the cerebellum, who never once drew in their legs after they firſt ſtretched them out, but died inſtantaneouſly; in a ſimilar manner the lungs ſeem to be rendered inſtantly inanimate by the fumes of burning ſulphur.

The

The lungs may be ſometimes primarily affected with contagious matter floating in the atmoſphere as well as the ſtomach, as mentioned in article 9. of this Supplement. But probably this may occur much leſs frequently, becauſe the oxygene of the atmoſphere does not appear to be taken into the blood by animal abſorption, as the ſaliva in the ſtomach, but paſſes through the moiſt membranes into the blood, like the ethereal fluids of electricity or heat, or by chemical attraction, and in conſequence the contagious matter may be left behind; except it may ſometimes be abſorbed along with the mucus; of which however in this caſe there appeared no ſymptoms.

The tonſils are other organs liable to receive contagious matter, as in the ſmall-pox, ſcarlet-fever, and in other ſenſitive inirritated fevers; but no ſymptom of this appeared here, as the tonſils were at no time of the fever inflamed, though they were in this child previouſly uncommonly large.

The pain of the forehead does not ſeem to have been of the internal parts of the head, becauſe the nerves, which ſerve the ſtomach, are not derived from the anterior part of the brain; but it ſeems to have been owing to a torpor of the external membranes about the forehead from their direct ſympathy with thoſe of the ſtomach; that is, from the deficient excitement of the ſenſorial power of aſſociation; and ſeemed in ſome meaſure to be relieved by the emetics and bliſters.

The pulſations of the heart were weaker and in conſequence quicker than natural, owing to their direct ſympathy with the torpid periſtaltic motions of the ſtomach; that is to the deficient excitement of the ſenſorial power of aſſociation.

The action of the cutaneous capillaries and abſorbents were ſtronger than natural, as appeared by the perpetual heat and dryneſs of the ſkin; which was owing to their reverſe ſympathy with the heart and arteries. This weaker and quicker action of the heart



and

and arteries, and the ſtronger action of the cutaneous capillaries and abſorbents, continued throughout the diſeaſe, and may be ſaid to have conſtituted the fever, of which the torpor of the ſtomach was the remote cauſe.

His tongue was not very much furred or very dry, nor his breath very hot; which ſhewed, that there was no great increaſe of the action of the mucous abſorbents, nor of the pulmonary capillaries, and yet ſufficient to produce great emaciation. His urine was nearly natural both in quantity and colour; which ſhewed, that there was no increaſe of action either of the kidnies, or of the urinary abſorbents.

The bathing his legs and hands and face for half an hour twice a day ſeemed to refreſh him, and ſometimes made his pulſe ſlower, and thence I ſuppoſe ſtronger. This ſeems to have been cauſed by the water, though ſubtepid, being much below the heat of his ſkin, and conſequently contributing to cool the capillaries, and by ſatiating the abſorbents to relieve the uneaſy ſenſation from the dryneſs of the ſkin.

He continued the uſe of three drops of tincture of opium from about the eighth day to the twenty-fourth, and for the three preceding days took along with it two large ſpoonfuls of an infuſion or bark in equal parts of wine and water. The former of theſe by its ſtimulus ſeemed to decreaſe his languor for a time, and the latter to ſtrengthen his returning power of digeſtion.

The daily exacerbations or remiſſions were obſcure, and not well attended to; but he appeared to be worſe on the fourteenth or fifteenth days, as his pulſe was then quickeſt, and his inattention greateſt; and he began to get better on the twentieth or twenty-firſt days of his diſeaſe; for the pulſe then became leſs frequent, and his ſkin cooler, and he took rather more food: theſe circumſtances ſeemed to obſerve the quarter periods of lunation.

XIV. *Termination*

## XIV. *Termination of continued fever.*

1. When the ſtomach is primarily affected with torpor not by defect of ſtimulus, but in conſequence of the previous exhauſtion of its ſenſorial power; and not ſecondarily by its aſſociation with other torpid parts; it ſeems to be the general cauſe of the weak pulſations of the heart and arteries, and the conſequent increaſed action of the capillaries, which conſtitute continued fever with weak pulſe. In this ſituation if the patient recovers, it is owing to the renovation of life in the torpid ſtomach, as happens to the whole ſyſtem in winter-ſleeping animals. If he periſhes, it is owing to the exhauſtion of the body for want of nouriſhment occaſioned by indigeſtion; which is haſtened by the increaſed actions of the capillaries and abſorbents.

2. When the ſtomach is primarily affected by defect of ſtimulus, as by cold or hunger; or ſecondarily by defect of the power of aſſociation, as in intermittent fevers; or laſtly in conſequence of the introduction of the ſenſorial power of ſenſation, as in inflammatory diſeaſes; the actions of the heart and arteries are not diminiſhed, as when the ſtomach is primarily affected with torpor by its previous exhauſtion of ſenſorial power; but become greatly increaſed, producing irritative or inflammatory fever. Where this fever is continued, though with ſome remiſſions and exacerbations, the exceſſive action is at length ſo much leſſened by expenditure of ſenſorial power, as to gradually terminate in health; or it becomes totally exhauſted, and death ſucceeds the deſtruction of the irritability and aſſociability of the ſyſtem.

3. There is alſo another termination of the diſeaſes in conſequence of great torpor of the ſtomach, which are not always termed fevers;

 one

one of theſe is attended with ſo great and univerſal torpor, that the patient dies in the firſt cold fit; that is, within twelve hours or leſs of the firſt ſeizure; this is commonly termed ſudden death. But the quickneſs of the pulſe, and the coldneſs with ſhuddering, and with ſick ſtomach, diſtinguiſhed a caſe, which I lately ſaw, from the ſudden deaths occaſioned by apoplexy, or ruptured blood-veſſels.

In hemicrania I believe the ſtomach is always affected ſecondarily, as no quickneſs of pulſe generally attends it, and as the ſtomach recovers its activity in about two whole days. But in the following caſe, which I ſaw laſt week, I ſuppoſe the ſtomach ſuddenly became paralytic, and cauſed in about a week the death of the patient. Miſs ———, a fine young lady about nineteen, had bathed a few times, about a month before, in a cold ſpring, and was always much indiſpoſed after it; ſhe was ſeized with ſickneſs, and cold ſhuddering, with very quick pulſe, which was ſucceeded by a violent hot fit; during the next cold paroxyſm ſhe had a convulſion fit; and after that ſymptoms of inſanity, ſo as to ſtrike and bite the attendants, and to ſpeak furious language; the ſame circumſtances occurred during a third fit, in which I believe a ſtrait waiſtcoat was put on, and ſome blood taken from her; during all this time her ſtomach would receive no nutriment, except once or twice a little wine and water. On the ſeventh day of the diſeaſe, when I ſaw her, the extremities were cold, the pulſe not to be counted, and ſhe was unable to ſwallow, or to ſpeak; a clyſter was uſed with turpentine and muſk and opium, with warm fomentations, but ſhe did not recover from that cold fit.

In this caſe the convulſion fit and the inſanity ſeem to have been violent efforts to relieve the diſagreeable ſenſation of the paralytic ſtomach; and the quick pulſe, and returning fits of torpor and of orgaſm, evinced the diſeaſe to be attended with fever, though it might have been called anorexia maniacalis, or epileptica.

4. Might

4. Might not many be ſaved in theſe fevers with weak pulſe for a few weeks by the introduction of blood into a vein, once in two or three days; which might thus give further time for the recovery of the torpid ſtomach? Which ſeems to require ſome weeks to acquire its former habits of action, like the muſcles of paralytic patients, who have all their habits of voluntary aſſociations to form afreſh, as in infancy.

If this experiment be again tried on the human ſubject, it ſhould be ſo contrived, that the blood in paſſing from the well perſon to the ſick one ſhould not be expoſed to the air; it ſhould not be cooled or heated; and it ſhould be meaſured; all which may be done in the following manner. Procure two ſilver pipes, each about an inch long, in the form of funnels, wide at top, with a tail beneath, the former ſomething wider than a ſwan-quill, and the latter leſs than a ſmall crow-quill. Fix one of theſe ſilver funnels by its wide end to one end of the gut of a chicken freſh killed about four or ſix inches long, and the other to the other end of the gut; then introduce the ſmall end of one funnel into the vein of the arm of a well perſon downwards towards the hand; and laying the gut with the other end on a water-plate heated to 98 degrees in a very warm room, let the blood run through it. Then preſſing the finger on the gut near the arm of the well perſon, ſlide it along ſo as to preſs out one gut-ful into a cup, in order to aſcertain the quantity by weight. Then introduce the other end of the other funnel into a ſimilar vein in the arm of the ſick perſon upwards towards the ſhoulder; and by ſliding one finger, and then another reciprocally, along the chicken's gut, ſo as to compreſs it, from the arm of the well perſon to the arm of the ſick one, the blood may be meaſured, and thus the exact quantity known which is given and received. See Claſs I. 2. 3. 25.

XV. *Inflammation*

## XV. *Inflammation excited in fever.*

1. When the actions of any part of the ſyſtem of capillaries are excited to a certain degree, ſenſation is produced, along with a greater quantity of heat, as mentioned in the fifth article of this ſupplement. When this increaſed capillary action becomes ſtill more energetic, by the combined ſenſorial powers of ſenſation with irritation, new fibres are ſecreted, or new fluids, (which harden into fibres like the mucus ſecreted by the ſilk-worm, or ſpider, or pinna,) from which new veſſels are conſtructed; it is then termed inflammation: if this exiſts in the capillary veſſels of the cellular membrane or ſkin only, with feeble pulſations of the heart and arteries, the febris ſenſitiva inirritata, or malignant fever, occurs; if the coats of the arteries are alſo inflamed, the febris ſenſitiva irritata, or inflammatory fever, exiſts.

In all theſe fevers the part inflamed is called a phlegmon, and by its violent actions excites ſo much pain, that is, ſo much of the ſenſorial power of ſenſation, as to produce more violent actions, and inflammation, throughout the whole ſyſtem. Whence great heat from the excited capillaries of the ſkin, large and quick pulſations of the heart, full and hard arteries, with great univerſal ſecretions and abſorptions. Theſe perpetually continue, though with exacerbations and remiſſions; which ſeem to be governed by ſolar or lunar influence.

2. In this ſituation there generally, I ſuppoſe, exiſts an increaſed activity of the ſecerning veſſels of the brain, and conſequently an increaſed production of ſenſorial power; in leſs violent quantity of this diſeaſe however the increaſe of the action of the heart and arteries may be owing ſimply to the accumulation of ſenſorial power of aſſociation in the ſtomach, when that organ is affected by ſympathy with ſome inflamed part. In the ſame manner as the capillaries are violently

lently and permanently actuated by the accumulation of the ſenſorial power of aſſociation in the heart and arteries, when the ſtomach is affected primarily by contagious matter, and the heart and arteries ſecondarily. Thus I ſuſpect, that in the diſtinct ſmall-pox the ſtomach is affected ſecondarily by ſympathy with the infected tonſils or inoculated arm; but that in the confluent ſmall-pox the ſtomach is affected primarily, as well as the tonſils, by contagious matter mixed with the ſaliva, and ſwallowed.

3. In inflammatory fevers with great arterial action, as the ſtomach is not always affected with torpor, and as there is a direct ſympathy between the ſtomach and heart, ſome people have believed, that nauſeating doſes of ſome emetic drug, as of antimonium tartarizatum, have been adminiſtered with advantage, abating by direct ſympathy the actions of the heart. This theory is not ill founded, and the uſe of digitalis, given in ſmall doſes, as from half a dram to a dram of the ſaturated tincture, two or three times a day, as well as other leſs violent emetic drugs, would be worth the attention of hoſpital phyſicians.

Sickneſs might alſo be produced probably with advantage by whirling the patient in a chair ſuſpended from the cieling by two parallel cords; which after being revolved fifty or one hundred times in one direction, would return with great circular velocity, and produce vertigo, ſimilar I ſuppoſe to ſea-ſickneſs. And laſtly the ſickneſs produced by reſpiring an atmoſphere mixed with one tenth of carbonated hydrogen, diſcovered by Mr. Watt, and publiſhed by Dr. Beddoes, would be well worthy exact and repeated experiment.

4. Cool air, cool fomentations, or ablutions, are alſo uſeful in this inflammatory fever; as by cooling the particles of blood in the cutaneous and pulmonary veſſels, they muſt return to the heart with leſs ſtimulus, than when they are heated above the natural degree of ninety-eight.

For

For this purpose snow and ice have been scattered on the patients in Italy; and cold bathing has been used at the eruption of the small pox in China, and both, it is said, with advantage. See Class III. 2. 1. 12, and Suppl. I. 8.

5. The lancet however with repeated mild cathartics is the great agent in destroying this enormous excitement of the system, so long as the strength of the patient will admit of evacuations. Blisters over the painful part, where the phlegmon or topical inflammation is situated, after great evacuation, is of evident service, as in pleurisy. Warm bathing for half an hour twice a day, when the patient becomes enfeebled, is of great benefit, as in peripneumony and rheumatism.

6. When other means fail of success in abating the violent excitement of the system in inflammatory diseases, might not the shaved head be covered with large bladders of cold water, in which ice or salt had been recently dissolved; and changed as often as necessary, till the brain is rendered in some degree torpid by cold?—Might not a greater degree of cold, as iced water, or snow, be applied to the cutaneous capillaries?

7. Another experiment I have frequently wished to try, which cannot be done in private practice, and which I therefore recommend to some hospital physician; and that is, to endeavour to still the violent actions of the heart and arteries, after due evacuations by venesection and cathartics, by gently compressing the brain. This might be done by suspending a bed, so as to whirl the patient round with his head most distant from the center of motion, as if he lay across a millstone, as described in Sect. XVIII. 20. For this purpose a perpendicular shaft armed with iron gudgeons might have one end pass

into the floor, and the other into a beam in the cieling, with an horizontal arm, to which a ſmall bed might be readily ſuſpended.

By thus whirling the patient with increaſing velocity ſleep might be produced, and probably the violence of the actions of the heart and arteries might be diminiſhed in inflammatory fevers; and, as it is believed, that no accumulation of ſenſorial power would ſucceed a torpor of the origin of the nerves, either thus procured by mechanical compreſſion, or by the bladder-cap of cold water above deſcribed, the lives of thouſands might probably be ſaved by thus extinguiſhing the exacerbations of febrile paroxyſms, or preventing the returns of them.

In fevers with weak pulſe ſleep, or a degree of ſtupor, thus produced, might prevent the too great expenditure of ſenſorial power, and thus contribute to preſerve the patient. See Claſs I. 2. 5. 10. on ſtupor. What might be the conſequence of whirling a perſon with his head next the center of motion, ſo as to force the blood from the brain into the other parts of the body, might be diſcovered by cautious experiment without danger, and might probably add to our ability of curing fever.

## XVI. *Recapitulation.*

1. The ſenſorial power cauſes the contraction of the fibres, and is excited into action by four different circumſtances, by the ſtimulus of external bodies, by pain or pleaſure, by deſire or averſion, or by the previous motions of other contracting fibres. In the firſt ſituation it is called the ſenſorial power of irritation, in the ſecond the ſenſorial power of ſenſation, in the third the ſenſorial power of volition, and in the fourth the ſenſorial power of aſſociation.

Many parts of the body are excited into perpetual action, as the ſanguiferous veſſels conſiſting of the heart, arteries, and veins; others into nearly perpetual action, as the conglomerate and capillary glands; and others into actions still somewhat less frequent, as the alimentary canal, and the lacteal and lymphatic absorbents with their conglobate glands: all these are principally actuated by the sensorial powers of irritation, and of association; but in some degree or at some times by those of sensation, and even of volition. There are three kinds of stimulus, which may easily be occasionally diminished, that of heat on the skin, of food in the stomach, and of the oxygenous part of the atmosphere, which mixes with the blood in respiration, and stimulates the heart and arteries.

2. When any parts, which are naturally excited into perpetual action by stimulus, become torpid or less active from decrease of that stimulus; there first occurs a decrease of the activity of the parts next catenated with them; thus going into cold water produces a torpor of the capillary vessels of the lungs, as is known by the difficult respiration, which immediately occurs; for the sensorial power of association, which naturally contributes to actuate the lungs, is now less excited by the decreased actions of the cutaneous vessels, with which they are catenated. This constitutes the cold fit of fever.

There next occurs an accumulation of the sensorial power of irritation in the parts, which were torpid from defect of stimulus, as the cutaneous vessels for instance when exposed to cold air; and a similar accumulation of the sensorial power of association occurs in the parts which were catenated with the former, as the vessels of the lungs in the example above mentioned. Whence, if the subduction of stimulus has not been too great, so as to impair the health of the part, the activity of the irritative motions returns, even though the stimulus continues less than usual; and those of the associate motions become considerably increased, because these latter are now excited by the previous fibrous

 motions,

motions, which now act as strong or stronger than formerly, and have also acquired an accumulation of the sensorial power of association. This accounts for the curious event of our becoming warm in a minute or two after remaining in water of about 80 degrees of heat, as in the bath at Buxton; or in the cold air of a frosty morning of about 30 degrees of heat.

But if the parts thus possessed of the accumulated sensorial powers of irritation and of association be exposed again to their natural quantity of stimulus, a great excess of activity supervenes; because the fibres, which possess accumulated irritation, are now excited by their usual quantity of stimulus; and those which possess accumulated association, are now excited by double or treble the quantity of the preceding irritative fibrous motions, with which they are catenated; this constitutes the hot fit of fever.

Another important circumstance occurs, when the parts, which are torpid from decreased stimulus, do not accumulate a quantity of sensorial power sufficient for the purpose of renewing their own natural quantity of action; but are nevertheless not so torpid, as to have the life of the part impaired. In this situation the superabundance of the accumulated power of irritation contributes to actuate the associate motions next catenated with them. Thus, when a person breathes air with less oxygene than natural, as by covering his head in bed, and thus respiring the same atmosphere repeatedly, the heart and arteries become less active by defect of the stimulus of oxygene; and then the accumulation of sensorial power of irritation becomes instantly very great, as these organs are subject to perpetual and energetic action. This accumulation nevertheless is not so great as to renew their own activity under this defect of stimulus, but yet is in sufficient abundance to increase the associability of the next link of catenation, that is, to actuate the capillaries of the skin with great and perpetual increase of energy. This resembles continued fever with weak pulse; in which the accumulation of the sensorial power

 caused

caused by the lessened motions of the heart and arteries, actuates the capillaries with increase of energy.

3. When the accumulation of the sensorial power of association, which is caused as above explained by deficient excitement owing to the lessened quantity of action of the irritative fibrous motions, with which the associate train is catenated, is not in quantity sufficient to renew the natural actions of the first link of an associate train of motions; it is nevertheless frequently so abundant as to actuate the next link of the associated train with unnatural energy by increasing its associability; and that in a still greater degree if that second link of the associated train was previously in a torpid state, that is, had previously acquired some accumulation of the sensorial power of association. This important circumstance of the animal economy is worthy our most accurate attention. Thus if the heart and arteries are deprived of their due quantity of the stimulus of oxygene in the blood, a weak and quick pulse ensues, with an accumulation of the sensorial power of irritation; next follows an increase of the action of the capillaries by the superabundance of this accumulated power of irritation; but there also exists an accumulation of the power of association in these acting capillaries, which is not now excited by the deficient actions of the heart and arteries; but which by its abundance contributes to actuate the next link of association, which is the sick stomach in the case related from Sydenham in Class IV. 1. 1. 2. and explained in this Supplement I. 4. And as this sick stomach was in a previous state of torpor, it might at the same time possess an accumulation of some sensorial power, which, if it was of association, would be thus more powerfully excited by the increased actions of the capillaries; which existed in consequence of the weak action of the heart and arteries. This also resembles in some respects the continued fevers with weak pulse, and with increased activity of the capillaries.

4. When

4. When a torpor of ſome irritative motions occurs from a previous exhauſtion of the ſenſorial power of irritation by the action of ſome very great ſtimulus, it is long before any accumulation of the ſenſorial power of irritation is produced; as is experienced in the ſickneſs and languor, which continues a whole day after a fit of drunkenneſs. But nevertheleſs there occurs an accumulation of the ſenſorial power of aſſociation in the firſt link of the aſſociate train of motions, which is catenated with theſe torpid irritative ones; which accumulation is owing to deficient excitement of that ſenſorial power in the firſt link of the aſſociate train. This firſt link therefore exiſts alſo in a leſs active or torpid ſtate, but the accumulation of the ſenſorial power of aſſociation by its ſuperabundance contributes to actuate the ſecond link of the aſſociate train with unnatural quantity of motion; and that though its own natural quantity of the power of aſſociation is not excited by the deficient action of preceding fibrous motions.

When this happens to the ſtomach, as after its irritative motions have been much exerted from the unnatural ſtimulus of wine, or opium, or of contagious matter mixed with the ſaliva, a torpor or inactivity of it ſucceeds for a greater or leſs length of time; as no accumulation of the ſenſorial power of irritation can occur, till the natural quantity, which has been previouſly expended, is firſt reſtored. Then the heart and arteries, which are next in catenation, become leſs active from the want of ſufficient excitement of the ſenſorial power of aſſociation, which previouſly contributed to actuate them. This ſenſorial power of aſſociation therefore becomes accumulated, and by its ſuperabundance contributes to actuate the link next in aſſociation, which has thus acquired ſo great a degree of aſſociability, as to overbalance the leſs quantity of the excitement of it by the torpid action of the previous or firſt aſſociate link. This happens to the capillaries, when the heart and arteries are affected as above by the torpor of the ſtomach, when it is occaſioned by previous great

 expenditure

expenditure of its sensorial power, and thus constitutes fever with weak pulse, which is here termed inirritative fever, typhus mitior.

5. When a deficiency of stimulus is too great or too long continued, so as to impair the life of the part, no further accumulation of sensorial power occurs; as when the skin is long exposed to cold and damp air. In that case the link in catenation, that is, the first of the associate train, is rendered torpid by defect of excitement of its usual quantity of the sensorial power of association, and from there being no accumulation of the sensorial power of irritation to increase its associability, and thus to contribute to actuate it by overbalancing the defect of the excitement of its association.

Thus on riding long and slowly on a cold and damp day, the exhalation of the vapour, which is impinged on the skin, as the traveller proceeds, carries away his warmth faster, than it is generated within the system; and thus the capillaries of the skin have their actions so much impaired after a time, that no accumulation of the sensorial power of irritation occurs; and then the stomach, whose motions are catenated with those of the capillaries, ceases to act from the deficient excitement of the power of association; and indigestion and flatulency succeed, instead of the increased digestion and hunger, which occur, when the cutaneous capillaries are exposed to a less degree of cold, and for a shorter time. In which latter situation the accumulation of the sensorial power of irritation increases by its superabundance the associability of the fibres of the stomach, so as to overbalance the defect of the excitement of their association.

6. The stomach is affected secondarily in fevers with strong pulse, as in those with weak pulse it is affected primarily. To illustrate this doctrine I shall relate the following case of Mr. Y———. He was a young man rather intemperate in the use of wine or beer, and was

was ſeized with a cold fit, and with a conſequent hot one with ſtrong pulſe; on examining his hypochondrium an oblong tumour was diſtinctly felt on the left ſide of the ſtomach, which extended ſix or eight inches downward, and was believed to be a tumour of the ſpleen, which thus occaſioned by its torpor the cold fit and conſequent hot fit of fever with ſtrong pulſe. This fever continued, though with remiſſions, for two or three weeks; and the patient repeatedly loſt blood, uſed cathartics with calomel and ſena, and had frequent antimonial and ſaline medicines. And after he was much weakened by evacuations, the peruvian bark and ſmall doſes of ſteel removed the fever, but the tumour remained many years during the remainder of his life.

In this caſe the tumour of the ſpleen was occaſioned by the torpor of the abſorbent veſſels; while the ſecerning veſſels continued ſomewhat longer to pour their fluids into the cells of it. Then the inactivity of this viſcus affected the whole ſyſtem with torpor by the deficient excitement of the ſenſorial power of aſſociation, which contributes along with the irritation cauſed by their ſpecific ſtimuli to actuate the whole ſanguiferous, ſecerning, and abſorbent veſſels; and along with theſe the ſtomach, which poſſeſſes perhaps greater mobility, or promptitude to torpor or to orgaſm, than any other part. And after a time all theſe parts recover their actions by the accumulation of their ſenſorial power of aſſociation. But the ſpleen not recovering its action from the accumulation of its power of irritation, as appeared from the continuance of the tumor, ſtill affects the ſtomach by its defective irritative motions ceaſing to excite the aſſociation, which ought to contribute to actuate it.

Hence the ſtomach continues torpid in reſpect to its motions, but accumulates its power of aſſociation; which is not excited into action by the defective motions of the ſpleen; this accumulation of the ſenſorial power of aſſociation now by its ſuperabundance actuates the next link of aſſociate motions, which conſiſts of the heart and arteries,

into

into greater energy of action than natural, and thus causes fever with strong pulse; which, as it was supposed to be most frequently excited by increase of irritation, is called irritative fever or synocha.

Similar to this in the small pox, which is given by inoculation, the stomach is affected secondarily, when the fever commences; and hence in this small-pox the pulsations of the heart and arteries are frequently stronger than natural, but never weaker, for the reasons above given. Whereas in that small-pox, which is caused by the stomach being primarily affected, by the contagious matter being swallowed with the saliva, whether the tonsils are at the same time affected or not, the pulsations of the heart and arteries become weak, and the inirritative fever is produced, as explained above, along with the confluent small-pox. This unfolds the cause of the mildness of the inoculated small-pox; because in this disease the stomach is affected secondarily, whereas in the natural small-pox it is frequently affected primarily by swallowing the contagious matter mixed with saliva.

In the measles I suppose the contagious matter to be dissolved in the air, and therefore not liable to be mixed with the saliva; whereas the variolous matter is probably only diffused in the air, and thence more readily mixed with the saliva in the mouth during respiration. This difference appears more probable, as the small-pox I believe is always taken at a less distance from the diseased person than is necessary to acquire the measles. The contagion of the measles affects the membranes of the nostrils, and the secretion of tears in consequence, but never I suspect the stomach primarily, but always secondarily; whence the pulsation of the heart and arteries is always stronger than natural, so as to bear the lancet at any period of the disease.

The great mildness sometimes, and fatality at other times, of the scarlet fever may depend on the same circumstance; that is, on the stomach being primarily or secondarily affected by the contagious matter,

matter, obſerving that the tonſils may be affected at the ſame time with the ſtomach. Should this prove to be the caſe, which future obſervations muſt determine, what certain advantage muſt ariſe from the inoculation of this diſeaſe! When it is received by the ſkin primarily I ſuppoſe no ſore throat attends it, nor fever with weak pulſe; when it is received by the ſtomach primarily, the tonſils are affected at the ſame time, and the torpor of the ſtomach produces inirritative fever, and the mortification of the tonſils ſucceeds.

We may hence conclude, that when the torpor of the ſtomach is either owing to defect of ſtimulus, which is not ſo great as to impair the life of the part, as in moderate hunger, or in ſwallowing iced water, or when its torpor is induced by its catenation or aſſociation with other torpid parts, as in the commencement of intermittent fevers, and inoculated ſmall-pox, that the ſubſequent action of the heart and arteries is generally increaſed, producing irritative fever. Which is owing to the accumulation of the ſenſorial power of irritation in one caſe, and of aſſociation in the other, contributing to actuate the next link of the catenated or aſſociated motions. But when the torpor of the ſtomach is induced by previous exhauſtion of its ſenſorial powers of irritation or of aſſociation by continued violent action, as by the ſtimulus of digitalis, or of contagious matter, or after intoxication from wine or opium, a weaker action of the heart and arteries ſucceeds, becauſe there is no accumulation of ſenſorial power, and a deficient excitement of aſſociation. And finally, as this weak action of the heart and arteries is not induced by exhauſtion of ſenſorial power, but by defect of the excitement of aſſociation, the accumulation of this power of aſſociation increaſes the action of the capillaries, and thus induces inirritative fever.

7. When any part of the ſyſtem acts very violently in fevers, the ſenſorial power of ſenſation is excited, which increaſes the actions of the moving ſyſtem; whereas the pain, which ariſes from decreaſed

irritative motions, as in hemicrania, ſeems to exhauſt a quantity of ſenſorial power, without producing or increaſing any fibrous actions.

When the ſtomach is primarily affected, as in inirritative fevers from contagion, and in ſuch a manner as to occaſion pain, the action of the capillaries ſeems to be increaſed by this additional ſenſorial power of ſenſation, whence extenſive inflammation or mortification; but when the ſtomach and conſequently the heart and arteries continue their torpidity of action; as in confluent ſmall-pox, and fatal ſcarlatina; this conſtitutes ſenſitive inirritative fever, or typhus gravior.

But when the ſtomach is ſecondarily affected, if the ſenſorial power of ſenſation is excited, as in pleuriſy or peripneumony, the actions of the heart and arteries are violently increaſed, and of all the moving ſyſtem along with them. Thus the peripneumony is generally induced by the patient reſpiring very cold air, and this eſpecially after being long confined to warm air, or after being much fatigued and heated by exceſſive labour or exerciſe. For we can cover the ſkin with more clothes, when we feel ourſelves cold; but the lungs not having the perception of cold, we do not think of covering them, nor have the power to cover them, if we deſired it; and the torpor, thus produced is greater, or of longer duration, in proportion to the previous expenditure of ſenſorial power by heat or exerciſe.

This torpor of the lungs affects the ſkin with ſhuddering, and the ſtomach is alſo ſecondarily affected; next follows the violent action of the lungs from the accumulation of the power of irritation, and an inflammation of them follows this violent action. While the ſtomach recovers its activity by the increaſe of the excitement of the ſenſorial power of aſſociation, and along with it the heart and arteries, and the whole moving ſyſtem. Hence this inflammation occurs during the hot fit of fever, and no cold fit ſucceeds, becauſe the exceſs

excess of the sensorial power of sensation prevents a succeeding torpor.

These new motions of certain parts of the system produce increased secretions of nutritious or organic mucus, which forms new vessels; these new vessels by their unusual motions produce new kinds of fluids; which are termed contagious, because they have the power, when introduced into a healthy body, of producing similar actions and effects, with or without fever, as in the small-pox and measles, or in the itch and venereal disease.

If any of these contagious matters affect the stomach with torpor either by their stimulus immediately applied, or by its sympathy with the parts first diseased, a fever is produced with sickness and want of appetite; as in small-pox, and scarlatina. If the stomach is not affected by contagious matter, no fever succeeds, as in itch, tinea, syphilis.

All these contagious matters are conceived to be harmless, till they have been exposed to the air, either openly or through a moist membrane; from which they are believed to acquire oxygene, and thence to become some kinds of animal acids. As the preparations of mercury cure venereal ulcers; as a quarter of a grain of sublimate dissolved in wine, and given thrice a day; this effect seems to be produced either by its stimulating the absorbents in the ulcer to absorb the venereal matter before it has acquired oxygene; or by afterwards uniting with it chemically, and again depriving it of its acquired acidity. On either supposition it might probably be given with advantage in small pox, and in all infectious diseases, both previous to their commencement, and during their whole progress.

8. The cold fits of intermittent fevers are caused by the torpor of some part owing to deficient irritation, and of the other parts of the system from deficient association. The hot fits are owing first to the accumulation of irritation in the part primarily affected, if it recovers

its action, which does not always happen; and secondly to the accumulation of association in the other parts of the system, which during health are subject to perpetual action; and lastly also to the greater excitement of the power of association, when the part primarily affected recovers its irritability, and acts with greater energy than natural.

The deficient secretions in the cold fit depend on the torpor of the glandular system; and the increased secretions in the hot fit on their more energetic action. The thirst in the cold fit is owing to the deficient absorption from the skin, cellular membrane, and bladder; the thirst in the hot fit is owing to the too great dissipation of the aqueous part of the blood. The urine is pale and in small quanity in the cold fit from deficient secretion of it, and from deficient absorption of its aqueous parts; it is high coloured, and sometimes deposits a sediment, in the hot fit from the greater secretion of it in the kidneys, and the greater absorption of its aqueous and saline part in the bladder. The dryness and scurf on the tongue and nostrils is owing to the increased heat of the air expired from the lungs, and consequent greater evaporation of the aqueous part of the mucus. The sweats appear in consequence of the declension of the hot fit, owing to the absorbent vessels of the skin losing their increased action sooner than the secerning ones; and to the evaporation lessening as the skin becomes cooler. The returns of the paroxysms are principally owing to the torpor of some less essential part of the system remaining after the termination of the last fit; and are also dependent on solar or lunar diurnal periods.

The torpor of the part, which induces the cold paroxysm, is owing to deficient irritation occasioned either by the subduction of the natural stimuli of food, or water, or pure air, or by deficiency of external influences, as of heat, or of solar or lunar gravitation. Or secondly, in consequence of the exhaustion of sensorial power by great previous exertions of some parts of the system, as of the limbs by great

great labour or exercise, or of the stomach by great stimulus, as by contagious matter swallowed with the saliva, or by much wine or opium previously taken into it. Or lastly a torpor of a part may be occasioned by some mechanic injury, as by a compression of the nerves of the part, or of their origin in the brain; as the sitting long with one leg crossed over the other occasions numbness, and as a torpor of the stomach with vomiting frequently precedes paralytic strokes of the limbs.

As sleep is produced, either by defect of stimulus, or by previous exhaustion of sensorial power; so the accumulation of the sensorial power of volition in those muscles and organs of sense, which are generally obedient to it, awakens the sleeping person; when it has increased the quantity of voluntarity so much as to overbalance the defect of stimulus in one case, and the exhaustion of sensorial power in the other; which latter requires a much longer time of sleep than the former. So the cold paroxysm of fever is produced either by defect of stimulus, or by previous exhaustion of the sensorial power of some part of the system; and the accumulation of the sensorial power of irritation in that part renews the action of it, when it has increased its irritability so much as to overbalance the defect of stimulus in one case, and the exhaustion of sensorial power in the other; which latter requires a much longer torpor or cold fit than the former.

But in the cold paroxysm of fever besides the torpor of one part of the system from defect of irritation, the remainder of it becomes torpid owing to defect of excitement of the sensorial power of association by the lessened action of the part first affected. This torpor of the general system remains, till the accumulation of the sensorial power of association has increased the associability so much as to overbalance the defect of the excitement of association; then the torpor ceases, and if the first affected part has recovered its activity the other parts are all thrown into excess of action by their increased associability, and the hot fit of fever is produced.

9. In

9. In the continued fevers with ſtrong pulſe the ſtomach is affected ſecondarily, and thus acts feebly from deficient excitement of the power of aſſociation; but the accumulation of the power of aſſociation thus produced in an organ ſubject to perpetual and energetic action, is ſo great as to affect the next link of the aſſociate train, which conſiſts of the heart and arteries; theſe therefore are exerted perpetually with increaſe of action.

In continued fevers with weak pulſe the torpid ſtomach is affected primarily by previous exhauſtion of its irritability by ſtimulus, as of contagious matter ſwallowed into it. The heart and arteries act feebly from deficient excitement of the power of aſſociation, owing to the torpor of the ſtomach, with which they are catenated; but the accumulation of the power of aſſociation, thus produced in organs ſubject to perpetual and energetic motion, is ſo great, as to affect the next link of the aſſociate train; which conſiſts of the capillaries of the ſkin or other glands; theſe therefore are exerted perpetually with great increaſe of action.

The continued fevers with ſtrong pulſe terminate by the reduction or exhauſtion of the ſenſorial power by violent action of the whole ſyſtem; which is followed either by return of health with the natural quantity of irritability, and of aſſociability, or by a total deſtruction of them both, and conſequent death.

In continued fevers with weak pulſe the ſtomach remains torpid during the whole courſe of the fever; and at length by the recovery of its irritability and ſenſibility effects the cure of it. Which generally happens about the firſt, ſecond, or third quarter of the lunar period, counted from the commencement of the diſeaſe, or continues a whole lunation, and ſometimes more; which gave riſe to what are termed critical days. See Sect. XXXVI. 4. on this ſubject. If the ſtomach does not recover from its torpor, the patient becomes emaciated, and dies exhauſted by the continuance of the increaſed action of the capillaries and abſorbents, and the want of nouriſhment.

The cure of continued fever with weak pulse consists first in weakening the undue action of the capillaries of the skin by ablution with cold water from 32 to 80 degrees of heat; or by exposing them to cool air. Secondly by invigorating the actions of the stomach, by decreasing them for a time, and thence accumulating the power of irritation, as by an emetic, or by iced water, or iced wine. Or by increase of stimulus, as by bark, wine, opium, and food, in small quantities frequently repeated. Or by renewing the action of the stomach by slight electric shocks. Or by fomenting it frequently with water heated to 96 or 100 degrees. Or lastly by exciting its power of association with other parts of the system, as by a blister; which succeeds best when the extremities are cool; or by swinging, as in vertigo rotatoria.

If by the stimulus of the Peruvian bark on the fibres of the stomach, they regain their due action, the heart and arteries also regain their due action; as their sensorial power of association is now excited, and expended as usual. And as there is then no accumulation of sensorial power in the heart and arteries, the capillaries cease to act with too great energy, and the fever is cured.

Thirdly. If the heart and arteries could be themselves stimulated into greater action, although the stomach remained torpid, they might probably by expending a greater quantity of the sensorial power of irritation, prevent an accumulation of the sensorial power of association, (for these may possibly be only different modes of action of the spirit of animation,) and thus the too great action of the capillaries might be prevented and the fever cease. This new mode of cure might possibly be accomplished, if the patient was to breathe a gallon or two of pure or diluted oxygene gas frequently in a day; which by passing through the moist membranes of the lungs and uniting with the blood might render it more stimulant, and thus excite the heart and arteries into greater action.

Fourthly. Greater energy might probably be given to the whole system, and particularly to those parts which act too feebly in fevers,

as

as the ſtomach and the heart and arteries, if the action of the ſecerning veſſels of the brain could be increaſed in energy; this is probably one effect of all thoſe drugs, which when given in large quantity induce intoxication, as wine and opium. And when given with great caution in ſmall quantities uniformly repeated, as from three drops to five of the tincture of opium, but not more, every ſix hours, I believe they ſupply an efficacious medicine in fevers with great arterial debility; and the more ſo, if the Peruvian bark be exhibited alternately every ſix hours along with them. There are other means of exciting the veſſels of the brain into action; as firſt by decreaſing the ſtimulus of heat by temporary cold fomentation; ſecondly, increaſing the ſtimulus of heat by long continued warm fomentation; thirdly, by electricity, as very ſmall ſhocks paſſed through it in all directions; and laſtly by bliſters on the head. All thoſe require to be uſed with great caution, and eſpecially where there exiſts an evident ſtupor, as the removing of that is I believe frequently injurious. See ſtupor, Claſs I. 2. 5. 10.

The cure of fever with ſtrong pulſe conſiſts in the repeated uſe of veneſection, gentle cathartics, diluents; medicines producing ſickneſs, as antimonials, digitalis; or the reſpiration of carbonated hydrogen; or by reſpiration of atmoſpheric air lowered by a mixture of hydrogen, azote, or carbonic acid gas, or by compreſſing the brain by whirling in a decumbent poſture, as if lying acroſs an horizontal mill-ſtone. See the former parts of this ſupplement for the methods of cure both of fevers with ſtrong and weak pulſe.

10. When any difficulty occurs in determining the weak pulſe from the ſtrong one, it may generally be aſſiſted by counting its frequency. For when an adult patient lies horizontally in a cool room, and is not hurried or alarmed by the approach of his phyſician, nor ſtimulated by wine or opium, the ſtrong pulſe ſeldom exceeds 118 or 120 in a minute; and the weak pulſe is generally not much below

130,

130, and often much above that number. Secondly in ſitting up in bed, or changing the horizontal to a perpendicular poſture, the quickneſs of the weak pulſe is liable immediately to increaſe 10 or 20 pulſations in a minute, which does not I believe occur in the ſtrong pulſe, when the patient has reſted himſelf after the exertion of riſing.

## XVII. *Concluſion.*

Thus have I given an outline of what may be termed the ſympathetic theory of fevers, to diſtinguiſh it from the mechanic theory of Boerhaave, the ſpaſmodic theory of Hoffman and of Cullen, and the putrid theory of Pringle. What I have thus delivered, I beg to be conſidered rather as obſervations and conjectures, than as things explained and demonſtrated; to be conſidered as a foundation and a ſcaffolding, which may enable future induſtry to erect a ſolid and a beautiful edifice, eminent both for its ſimplicity and utility, as well as for the permanency of its materials,—which may not moulder, like the ſtructures already erected, into the ſand of which they were compoſed; but which may ſtand unimpaired, like the Newtonian philoſophy, a rock amid the waſte of ages!

# ADDITIONS.

## ADDITION I.

AT the end of the article Canities, in Clafs I. 2. 2. 11. pleafe to add the following:

As mechanical injury from a percuffion, or a wound, or a cauftic, is liable to occafion the hair of the part to become grey; fo I fufpect the compreffion of parts againft each other of fome animals in the womb is liable to render the hair of thofe parts of a lighter colour; as feems often to occur in black cats and dogs. A fmall terrier bitch now ftands by me, which is black on all thofe parts, which were external, when fhe was wrapped up in the uterus, teres atque rotunda; and thofe parts white, which were moft conftantly preffed together; and thofe parts tawny, which were generally but lefs conftantly preffed together. Thus the hair of the back from the forehead to the end of the tail is black, as well as that of the fides, and external parts of the legs, both before and behind.

As in the uterus the chin of the whelp is bent down, and lies in contact with the fore part of the neck and breaft; the tail is applied clofe againft the divifion of the thighs behind; the infide of the hinder

hinder thighs are preſſed cloſe to the ſides of the belly, all theſe parts have white hairs.

The fore-legs in the uterus lie on each ſide of the face; ſo that the feet cover part of the temples, and compreſs the prominent part of the upper eye brows, but are ſo placed as to defend the eye-balls from preſſure; it is curious to obſerve, that the hair of the ſides of the face, and of the prominent upper eye-brows, are tawny, and of the inſide of the feet and legs, which covered them; for as this poſture admitted of more change in the latter weeks of geſtation, the colour of theſe parts is not ſo far removed from black, as of thoſe parts, where the contact or compreſſion was more uniform.

Where this uterine compreſſion of parts has not been ſo great as to render the hair white in other animals, it frequently happens, that the extremities of the body are white, as the feet, and noſes, and tips of the ears of dogs and cats and horſes, where the circulation is naturally weaker; whence it would ſeem, that the capillary glands, which form the hair, are impeded in the firſt inſtance by compreſſion, and in the laſt by the debility of the circulation in them. See Claſs I. 1. 2. 15.

This day, Auguſt 8th, 1794, I have ſeen a negro, who was born (as he reports) of black parents, both father and mother, at Kingſton in Jamaica, who has many large white blotches on the ſkin of his limbs and body; which I thought felt not ſo ſoft to the finger, as the black parts. He has a white divergent blaze from the ſummit of his noſe to the vertex of his head; the upper part of which, where it extends on the hairy ſcalp, has thick curled hair, like the other part of his head, but quite white. By theſe marks I ſuppoſed him to be the ſame black, who is deſcribed, when only two years old, in the Tranſactions of the American Philoſophical Society, Vol. II. page 292, where a female one is likewiſe deſcribed with nearly ſimilar marks.

The joining of the frontal bones, and the bregma, having been

 later

later than that of the other futures of the cranium, probably gave cauſe to the whiteneſs of the hair on theſe parts by delaying or impeding its growth.

## ADDITION II.

The following extract from a letter of Dr. Beddoes on hydrocephalus internus, I eſteem a valuable addition to the article on that ſubject at Claſs I. 2. 3. 12.

"Maſter L———, aged 9 years, became ſuddenly ill in the night about a week before I ſaw him. On the day before the attack, he had taken opening medicines, and had bathed afterwards. He had complained of violently acute pain in his head, ſhrieked frequently, ground his teeth hard, could not bear to have his head raiſed from the pillow, and was torpid or deaf. His tongue was white, pulſe 110 in the evening and full. As yet the pupil of the eye was irritable, and he had no ſtrabiſmus. He had been bled with leeches about the head, and bliſtered. I directed mercurial inunction, and calomel from 3 to 6 grains to be taken at firſt every ſix, and afterwards every three hours. This plan produced no ſenſible effect, and the patient died on the 18th day after the ſeizure. He had convulſion fits two days preceding his death, and the well-known ſymptoms of hydrocephalus internus all made their appearance. From what I had ſeen and read of this diſeaſe, I believed it to belong to inflammations, and at an earlier period I ſhould be tempted to bleed as largely as for pneumonia. The fluid found after death in the ventricules of the brain I impute to debility of the abſorbents induced by inflammation. My reaſons are briefly theſe; 1. The acuteneſs of the pain. 2. The ſtate of the pulſe. In the above caſe for the firſt 9 or 10 days it did not exceed 110, and was full and ſtrong.

3. To

3. To find out whether any febrile alternations took place, Master L.'s feet were frequently felt, and they were found at times cold, and at other times of a dry heat. I have many times seen this disease, but the patients were too young, or too far advanced, to inform me, whether they had chillness succeeded by heat at its onset.

4. The disorders to which the young are more peculiarly liable afford a presumption, that hydrocephalus internus is an inflammatory disease; and this is confirmed by the regularity of the period, within which it finishes its course. And lastly, does it not happen more frequently than is suspected from external injury?

I have just now been well informed, that Dr. Rush has lately cured five out of six patients by copious bleedings. I relate here the reasons for an opinion without pretending to a discovery. Something like this doctrine may be found in certain modern publications, but it is delivered in that vague and diffuse style, which I trust your example will banish from medical literature."

Clifton, near Bristol,
*July* 28, 1795.

To this idea of Dr. Beddoes may be added, that the hydrocele generally succeeds an injury, and consequent inflammation of the bag, which contains it. And that other dropsies, which principally attend inebriates, are consequent to too great action of the mucous membranes by the stimulus of beer, wine, and spirits. And lastly, that as these cases of hydrocephalus end so fatally, a new mode of treating them is much to be desired, and deserves to be seriously attended to.

## ADDITION III. ON VERTIGO.

*To be placed after the additional Note at the end of Vol. I. on this Subject.*

Having reperused the ingenious Essay of Dr. Wells on Single Vision, and his additional observations in the Gentleman's Magazine on the apparent retrogression of objects in vertigo, I am induced to believe, that this apparent retrogression of objects is not always owing to the same cause.

When a person revolves with his eyes closed, till he becomes vertiginous, and then stands still without opening them, he seems for a while to go forward in the same direction. This hallucination of his ideas cannot be owing to ocular spectra, because, as Dr. Wells observes, no such can have been formed; but it must arise from a similar continuance or repetition of ideas belonging to the sense of touch, instead of to the sense of vision; and should therefore be called a tangible, not a visual, vertigo. In common language this belief of continuing to revolve for some time, after he stands still, when a person has turned round for a minute in the dark, would be called a deception of imagination.

Now at this time if he opens his eyes upon a gilt book, placed with other books on a shelf about the height of his eye, the gilt book seems to recede in the contrary direction; though his eyes are at this time kept quite still, as well as the gilt book. For if his eyes were not kept still, other books would fall on them in succession; which, when I repeatedly made the experiment, did not occur; and which thus evinces, that no motion of the eyes is the cause of the apparent retrocession of the gilt book. Why then does it happen?—Certainly

 from

from an hallucination of ideas, or in common language the deception of imagination.

The vertiginous perſon ſtill imagines, that he continues to revolve forwards, after he has opened his eyes; and in conſequence that the objects, which his eyes happen to fall upon, are revolving backward; as they would appear to do, if he was actually turning round with his eyes open. For he has been accuſtomed to obſerve the motions of bodies, whether apparent or real, ſo much more frequently by the eye than by the touch; that the preſent belief of his gyration, occaſioned by the hallucinations of the ſenſe of touch, is attended with ideas of ſuch imagined motions of viſible objects, as have always accompanied his former gyrations, and have thus been aſſociated with the muſcular actions and perceptions of touch, which occurred at the ſame time.

When the remains of colours are ſeen in the eye, they are termed ocular ſpectra; when remaining ſounds are heard in the ear, they may be called auricular murmurs; but when the remaining motions, or ideas, of the ſenſe of touch continue, as in this vertigo of a blindfolded perſon, they have acquired no name, but may be termed evaneſcent titillations, or tangible hallucinations.

Whence I conclude, that vertigo may have for its cauſe either the ocular ſpectra of the ſenſe of viſion, when a perſon revolves with his eyes open; or the auricular murmurs of the ſenſe of hearing, if he is revolved near a caſcade; or the evaneſcent titillations of the ſenſe of touch, if he revolves blindfold. All theſe I ſhould wiſh to call vaniſhing ideas, or ſenſual motions, of thoſe organs of ſenſe; which ideas, or ſenſual motions, have lately been aſſociated in a circle, and therefore for a time continue to be excited. And what are the ideas of colours, when they are excited by imagination or memory, but the repetition of finer ocular ſpectra? What the idea of ſounds, but the repetition of finer auricular murmurs? And what the ideas of tangible objects, but the repetition of finer evaneſcent titillations?

The

The tangible, and the auricular, and the viſual vertigo, are all perceived by many people for a day or two after long travelling in a boat or coach; the motions of the veſſel, or vehicle, or of the ſurrounding objects, and the noiſe of the wheels and oars, occur at intervals of reverie, or at the commencement of ſleep. See Sect. XX. 5. Theſe ideas, or ſenſual motions, of ſight, of hearing, and of touch, are ſucceeded by the ſame effects as the ocular ſpectra, the auricular murmurs, and the evaneſcent titillations above mentioned; that is, by a kind of vertigo, and cannot in that reſpect be diſtinguiſhed from them. Which is a further confirmation of the truth of the doctrine delivered in Sect. III. of this work, that the colours remaining in the eyes, which are termed ocular ſpectra, are ideas, or ſenſual motions, belonging to the ſenſe of viſion, which for too long a time continue their activity.

## ADDITION IV. Of Voluntary Motions.

A correſpondent acquaints me, that he finds difficulty in underſtanding how the convulſions of the limbs in epilepſy can be induced by voluntary exertions. This I ſuſpect firſt to have ariſen from the double meaning of the words "involuntary motions;" which are ſometimes uſed for thoſe motions, which are performed without the interference of volition, as the pulſations of the heart and arteries; and at other times for thoſe actions, which occur, where two counter volitions oppoſe each other, and the ſtronger prevails; as in endeavouring to ſuppreſs laughter, and to ſtop the ſhudderings, when expoſed to cold. Thus when the poet writes,

——video meliora, proboque,
Deteriora ſequor.——

 The

The ſtronger volition actuates the ſyſtem, but not without the counteraction of unavailing ſmaller ones; which conſtitute deliberation.

A ſecond difficulty may have ariſen from the confined uſe of the words "to will," which in common diſcourſe generally mean to chooſe after deliberation; and hence our will or volition is ſuppoſed to be always in our own power. But the will or voluntary power, acts always from motive, as explained in Sect. XXXIV. 1. and in Claſs IV. 1. 3. 2. and III. 2. 1. 12. which motive can frequently be examined previous to action, and balanced againſt oppoſite motives, which is called deliberation; at other times the motive is ſo powerful as immediately to excite the ſenſorial power of volition into action, without a previous balancing of oppoſite motives, or counter volitions. The former of theſe volitions is exerciſed in the common purpoſes of life, and the latter in the exertions of epilepſy and inſanity.

It is difficult *to think without words*, which however all thoſe muſt do, who diſcover new truths by reaſoning; and ſtill more difficult, when the words in common uſe deceive us by their twofold meanings, or by the inaccuracy of the ideas, which they ſuggeſt.

## ADDITION V. OF FIGURE.

I feel myſelf much obliged by the accurate attention given to the firſt volume of Zoonomia, and by the ingenious criticiſms beſtowed on it, by the learned writers of that article both in the Analytical and Engliſh Reviews. Some circumſtances, in which their ſentiments do not accord with thoſe expreſſed in the work, I intend to reconſider, and to explain further at ſome future time. One thing, in which both theſe gentlemen ſeem to diſſent from me, I ſhall now mention, it is concerning the manner, in which we acquire the idea of figure; a circumſtance of great importance in the knowledge of our intellect,



as

as it ſhews the cauſe of the accuracy of our ideas of motion, time, ſpace, number, and of the mathematical ſciences, which are concerned in the menſurations or proportions of figure.

This I imagine may have in part ariſen from the prepoſſeſſion, which has almoſt univerſally prevailed, that ideas are immaterial beings, and therefore poſſeſs no properties in common with ſolid matter. Which I ſuppoſe to be a fanciful hypotheſis, like the ſtories of ghoſts and apparitions, which have ſo long amuſed, and ſtill amuſe, the credulous without any foundation in nature.

The exiſtence of our own bodies, and of their ſolidity, and of their figure, and of their motions, is taken for granted in my account of ideas; becauſe the ideas themſelves are believed to conſiſt of motions or configurations of ſolid fibres; and the queſtion now propoſed is, how we become acquainted with the figures of bodies external to our organs of ſenſe? Which I can only repeat from what is mentioned in Sect. XIV. 2. 2. that if part of an organ of ſenſe be ſtimulated into action, as of the ſenſe of touch, that part ſo ſtimulated into action muſt poſſeſs figure, which muſt be ſimilar to the figure of the body, which ſtimulates it.

Another previous prepoſſeſſion of the mind, which may have rendered the manner of our acquiring the knowledge of figure leſs intelligible, may have ariſen from the common opinion of the perceiving faculty reſiding in the head; whereas our daily experience ſhews, that our perception (which conſiſts of an idea, and of the pleaſure or pain it occaſions) exiſts principally in the organ of ſenſe, which is ſtimulated into action; as every one, who burns his finger in the candle, muſt be bold to deny.

When an ivory triangle is preſſed on the palm of the hand, the figure of the ſurface of the part of the organ of touch thus compreſſed is a triangle, reſembling in figure the figure of the external body, which compreſſes it. The action of the ſtimulated fibres, which conſtitute the idea of hardneſs and of figure, remains in this part of the ſenſorium,

ſenſorium, which forms the ſenſe of touch; but the ſenſorial motion, which conſtitutes pleaſure or pain, and which is excited in conſequence of theſe fibrous motions of the organ of ſenſe, is propagated to the central parts of the ſenſorium, or to the whole of it; though this generally occurs in leſs degree of energy, than it exiſts in the ſtimulated organ of ſenſe; as in the inſtance above mentioned of burning a finger in the candle.

Some, who have eſpouſed the doctrine of the immateriality of ideas, have ſeriouſly doubted the exiſtence of a material world, with which only our ſenſes acquaint us; and yet have aſſented to the exiſtence of ſpirit, with which our ſenſes cannot acquaint us; and have finally allowed, that all our knowledge is derived through the medium of our ſenſes! They forget, that if the ſpirit of animation had no properties in common with matter, it could neither affect nor be affected by the material body. But the knowledge of our own material exiſtence being granted, which I ſuſpect few rational perſons will ſeriouſly deny, the exiſtence of a material external world follows in courſe; as our perceptions, when we are awake and not inſane, are diſtinguiſhed from thoſe excited by ſenſation, as in our dreams, and from thoſe excited by volition or by aſſociation as in inſanity and reverie, by the power we have of comparing the preſent perceptions of one ſenſe with thoſe of another, as explained in Sect. XIV. 2. 5. And alſo by comparing the tribes of ideas, which the ſymbols of pictures, or of languages, ſuggeſt to us, by intuitive analogy with our previous experience, that is, with the common courſe of nature. See Claſs III. 2. 2. 3. on Credulity.

## ADDITION VI.

*Please to add the following in page 14, after line 20.*

### *Cold and hot Fit.*

As the torpor, with which a fit of fever commences, is sometimes owing to defect of stimulus, as in going into the cold-bath; and sometimes to a previous exhaustion of the sensorial power by the action of some violent stimulus, as after coming out of a hot room into cold air; a longer time must elapse, before there can be a sufficient accumulation of sensorial power to produce a hot fit in one case than in the other. Because in the latter case the quantity of sensorial power previously expended must be supplied, before an accumulation can begin.

The cold paroxysm commences, when the torpor of a part becomes so great, and its motions in consequence so slow or feeble, as not to excite the sensorial power of association; which in health contributes to move the rest of the system, which is catenated with it. And the hot fit commences by the accumulation of the sensorial power of irritation of the part first affected, either so as to counteract its deficient stimulus, or its previous waste of sensorial power; and it becomes general by the accumulation of the sensorial power of association; which is excited by the renovated actions of the part first affected; or becomes so great as to overbalance the deficient excitement of it. On all these accounts the hot fit cannot be supposed to bear any proportion to the cold one in length of time, though the latter may be the consequence of the former. See Suppl. I. 16. 8.

## ADDITION VII. ON WARMTH.

*To be added at the end of the Species Sudor Calidus, in Clafs* I. 1. 2. 3.

WHEN the heat of the body in weak patients in fevers is increafed by the ftimulus of the points of flannel, a greater confequent debility fucceeds, than when it is produced by the warmth of fire; as in the former the heat is in part owing to the increafed activity of the fkin, and confequent expenditure of fenforial power; whereas in the latter cafe it is in part owing to the influx of the fluid matter of heat.

So the warmth produced by equitation, or by rubbing the body and limbs with a fmooth brufh or hand, as is done after bathing in fome parts of the Eaft, does not expend nearly fo much fenforial power, as when the warmth is produced by the locomotion of the whole weight of the body by mufcular action, as in walking, or running, or fwimming. Whence the warmth of a fire is to be preferred to flannel fhirts for weak people, and the agitation of a horfe to exercife on foot. And I fuppofe thofe, who are unfortunately loft in fnow, who are on foot, are liable to perifh fooner by being exhaufted by their mufcular exertions; and might frequently preferve themfelves by lying on the ground, and covering themfelves with fnow, before they were too much exhaufted by fatigue. See Botan. Garden, Vol. II. the note on Barometz.

## ADDITION VIII. PUERPERAL FEVER.

*To be added to Class* II. 1. 6. 16.

A very interesting account of the puerperal fever, which was epidemic at Aberdeen, has been lately published by Dr. Alexander Gordon. (Robinson, London.) In several dissections of those, who died of this disease, purulent matter was found in the cavity of the abdomen; which he ascribes to an erysipelatous inflammation of the peritonæum, as its principal seat, and of its productions, as the omentum, mesentery, and peritonæal coat of the intestines.

He believes, that it was infectious, and that the contagion was always carried by the accoucheur or the nurse from one lying-in woman to another.

The disease began with violent unremitting pain of the abdomen on the day of delivery, or the next day, with shuddering, and very quick pulse, often 140 in a minute. In this situation, if he saw the patient within 12 or 24 hours of her seizure, he took away from 16 to 24 ounces of blood, which was always sizy. He then immediately gave a cathartic consisting of three grains of calomel, and 40 grains of powder of jalap. After this had operated, he gave an opiate at night; and continued the purging and the opiate for several days.

He asserts, that almost all those, whom he was permitted to treat in this manner early in the disease, recovered to the number of 50; and that almost all the rest died. But that when two or three days were elapsed, the patient became too weak for this method; and the matter was already formed, which destroyed them. Except that he saw two patients, who recovered after discharging a large quantity of matter at the navel. And a few, who were relieved by the appearance of external erysipelas on the extremities.

This

This diſeaſe, conſiſting of an eryſipelatous inflammation, may occaſion the great debility ſooner to occur than in inflammation of the uterus; which latter is neither eryſipelatous, I ſuppoſe, nor contagious. And the ſucceſs of Dr. Gordon's practice ſeems to correſpond with that of Dr. Ruſh in the contagious fever or plague at Philadelphia; which appeared to be much aſſiſted by early evacuations. One caſe I ſaw ſome time ago, where violent unceaſing pain of the whole abdomen occurred a few hours after delivery, with quick pulſe; which ceaſed after the patient had twice loſt about eight ounces of blood, and had taken a moderate cathartic with calomel.

This caſe induces me to think, that it might be ſafer and equally efficacious, to take leſs blood at firſt, than Dr. Gordon mentions, and to repeat the operation in a few hours, if the continuance of the ſymptoms ſhould require it. And the ſame in reſpect to the cathartic, which might perhaps be given in leſs quantity, and repeated every two or three hours.

Nor ſhould I wiſh to give an opiate after the firſt veneſection and cathartic; as I ſuſpect that this might be injurious, except thoſe evacuations had emptied the veſſels ſo much, that the ſtimulus of the opiate ſhould act only by increaſing the abſorption of the new veſſels or fluids produced on the ſurfaces of the inflamed membranes. In other inflammations of the bowels, and in acute rheumatiſm, I have ſeen the diſeaſe much prolonged, and I believe ſometimes rendered fatal, by the too early adminiſtration of opiates, either along with cathartics, or at their intervals; while a ſmall doſe of opium given after ſufficient evacuations produces abſorption only by its ſtimulus, and much contributes to the cure of the patient. We may have viſible teſtimony of this effect of opium, when a ſolution of it is put into an inflamed eye; if it be thus uſed previous to ſufficient evacuation, it increaſes the inflammation; if it be uſed after ſufficient evacuation, it increaſes abſorption only, and clears the eye in a very ſmall time.

6 I cannot

I cannot omit obſerving, from conſidering theſe circumſtances, how unwiſe is the common practice of giving an opiate to every woman immediately after her delivery, which muſt often have been of dangerous conſequence.

END OF THE SECOND PART.

---

# ZOONOMIÆ AUCTORI

*S. P. D.*

AMICUS.

---

*CURRUS TRIUMPHALIS MEDICINÆ.*

---

Currus it Hygeiæ, Medicus movet arma triumphans,
Undique victa fugit lurida turma mali.——
Laurea dum Phœbi viridis tua tempora cingit,
Nec mortale ſonans Fama coronat opus;
Poſt equitat trepidans, repetitque Senectus in aurem,
Voce canens ſtridulâ, " ſis memor ipſe mori!"

# INDEX

OF THE

# CLASSES.

## B.



## C.

## D.

E.



F.

## G.



## H.

## I.



## L.



Leg,

## P.



Placenta,

R.



S.

## T.

## U.



## V.

W.



Y.



Z.

# ZOONOMIA;

OR,

# THE LAWS OF ORGANIC LIFE.

## PART III.

CONTAINING

THE ARTICLES OF THE MATERIA MEDICA,

WITH AN ACCOUNT OF THE

OPERATION OF MEDICINES.

---

IN VIVUM CORPUS
AGUNT MEDICAMENTA.

---

# PREFACE.

THE MATERIA MEDICA includes all thoſe ſubſtances, which may contribute to the reſtoration of health. Theſe may be conveniently diſtributed under ſeven articles according to the diverſity of their operations.

1. NUTRIENTIA, or thoſe things which preſerve in their natural ſtate the due exertions of all the irritative motions.

2. INCITANTIA, or thoſe things which increaſe the exertions of all the irritative motions.

3. SECERNENTIA, or thoſe things which increaſe the irritative motions, which conſtitute ſecretion.

4. SORBENTIA, or thoſe things which increaſe the irritative motions, which conſtitute abſorption.

5. INVERTENTIA, or thoſe things which invert the natural order of the ſucceſſive irritative motions.

6. REVERTENTIA, or thoſe things which reſtore the natural order of the inverted irritative motions.

7. Torpentia, thoſe things which diminiſh the exertions of all the irritative motions.

It is neceſſary to apprize the reader, that in the following account of the virtues of Medicines their uſual doſes are always ſuppoſed to be exhibited; and the patient to be expoſed to the degree of exterior heat, which he has been accuſtomed to, (where the contrary is not mentioned), as any variation of either of theſe circumſtances varies their effects.

# ARTICLES

OF THE

# MATERIA MEDICA.

## Art. I.

## NUTRIENTIA.

I. 1. Those things, which preſerve in their natural ſtate the due exertions of all the irritative motions, are termed nutrientia; they produce the growth, and reſtore the waſte, of the ſyſtem. Theſe conſiſt of a variety of mild vegetable and animal ſubſtances, water, and air.

2. Where ſtronger ſtimuli have been long uſed, they become neceſſary for this purpoſe, as muſtard, ſpice, ſalt, beer, wine, vinegar, alcohol, opium. Which however, as they are unnatural ſtimuli, and difficult to manage in reſpect to quantity, are liable to ſhorten the ſpan of human life, ſooner rendering the ſyſtem incapable of being ſtimulated into action by the nutrientia. See Sect. XXXVII. 4. On the ſame account life is ſhorter in warmer climates than in more temperate ones.

 II. Obser-

## II. OBSERVATIONS ON THE NUTRIENTIA.

I. 1. The flesh of animals contains more nourishment, and stimulates our absorbent and secerning vessels more powerfully, than the vegetable productions, which we use as food; for the carnivorous animals can fast longer without injury than the graminivorous; and we feel ourselves warmer and stronger after a meal of flesh than of grain. Hence in diseases attended with cold extremities and general debility this kind of diet is preferred; as in rickets, dropsy, scrophula, and in hysteric and hypochondriac cases, and to prevent the returns of agues. Might not flesh in small quantities bruised to a pulp be more advantageously used in fevers attended with debility than vegetable diet?

That flesh, which is of the darkest colour, generally contains more nourishment, and stimulates our vessels more powerfully, than the white kinds. The flesh of the carnivorous and piscivorous animals is so stimulating, that it seldom enters into the food of European nations, except the swine, the Soland goose (Pelicanus Bassanus), and formerly the swan. Of these the swine and the swan are fed previously upon vegetable aliment; and the Soland goose is taken in very small quantity, only as a whet to the appetite. Next to these are the birds, that feed upon insects, which are perhaps the most stimulating and the most nutritive of our usual food.

It is said that a greater quantity of volatile alkali can be obtained from this kind of flesh, to which has been ascribed its stimulating quality. But it is more probable, that fresh flesh contains only the elements of volatile alkali.

2. Next to the dark coloured flesh of animals, the various tribes of shell-fish seem to claim their place, and the wholesome kinds of mushrooms,

mushrooms, which must be esteemed animal food, both for their alkalescent tendency, their stimulating quality, and the quantity of nourishment, which they afford; as oysters, lobsters, crabfish, shrimps; mushrooms; to which perhaps might be added some of the fish without scales; as the eel, barbolt, tench, smelt, turbot, turtle.

The flesh of many kinds of fish, when it is supposed to have undergone a beginning putrefaction, becomes luminous in the dark. This seems to shew a tendency in the phosphorus to escape, and combine with the oxygen of the atmosphere; and would hence shew, that this kind of flesh is not so perfectly animalized as those before mentioned. This light, as it is frequently seen on rotten wood, and sometimes on veal, which has been kept too long, as I have been told, is commonly supposed to have its cause from putrefaction; but is nevertheless most probably of phosphoric origin, like that seen in the dark on oyster-shells, which have previously been ignited, and afterwards exposed to the sunshine, and on the Bolognian stone. See Botan. Gard. Vol. I. Cant. I. line 1 and 2, the note.

3. The flesh of young animals, as of lamb, veal, and sucking pigs, supplies us with a still less stimulating food. The broth of these is said to become sour, and continues so a considerable time before it changes into putridity; so much does their flesh partake of the chemical properties of the milk, with which these animals are nourished.

4. The white meats, as of turkey, partridge, pheasant, fowl, with their eggs, seem to be the next in mildness; and hence are generally first allowed to convalescents from inflammatory diseases.

5. Next to those should be ranked the white river-fish, which have scales, as pike, perch, gudgeon.

II. 1. Milk unites the animal with the vegetable ſource of our nouriſhment, partaking of the properties of both. As it contains ſugar, and will therefore ferment and produce a kind of wine or ſpirit, which is a common liquor in Siberia; or will run into an acid by ſimple agitation, as in the churning of cream; and laſtly, as it contains coagulable lymph, which will undergo the proceſs of putrefaction like other animal ſubſtances, as in old cheeſe.

2. Milk may be ſeparated by reſt or by agitation into cream, butter, butter-milk, whey, curd. The cream is eaſier of digeſtion to adults, becauſe it contains leſs of the coagulum or cheeſy part, and is alſo more nutritive. Butter conſiſting of oil between an animal and vegetable kind contains ſtill more nutriment, and in its recent ſtate is not difficult of digeſtion if taken in moderate quantity. See Art. I. 2. 3. 2. Butter-milk if it be not bitter is an agreeable and nutritive fluid, if it be bitter it has ſome putrid parts of the cream in it, which had been kept too long; but is perhaps not leſs wholeſome for being ſour to a certain degree: as the inferior people in Scotland chooſe ſour milk in preference to ſkimmed milk before it is become ſour. Whey is the leaſt nutritive and eaſieſt of digeſtion. And in the ſpring of the year, when the cows feed on young graſs, it contains ſo much of vegetable properties, as to become a ſalutary potation, when drank to about a pint every morning to thoſe, who during the winter have taken too little vegetable nouriſhment, and who are thence liable to bilious concretions.

3. Cheeſe is of various kinds, according to the greater or leſs quantity of cream, which it contains, and according to its age. Thoſe cheeſes, which are eaſieſt broken to pieces in the mouth, are generally eaſieſt of digeſtion, and contain moſt nutriment. Some kinds of cheeſe, though ſlow of digeſtion, are alſo ſlow in changing by chemical

mical processes in the stomach, and therefore will frequently agree well with those, who have a weak digestion; as I have seen toasted cheese vomited up a whole day after it was eaten without having undergone any apparent change, or given any uneasiness to the patient. It is probable a portion of sugar, or of animal fat, or of the gravy of boiled or roasted meat, mixed with cheese at the time of making it, might add to its pleasant and nutritious quality.

4. The reason, why autumnal milk is so much thicker or coagulable than vernal milk, is not easy to understand, but as new milk is in many respects similar to chyle, it may be considered as food already in part digested by the animal it is taken from, and thence supplies a nutriment of easy digestion. But as it requires to be curdled by the gastric acid, before it can enter the lacteals, as is seen in the stomachs of calves, it seems more suitable to children, whose stomachs abound more with acidity, than to adults; but nevertheless supplies good nourishment to many of the latter, and particularly to those, who use vegetable food, and whose stomachs have not been much accustomed to the unnatural stimulus of spice, salt, and spirit. See Class I. 1. 2. 5.

III. 1. The seeds, roots, leaves, and fruits of plants, constitute the greatest part of the food of mankind; the respective quantities of nourishment, which these contain, may perhaps be estimated from the quantity of starch, or of sugar, they can be made to produce: in farinaceous seeds, the mucilage seems gradually to be converted into starch, while they remain in our granaries; and the starch by the germination of the young plant, as in making malt from barley, or by animal digestion, is converted into sugar. Hence old wheat and beans contain more starch than new; and in our stomachs other ve-

 getable

getable and animal materials are converted into ſugar; which conſtitutes in all creatures a part of their chyle.

Hence it is probable, that ſugar is the moſt nutritive part of vegetables; and that they are more nutritive, as they are convertible in greater quantity into ſugar by the power of digeſtion; as appears from ſugar being found in the chyle of all animals, and from its exiſting in great quantity in the urine of patients in the diabætes, of which a curious caſe is related in Sect. XXIX. 4. where a man labouring under this malady eat and drank an enormous quantity, and ſometimes voided ſixteen pints of water in a day, with an ounce of ſugar in each pint.

2. Oil, when mixed with mucilage or coagulable lymph, as in cream or new milk, is eaſy of digeſtion, and conſtitutes probably the moſt nutritive part of animal diet; as oil is another part of the chyle of all animals. As theſe two materials, ſugar and butter, contain much nutriment under a ſmall volume, and readily undergo ſome chemical change ſo as to become acid or rancid; they are liable to diſturb weak ſtomachs, when taken in large quantity, more than aliment, which contains leſs nouriſhment, and is at the ſame time leſs liable to chemical changes; becauſe the chyle is produced quicker than the torpid lacteals can abſorb it, and thence undergoes a further chemical proceſs. Sugar and butter therefore are not ſo eaſily digeſted, when taken in large quantity, as thoſe things, which contain leſs nutriment; hence, where the ſtomach is weak, they muſt be uſed in leſs quantity. But the cuſtom of ſome people in reſtraining children entirely from them, is depriving them of a very wholeſome, agreeable, and ſubſtantial part of their diet. Honey, manna, ſap-juice, are different kinds of leſs pure ſugar.

3. All the eſculent vegetables contain a bland oil, or mucilage, or ſtarch,

ſtarch, or ſugar, or acid; and, as their ſtimulus is moderate, are properly given alone as food in inflammatory diſeaſes; and mixed with milk conſtitute the food of thouſands. Other vegetables poſſeſs various degrees and various kinds of ſtimulus; and to theſe we are beholden for the greater part of our Materia Medica, which produce nauſea, ſickneſs, vomiting, catharſis, intoxication, inflammation, and even death, if unſkilfully adminiſtered.

The acrid or intoxicating, and other kinds of vegetable juices, ſuch as produce ſickneſs, or evacuate the bowels, or ſuch even as are only diſagreeable to the palate, appear to be a part of the defence of thoſe vegetables, which poſſeſs them, from the aſſaults of larger animals or of inſects. As mentioned in the Botanic Garden, Part II. Cant. I. line 161, note. This appears in a forcible manner from the peruſal of ſome travels, which have been publiſhed of thoſe unfortunate people, who have ſuffered ſhipwreck on uncultivated countries, and have with difficulty found food to ſubſiſt, in otherwiſe not inhoſpitable climates.

4. As theſe acrid and intoxicating juices generally reſide in the mucilage, and not in the ſtarch of many roots, and ſeeds, according to the obſervation of M. Parmentier, the wholeſome or nutritive parts of ſome vegetables may be thus ſeparated from the medicinal parts of them. Thus if the root of white briony be raſped into cold water, by means of a bread-grater made of a tinned iron plate, and agitated in it, the acrid juice of the root along with the mucilage will be diſſolved, or ſwim, in the water; while a ſtarch perfectly wholeſome and nutritious will ſubſide, and may be uſed as food in times of ſcarcity.

M. Parmentier further obſerves, that potatoes contain too much mucilage in proportion to their ſtarch, which prevents them from being converted into good bread. But that if the ſtarch be collected from ten pounds of raw potatoes by grating them into cold water, and agitating them, as above mentioned; and if the ſtarch thus procured be mixed with other ten pounds of boiled potatoes, and properly ſubjected to fermentation like wheat flour, that it will make as good bread as the fineſt wheat.

Good bread may alſo be made by mixing wheat-flour with boiled potatoes. Eighteen pounds of wheat flour are ſaid to make twenty-two pounds and a half of bread. Eighteen pounds of wheat-flour mixed with nine pounds of boiled potatoes, are ſaid to make twenty-nine pounds and a half of bread. This difference of weight muſt ariſe from the difference of the previous dryneſs of the two materials. The potatoes might probably make better flour, if they were boiled in ſteam, in a cloſe veſſel, made ſome degrees hotter than common boiling water.

Other vegetable matters may be deprived of their too great acrimony by boiling in water, as the great variety of the cabbage, the young tops of white briony, water-creſſes, aſparagus, with innumerable roots, and ſome fruits. Other plants have their acrid juices or bitter particles diminiſhed by covering them from the light by what is termed blanching them, as the ſtems and leaves of cellery, endive, ſea-kale. The former method either extracts or decompoſes the acrid particles, and the latter prevents them from being formed. See Botanic Garden, Vol. I. additional note XXXIV. on the Etiolation of vegetables.

5. The art of cookery, by expoſing vegetable and animal ſubſtances to heat, has contributed to increaſe the quantity of the food of mankind by other means beſides that of deſtroying their acrimony. One of theſe is by converting the acerb juices of ſome fruits into ſugar, as in the baking of unripe pears, and the bruiſing of unripe apples; in both which ſituations the life of the vegetable is deſtroyed, and the converſion

conversion of the harsh juice into a sweet one must be performed by a chemical process; and not by a vegetable one only, as the germination of barley in making malt has generally been supposed.

Some circumstances, which seem to injure the life of several fruits, seem to forward the saccharine process of their juices. Thus if some kinds of pears are gathered a week before they would ripen on the tree, and are laid on a heap and covered, their juice becomes sweet many days sooner. The taking off a circular piece of the bark from a branch of a pear-tree causes the fruit of that branch to ripen sooner by a fortnight, as I have more than once observed. The wounds made in apples by insects occasion those apples to ripen sooner; caprification, or the piercing of figs, in the island of Malta, is said to ripen them sooner; and I am well informed, that when bunches of grapes in this country have acquired their expected size, that if the stalk of each bunch be cut half through, that they will sooner ripen.

The germinating barley in the malt-house I believe acquires little sweetness, till the life of the seed is destroyed, and the saccharine process then continued or advanced by the heat in drying it. Thus in animal digestion, the sugar produced in the stomach is absorbed by the lacteals as fast as it is made, otherwise it ferments, and produces flatulency; so in the germination of barley in the malt-house, so long as the new plant lives, the sugar, I suppose, is absorbed as fast as it is made; but that, which we use in making beer, is the sugar produced by a chemical process after the death of the young plant, or which is made more expeditiously, than the plant can absorb it.

It is probably this saccharine process, which obtains in new haystacks too hastily, and which by immediately running into fermentation produces so much heat as to set them on fire. The greatest part of the grain, or seeds, or roots, used in the distilleries, as wheat, canary seed, potatoes, are not I believe previously subjected to germination, but are in part by a chemical process converted into sugar, and immediately subjected to vinous fermentation; and it is probable a process may sometime be discovered of producing sugar

 from

from ſtarch or meal; and of ſeparating it from them for domeſtic purpoſes by alcohol, which diſſolves ſugar but not mucilage; or by other means.

Another method of increaſing the nutriment of mankind by cookery, is by diſſolving cartilages and bones, and tendons, and probably ſome vegetables, in ſteam or water at a much higher degree of heat than that of boiling. This is to be done in a cloſe veſſel, which is called Papin's digeſter; in which, it is ſaid, that water may be made red-hot, and will then diſſolve all animal ſubſtances; and might thus add to our quantity of food in times of ſcarcity. This veſſel ſhould be made of iron, and ſhould have an oval opening at top, with an oval lid of iron larger than the aperture; this lid ſhould be ſlipped in endways, when the veſſel is filled, and then turned, and raiſed by a ſcrew above it into contact with the under edges of the aperture. There ſhould alſo be a ſmall tube or hole covered with a weighted valve to prevent the danger of burſting the digeſter.

Where the powers of digeſtion are weakened, broths made by boiling animal and vegetable ſubſtances in water afford a nutriment; though I ſuppoſe not ſo great as the fleſh and vegetables would afford, if taken in their ſolid form, and mixed with ſaliva in the act of maſtication. The aliment thus prepared ſhould be boiled but a ſhort time, nor ſhould be ſuffered to continue in our common kitchen-utenſils afterwards, as they are lined with a mixture of half lead and half tin, and are therefore unwholeſome, though the copper is completely covered. And thoſe ſoups, which have any acid or wine boiled in them, unleſs they be made in ſilver, or in china, or in thoſe pot-veſſels, which are not glazed by the addition of lead, are truly poiſonous; as the acid, as lemon-juice or vinegar, when made hot, erodes or diſſolves the lead and tin lining of the copper-veſſels, and the leaden glaze of the porcelain ones. Hence, where ſilver cannot be had, iron veſſels are preferable to tinned copper ones; or thoſe made of tinned iron-plates in the common tin-ſhops, which are ſaid to be covered with pure or block tin.

6 6. Another

6. Another circumſtance, which facilitates the nouriſhment of mankind, is the mechanic art of grinding farinaceous ſeeds into powder between mill-ſtones; which may be called the artificial teeth of ſociety. It is probable, that ſome ſoft kinds of wood, eſpecially when they have undergone a kind of fermentation, and become of looſer texture, might be thus uſed as food in times of famine.

Nor is it improbable, that hay, which has been kept in ſtacks, ſo as to undergo the ſaccharine proceſs, may be ſo managed by grinding and by fermentation with yeaſt like bread, as to ſerve in part for the ſuſtenance of mankind in times of great ſcarcity. Dr. Prieſtley gave to a cow for ſome time a ſtrong infuſion of hay in large quantity for her drink, and found that ſhe produced during this treatment above double the quantity of milk. Hence if bread cannot be made from ground hay, there is great reaſon to ſuſpect, that a nutritive beverage may be thus prepared either in its ſaccharine ſtate, or fermented into a kind of beer.

In times of great ſcarcity there are other vegetables, which though not in common uſe, would moſt probably afford wholeſome nouriſhment, either by boiling them, or drying and grinding them, or by both thoſe proceſſes in ſucceſſion. Of theſe are perhaps the tops and the bark of all thoſe vegetables, which are armed with thorns or prickles, as gooſeberry trees, holly, gorſe, and perhaps hawthorn. The inner bark of the elm tree makes a kind of gruel. And the roots of fern, and probably of very many other roots, as of graſs and of clover taken up in winter, might yield nouriſhment either by boiling or baking, and ſeparating the fibres from the pulp by beating them; or by getting only the ſtarch from thoſe, which poſſeſs an acrid mucilage, as the white briony.

7. However the arts of cookery and of grinding may increaſe or facilitate the nouriſhment of mankind, the great ſource of it is from agriculture. In the ſavage ſtate, where men live ſolely by hunting,

I. was

I was informed by Dr. Franklin, that there was ſeldom more than one family exiſted in a circle of five miles diameter; which in a ſtate of paſturage would ſupport ſome hundred people, and in a ſtate of agriculture many thouſands. The art of feeding mankind on ſo ſmall a grain as wheat, which ſeems to have been diſcovered in Egypt by the immortal name of Ceres, ſhewed greater ingenuity than feeding them with the large roots of potatoes, which ſeem to have been a diſcovery of ill-fated Mexico.

This greater production of food by agriculture than by paſturage, ſhews that a nation nouriſhed by animal food will be leſs numerous than if nouriſhed by vegetable; and the former will therefore be liable, if they are engaged in war, to be conquered by the latter, as Abel was ſlain by Cain. This is perhaps the only valid argument againſt incloſing open arable fields. The great production of human nouriſhment by agriculture and paſturage evinces the advantage of ſociety over the ſavage ſtate; as the number of mankind becomes increaſed a thouſand fold by the arts of agriculture and paſturage; and their happineſs is probably under good governments improved in as great a proportion, as they become liberated from the hourly fear of beaſts of prey, from the daily fear of famine, and of the occaſional incurſions of their cannibal neighbours.

But paſturage cannot exiſt without property both in the ſoil, and the herds which it nurtures; and for the invention of arts, and production of tools neceſſary to agriculture, ſome muſt think, and others labour; and as the efforts of ſome will be crowned with greater ſucceſs than that of others, an inequality of the ranks of ſociety muſt ſucceed; but this inequality of mankind in the preſent ſtate of the world is too great for the purpoſes of producing the greateſt quantity of human nouriſhment, and the greateſt ſum of human happineſs; there ſhould be no ſlavery at one end of the chain of ſociety, and no deſpotiſm at the other.—By the future improvements of human reaſon ſuch governments may poſſibly hereafter be eſtabliſhed, as may a hundred-

hundred-fold increaſe the numbers of mankind, and a thouſand-fold their happineſs.

IV. 1. Water muſt be conſidered as a part of our nutriment, becauſe ſo much of it enters the compoſition of our ſolids as well as of our fluids; and becauſe vegetables are now believed to draw almoſt the whole of their nouriſhment from this ſource. As in them the water is decompoſed, as it is perſpired by them in the ſunſhine, the oxygen gas increaſes the quantity and the purity of the atmoſphere in their vicinity, and the hydrogen ſeems to be retained, and to form the nutritive juices, and conſequent ſecretions of roſin, gum, wax, honey, oil, and other vegetable productions. See Botanic Garden, Part I. Cant. IV. line 25, note. It has however other uſes in the ſyſtem, beſides that of a nouriſhing material, as it dilutes our fluids, and lubricates our ſolids; and on all theſe accounts a daily ſupply of it is required.

2. River-water is in general purer than ſpring-water; as the neutral ſalts waſhed down from the earth decompoſe each other, except perhaps the marine ſalt; and the earths, with which ſpring-water frequently abounds, is precipitated; yet it is not improbable, that the calcareous earth diſſolved in the water of many ſprings may contribute to our nouriſhment, as the water from ſprings, which contain earth, is ſaid to conduce to enrich thoſe lands, which are flooded with it, more than river water.

3. Many arguments ſeem to ſhew, that calcareous earth contributes to the nouriſhment of animals and vegetables. Firſt becauſe calcareous earth conſtitutes a conſiderable part of them, and muſt therefore either be received from without, or formed by them, or both, as milk, when taken as food by a lacteſcent woman, is decompoſed in the ſtomach by the proceſs of digeſtion, and again in part converted

into

into milk by the pectoral glands. Secondly, becaufe from the analogy of all organic life, whatever has compofed a part of a vegetable or animal may again after its chemical folution become a part of another vegetable or animal, fuch is the general tranfmigration of matter. And thirdly, becaufe the great ufe of lime in agriculture on almoft all kinds of foil and fituation cannot be fatisfactorily explained from its chemical properties alone. Though thefe may alfo in certain foils and fituations have confiderable effect.

The chemical ufes of lime in agriculture may be, 1. from its deftroying in a fhort time the cohefion of dead vegetable fibres, and thus reducing them to earth, which otherwife is effected by a flow procefs either by the confumption of infects or by a gradual putrefaction. Thus I am informed that a mixture of lime with oak bark, after the tanner has extracted from it whatever is foluble in water, will in two or three months reduce it to a fine black earth, which, if only laid in heaps, would require as many years to effect by its own fpontaneous fermentation or putrefaction. This effect of lime muft be particularly advantageous to newly inclofed commons when firft broken up.

Secondly, lime for many months continues to attract moifture from the air or earth, which it deprives I fuppofe of carbonic acid, and then fuffers it to exhale again, as is feen on the plaftered walls of new houfes. On this account it muft be advantageous when mixed with dry or fandy foils, as it attracts moifture from the air above or the earth beneath, and this moifture is then abforbed by the lymphatics of the roots of vegetables. Thirdly, by mixing lime with clays it is believed to make them lefs cohefive, and thus to admit of their being more eafily penetrated by vegetable fibres. A mixture of lime with clays deftroys their fuperabundancy of acid, if fuch exifts, and by uniting with it converts it into gypfum or alabafter. And laftly, frefh lime deftroys worms, fnails, and other infects, with which it happens to come in contact.

Yet do not all thefe chemical properties feem to account for the great

great uſes of lime in almoſt all ſoils and ſituations, as it contributes ſo much to the melioration of the crops, as well as to their increaſe in quantity. Wheat from land well limed is believed by farmers, millers, and bakers, to be, as they ſuppoſe, thinner ſkinned; that is, it turns out more and better flour; which I ſuppoſe is owing to its containing more ſtarch and leſs mucilage. In reſpect to graſs-ground I am informed, that if a ſpadeful of lime be thrown on a tuſſock, which horſes or cattle have refuſed to touch for years, they will for many ſucceeding ſeaſons eat it quite cloſe to the ground.

One property of lime is not perhaps yet well underſtood, I mean its producing ſo much heat, when it is mixed with water; which may be owing to the elementary fluid of heat conſolidated in the lime. It is the ſteam occaſioned by this heat, when water is ſprinkled upon lime, if the water be not in too great quantity or too cold, which breaks the lime into ſuch fine powder as almoſt to become fluid, which cannot be effected perhaps by any other means, and which I ſuppoſe muſt give great preference to lime in agriculture, and to the ſolutions of calcareous earth in water, over chalk or powdered lime-ſtone, when ſpread upon the land.

4. It was formerly believed that waters replete with calcareous earth, ſuch as incruſt the inſide of tea-kettles, or are ſaid to petrify moſs, were liable to produce or to increaſe the ſtone in the bladder. This miſtaken idea has lately been exploded by the improved chemiſtry, as no calcareous earth, or a very minute quantity, was found in the calculi analyſed by Scheel and Bergman. The waters of Matlock and of Carlſbad, both which cover the moſs, which they paſs through, with a calcareous cruſt, are ſo far from increaſing the ſtone of the bladder or kidnies, that thoſe of Carlſbad are celebrated for giving relief to thoſe labouring under theſe diſeaſes. Philoſ. Tranſ. Thoſe of Matlock are drank in great quantities without any ſuſpicion of injury; and I well know a perſon who for above ten years has

 drank

drank about two pints a day of cold water from a ſpring, which very much incruſts the veſſels, it is boiled in, with calcareous earth, and affords a copious calcareous ſediment with a ſolution of ſalt of tartar, and who enjoys a ſtate of uninterrupted health.

V. 1. As animal bodies conſiſt much both of oxygen and azote, which make up the compoſition of atmoſpheric air, theſe ſhould be counted amongſt nutritious ſubſtances. Beſides that by the experiments of Dr. Prieſtley it appears, that the oxygen gains admittance into the blood through the moiſt membranes of the lungs; and ſeems to be of much more immediate conſequence to the preſervation of our lives than the other kinds of nutriment above ſpecified.

As the baſis of fixed air, or carbonic acid gas, is carbone, which alſo conſtitutes a great part both of vegetable and animal bodies; this air ſhould likewiſe be reckoned amongſt nutritive ſubſtances. Add to this, that when this carbonic acid air is ſwallowed, as it eſcapes from beer or cyder, or when water is charged with it as detruded from limeſtone by vitriolic acid, it affords an agreeable ſenſation both to the palate and ſtomach, and is therefore probably nutritive.

The immenſe quantity of carbone and of oxygen which conſtitute ſo great a part of the limeſtone countries is almoſt beyond conception, and, as it has been formed by animals, may again become a part of them, as well as the calcareous matter with which they are united. Whence it may be conceived, that the waters, which abound with limeſtone in ſolution, may ſupply nutriment both to animals and to vegetables, as mentioned above.

VI. 1. The manner, in which nutritious particles are ſubſtituted in the place of thoſe, which are mechanically abraded, or chemically decompoſed, or which vaniſh by animal abſorption, muſt be owing to animal appetency, as deſcribed in Sect. XXXVII. 3. and is probably ſimilar to the proceſs of inflammation, which produces new veſſels and new fluids; or to that which conſtitutes the growth of the body to maturity. Thus the granulations of new fleſh to repair the injuries of wounds are viſible to the eye; as well as the callous matter, which cements broken bones; the calcareous matter, which repairs injured ſnail-ſhells; and the threads, which are formed by ſilk-worms and ſpiders; which are all ſecreted in a ſofter ſtate, and harden by exſiccation, or by the contact of the air, or by abſorption of their more fluid parts.

Whether the materials, which thus ſupply the waſte of the ſyſtem, can be given any other way than by the ſtomach, ſo as to preſerve the body for a length of time, is worth our inquiry; as caſes ſometimes occur, in which food cannot be introduced into the ſtomach, as in obſtructions of the œſophagus, inflammations of the throat, or in hydrophobia; and other caſes are not unfrequent in which the power of digeſtion is nearly or totally deſtroyed, as in anorexia epileptica, and in many fevers.

In the former of theſe circumſtances liquid nutriment may ſometimes be got into the ſtomach through a flexible catheter; as deſcribed in Claſs III. 1. 1. 15. In the latter many kinds of mild aliment, as milk or broth, have frequently been injected as clyſters, together with a ſmall quantity of opium, as ten drops of the tincture, three or four times a day; to which alſo might be added very ſmall quantities of vinous ſpirit. But theſe, as far as I have obſerved, will not long ſuſtain a perſon, who cannot take any ſuſtenance by the ſtomach.

2. Another mode of applying nutritive fluids might be by extenſive fomentations, or by immerging the whole body in a bath of broth, or of warm milk, which might at the ſame time be coagulated by rennet, or the acid of the calf's ſtomach; broth or whey might thus probably be introduced, in part at leaſt, into the circulation, as a ſolution of nitre is ſaid to have been abſorbed in a pediluvium, which was afterwards diſcovered by the manner in which paper dipped fre-

 quently

quently in the urine of the patient and dried, burnt and ſparkled like touch-paper. Great quantity of water is alſo known to be abſorbed by thoſe, who have bathed in the warm bath after exerciſe and abſtinence from liquids. Cleopatra was ſaid to travel with 4000 milch-aſſes in her train, and to bathe every morning in their milk, which ſhe probably might uſe as a coſmetic rather than a nutritive.

3. The transfuſion of blood from another animal into the vein of one, who could take no ſuſtenance by the throat, or digeſt none by the ſtomach, might long continue to ſupport him; and perhaps other nutriment, as milk or mucilage, might be this way introduced into the ſyſtem, but we have not yet ſufficient experiments on this ſubject. See Sect. XXXII. 4. and Claſs I. 2. 3. 25. and Sup. I. 14. 2.

VII. Various kinds of condiments, or ſauces, have been taken along with vegetable or animal food, and have been thought by ſome to ſtrengthen the proceſs of digeſtion and conſequent proceſs of nutrition. Of theſe wine, or other fermented liquors, vinegar, ſalt, ſpices, and muſtard, have been in moſt common uſe, and I believe to the injury of thouſands. As the ſtomach by their violent ſtimulus at length loſes its natural degree of irritability, and indigeſtion is the conſequence; which is attended with flatulency and emaciation. Where any of theſe have been taken ſo long as to induce a habit, they muſt either be continued, but not increaſed; or the uſe of them ſhould be gradually and cautiouſly diminiſhed or diſcontinued, as directed in Sect. XII. 7. 8.

III. CATALOGUE

### III. Catalogue of the Nutrientia.

I. 1. Veniſon, beef, mutton, hare, gooſe, duck, woodcock, ſnipe, moor-game.

2. Oyſters, lobſters, crabs, ſhrimps, muſhrooms, eel, tench, barbolt, ſmelt, turbot, ſole, turtle.

3. Lamb, veal, ſucking-pig.

4. Turkey, partridge, pheaſant, fowl, eggs.

5. Pike, perch, gudgeon, trout, grayling.

II. Milk, cream, butter, buttermilk, whey, cheeſe.

III. Wheat, barley, oats, peas, potatoes, turnips, carrots, cabbage, aſparagus, artichoke, ſpinach, beet, apple, pear, plum, apricot, nectarine, peach, ſtrawberry, grape, orange, melon, cucumber, dried figs, raiſins, ſugar, honey. With a great variety of other roots, ſeeds, leaves, and fruits.

IV. Water, river-water, ſpring-water, calcareous earth.

V. Air, oxygene, azote, carbonic acid gas.

VI. Nutritive baths and clyſters, transfuſion of blood.

VII. Condiments.

Art.

## Art. II.

# INCITANTIA.

I. 1. Those things, which increaſe the exertions of all the irritative motions, are termed incitantia. As alcohol, or the ſpirituous part of fermented liquors, opium, and many drugs, which are ſtill eſteemed poiſons, their proper doſes not being aſcertained. To theſe ſhould be added the exhilarating paſſions of the mind, as joy, love: and externally the application of heat, electricity, æther, eſſential oils, friction, and exerciſe.

2. Theſe promote both the ſecretions and abſorptions, increaſe the natural heat, and remove thoſe pains, which originate from the defect of irritative motions, termed nervous pains; and prevent the convulſions conſequent to them. When given internally they induce coſtiveneſs, and deep coloured urine; and by a greater doſe intoxication, and its conſequences.

## II. Observations on the Incitantia.

I. 1. Opium and alcohol increaſe all the ſecretions and abſorptions. The increaſe of the ſecretion of ſenſorial power appears from the violent exertions of drunken people; the ſecretion of ſweat is more certainly excited by opium or wine than by any other medicine; and the increaſe of general heat, which theſe drugs produce, is an evidence of their effect in promoting all the ſecretions; ſince an increaſe of of ſecretion is always attended with increaſe of heat in the part, as in hepatic and other inflammations.

2. But as they at the ſame time promote abſorption; thoſe fluids, which are ſecreted into receptacles, as the urine, bile, inteſtinal and pulmonary mucus, have again their thinner parts abſorbed; and hence, though the quantity of ſecreted fluid was increaſed, yet as the abſorption was alſo increaſed, the excretion from theſe receptacles is leſſened; at the ſame time that it is deeper coloured or of thicker conſiſtence, as the urine, alvine feces, and pulmonary mucus. Whereas the perſpiration being ſecreted on the ſurface of the body is viſible in its increaſed quantity, before it can be reabſorbed; whence ariſes that erroneous opinion, that opium increaſes the cutaneous ſecretion, and leſſens all the others.

3. It muſt however be noted, that after evacuations opium ſeems to promote the abſorptions more than the ſecretions; if you except that of the ſenſorial power in the brain, which probably ſuffers no abſorption. Hence its efficacy in reſtraining hæmorrhages, after the veſſels are emptied, by promoting venous abſorption.

4. In ulcers the matter is thickened by the exhibition of opium from the increaſed abſorption of the thinner parts of it; but it is probable, that the whole ſecretion, including the part which is abſorbed, is increaſed; and hence new fibres are ſecreted along with the matter, and the ulcer fills with new granulations of fleſh. But as no ulcer can heal, till it ceaſes to diſcharge; that is, till the abſorption becomes as great as the excretion; thoſe medicines, which promote abſorption only, are more advantageous for the healing an ulcer after it is filled with new fleſh; as the Peruvian bark internally; with bandages and ſolutions of lead externally.

5. There

5. There are many pains which originate from a want of due motion in the part, as thoſe occaſioned by cold; and all thoſe pains which are attended with cold extremities, and are generally termed nervous. Theſe are relieved by whatever excites the part into its proper actions, and hence by opium and alcohol; which are the moſt univerſal ſtimulants we are acquainted with. In theſe caſes the effect of opium is produced, as ſoon as the body becomes generally warm; and a degree of intoxication or ſleep follows the ceſſation of the pain.

Theſe nervous pains (as they are called) frequently return at certain periods of time, and are alſo frequently ſucceeded by convulſions; in theſe caſes if opium removes the pain, the convulſions do not come on. For this purpoſe it is beſt to exhibit it gradually, as a grain every hour, or half hour, till it intoxicates. Here it muſt be noted, that a much leſs quantity will prevent the periods of theſe cold pains, than is neceſſary to relieve them after their acceſs. As a grain and half of opium given an hour before the expected paroxyſm will prevent the cold fit of an intermittent fever, but will not ſoon remove it, when it is already formed. For in the former caſe the uſual or healthy aſſociations or catenations of motion favour the effect of the medicine; in the latter caſe theſe aſſociations or catenations are diſordered, or interrupted, and new ones are formed, which ſo far counteract the effect of the medicine.

When opium has been required in large doſes to eaſe or prevent convulſions, ſome have adviſed the patient to omit the uſe of wine, as a greater quantity of opium might then be exhibited; and as opium ſeems to increaſe abſorption more, and ſecretion leſs, than vinous ſpirit; it may in ſome caſes be uſeful to exchange one for the other; as in diſeaſes attended with too great evacuation, as diarrhœa, and dyſentery, opium may be preferable; on the contrary in tetanus, or locked-jaw, where inflammation of the ſyſtem might be of ſervice, wine may be preferable to opium; ſee Claſs III. 1. 1. 12. I have generally obſerved, that a mixture of ſpirit of wine and warm water,

given

given alternately with the doſes of opium, has ſooneſt and moſt certainly produced that degree of intoxication, which was neceſſary to relieve the patient in the epilepſia doloroſica.

6. There is likewiſe ſome relief given by opium to inflammatory pains, or thoſe from exceſs of motion in the affected part; but with this difference, that this relief from the pains, and the ſleep, which it occaſions, does not occur till ſome hours after the exhibition of the opium. This requires to be explained; after the ſtimulus of opium or of alcohol ceaſes, as after common drunkenneſs, a conſequent torpor comes on; and the whole habit becomes leſs irritable by the natural ſtimuli. Hence the head-achs, ſickneſs, and languor, on the next day after intoxication, with cold ſkin, and general debility. Now in pains from exceſs of motion, called inflammatory pains, when opium is given, the pain is not relieved, till the debility comes on after the ſtimulus ceaſes to act; for then after the greater ſtimulus of the opium has exhauſted much of the ſenſorial power; the leſs ſtimulus, which before cauſed the pain, does not now excite the part into unnatural action.

In theſe caſes the ſtimulus of the opium firſt increaſes the pain; and it ſometimes happens, that ſo great a torpor follows, as to produce the death or mortification of the affected part; whence the danger of giving opium in inflammatory diſeaſes, eſpecially in inflammation of the bowels; but in general the pain returns with its former violence, when the torpor above mentioned ceaſes. Hence theſe pains attended with inflammation are beſt relieved by copious veneſection, other evacuations, and the claſs of medicines called torpentia.

7. Theſe pains from exceſs of motion are attended with increaſed heat of the whole, or of the affected part, and a ſtrong quick pulſe; the pains from defect of motion are attended with cold extremities,

and a weak pulse; which is also generally more frequent than natural, but not always so.

8. Opium and alcohol are the only two drugs, we are much acquainted with, which intoxicate; and by this circumstance are easily distinguished from the secernentia and sorbentia. Camphor, and cicuta, and nicotiana, are thought to induce a kind of intoxication; and there are many other drugs of this class, whose effects are less known, or their doses not ascertained; as atropa belladonna, hyocyamus, stramonium, prunus laurocerasus, menispermum, cynoglossum, some fungi, and the water distilled from black cherry-stones; the last of which was once much in use for the convulsions of children, and was said to have good effect; but is now improvidently left out of our pharmacopias. I have known one leaf of the laurocerasus, shred and made into tea, given every morning for a week with no ill consequence to a weak hysteric lady, but rather perhaps with advantage.

9. The pernicious effects of a continued use of much vinous spirit is daily seen and lamented by physicians; not only early debility, like premature age, but a dreadful catalogue of diseases is induced by this kind of intemperance; as dropsy, gout, leprosy, epilepsy, insanity, as described in Botanic Garden, Part II. Canto III. line 357. The stronger or less diluted the spirit is taken, the sooner it seems to destroy, as in dram-drinkers; but still sooner, when kernels of apricots, or bitter almonds, or laurel-leaf, are infused in the spirit, which is termed ratafia; as then two poisons are swallowed at the same time. And vinegar, as it contains much vinous spirit, is probably a noxious part of our diet. And the distilled vinegar, which is commonly sold in the shops, is truly poisonous, as it is generally distilled by means of a pewter or leaden alembic-head or worm-tube, and abounds with lead;

lead; which any one may detect by mixing with it a solution of liver of sulphur. Opium, when taken as a luxury, not as a medicine, is as pernicious as alcohol; as Baron de Tott relates in his account of the opium-eaters in Turkey.

10. It must be observed, that a frequent repetition of the use of this class of medicines so habituates the body to their stimulus, that their dose may gradually be increased to an astonishing quantity, such as otherwise would instantly destroy life; as is frequently seen in those, who accustom themselves to the daily use of alcohol and opium; and it would seem, that these unfortunate people become diseased as soon as they omit their usual potations; and that the consequent gout, dropsy, palsy, or pimpled face, occur from the debility occasioned from the want of accustomed stimulus, or to some change in the contractile fibres, which requires the continuance or increase of it. Whence the cautions necessary to be observed are mentioned in Sect. XII. 7. 8.

11. It is probable, that some of the articles in the subsequent catalogue do not induce intoxication, though they have been esteemed to do so; as tobacco, hemlock, nux vomica, stavisagria; and on this account should rather belong to other arrangements, as to the secernentia, or sorbentia, or invertentia.

II. 1. Externally the application of heat, as the warm bath, by its stimulus on the skin excites the excretory ducts of the perspirative glands, and the mouths of the lymphatics, which open on its surface, into greater action; and in consequence many other irritative motions, which are associated with them. To this increased action is added pleasurable sensation, which adds further activity to the system; and thus many kinds of pain receive relief from this additional atmosphere of heat.

The uſe of a warm bath of about 96 or 98 degrees of heat, for half an hour once a day for three or four months, I have known of great ſervice to weak people, and is perhaps the leaſt noxious of all unnatural ſtimuli; which however, like all other great excitement, may be carried to exceſs, as complained of by the ancients. The unmeaning application of the words relaxation and bracing to warm and cold baths has much prevented the uſe of this grateful ſtimulus; and the miſuſe of the term warm-bath, when applied to baths colder than the body, as to thoſe of Buxton and Matlock, and to artificial baths of leſs than 90 degrees of heat, which ought to be termed cold ones, has contributed to miſlead the unwary in their application.

The ſtimulus of wine, or ſpice, or ſalt, increaſes the heat of the ſyſtem by increaſing all or ſome of the ſecretions; and hence the ſtrength is diminiſhed afterwards by the loſs of fluids, as well as by the increaſed action of the fibres. But the ſtimulus of the warm-bath ſupplies heat rather than produces it; and rather fills the ſyſtem by increaſed abſorption, than empties it by increaſed ſecretion; and may hence be employed with advantage in almoſt all caſes of debility with cold extremities, perhaps even in anaſarca, and at the approach of death in fevers. In theſe caſes a bath much beneath 98 degrees, as of 80 or 85, might do injury, as being a cold-bath compared with the heat of the body, though ſuch a bath is generally called a warm one.

The activity of the ſyſtem thus produced by a bath of 98 degrees of heat, or upwards, does not ſeem to render the patients liable to take cold, when they come out of it; for the ſyſtem is leſs inclined to become torpid than before, as the warmth thus acquired by communication, rather than by increaſed action, continues long without any conſequent chillneſs. Which accords with the obſervation of Dr. Fordyce, mentioned in Sup. I. 5. 1. who ſays, that thoſe who are confined ſome time in an atmoſphere of 120 or 130 degrees of heat, do not feel cold or look pale on coming into a temperature of

30

30 or 40 degrees; which would produce great paleneſs and ſenſation of coldneſs in thoſe, who had been ſome time confined in an atmoſphere of only 86 or 90 degrees of heat. Treatiſe on Simple Fever, p. 168.

Hence heat, where it can be confined on a torpid part along with moiſture, as on a ſcrophulous tumour, will contribute to produce ſuppuration or reſolution. This is done by applying a warm poultice, which ſhould be frequently repeated; or a plaſter of reſin, wax, or fat; or by covering the part with oiled ſilk; both which laſt prevent the perſpirable matter from eſcaping as well as the heat of the part, as theſe ſubſtances repel moiſture, and are bad conductors of heat. Another great uſe of the ſtimulus of heat is by applying it to torpid ulcers, which are generally termed ſcrophulous or ſcorbutic, and are much eaſier inclined to heal, when covered with ſeveral folds of flannel.

Mr. —— had for many months been afflicted with an ulcer in perinæo, which communicated with the urethra, through which a part of his urine was daily evacuated with conſiderable pain; and was reduced to a great degree of debility. He uſed a hot-bath of 96 or 98 degrees of heat every day for half an hour during about ſix months. By this agreeable ſtimulus repeated thus at uniform times not only the ulcer healed, contrary to the expectation of his friends, but he acquired greater health and ſtrength, than he had for ſome years previouſly experienced.

Mrs. —— was affected with tranſient pains, which were called nervous ſpaſms, and with great fear of diſeaſes, which ſhe did not labour under, with cold extremities, and general debility. She uſed a hot-bath every other day of 96 degrees of heat for about four months, and recovered a good ſtate of health, with greater ſtrength and courage, than ſhe had poſſeſſed for many months before.

Mr. Z. a gentleman about 65 years of age, who had lived rather intemperately in reſpect to vinous potation, and had for many years

had

had annual visits of the gout, which now became irregular, and he appeared to be losing his strength, and beginning to feel the effects of age. He used a bath, as hot as was agreeable to his sensations, twice a week for about a year and half, and greatly recovered his health and strength with less frequent and less violent returns of regular gout, and is now near 80 years of age.

When Dr. Franklin, the American philosopher, was in England many years ago, I recommended to him the use of a warm-bath twice a week to prevent the too speedy access of old age, which he then thought that he felt the approach of, and I have been informed, that he continued the use of it till near his death, which was at an advanced age.

All these patients were advised not to keep themselves warmer than their usual habits, after they came out of the bath, whether they went into bed or not; as the design was not to promote perspiration, which weakens all constitutions, and seldom is of service to any. Thus a flannel shirt, particularly if it be worn in warm weather, occasions weakness by stimulating the skin by its points into too great action, and producing heat in consequence; and occasions emaciation by increasing the discharge of perspirable matter; and in both these respects differs from the effect of warm bathing, which communicates heat to the system at the same time that it stimulates it, and causes absorption more than exhalation.

2. The effect of the passage of an electric shock through a paralytic limb in causing it to contract, besides the late experiments of Galvani and Volta on frogs, intitle it to be classed amongst universal stimulants. Electric shocks frequently repeated daily for a week or two remove chronical pains, as the pleurodyne chronica, Class I. 2. 4. 14. and other chronic pains, which are termed rheumatic, probably by promoting the absorption of some extravasated material. Scrophulous tumours

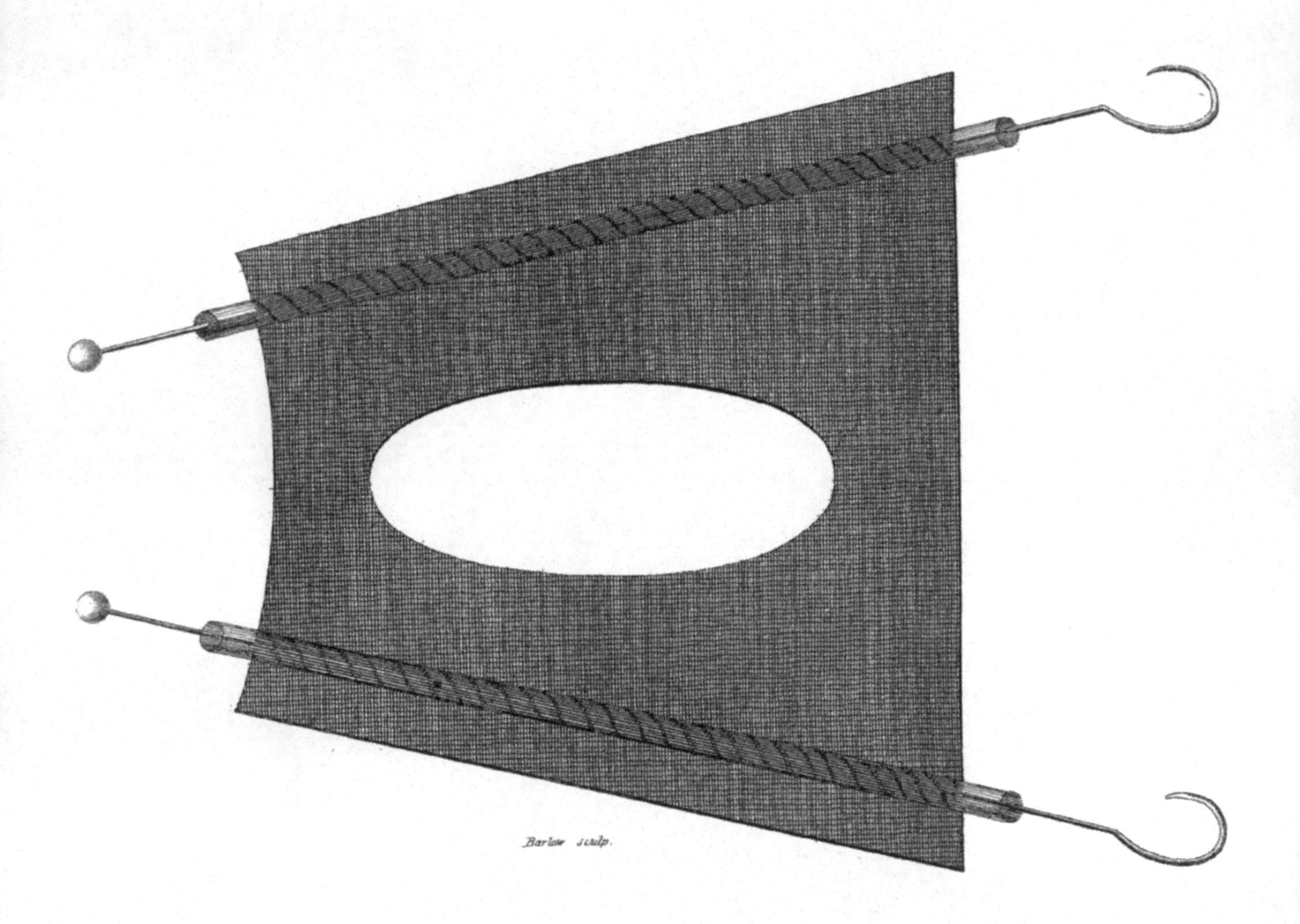

Barlow sculp.

tumours are ſometimes abſorbed, and ſometimes brought to ſuppurate by paſſing electric ſhocks through them daily for two or three weeks.

Miſs ——, a young lady about eight years of age, had a ſwelling about the ſize of a pigeon's egg on her neck a little below her ear, which long continued in an indolent ſtate. Thirty or forty ſmall electric ſhocks were paſſed through it once or twice a day for two or three weeks, and it then ſuppurated and healed without difficulty. For this operation the coated jar of the electric machine had on its top an electrometer, which meaſured the ſhocks by the approach of a braſs knob, which communicated with the external coating to another, which communicated with the internal one, and their diſtance was adjuſted by a ſcrew. So that the ſhocks were ſo ſmall as not to alarm the child, and the accumulated electricity was frequently diſcharged, as the wheel continued turning. The tumour was incloſed between two other braſs knobs, which were fixed on wires, which paſſed through glaſs tubes, the tubes were cemented in two grooves on a board, ſo that at one end they were nearer each other than at the other, and the knobs were puſhed out ſo far as exactly to include the tumour, as deſcribed in the annexed plate, which is about half the ſize of the original apparatus.

Inflammations of the eyes without fever are frequently cured by taking a ſtream of very ſmall electric ſparks from them, or giving the electric ſparks to them, once or twice a day for a week or two; that is, the new veſſels, which conſtitute inflammation in theſe inirritable conſtitutions, are abſorbed by the activity of the abſorbents induced by the ſtimulus of the electric aura. For this operation the eaſieſt method is to fix a pointed wire to a ſtick of ſealing wax, or to an inſulating handle of glaſs, one end of this wire communicates with the prime conductor, and the point is approached near the inflamed eye in every direction.

III. Externally

III. Externally the application of ether, and of essential oils, as of cloves or cinnamon, seem to possess a general stimulating effect. As they instantly relieve tooth-ach, and hiccough, when these pains are not in violent degree; and camphor in large doses is said to produce intoxication; this effect however I have not been witness to, and have reason to doubt.

The manner in which ether and the essential oil operate on the system when applied externally, is a curious question, as pain is so immediately relieved by them, that they must seem to penetrate by the great fluidity or expansive property of a part of them, as of their odoriferous exhalation or vapour, and that they thus stimulate the torpid part, and not by their being taken up by the absorbent vessels, and carried thither by the long course of circulation; nor is it probable, that these pains are relieved by the sympathy of the torpid membrane with the external skin, which is thus stimulated into action; as it does not succeed, unless it is applied over the pained part. Thus there appears to be three different modes by which extraneous bodies may be introduced into the system, besides that of absorption. 1st. By ethereal transition, as heat and electricity; 2d. by chemical attraction, as oxygen; and 3d. by expansive vapour, as ether and essential oils.

IV. The perpetual necessity of the mixture of oxygen gas with the blood in the lungs evinces, that it must act as a stimulus to the sanguiferous system, as the motions of the heart and arteries presently cease, when animals are immersed in airs which possess no oxygen. It may also subsequently answer another important purpose, as it probably affords the material for the production of the sensorial power; which is supposed to be secreted in the brain or medullary part of the nerves; and that the perpetual demand of this fluid in respiration is occasioned by the sensorial power, which is supposed to be produced

produced from it, being too ſubtle to be long confined in any part of the ſyſtem.

Another proof of the ſtimulant quality of oxygen appears from the increaſed acrimony, which the matter of a common abſceſs poſſeſſes, after it has been expoſed to the air of the atmoſphere, but not before; and probably all other contagious matters owe their fever-producing property to having been converted into acids by their union with oxygen.

As oxygen penetrates the fine moiſt membranes of the air-veſſels of the lungs, and unites with the blood by a chemical attraction, as is ſeen to happen, when blood is drawn into a baſon, the lower ſurface of the craſſamentum is of a very dark red ſo long as it is covered from the air by the upper ſurface, but becomes florid in a ſhort time on its being expoſed to the atmoſphere; the manner of its introduction into the ſyſtem is not probably by animal abſorption but by chemical attraction, in which circumſtance it differs from the fluids before mentioned both of heat and electricity, and of ether and eſſential oils.

As oxygen has the property of paſſing through moiſt animal membranes, as firſt diſcovered by the great Dr. Prieſtley, it is probable it might be of uſe in vibices, and petechiæ in fevers, and in other bruiſes; if the ſkin over thoſe parts was kept moiſt by warm water, and covered with oxygen gas by means of an inverted glaſs, or even by expoſing the parts thus moiſtened to the atmoſphere, as the dark coloured extravaſated blood might thus become florid, and by its increaſe of ſtimulus facilitate its reabſorption.

Two weak patients, to whom I gave oxygen gas in as pure a ſtate as it can eaſily be procured from Exeter mangane ſe, and in the quantity of about four gallons a day, ſeemed to feel refreſhed, and ſtronger, and to look better immediately after reſpiring it, and gained ſtrength in a ſhort time. Two others, one of whom laboured under confirmed hydrothorax, and the other under a permanent and uniform difficulty of respiration, were not refreshed, or in any way served by the use of oxygen in the above quantity of four gallons a day for a fortnight, which I ascribed to the inirritability of the diseased lungs. For other cases the reader is referred to the publications of Dr. Beddoes; Considerations on the Use of Factitious Airs, sold by Johnson, London.

Its effects would probably have been greater in respect to the quantity breathed, if it had been given in a dilute state, mixed with 10 or 20 times its quantity of atmospheric air, as otherwise much of it returns by expiration without being deprived of its quality, as may be seen by the person breathing on the flame of a candle, which it enlarges. See the Treatise of Dr. Beddoes above mentioned.

V. Those passions, which are attended with pleasurable sensation, excite the system into increased action in consequence of that sensation, as joy, and love, as is seen by the flush of the skin. Those passions, which are attended with disagreeable sensation, produce torpor in general by the expence of sensorial power occasioned by inactive pain; unless volition be excited in consequence of the painful sensation; and in that case an increased activity of the system occurs; thus paleness and coldness are the consequence of fear, but warmth and redness are the consequence of anger.

VI. Besides the exertions of the system occasioned by increased stimuli, and consequent irritation, and by the passions of the mind above described, the increased actions occasioned by exercise belong to this article. These may be divided into the actions of the body in consequence of volition, which is generally termed labour; or secondly, in consequence of agreeable sensation, which is termed play or ſport; thirdly, the exerciſe occaſioned by agitation, as in a carriage or on horſeback; fourthly, that of friction, as with a bruſh or hand, ſo much uſed in the baths of Turkey; and laſtly, the exerciſe of ſwinging.

The firſt of theſe modes of exerciſe is frequently carried to great exceſs even amongſt our own labourers, and more ſo under the laſh of ſlavery; ſo that the body becomes emaciated and ſinks under either the preſent hardſhips, or by a premature old age. The ſecond mode of exerciſe is ſeen in the play of all young animals, as kittens, and puppies, and children; and is ſo neceſſary to their health as well as to their pleaſure, that thoſe children, which are too much confined from it, not only become pale-faced and bloated, with tumid bellies, and conſequent worms, but are liable to get habits of unnatural actions, as twitching of their limbs, or of ſome parts of their countenance; together with an ill-humoured or diſcontented mind.

Agitation in a carriage or on horſeback, as it requires ſome little voluntary exertion to preſerve the body perpendicular, but much leſs voluntary exertion than in walking, ſeems the beſt adapted to invalids; who by theſe means obtain exerciſe principally by the ſtrength of the horſe, and do not therefore too much exhauſt their own ſenſorial power. The uſe of friction with a bruſh or hand, for half an hour or longer morning and evening, is ſtill better adapted to thoſe, who are reduced to extreme debility; and none of their own ſenſorial power is thus expended, and affords ſomewhat like the warm-bath activity without ſelf-exertion, and is uſed as a luxury after warm bathing in many parts of Aſia.

Another kind of exerciſe is that of ſwinging, which requires ſome exertion to keep the body perpendicular, or pointing towards the center of the ſwing, but is at the ſame time attended with a degree of vertigo; and is deſcribed in Claſs II. 1. 6. 7. IV. 2. 1. 10. Sup. I. 3. and 15.

The neceſſity of much exerciſe has perhaps been more inſiſted upon by phyſicians, than nature ſeems to demand. Few animals exerciſe themſelves ſo as to induce viſible ſweat, unleſs urged to it by mankind, or by fear, or hunger. And numbers of people in our market towns, of ladies particularly, with ſmall fortunes, live to old age in health, without any kind of exerciſe of body, or much activity of mind.

In ſummer weak people cannot continue too long in the air, if it can be done without fatigue; and in winter they ſhould go out ſeveral times in a day for a few minutes, uſing the cold air like a cold-bath, to invigorate and render them more hardy.

### III. CATALOGUE OF THE INCITANTIA.

I. Papaver ſomniferum; poppy, opium.

Alcohol, wine, beer, cyder.

Prunus lauro-ceraſus; laurel, diſtilled water from the leaves.

Prunus ceraſus; black cherry, diſtilled water from the kernels.

Nicotiana tabacum; tobacco? the eſſential oil, decoction of the leaf.

Atropa belladona; deadly nightſhade, the berries.

Datura ſtramoneum; thorn-apple, the fruit boiled in milk.

Hyoſcyamus reticulatus; henbane, the ſeeds and leaves.

Cynogloſſum; hounds tongue.

Meniſpermum, cocculus; Indian berry.

Amygdalus amarus; bitter almond.

Cicuta;

Cicuta; hemlock. Conium maculatum?

Strychnos nuc vomica?

Delphinium ſtaviſagria?

II. Externally, heat, electricity.

III. Ether, eſſential oils.

IV. Oxygen gas.

V. Paſſions of love, joy, anger.

VI. Labour, play, agitation, friction.

## ART. III.

## SECERNENTIA.

I. THOSE THINGS which increafe the irritative motions, which conftitute fecretion, are termed fecernentia; which are as various as the glands, which they ftimulate into action.

1. Diaphoretics, as aromatic vegetables, effential oils, ether, volatile alcali, neutral falts, antimonial preparations, external heat, exercife, friction, cold water for a time with fubfequent warmth, blifters, electric fluid.

2. Sialagogues, as mercury internally, and pyrethrum externally.

3. Expectorants, as fquill, onions, gum ammoniac, feneka root, mucilage: fome of thefe increafe the pulmonary perfpiration, and perhaps the pulmonary mucus.

4. Diuretics, as neutral falts, fixed alcali, balfams, refins, afparagus, cantharides.

5. Cathartics of the mild kind, as fena, jalap, neutral falts, manna. They increafe the fecretions of bile, pancreatic juice, and inteftinal mucus.

6. The mucus of the bladder is increafed by cantharides, and perhaps by oil of turpentine.

7. The

7. The mucus of the rectum by aloe internally, by clysters and suppositories externally.

8. The mucus of the cellular membrane is increased by blisters and sinapisms.

9. The mucus of the nostrils is increased by errhines of the milder kind, as marum, common snuff.

10. The secretion of tears is increased by volatile salts, the vapour of onions, by grief, and joy.

11. All those medicines increase the heat of the body, and remove those pains, which originate from a defect of motion in the vessels, which perform secretion; as pepper produces a glow on the skin, and balsam of Peru is said to relieve the flatulent cholic. But these medicines differ from the preceding class, as they neither induce costiveness nor deep coloured urine in their usual dose, nor intoxication in any dose.

12. Yet if any of these are used unnecessarily, it is obvious, like the incitantia, that they must contribute to shorten our lives by sooner rendering peculiar parts of the system disobedient to their natural stimuli. Of those in daily use the great excess of common salt is probably the most pernicious, as it enters all our cookery, and is probably one cause of scrophula, and of sea-scurvy, when joined with other causes of debility. See Botanic Garden, Part II. Canto IV. line 221. Spices taken to excess by stimulating the stomach, and the vessels of the skin by association, into unnecessary action, contribute to weaken these parts of the system, but are probably less noxious than the general use of so much salt.

II. Observations

## II. Observations on the Secernentia.

I. 1. Some of the medicines of this claſs produce abſorption in ſome degree, though their principal effect is exerted on the ſecerning part of our ſyſtem. We ſhall have occaſion to obſerve a ſimilar circumſtance in the next claſs of medicines termed Sorbentia; as of theſe ſome exert their effects in a ſmaller degree on the ſecerning ſyſtem. Nor will this ſurpriſe any one, who has obſerved, that all natural objects are preſented to us in a ſtate of combination; and that hence the materials, which produce theſe different effects, are frequently found mingled in the ſame vegetable. Thus the pure aromatics increaſe the action of the veſſels, which ſecrete the perſpirable matter; and the pure aſtringents increaſe the action of the veſſels, which abſorb the mucus from the lungs, and other cavities of the body; hence it muſt happen, that nutmeg, which poſſeſſes both theſe qualities, ſhould have the double effect above mentioned.

Other drugs have this double effect, and belong either to the claſs of Secernentia or Sorbentia, according to the doſe in which they are exhibited. Thus a ſmall doſe of alum increaſes abſorption, and induces coſtiveneſs; and a large one increaſes the ſecretions into the inteſtinal canal, and becomes cathartic. And this accounts for the conſtipation of the belly left after the purgative quality of rhubarb ceaſes, for it increaſes abſorption in a ſmaller doſe, and ſecretion in a greater. Hence when a part of the larger doſe is carried out of the habit by ſtools, the ſmall quantity which remains induces coſtiveneſs. Hence rhubarb exhibited in ſmall doſes, as 2 or 3 grains twice a day, ſtrengthens the ſyſtem by increaſing the action of the abſorbent veſſels, and of the inteſtinal canal.

2. Diaphoretics. The perſpiration is a ſecretion from the blood in its paſſage through the capillary veſſels, as other ſecretions are produced in the termination of the arteries in the various glands. After this ſecretion the blood loſes its florid colour, which it regains in its paſſage through the lungs; which evinces that ſomething beſides water is ſecreted on the ſkins of animals.

No ſtatical experiments can aſcertain the quantity of our perſpiration; as a continued abſorption of the moiſture of the atmoſphere exiſts at the ſame time both by the cutaneous and pulmonary lymphatics.

3. Every gland is capable of being excited into greater exertions by an appropriated ſtimulus applied either by its mixture with the blood immediately to the ſecerning veſſel, or applied externally to its excretory duct. Thus mercury internally promotes an increaſed ſalivation, and pyrethrum externally applied to the excretory ducts of the ſalival glands. Aloes ſtimulate the rectum internally mixed with the circulating blood; and ſea-ſalt by injection externally. Now as the capillaries, which ſecrete the perſpirable matter, lie near the ſurface of the body, the application of external heat acts immediately on their excretory ducts, and promotes perſpiration; internally thoſe drugs which poſſeſs a fragrant eſſential oil, or ſpiritus rector, produce this effect, as the aromatic vegetables, of which the number is very great.

4. It muſt be remembered, that a due quantity of ſome aqueous vehicle muſt be given to ſupport this evacuation; otherwiſe a burning heat without much viſible ſweat muſt be the conſequence. When the ſkin acquires a degree of heat much above 108, as appears by Dr. Alexander's experiments, no viſible ſweat is produced; which is owing to the great heat of the ſkin evaporating it as haſtily, as it is

 ſecreted;

ſecreted; and, where the ſweat is ſecreted in abundance, its evaporation cannot carry off the exuberant heat, like the vapour of boiling water; becauſe a great part of it is wiped off, or abſorbed by the bed-clothes; or the air about the patient is not changed ſufficiently often, as it becomes ſaturated with the perſpirable matter. And hence it is probable, that the waſte of perſpirable matter is as great, or greater, when the ſkin is hot and dry, as when it ſtands in drops on the ſkin; as appears from the inextinguiſhable thirſt.

Hence Dr. Alexander found, that when the heat of the body was greater than 108, nothing produced ſweats but repeated draughts of cold water; and of warm fluids, when the heat was much below that degree. And that cold water which procured ſweats inſtantaneouſly when the heat was above 108, ſtopped them as certainly when it was below that heat; and that flannels, wrung out of warm water and wrapped round the legs and thighs, were then moſt certainly productive of ſweats.

5. The diaphoretics are all ſaid to ſucceed much better, if given early in the morning, about an hour before ſun-riſe, than at any other time; which is owing to the great excitability of every part of the ſyſtem after the ſenſorial power has been accumulated during ſleep. In thoſe, who have hectic fever, or the febricula, or nocturnal fever of debility, the morning ſweats are owing to the decline of the fever-fit, as explained in Sect. XXXII. 9. In ſome of theſe patients the ſweat does not occur till they awake; becauſe then the ſyſtem is ſtill more excitable than during ſleep, becauſe the aſſiſtance of the voluntary power in reſpiration facilitates the general circulation. See Claſs I. 2. 1. 3.

6. It muſt be obſerved, that the ſkin is very dry and hard to the touch, where the abſorbents, which open on its ſurface, do not act; as in

in ſome dropſies, and other diſeaſes attended with great thirſt. This dryneſs, and ſhrivelled appearance, and roughneſs, are owing to the mouths of the abſorbents being empty of their accuſtomed fluid, and is diſtinguiſhable from the dryneſs of the ſkin above mentioned in the hot fits of fever, by its not being attended with heat.

As the heat of the ſkin in the uſual temperature of the air always evinces an increaſed perſpiration, whether viſible or not, the heat being produced along with the increaſe of ſecretion; it follows, that a defect of perſpiration can only exiſt, when the ſkin is cold.

7. Volatile alcali is a very powerful diaphoretic, and particularly if exhibited in wine-whey; 20 drops of ſpirit of hartſhorn every half hour in half a pint of wine-whey, if the patient be kept in a moderately warm bed, will in a few hours elicit moſt profuſe ſweats.

Neutral ſalts promote inviſible perſpiration, when the ſkin is not warmed much externally, as is evinced from the great thirſt, which ſucceeds a meal of ſalt proviſions, as of red herrings. When theſe are ſufficiently diluted with water, and the ſkin kept warm, copious ſweats without inflaming the habit, are the conſequence. Half an ounce of vinegar ſaturated with volatile alcali, taken every hour or two hours, well anſwers this purpoſe; and is preferable perhaps in general to all others, where ſweating is advantageous. Boerhaave mentions one cured of a fever by eating red-herrings or anchovies, which, with repeated draughts of warm water or tea, would I ſuppoſe produce copious perſpiration.

Antimonial preparations have alſo been of late much uſed with great advantage as diaphoretics. For the hiſtory and uſe of theſe preparations I ſhall refer the reader to the late writers on the Materia Medica, only obſerving that the ſtomach becomes ſo ſoon habituated to its ſtimulus, that the ſecond doſe may be conſiderably increaſed, if the firſt had no operation.

Where it is adviſable to procure copious ſweats, the emetics, as ipecacuanha, joined with opiates, as in Dover's powder, produce this effect with greater certainty than the above.

8. We muſt not diſmiſs this ſubject without obſerving, that perſpiration is deſigned to keep the ſkin flexile, as the tears are intended to clean and lubricate the eye; and that neither of theſe fluids can be conſidered as excretions in their natural ſtate, but as ſecretions. See Claſs I. 1. 2. 3. And that therefore the principal uſe of diaphoretic medicines is to warm the ſkin, and thence in conſequence to produce the natural degree of inſenſible perſpiration in languid habits.

9. When the ſkin of the extremities is cold, which is always a ſign of preſent debility, the digeſtion becomes frequently impaired by aſſociation, and cardialgia or heartburn is induced from the vinous or acetous fermentation of the aliment. In this diſeaſe diaphoretics, which have been called cordials, by their action on the ſtomach reſtore its exertion, and that of the cutaneous capillaries by their aſſociation with it, and the ſkin becomes warm, and the digeſtion more vigorous.

10. But a bliſter acts with more permanent and certain effect by ſtimulating a part of the ſkin, and thence affecting the whole of it, and of the ſtomach by aſſociation, and thence removes the moſt obſtinate heartburns and vomitings. From this the principal uſe of bliſters is underſtood, which is to invigorate the exertions of the arterial and lymphatic veſſels of the ſkin, producing an increaſe of inſenſible perſpiration, and of cutaneous abſorption; and to increaſe the action of the ſtomach, and the conſequent power of digeſtion; and thence by ſympathy to excite all the other irritative motions: hence they relieve pains of the cold kind, which originate from defect of motion;

not

not from their introducing a greater pain, as ſome have imagined, but by ſtimulating the torpid veſſels into their uſual action; and thence increaſing the action and conſequent warmth of the whole ſkin, and of all the parts which are aſſociated with it.

II. 1. *Sialagogues*. The preparations of mercury conſiſt of a ſolution or corroſion of that metal by ſome acid; and, when the doſe is known, it is probable that they are all equally efficacious. As their principal uſe is in the cure of the venereal diſeaſe, they will be mentioned in the catalogue amongſt the ſorbentia. Where ſalivation is intended, it is much forwarded by a warm room and warm clothes; and prevented by expoſing the patient to his uſual habits of cool air and dreſs, as the mercury is then more liable to go off by the bowels.

2. Any acrid drug, as pyrethrum, held in the mouth acts as a ſialagogue externally by ſtimulating the excretory ducts of the ſalivary glands; and the ſiliqua hirſuta applied externally to the parotid gland, and even hard ſubſtances in the ear, are ſaid to have the ſame effect. Maſtich chewed in the mouth emulges the ſalivary glands.

3. The unwiſe cuſtom of chewing and ſmoking tobacco for many hours in a day not only injures the ſalivary glands, producing dryneſs in the mouth when this drug is not uſed, but I ſuſpect that it alſo produces ſchirrhus of the pancreas. The uſe of tobacco in this immoderate degree injures the power of digeſtion, by occaſioning the patient to ſpit out that ſaliva, which he ought to ſwallow; and hence produces that flatulency, which the vulgar unfortunately take it to prevent. The mucus, which is brought from the fauces by hawking, ſhould be ſpit out, as well as that coughed up from the lungs; but that which comes ſpontaneouſly into the mouth from the ſalivary glands,

glands, ſhould be ſwallowed mixed with our food or alone for the purpoſes of digeſtion. See Claſs I. 2. 2. 7.

III. 1. Expectorants are ſuppoſed to increaſe the ſecretion of mucus in the branches of the windpipe, or to increaſe the perſpiration of the lungs ſecreted at the terminations of the bronchial artery.

2. If any thing promotes expectoration toward the end of peripneumonies, when the inflammation is reduced by bleeding and gentle cathartics, ſmall repeated bliſters about the cheſt, with tepid aqueous and mucilaginous or oily liquids, are more advantageous than the medicines generally enumerated under this head; the bliſters by ſtimulating into action the veſſels of the ſkin produce by aſſociation a greater activity of thoſe of the mucous membrane, which lines the branches of the windpipe, and air-cells of the lungs; and thus after evacuation they promote the abſorption of the mucus and conſequent healing of the inflamed membrane, while the diluting liquids prevent this mucus from becoming too viſcid for this purpoſe, or facilitate its expuition.

Bliſters, one at a time, on the ſides or back, or on the ſternum, are alſo uſeful towards the end of peripneumonies, by preventing the evening acceſs of cold fit, and thence preventing the hot fit by their ſtimulus on the ſkin; in the ſame manner as five drops of laudanum by its ſtimulus on the ſtomach. For the increaſed actions of the veſſels of the ſkin or ſtomach excite a greater quantity of the ſenſorial power of aſſociation, and thus prevent the torpor of the other parts of the ſyſtem; which, when patients are debilitated, is ſo liable to return in the evening.

3. Warm bathing is of great ſervice towards the end of peripneumony to promote expectoration, eſpecially in thoſe children who drink too little aqueous fluids, as it gently increaſes the action of the pulmonary

pulmonary capillaries by their conſent with the cutaneous ones, and ſupplies the ſyſtem with aqueous fluid, and thus dilutes the ſecreted mucus.

Some have recommended oil externally around the cheſt, as well as internally, to promote expectoration; and upon the noſe, when its mucous membrane is inflamed, as in common catarrh.

IV. 1. Diuretics. If the ſkin be kept warm, moſt of theſe medicines promote ſweat inſtead of urine; and if their doſe is enlarged, moſt of them become cathartic. Hence the neutral ſalts are uſed in general for all theſe purpoſes. Thoſe indeed, which are compoſed of the vegetable acid, are moſt generally uſed as ſudorifics; thoſe with the nitrous acid as diuretics; and thoſe with the vitriolic acid as cathartics; while thoſe united with the marine acid enter our common nutriment, as a more general ſtimulus. All theſe increaſe the acrimony of the urine, hence it is retained a leſs time in the bladder; and in conſequence leſs of it is reabſorbed into the ſyſtem, and the apparent quantity is greater, as more is evacuated from the bladder; but it is not certain from thence, that a greater quantity is ſecreted by the kidnies. Hence nitre, and other neutral ſalts, are erroneouſly given in the gonorrhœa; as they augment the pain of making water by their ſtimulus on the excoriated or inflamed urethra. They are alſo erroneouſly given in catarrhs or coughs, where the diſcharge is too thin and ſaline, as they increaſe the frequency of coughing.

2. Balſam of Copaiva is thought to promote urine more than the other native balſams; and common reſin is ſaid to act as a powerful diuretic in horſes. Theſe are alſo much recommended in gleets, and in fluor albus, perhaps more than they deſerve; they give a violet ſmell to the urine, and hence probably increaſe the ſecretion of it.

4 Calcined

Calcined egg-ſhells are ſaid to promote urine, perhaps from the phoſphoric acid they contain.

3. Cold air and cold water will increaſe the quantity of urine by decreaſing the abſorption from the bladder; and neutral and alcalious ſalts and cantharides by ſtimulating the neck of the bladder to diſcharge the urine as ſoon as ſecreted; and alcohol as gin and rum at the beginning of intoxication, if the body be kept cool, occaſion much urine by inverting the urinary lymphatics, and thence pouring a fluid into the bladder, which never paſſed the kidnies. But it is probable, that thoſe medicines, which give a ſcent to the urine, as the balſams and reſins, but particularly aſparagus and garlic, are the only drugs, which truly increaſe the ſecretion of the kidnies. Alcohol however, uſed as above mentioned, and perhaps great doſes of tincture of cantharides, may be conſidered as draſtic diuretics, as they pour a fluid into the bladder by the retrograde action of the lymphatics, which are in great abundance ſpread about the neck of it. See Sect. XXIX. 3.

V. Mild cathartics. The ancients believed that ſome purges evacuated the bile, and hence were termed Cholagogues; others the lymph, and were termed Hydragogues; and that in ſhort each cathartic ſelected a peculiar humour, which it diſcharged. The moderns have too haſtily rejected this ſyſtem; the ſubject well deſerves further obſervation.

Calomel given in the doſe from ten to twenty grains, ſo as to induce purging without the aſſiſtance of other drugs, appears to me to particularly increaſe the ſecretion of bile, and to evacuate it; aloe ſeems to increaſe the ſecretion of the inteſtinal mucus; and it is probable that the pancreas and ſpleen may be peculiarly ſtimulated into action by ſome other of this tribe of medicines; whilſt others of them

them may ſimply ſtimulate the inteſtinal canal to evacuate its contents; as the bile of animals. It muſt be remarked, that all theſe cathartic medicines are ſuppoſed to be exhibited in their uſual doſes, otherwiſe they become draſtic purges, and are treated of in the Claſs of Invertentia.

VI. The mucus of the bladder is ſeen in the urine, when cantharides have been uſed, either internally or externally, in ſuch doſes as to induce the ſtrangury. Spirit of turpentine is ſaid to have the ſame effect. I have given above a dram of it twice a day floating on a glaſs of water in chronic lumbago without this effect, and the patient gradually recovered.

VII. Aloe given internally ſeems to act chiefly on the rectum and ſpincter ani, producing teneſmus and piles. Externally in clyſters or ſuppoſitories, common ſalt ſeems to act on that bowel with greater certainty. But where the thread-worm or aſcarides exiſt, 60 or 100 grains of aloes reduced to powder and boiled in a pint of gruel, and uſed as a clyſter twice a week for three months, has frequently deſtroyed them.

VIII. The external application of cantharides by ſtimulating the excretory ducts of the capillary glands produces a great ſecretion of ſubcutaneous mucus with pain and inflammation; which mucaginous fluid, not being able to permeate the cuticle, raiſes it up; a ſimilar ſecretion and elevation of the cuticle is produced by actual fire; and by cauſtic materials, as by the application of the juice of the root of white briony, or bruiſed muſtard-ſeed. Experiments are wanting to introduce ſome acrid application into practice inſtead of cantharides, which might not induce the ſtrangury.

Muſtard-ſeed alone is too acrid, and if it be ſuffered to lie on the ſkin many minutes is liable to produce a ſlough and conſequent ulcer, and ſhould therefore be mixed with flour when applied to cold extremities. Volatile alkali properly diluted might ſtimulate the ſkin without inducing ſtrangury.

IX. The mild errhines are ſuch as moderately ſtimulate the membrane of the noſtrils, ſo as to increaſe the ſecretion of naſal mucus; as is ſeen in thoſe, who are habituated to take ſnuff. The ſtronger errhines are mentioned in Art. V. 2. 3.

X. The ſecretion of tears is increaſed either by applying acrid ſubſtances to the eye; or acrid vapours, which ſtimulate the excretory duct of the lacrymal gland; or by applying them to the noſtrils, and ſtimulating the excretory duct of the lacrymal ſack, as treated of in the Section on Inſtinct.

Or the ſecretion of tears is increaſed by the aſſociation of the motions of the excretory duct of the lacrymal ſack with ideas of tender pleaſure, or of hopeleſs diſtreſs, as explained in Sect. XVI. 8. 2. and 3.

XI. The ſecretion of ſenſorial power in the brain is probably increaſed by opium or wine, becauſe when taken in certain quantity an immediate increaſe of ſtrength and activity ſucceeds for a time, with conſequent debility if the quantity taken be ſo great as to intoxicate in the leaſt degree. The neceſſity of perpetual reſpiration ſhews, that the oxygen of the atmoſphere ſupplies the ſource of the ſpirit of animation; which is conſtantly expended, and is probably too fine to be long contained in the nerves after its production in the brain. Whence it is probable, that the reſpiration of oxygen

gas

gas mixed with common air may increaſe the ſecretion of ſenſorial power; as indeed would appear from its exhilarating effect on moſt patients.

### III. CATALOGUE OF THE SECERNENTIA.

I. Diaphoretics.

1. Amomum zinziber, ginger. Caryophyllus aromaticus, cloves. Piper indicum, pepper. Capſicum. Cardamomum. Pimento, myrtus pimenta. Canella alba. Serpentaria virginiana, ariſtolochia ſerpentaria, guaiacum. Saſſafras, laurus ſaſſafras. Opium. Wine.
2. Eſſential oils of cinnamon, laurus cinnamomum. Nutmeg, myriſtica moſchata. Cloves, caryophyllus aromaticus. Mint, mentha. Camphor, laurus camphora. Ether.
3. Volatile ſalts, as of ammoniac and of hartſhorn. Sal cornu cervi.
4. Neutral ſalts, as thoſe with vegetable acid; or with marine acid, as common ſalt. Hálex. Red-herring, anchovy.
5. Preparations of antimony, as emetic tartar, antimonium tartarizatum, wine of antimony. James's powder.
6. External applications. Bliſters. Warm bath. Warm air. Exerciſe. Friction.
7. Cold water with ſubſequent warmth.

II. Sialagogues. Preparations of mercury, hydrargyrus. Pyrethrum, anthemis pyrethrum, tobacco, cloves, pepper, cowhage, ſtizolobium ſiliqua hirſuta. Maſtich, piſtacia lentiſcus.

III. Expectorants:

1. Squill, ſcilla maritima, garlic, leek, onion, allium, aſafœtida, ferula aſafœtida, gum ammoniac, benzoin, tar, pix liquida, balſam of Tolu.

2. Root of ſeneka, polygala ſeneka, of elicampane, inula helenium.

3. Marſh-mallow, althæa, coltsfoot, tuſſilago farfara, gum arabic, mimoſa nilotica, gum tragacanth, aſtragalus tragacantha. Decoction of barley, hordeum diſtichon. Expreſſed oils. Spermaceti, ſoap. Extract of liquorice, glycyrrhiza glabra. Sugar. Honey.

4. Externally bliſters. Oil. Warm bath.

IV. Mild diuretics.

1. Nitre, kali acetatum, other neutral ſalts.
2. Fixed alkali, ſoap, calcined egg-ſhells.
3. Turpentine. Balſam of Copaiva. Reſin. Olibanum.
4. Aſparagus, garlic, wild daucus. Parſley, apium. Fennel fæniculum, pareira brava, Ciſſampelos?
5. Externally cold air, cold water.
6. Alcohol. Tincture of cantharides. Opium.

V. Mild cathartics.

1. Sweet ſubacid fruits. Prunes, prunus domeſtica. Caſſia fiſtula. Tamarinds, cryſtals of tartar, unrefined ſugar. Manna. Honey.

2. Whey of milk, bile of animals.

3. Neutral ſalts, as Glauber's ſalt, vitriolated tartar, ſea-water, magneſia alba, ſoap.

4. Gum guaiacum. Balsam of Peru. Oleum ricini, castor-oil, oil of almonds, oil of olives, sulphur.

5. Senna, cassia senna, jalap, aloe, rhubarb, rheum palmatum.

6. Calomel. Emetic tartar, antimonium tartarizatum.

VI. Secretion of mucus of the bladder is increased by cantharides, by spirit of turpentine?

VII. Secretion of mucus of the rectum is increased by aloe internally, by various clysters and suppositories externally.

VIII. Secretion of subcutaneous mucus is increased by blisters of cantharides, by application of a thin slice of the fresh root of white briony, by sinapisms, by root of horse-radish, cochlearia armoracia. Volatile alcali.

IX. Mild errhines. Marjoram. Origanum. Marum, tobacco.

X. Secretion of tears is increased by vapour of sliced onion, of volatile alcali. By pity, or ideas of hopeless distress.

XI. Secretion of sensorial power in the brain is probably increased by opium, by wine, and perhaps by oxygen gas added to the common air in respiration.

ART.

## Art. IV.
## SORBENTIA.

I. Those things which increaſe the irritative motions, which conſtitute abſorption, are termed ſorbentia; and are as various as the abſorbent veſſels, which they ſtimulate into action.

1. Cutaneous abſorption is increaſed by auſtere acids, as of vitriol; hence they are believed to check colliquative ſweats, and to check the eruption of ſmall-pox, and contribute to the cure of the itch, and tinea; hence they thicken the ſaliva in the mouth, as lemon-juice, crab-juice, ſloes.

2. Abſorption from the mucous membrane is increaſed by opium, and Peruvian bark, internally; and by blue vitriol externally. Hence the expectoration in coughs, and the mucous diſcharge from the urethra, are thickened and leſſened.

3. Abſorption from the cellular membrane is promoted by bitter vegetables, and by emetics, and cathartics. Hence matter is thickened and leſſened in ulcers by opium and Peruvian bark; and ſerum is abſorbed in anaſarca by the operation of emetics and cathartics.

4. Venous abſorption is increaſed by acrid vegetables; as watercreſs, cellery, horſe-radiſh, muſtard. Hence their uſe in ſeaſcurvy, the vibices of which are owing to a defect of venous abſorption; and by external ſtimulants, as vinegar, and by electricity, and perhaps by oxygen.

5. Inteſtinal

5. Inteſtinal abſorption is increaſed by aſtringent vegetables, as rhubarb, galls; and by earthy ſalts, as alum; and by argillaceous and calcareous earth.

6. Hepatic abſorption is increaſed by metallic ſalts, hence calomel and ſal martis are ſo efficacious in jaundice, worms, chloroſis, dropſy.

7. Venereal virus in ulcers is abſorbed by the ſtimulus of mercury; hence they heal by the uſe of this medicine.

8. Veneſection, hunger, thirſt, and violent evacuations, increaſe all abſorptions; hence ſweating produces coſtiveneſs.

9. Externally bitter aſtringent vegetables, earthy and metallic ſalts, and bandages, promote the abſorption of the parts on which they are applied.

10. All theſe in their uſual doſes do not increaſe the natural heat; but they induce coſtiveneſs, and deep coloured urine with earthy ſediment.

In greater doſes they invert the motions of the ſtomach and lacteals; and hence vomit or purge, as carduus benedictus, rhubarb. They promote perſpiration, if the ſkin be kept warm; as camomile tea, and teſtaceous powders, have been uſed as ſudorifics.

The preparations of antimony vomit, purge, or ſweat, either according to the quantity exhibited, or as a part of what is given is evacuated. Thus a quarter of a grain of emetic tartar (if well prepared) will promote a diaphoreſis, if the ſkin be kept warm; half a grain will procure a ſtool or two firſt, and ſweating afterwards; and a grain will generally vomit, and then purge, and laſtly ſweat the patient. In leſs quantity it is probable, that this medicine acts like other metallic

tallic ſalts, as ſteel, zinc, or copper in ſmall doſes; that is, that it ſtrengthens the ſyſtem by its ſtimulus. As camomile or rhubarb in different doſes vomit, or purge, or act as ſtimulants ſo as to ſtrengthen the ſyſtem.

## II. Observations on the Sorbentia.

I. 1. As there is great difference in the apparent ſtructure of the various glands, and of the fluids which they ſelect from the blood, theſe glands muſt poſſeſs different kinds of irritability, and are therefore ſtimulated into ſtronger or unnatural actions by different articles of the materia medica, as ſhewn in the ſecernentia. Now as the abſorbent veſſels are likewiſe glands, and drink up or ſelect different fluids; as chyle, water, mucus, with a part of every different ſecretion, as a part of the bile, a part of the ſaliva, a part of the urine, &c. it appears, that theſe abſorbent veſſels muſt likewiſe poſſeſs different kinds of irritability, and in conſequence muſt require different articles of the materia medica to excite them into unuſual action. This part of the ſubject has been ſo little attended to, that the candid reader will find in this article a great deal to excuſe.

It was obſerved, that ſome of the ſecernentia did in a leſs degree increaſe abſorption, from the combination of different properties in the ſame vegetable body; for the ſame reaſon ſome of the claſs of ſorbentia produce ſecretion in a leſs degree, as thoſe bitters which have alſo an aroma in their compoſition; theſe are known from their increaſing the heat of the ſyſtem above its uſual degree.

It muſt alſo be noted, that the actions of every part of the abſorbent ſyſtem are ſo aſſociated with each other, that the drugs which ſtimulate one branch increaſe the action of the whole; and the torpor or quieſcence of one branch weakens the exertions of the whole; or when

when one branch is excited into ſtronger action, ſome other branch has its actions weakened or inverted. Yet though peculiar branches of the abſorbent ſyſtem are ſtimulated into action by peculiar ſubſtances, there are other ſubſtances which ſeem to ſtimulate the whole ſyſtem, and that without immediately increaſing any of the ſecretions; as thoſe bitters which poſſeſs no aromatic ſcent, at the head of which ſtands the famed Peruvian bark, or cinchona.

2. Cutaneous abſorption. I have heard of ſome experiments, in which the body was kept cold, and was thought to abſorb more moiſture from the atmoſphere than at any other time. This however cannot be determined by ſtatical experiments; as the capillary veſſels, which ſecrete the perſpirable matter, muſt at the ſame time have been benumbed by the cold; and from their inaction there could not have been the uſual waſte of the weight of the body; and as all other muſcular exertions are beſt performed, when the body poſſeſſes its uſual degree of warmth, it is concluſive, that the abſorbent ſyſtem ſhould likewiſe do its office beſt, when it is not benumbed by external cold.

The auſtere acids, as of vitriol, lemon-juice, juice of crabs and ſloes, ſtrengthen digeſtion, and prevent that propenſity to ſweat ſo uſual to weak convaleſcents, and diminiſh the colliquative ſweats in hectic fevers; all which are owing to their increaſing the action of the external and internal cutaneous abſorption. Hence vitriolic acid is given in the ſmall-pox to prevent the too haſty or too copious eruption, which it effects, by increaſing the cutaneous abſorption. Vinegar, from the quantity of alcohol which it contains, exerts a contrary effect to that here deſcribed, and belongs to the incitantia; as an ounce of it promotes ſweat, and a fluſhing of the ſkin; at the ſame time externally it acts as a venous abſorbent, as the lips become pale by moiſtening them with it. And it is ſaid, when taken internally in

 great

great and continued quantity, to induce palenefs of the fkin, and foftnefs of the bones.

The fweet vegetable acids, as of feveral ripe fruits, are among the torpentia; as they are lefs ftimulating than the general food of this climate, and are hence ufed in inflammatory difeafes.

Where the quantity of fluids in the fyftem is much leffened, as in hectic fever, which has been of fome continuance, or in fpurious peripneumony, a grain of opium given at night will fometimes prevent the appearance of fweats; which is owing to the ftimulus of opium increafing the actions of the cutaneous abforbents, more than thofe of the fecerning veffels of the fkin. Whence the fecretion of perfpirable matter is not decreafed, but its appearance on the fkin is prevented by its more facile abforption.

3. There is one kind of itch, which feldom appears between the fingers, is the leaft infectious, and moft difficult to eradicate, and which has its cure much facilitated by the internal ufe of acid of vitriol. This difeafe confifts of fmall ulcers in the fkin, which are healed by whatever increafes the cutaneous abforption. The external application of fulphur, mercury, and acrid vegetables, acts on the fame principle; for the animalcula, which are feen in thefe puftules, are the effect, not the caufe, of them; as all other ftagnating animal fluids, as the femen itfelf, abounds with fimilar microfcopic animals.

4. Young children have fometimes an eruption upon the head called Tinea, which difcharges an acrimonius ichor inflaming the parts, on which it falls. This eruption I have feen fubmit to the internal ufe of vitriolic acid, when only wheat-flour was applied externally. This kind of eruption is likewife frequently cured by teftaceous powders; two materials fo widely different in their

chemical properties, but agreeing in their power of promoting cutaneous absorption.

II. Absorption from the mucous membrane is increased by applying to its surface the austere acids, as of vitriol, lemon-juice, crab-juice, sloes. When these are taken into the mouth, they immediately thicken, and at the same time lessen the quantity of the saliva; which last circumstance cannot be owing to their coagulating the saliva, but to their increasing the absorption of the thinner parts of it. So alum applied to the tip of the tongue does not stop in its action there, but independent of its diffusion it induces cohesion and corrugation over the whole mouth. (Cullen's Mat. Med. Art. Astringentia.) Which is owing to the association of the motions of the parts or branches of the absorbent system with each other.

Absorption from the mucous membrane is increased by opium taken internally in small doses more than by any other medicine, as is seen in its thickening the expectoration in coughs, and the discharge from the nostrils in catarrh, and perhaps the discharge from the urethra in gonorrhoea. The bark seems next in power for all these purposes.

Externally slight solutions of blue vitriol, as two or three grains to an ounce of water, applied to ulcers of the mouth, or to chancres on the glans penis, more powerfully induces them to heal than any other material.

Where the lungs or urethra are inflamed to a considerable degree, and the absorption is so great, that the mucus is already too thick, and adheres to the membrane from its viscidity, opiates and bitter vegetable and austere acids are improper; and mucilaginous diluents should be used in their stead with venesection and torpentia.

III. 1. Abſorption from the cellular membrane, and from all the other cavities of the body, is too ſlowly performed in ſome conſtitutions; hence the bloated pale complexion; and when this occurs in its greateſt degree, it becomes an univerſal dropſy. Theſe habits are liable to intermittent fevers, hyſteric paroxyſms, cold extremities, indigeſtion, and all the ſymptoms of debility.

The abſorbent ſyſtem is more ſubject to torpor or quieſcence than the ſecerning ſyſtem, both from the coldneſs of the fluids which are applied to it, as the moiſture of the atmoſphere, and from the coldneſs of the fluids which we drink; and alſo from its being ſtimulated only by intervals, as when we take our food; whereas the ſecerning ſyſtem is perpetually excited into action by the warm circulating blood; as explained in Sect. XXXII.

2. The Peruvian bark, camomile flowers, and other bitter drugs, by ſtimulating this cellular branch of the abſorbent ſyſtem prevents it from becoming quieſcent; hence the cold paroxyſms of thoſe agues, which ariſe from the torpor of the cellular lymphatics, are prevented, and the hot fits in conſequence. The patient thence preſerves his natural heat, regains his healthy colour, and his accuſtomed ſtrength.

Where the cold paroxyſm of an ague originates in the abſorbents of the liver, ſpleen, or other internal viſcus, the addition of ſteel to vegetable bitters, and eſpecially after the uſe of one doſe of calomel, much advances the cure.

And where it originates in any part of the ſecerning ſyſtem, as is probably the caſe in ſome kinds of agues, the addition of opium in the doſe of a grain and half, given about an hour before the acceſs of the paroxyſm, or mixed with chalybeate and bitter medicines, enſures the cure. Or the ſame may be effected by wine given inſtead of opium before the paroxyſm, ſo as nearly to intoxicate.

Theſe

Theſe three kinds of agues are thus diſtinguiſhed; the firſt is not attended with any tumid or indurated viſcus, which the people call an ague cake, and which is evident to the touch. The ſecond is accompanied with a tumid viſcus; and the laſt has generally, I believe, the quartan type, and is attended with ſome degree of arterial debility.

3. This claſs of abſorbent medicines are ſaid to decreaſe irritability. After any part of our ſyſtem has been torpid or quieſcent, by whatever cauſe that was produced, it becomes afterwards capable of being excited into greater motion by ſmall ſtimuli; hence the hot fit of fever ſucceeds the cold one. As theſe medicines prevent torpor or quieſcence of parts of the ſyſtem, as cold hands or feet, which perpetually happen to weak conſtitutions, the ſubſequent increaſe of irritability of theſe parts is likewiſe prevented.

4. Theſe abſorbent medicines, including both the bitters, and metallic ſalts, and opiates, are of great uſe in the dropſy by their promoting univerſal abſorption; but here evacuations are likewiſe to be produced, as will be treated of in the Invertentia.

5. The matter in ulcers is thickened, and thence rendered leſs corroſive, the ſaline part of it being reabſorbed by the uſe of bitter medicines; hence the bark is uſed with advantage in the cure of ulcers.

6. Bitter medicines ſtrengthen digeſtion by promoting the abſorption of chyle; hence the introduction of hop into the potation uſed at our meals, which as a medicine may be taken advantageouſly, but, like other unneceſſary ſtimuli, muſt be injurious as an article of our daily diet.

The

The hop may perhaps in ſome degree contribute to the production of gravel in the kidnies, as our intemperate wine-drinkers are more ſubject to the gout, and ale-drinkers to the gravel; in the formation of both which diſeaſes, there can be no doubt, but that the alcohol is the principal, if not the only agent.

7. Vomits greatly increaſe the abſorption from the cellular membrane, as ſquill, and foxglove. The ſquill ſhould be given in the doſe of a grain of the dried root every hour, till it operates upwards and downwards. Four ounces of the freſh leaves of the foxglove ſhould be boiled from two pounds of water to one, and half an ounce of the decoction taken every two hours for four or more doſes. This medicine by ſtimulating into inverted action the abſorbents of the ſtomach, increaſes the direct action of the cellular lymphatics.

Another more convenient way of aſcertaining the doſe of foxglove is by making a ſaturated tincture of it in proof ſpirit; which has the twofold advantage of being invariable in its original ſtrength, and of keeping a long time as a ſhop-medicine without loſing any of its virtue. Put two ounces of the leaves of purple foxglove, digitalis purpurea, nicely dried, and coarſely powdered, into a mixture of four ounces of rectified ſpirit of wine and four ounces of water; let the mixture ſtand by the fire-ſide twenty-four hours frequently ſhaking the bottle, and thus making a ſaturated tincture of digitalis; which muſt be poured from the ſediment or paſſed through filtering paper.

As the ſize of a drop is greater or leſs according to the ſize of the rim of the phial from which it is dropped, a part of this ſaturated tincture is then directed to be put into a two-ounce phial, for the purpoſe of aſcertaining the ſize of the drop. Thirty drops of this tincture is directed to be put into an ounce of mint-water for a draught to be taken twice or thrice a day, till it reduces the anaſarca of the limbs,

or

or removes the difficulty of breathing in hydrothorax, or till it induces ſickneſs. And if theſe do not occur in two or three days, the doſe muſt be gradually increaſed to forty or ſixty drops, or further.

From the great ſtimulus of this medicine the ſtomach is rendered torpid with conſequent ſickneſs, which continues many hours and even days, owing to the great exhauſtion of its ſenſorial power of irritation; and the action of the heart and arteries becomes feeble from the deficient excitement of the ſenſorial power of aſſociation; and laſtly, the abſorbents of the cellular membrane act more violently in conſequence of the accumulation of the ſenſorial power of aſſociation in the torpid heart and arteries, as explained in Suppl. I. 12.

A circumſtance curiouſly ſimilar to this occurs to ſome people on ſmoking tobacco for a ſhort time, who have not been accuſtomed to it. A degree of ſickneſs is preſently induced, and the pulſations of the heart and arteries become feeble for a ſhort time, as in the approach to fainting, owing to the direct ſympathy between theſe and the ſtomach, that is from defect of the excitement of the power of aſſociation. Then there ſucceeds a tingling, and heat, and ſometimes ſweat, owing to the increaſed action of the capillaries, or perſpirative and mucous glands; which is occaſioned by the accumulation of the ſenſorial power of aſſociation by the weaker action of the heart and arteries, which now increaſes the action of the capillaries.

8. Another method of increaſing abſorption from the cellular membrane is by warm air, or by warm ſteam. If the ſwelled legs of a dropſical patient are incloſed in a box, the air of which is made warm by a lamp or two, copious ſweats are ſoon produced by the increaſed action of the capillary glands, which are ſeen to ſtand on the ſkin, as it cannot readily exhale in ſo ſmall a quantity of air, which is only changed ſo faſt as may be neceſſary to permit the lamps to burn. At the ſame time the lymphatics of the cellular membrane are ſtimulated

ſtimulated by the heat into greater action, as appears by the ſpeedy reduction of the tumid legs.

It would be well worth trying an experiment upon a perſon labouring under a general anaſarca by putting him into a room filled with air heated to 120 or 130 degrees, which would probably excite a great general diaphoreſis, and a general cellular abſorption both from the lungs and every other part. And that air of ſo great heat may be borne for many minutes without great inconvenience was ſhewn by the experiments made in heated rooms by Dr. Fordyce and others. Philoſ. Tranſ.

Another experiment of uſing warmth in anaſarca, or in other diſeaſes, might be by immerſing the patient in warm air, or in warm ſteam, received into an oil-ſkin bag, or bathing-tub of tin, ſo managed, that the current of warm air or ſteam ſhould paſs round and cover the whole of the body except the head, which might not be expoſed to it; and thus the abſorbents of the lungs might be induced to act more powerfully by ſympathy with the ſkin, and not by the ſtimulus of heat. See Uſes of Warm Bath, Claſs IV. 2. 2. 1.

IV. 1. Venous abſorption. Cellary, water-creſſes, cabbages, and many other vegetables of the Claſs Tetradynamia, do not increaſe the heat of the body (except thoſe whoſe acrimony approaches to corroſion), and hence they ſeem alone, or principally, to act on the venous ſyſtem; the extremities of which we have ſhewn are abſorbents of the red blood, after it has paſſed the capillaries and glands.

2. In the ſea-ſcurvy and petechial fever the veins do not perfectly perform this office of abſorption; and hence the vibices are occaſioned by blood ſtagnating at their extremities, or extravaſated into the cellular membrane. And this claſs of vegetables, ſtimulating the veins to perform their natural abſorption, without increaſing the energy of the

the arterial action, prevents future petechiæ, and may assist the absorption of the blood already stagnated, as soon as its chemical change renders it proper for that operation.

3. The fluids, which are extravasated, and received into the cells of the cellular membrane, seem to continue there for many days, so as to undergo some chemical change, and are then taken up again by the mouths of the cellular absorbents. But the new vessels produced in inflamed parts, as they communicate with the veins, are probably absorbed again by the veins along with the blood which they contain in their cavities. Hence the blood, which is extravasated in bruises or vibices, is gradually many days in disappearing; but after due evacuations the inflamed vessels on the white of the eye, if any stimulant lotion is applied, totally disappear in a few hours.

Amongst absorbents affecting the veins we should therefore add the external application of stimulant materials; as of vinegar, which makes the lips pale on touching them. Friction, and electricity.

4. Hæmorrhages are of two kinds, either arterial, which are attended with inflammation; or venous, from a deficiency in the absorbent power of this set of vessels. In the former case the torpentia are efficacious; in the latter steel, opium, alum, and all the tribe of sorbentia, are used with success.

5. Sydenham recommends vegetables of the class Tetradynamia in rheumatic pains left after the cure of intermittents. These pains are perhaps similar to those of the sea-scurvy, and seem to arise from want of absorption in the affected part, and hence are relieved by the same medicines.

V. 1. Inteſtinal abſorption. Some aſtringent vegetables, as rhubarb, may be given in ſuch doſes as to prove cathartic; and, after a part of it is evacuated from the body, the remaining part augments the abſorption of the inteſtines; and acts, as if a ſimilar doſe had been exhibited after the operation of any other purgative. Hence 4 grains of rhubarb ſtrengthen the bowels, 30 grains firſt empty them.

2. The earthy ſalts, as alum, increaſe the inteſtinal abſorption, and hence induce conſtipation in their uſual doſe; alum is ſaid ſometimes to cure intermittents, perhaps when their ſeat is in the inteſtines, when other remedies have failed. It is uſeful in the diabætes by exciting the abſorbents of the bladder into their natural action; and combined with reſin is eſteemed in the fluor albus, and in gleets. Lime-ſtone or chalk, and probably gypſum, poſſeſs effects in ſome degree ſimilar, and increaſe the abſorption of the inteſtines; and thus in certain doſes reſtrain ſome diarrhœas, but in greater doſes alum I ſuppoſe will act as a cathartic. Five or ten grains produce conſtipation, 20 or 30 grains are either emetic or cathartic.

3. Earth of alum, tobacco-pipe clay, marl, Armenian bole, lime, crab's eyes or claws, and calcined hartſhorn, or bone aſhes, reſtrain fluxes; either mechanically by ſupplying ſomething like mucilage, or oil, or rollers to abate the friction of the aliment over inflamed membranes; or by increaſing their abſorption. The two laſt conſiſt of calcareous earth united to phoſphoric acid, and the Armenian bole and marl may contain iron. By the conſent between the inteſtines and the ſkin 20 grains of Armenian bole given at going into bed to hectic patients will frequently check their tendency to ſweat as well as to purge, and the more certainly if joined with one grain of opium.

VI. 1. Ab-

VI. 1. Abſorption from the liver, ſtomach, and other viſcera. When inflammations of the liver are ſubdued to a certain degree by veneſection, with calomel and other gentle purges, ſo that the arterial energy becomes weakened, four or eight grains of iron-filings, or of ſalt of ſteel, with the Peruvian bark, have wonderful effect in curing the cough, and reſtoring the liver to its uſual ſize and ſanity; which it ſeems to effect by increaſing the abſorption of this viſcus. The ſame I ſuppoſe happens in reſpect to the tumours of other viſcera, as of the ſpleen, or pancreas, ſome of which are frequently enlarged in agues.

2. Hæmorrhages from the noſe, rectum, kidnies, uterus, and other parts, are frequently attendant on diſeaſed livers; the blood being impeded in the vena portarum from the decreaſed power of abſorption, and in conſequence of the increaſed ſize of this viſcus. Theſe hæmorrhages after veneſection, and a mercurial cathartic, are moſt certainly reſtrained by ſteel alone, or joined with an opiate; which increaſe the abſorption, and diminiſh the ſize of the liver.

Chalybeates may alſo reſtrain theſe hæmorrhages by their promoting venous abſorption, though they exert their principal effect upon the liver. Hence alſo opiates, and bitters, and vitriolic acid, are advantageouſly uſed along with them. It muſt be added that ſome hæmorrhages recur by periods like the paroxyſms of intermittent fevers, and are thence cured by the ſame treatment.

3. The jaundice is frequently cauſed by the inſipidity of the bile, which does not ſtimulate the gall-bladder and bile-ducts into their due action; hence it ſtagnates in the gall-bladder, and produces a kind of cryſtallization, which is too large to paſs into the inteſtines, blocks up the bile-duct, and occaſions a long and painful diſeaſe. A paralyſis of the bile-duct produces a ſimilar jaundice, but without pain.

4. Worms in ſheep called flukes are owing to the dilute ſtate of the bile; hence they originate in the inteſtines, and thence migrate into the biliary ducts, and corroding the liver produce ulcers, cough, and hectic fever, called the rot. In human bodies it is probable the inert ſtate of the bile is one cauſe of the production of worms; which inſipid ſtate of the bile is owing to deficient abſorption of the thinner parts of it; hence the pale and bloated complexion, and ſwelled upper lip, of wormy children, is owing to the concomitant deficiency of abſorption from the cellular membrane. Salt of ſteel, or the ruſt of it, or filings of it, with bitters, increaſe the acrimony of the bile by promoting the abſorption of its aqueous part; and hence deſtroy worms, as well as by their immediate action on the inteſtines, or on the worms themſelves. The cure is facilitated by premiſing a purge with calomel. See Claſs I. 2. 3. 9.

5. The chloroſis is another diſeaſe owing to the deficient action of the abſorbents of the liver, and perhaps in ſome degree alſo to that of the ſecretory veſſels, or glands, which compoſe that viſcus. Of this the want of the catameniæ, which is generally ſuppoſed to be a cauſe, is only a ſymptom or conſequence. In this complaint the bile is deficient perhaps in quantity, but certainly in acrimony, the thinner parts not being abſorbed from it. Now as the bile is probably of great conſequence in the proceſs of making the blood; it is on this account that the blood is ſo deſtitute of red globules; which is evinced by the great paleneſs of theſe patients. As this ſerous blood muſt exert leſs ſtimulus on the heart, and arteries, the pulſe in conſequence becomes quick as well as weak, as explained in Sect. XII. 1. 4.

The quickneſs of the pulſe is frequently ſo great and permanent, that when attended by an accidental cough, the diſeaſe may be miſtaken for hectic fever; but is cured by chalybeates, and bitters exhibited twice a day; with half a grain of opium, and a grain of aloe

 every

every night; and the expected catamenia appears in consequence of a restoration of the due quantity of red blood. This and the two former articles approach to the disease termed paralysis of the liver. Sect. XXX. 4.

6. It seems paradoxical, that the same treatment with chalybeates, bitters, and opiates, which produces menstruation in chlorotic patients, should repress the too great or permanent menstruation, which occurs in weak constitutions at the time of life when it should cease. This complaint is an hæmorrhage owing to the debility of the absorbent power of the veins, and belongs to the paragraph on venous absorption above described, and is thence curable by chalybeates, alum, bitters, and particularly by the exhibition of a grain of opium every night with five grains of rhubarb.

7. Metallic salts supply us with very powerful remedies for promoting absorption in dropsical cases; which frequently are caused by enlargement of the liver. First, as they may be given in such quantities as to prove strongly cathartic, of which more will be said in the article on invertentia; and then, when their purgative quality ceases, like the effect of rhubarb, their absorbent quality continues to act. The salts of mercury, silver, copper, iron, zinc, antimony, have all been used in the dropsy; either singly for the former purpose, or united with bitters for the latter, and occasionally with moderate but repeated opiates.

8. From a quarter of a grain to half a grain of blue vitriol given every four or six hours, is said to be very efficacious in obstinate intermittents; which also frequently arise from an enlarged viscus, as the liver or spleen, and are thence owing to the deficient absorption of the lymphatics of that viscus. A quarter of a grain of white arsenic, as I was informed by a surgeon of the army, cures a quartan ague with

great

great certainty, if it be given an hour before the expected fit. This dose he said was for a robust man, perhaps one eighth of a grain might be given and repeated with greater safety and equal efficacy.

Dr. Fowler has given many successful cases in his treatise on this subject. He prepares it by boiling sixty-four grains of white arsenic in a Florence flask along with as much pure vegetable fixed alcali in a pint of distilled water, till it is dissolved, and then adding to it as much distilled water as will make the whole exactly sixteen ounces. Hence there are four grains of arsenic in every ounce of the solution. This should be put into a phial of such a size of the edge of its aperture, that sixty drops may weigh one dram, which will contain half a grain of arsenic. To children from two years old to four he gives from two to five drops three or four times a day. From five years old to seven, he directs seven or eight drops. From eight years old to twelve, he directs from seven to ten drops. From thirteen years old to eighteen he directs from ten to twelve drops. From eighteen upwards, twelve drops. In so powerful a medicine it is always prudent to begin with smaller doses, and gradually to increase them.

A saturated solution of arsenic in water is preferable I think to the above operose preparation of it; as no error can happen in weighing the ingredients, and it more certainly therefore possesses an uniform strength. Put much more white arsenic reduced to powder into a given quantity of distilled water, than can be dissolved in it. Boil it for half an hour in a Florence flask, or in a tin sauce-pan; let it stand to subside, and filter it through paper. My friend Mr. Greene, a surgeon at Brewood in Staffordshire, assured me, that he had cured in one season agues without number with this saturated solution; that he found ten drops from a two-ounce phial given thrice a day was a full dose for a grown person, but that he generally began with five.

9. The

9. The manner, in which arſenic acts in curing intermittent fevers, cannot be by its general ſtimulus, becauſe no intoxication or heat follows the uſe of it; nor by its peculiar ſtimulus on any part of the ſecreting ſyſtem, ſince it is not in ſmall doſes ſucceeded by any increaſed evacuation, or heat, and muſt therefore exert its power, like other articles of the ſorbentia, on the abſorbent ſyſtem. In what manner it deſtroys life ſo ſuddenly is difficult to underſtand, as it does not intoxicate like many vegetable poiſons, nor produce fevers like contagious matter. When applied externally it ſeems chemically to deſtroy the part like other cauſtics. Does it chemically deſtroy the ſtomach, and life in conſequence? or does it deſtroy the action of the ſtomach by its great ſtimulus, and life in conſequence of the ſympathy between the ſtomach and the heart? This laſt appears to be the moſt probable mode of its operation.

The ſucceſs of arſenic in the cure of intermittent fevers I ſuſpect to depend on its ſtimulating the ſtomach into ſtronger action, and thus, by the aſſociation of this viſcus with the heart and arteries, preventing the torpor of any part of the ſanguiferous ſyſtem. I was led to this concluſion from the following conſiderations.

Firſt. The effects of arſenic given a long time internally in ſmall doſes, or when uſed in larger quantities externally, ſeem to be ſimilar to thoſe of other great ſtimuli, as of wine or alcohol. Theſe are a bloated countenance, ſwelled legs, hepatic tumours, and dropſy, and ſometimes eruptions on the ſkin. The former of theſe I have ſeen, where arſenic has been uſed externally for curing the itch; and the latter appears on evidence in the famous trial of Miſs Blandy at Chelmſford, about forty years ago.

Secondly. I ſaw an ague cured by arſenic in a child, who had in vain previouſly taken a very large quantity of bark with great regularity. And another caſe of a young officer, who had lived intemperately, and laboured under an intermittent fever, and had taken the bark repeatedly in conſiderable quantities, with a grain of opium at night,

night, and though the paroxyſms had been thrice thus for a time prevented, they recurred in about a week. On taking five drops of a ſaturated ſolution of arſenic thrice a day the paroxyſms ceaſed, and returned no more, and at the ſame time his appetite became much improved.

Thirdly. A gentleman about 65 years of age had for about ten years been ſubject to an intermittent pulſe, and to frequent palpitations of his heart. Lately the palpitations ſeemed to obſerve irregular periods, but the intermiſſion of every third or fourth pulſation was almoſt perpetual. On giving him four drops of a ſaturated ſolution of arſenic from a two-ounce phial about every four hours for one day, not only the palpitation did not return, but the intermiſſion ceaſed entirely, and did not return ſo long as he took the medicine, which was three or four days.

Now as when the ſtomach has its action much weakened by an over-doſe of digitalis, the pulſe is liable to intermit, this evinces a direct ſympathy between theſe parts of the ſyſtem, and as I have repeatedly obſerved, that when the pulſe begins to intermit in elderly people, that an eructation from the ſtomach, voluntarily produced, will prevent the threatened ſtop of the heart; I am induced to think, that the torpid ſtate of the ſtomach, at the inſtant of the production of air occaſioned by its weak action, cauſed the intermiſſion of the pulſe. And that arſenic in this caſe, as well as in the caſes of agues above mentioned, produced its effects by ſtimulating the ſtomach into more powerful action; and that the equality of the motions of the heart was thus reſtored by increaſing the excitement of the ſenſorial power of aſſociation. See Sect. XXV. 17. Claſs IV. 2. 1. 18.

10. Where arſenic has been given as a poiſon, it may be diſcovered in the contents of the ſtomach by the ſmell like garlic, when a few grains of it are thrown on a red-hot iron. 2. If a few grains are placed between two plates of copper, and ſubjected to a red heat, the

the copper becomes whitened. 3. Diſſolve arſenic in water along with vegetable alcali, add to this a ſolution of blue vitriol in water, and the mixture becomes of a fine green, which gradually precipitates, as diſcovered by Bergman. 4. Where the quantity is ſufficient, ſome wheat may be ſteeped in a ſolution of it, which given to ſparrows or chickens will deſtroy them.

VII. Abſorption of the matter from venereal ulcers. No ulcer can heal, unleſs the abſorption from it is as great as the depoſition in it. The preparations or oxydes of mercury in the cure of the venereal diſeaſe ſeem to act by their increaſing the abſorption of the matter in the ulcers it occaſions; and that whether they are taken into the ſtomach, or applied on the ſkin, or on the ſurface of the ulcers. And this in the ſame manner as ſugar of lead, or other metallic oxydes, promote ſo rapidly the healing of other ulcers by their external application; and probably when taken internally, as ruſt of iron given to children affected with ſcrophulous ulcers contributes to heal them, and ſolutions of lead were once famous in phthiſis.

The matter depoſited in large abſceſſes does not occaſion hectic fever, till it has become oxygenated by being expoſed to the open air, or to the air through a moiſt membrane; the ſame ſeems to happen to other kinds of matter, which produce fever, or which occaſion ſpreading ulcers, and are thence termed contagious. See Claſs II. 1. 3. II. 1. 5. II. 1. 6. 6. This may perhaps occur from theſe matters not being generally abſorbed, till they become oxygenated; and that it is the ſtimulus of the acid thus formed by their union with oxygen, which occaſions their abſorption into the circulation, and the fever, which they then produce. For though collections of matter, and milk, and mucus, are ſometimes ſuddenly abſorbed during the action of emetics or in ſea-ſickneſs, they are probably eliminated from the body without entering the circulation; that is, they are taken up by the increaſed action of one lymphatic branch, and evacuated by the inverted action of ſome other lymphatic branch, and thus carried off by ſtool or urine.

But as the matter in large abſceſſes is in general not abſorbed, till it becomes by ſome means expoſed to air, there is reaſon to conclude, that the ſtimulus of this new combination of the matter with oxygen occaſions its abſorption; and that hence the abſorption of matter in ulcers of all kinds, is ſtill more powerfully effected by the external application or internal uſe of metallic oxydes; which are alſo acids conſiſting of the metal united with oxygen; and laſtly, becauſe venereal ulcers, and thoſe of itch, and tinea, will not heal without ſome ſtimulant application; that is, the ſecretion of matter in them continues to be greater, than the abſorption of it; and the ulcers at the ſame time continue to enlarge, by the contagion affecting the edges of them; that is, by the ſtimulus of the oxygenated matter ſtimulating the capillary veſſels in its vicinity into actions ſimilar to thoſe of the ulcer, which produces it.

This effect of the oxydes of mercury occurs, whether ſalivation attends its uſe or not. Salivation is much forwarded by external warmth, when mercury is given to promote this ſecretion; but as the cure of venereal complaints depends on its abſorbent quality, the act of ſalivation is not neceſſary or uſeful. A quarter of a grain of good corroſive ſublimate twice a day will ſeldom fail of curing the moſt confirmed pox; and will as ſeldom ſalivate, if the patient be kept cool. A quarter of a grain thrice a day I believe to be infallible, if it be good ſublimate.

Mercury alone when ſwallowed does not act beyond the inteſtines, its active preparations are the ſalts formed by its union with the various acids, as mentioned in the catalogue. Its union with the vegetable acid, when triturated with manna, is ſaid to compoſe Keyſer's Pill. Triturated with gum arabic it is much recommended by Plenk; and triturated with ſugar and a little eſſential oil, as directed in a former Edinburgh Difpenfatory, it probably forms fome of the fyrups fold as noftrums.

United with fulphur it feldom enters the circulation, as when cinnabar, or Æthiop's mineral, are taken inwardly. But united with fat and rubbed on the fkin, it is readily abforbed. I know not whether it can be united to charcoal, nor whether it has been given internally when united with animal fat.

VIII. 1. Abforptions in general are increafed by inanition; hence the ufe of evacuations in the cure of ulcers. Dr. Jurin abforbed in one night, after a day's abftinence and exercife, eighteen ounces from the atmofphere in his chamber; and every one muft have obferved, how foon his fheets became dry, after having been moiftened by fweat, if he throws off part of the bed-clothes to cool himfelf; which is owing to the increafed cutaneous abforption after the evacuation by previous fweat.

2. Now as opium is an univerfal ftimulant, as explained in the article on Incitantia, it muft ftimulate into increafed action both the fecretory fyftem, and the abforbent one; but after repeated evacuation by venefection, and cathartics, the abforbent fyftem is already inclined to act more powerfully; as the blood-veffels being lefs diftended, there is lefs refiftance to the progrefs of the abforbed fluids into them. Hence after evacuations opium promotes abforption, if given in fmall dofes, much more than it promotes fecretion; and is thus eminently of fervice at the end of inflammations, as in pleurify, or peripneumony, in the dofe of four or five drops of the tincture, given before the accefs of the evening paroxyfm; which I have feen fucceed even when the rifus fardonicus has exifted. Some convulfions may originate in the want of the abforption of fome acrid

 fecretion.

ſecretion, which occaſions pain; hence theſe diſeaſes are ſo much more certainly relieved by opium after veneſection or other evacuations.

IX. 1. Abſorption is increaſed by the calces or ſolutions of mercury, lead, zinc, copper, iron, externally applied; and by arſenic, and by ſulphur, and by the application of bitter vegetables in fine powder. Thus an ointment conſiſting of mercury and hog's fat rubbed on the ſkin cures venereal ulcers; and many kinds of herpetic eruptions are removed by an ointment conſiſting of 60 grains of white precipitate of mercury and an ounce of hog's fat.

2. The tumours about the necks of young people are often produced by the abſorption of a ſaline or acrid material, which has been depoſited from eruptions behind the ears, owing to deficient abſorption in the ſurface of the ulcer, but which on running down on the ſkin below becomes abſorbed, and ſwells the lymphatic glands of the neck; as the variolous matter, when inſerted into the arm, ſwells the gland of the axilla. Sometimes the perſpirative matter produced behind the ears becomes putrid from the want of daily waſhing them, and may alſo cauſe by its abſorption the tumours of the lymphatics of the neck. In the former caſe the application of a cerate of lapis calaminaris, or of ceruſſa applied in dry powder, or of rags dipped in a ſolution of ſugar of lead, increaſes the abſorption in the ulcers, and prevents the effuſion of the ſaline part of the ſecreted material. The latter is to be prevented by cleanlineſs.

After the eruptions or ulcers are healed a ſolution of corroſive ſublimate of one grain to an ounce of water applied for ſome weeks behind the ear, and amongſt the roots of the hair on one ſide of the head, where the mouths of the lymphatics of the neck open themſelves, frequently removes theſe tumours.

3. Linen

3. Linen rags moiſtened with a ſolution of half an ounce of ſugar of lead to a pint of water applied on the eryſipelas on anaſarcous legs, which have a tendency to mortification, is more efficacious than other applications. White vitriol ſix grains diſſolved in one ounce of roſe-water removes inflammations of the eyes after evacuation more certainly than ſolutions of lead. Blue vitriol two or three grains diſſolved in an ounce of water cures ulcers in the mouth, and other mucous membranes, and a ſolution of arſenic externally applied cures the itch, but requires great caution in the uſe of it. See Claſs II. 1. 5. 6.

4. Bitter vegetables, as the Peruvian bark, quilted between two ſhirts, or ſtrewed in their beds, will cure the ague in children ſometimes. Iron in ſolution, and ſome bitter extract, as in the form of ink, will cure one kind of herpes called the ringworm. And I have ſeen ſeven parts of bark in fine powder mixed with one part of ceruſs, or white lead, in fine powder, applied dry to ſcrophulous ulcers, and renewed daily, with great advantage.

5. To theſe ſhould be added electric ſparks and ſhocks, which promote the abſorption of the veſſels in inflamed eyes of ſcrophulous children; and diſperſe, or bring to ſuppuration, ſcrophulous tumours about the neck. For this laſt purpoſe ſmart ſhocks ſhould be paſſed through the tumours only, by incloſing them between two braſs knobs, communicating with the external and internal coating of a charged phial. See Art. II. 2. 2. 2.

X. 1. Bandages increaſe abſorption, if they are made to fit nicely on the part; for which purpoſe it is neceſſary to ſpread ſome moderately adheſive plaſter on the bandage, and to cut it into tails, or into ſhreds two inches wide; the ends are to be wrapped over each other; and it muſt be applied when the part is leaſt tumid, as in the morn-

8 ing

ing before the patient riſes, if on the lower extremities. The emplaſtrum de minio made to cover the whole of a ſwelled leg in this manner, whether the ſwelling is hard, which is uſually termed ſcor butic; or more eaſily compreſſible, as in anaſarca, reduces the limb in two or three days to its natural ſize; for this purpoſe I have ſometimes uſed carpenter's glue, mixed with one twentieth part of honey to prevent its becoming too hard, inſtead of a reſinous plaſter; but the minium plaſter of the ſhops is in general to be preferred. Nothing ſo much facilitates the cure of ulcers in the legs, as covering the whole limb from the toes to the knee with ſuch a plaſter-bandage; which increaſes the power of abſorption in the ſurface of the ſore.

2. The lymph is carried along the abſorbent veſſels, which are replete with valves, by the intermitted preſſure of the arteries in their neighbourhood. Now if the external ſkin of the limb be lax, it riſes, and gives way to the preſſure of the arteries at every pulſation; and thence the lymphatic veſſels are ſubject to the preſſure of but half the arterial force. But when the external ſkin is tightened by the ſurrounding bandage, and thence is not elevated by the arterial diaſtole, the whole of this power is exerted in compreſſing the lymphatic veſſels, and carrying on the lymph already abſorbed; and thence the abſorbent power is ſo amazingly increaſed by bandage nicely applied. Pains are ſometimes left in the fleſhy parts of the thighs or arms, after the inflammation is gone, in the acute rheumatiſm, or after the patient is too weak for further evacuation; in this caſe after internal abſorbent medicines, as the bark, and opiates, have been uſed in vain, I have ſucceſsfully applied a plaſter-bandage, as above deſcribed, ſo as to compreſs the pained part.

XI. 1. We ſhall conclude by obſerving, that the ſorbentia ſtrengthen the whole habit by preventing the eſcape of the fluid part of of the ſecretions out of the body, before it has given up as much nouriſhment, as it is capable; as the liquid part of the ſecretion of urine, ſweat, ſaliva, and of all other ſecretions, which are poured into receptacles. Hence they have been ſaid to brace the body, and been called tonics, which are mechanical terms not applicable to the living bodies of animals; as explained in Sect. XXXII. 3. 2.

2. A continued uſe of bitter medicines for years together, as of Portland's powder, or of the bark, is ſuppoſed to induce apoplexy, or other fatal diſeaſes. Two caſes of this kind have fallen under my obſervation; the patients were both rather intemperate in reſpect to the uſe of fermented liquors, and one of them had been previouſly ſubject to the gout. As I believe the gout generally originates from a torpor of the liver, which inſtead of being ſucceeded by an inflammation of it, is ſucceeded by an inflammation of ſome of the joints; or by a pimpled face, which is another mode, by which the diſeaſe of the liver is terminated. I conceive, that the daily uſe of bitter medicine had in theſe patients prevented the removal of a gouty inflammation from the liver to the membranes of the joints of the extremities, or to the ſkin of the face, by preventing the neceſſary torpor of theſe parts previous to the inflammation of them; in the ſame manner as cold fits of fever are prevented by the ſame medicines; and, as I believe, the returns of the gout have ſometimes for two or three years been prevented by them.

One of theſe patients died of the apoplexy in a few hours; and the other of an inflammation of the liver, which I believe was called the gout, and in conſequence was not treated by veneſection, and other evacuations. From hence it appears, that the daily uſe of hop in our malt liquor muſt add to the noxious quality of the ſpirit in it, when taken to exceſs, and contribute to the production of apoplexy, or inflammation of the liver.

III. CATALOGUE

## III. Catalogue of the Sorbentia.

I. Sorbentia affecting the ſkin.

1. Acid of vitriol, of ſea-ſalt, lemons, ſloes, prunus ſpinoſa, crabs, pyrus, quince, pyrus cydonia, opium.
2. Externally calx of zinc, of lead, of mercury.

II. Sorbentia affecting the mucous membranes.

1. Juice of ſloes, crabs, Peruvian bark, cinchona, opium.
2. Externally blue vitriol.

III. Sorbentia affecting the cellular membrane.

1. Peruvian bark, wormwoods, artemiſia maritima, artemiſia abſynthium, worm-ſeed, artemiſia ſantonicum, chamomile, anthemis nobilis, tanſey tanacetum, bogbean, menyanthes trifoliata, centaury, gentiana centaurium, gentian, gentiana lutea, artichoke-leaves, cynara ſcolymus, hop, humulus lupulus.
2. Orange-peel, cinnamon, nutmeg, mace.
3. Vomits, ſquill, digitalis, tobacco.
4. Bath of warm air, of ſteam.

IV. Sorbentia affecting the veins.

1. Water-creſs, ſiſymbrium naſturtium aquaticum, muſtard, ſinapis, ſcurvy-graſs cochlearia hortenſis, horſe-radiſh cochlearia armoracia, cuckoo-flower, cardamine, dog's-graſs, dandelion, leontodon taraxacon, cellery apium, cabbage braſſica.

2. Chalybeates,

2. Chalybeates, bitters, and opium, after ſufficient evacuation.

3. Externally vinegar, friction, electricity.

V. Sorbentia affecting the inteſtines.

1. Rhubarb, rheum palmatum, oak-galls, gallæ quercinæ, tormentil, tormentilla erecta, cinquefoil potentilla, red-roſes, uva urſi, ſimarouba.

2. Logwood, hæmatoxylum campechianum, ſuccus acaciæ, dragon's blood, terra japonica, mimoſa catechu.

3. Alum, earth of alum, Armenian bole, chalk, creta, crab's claws, chelæ cancrorum, white clay, cimolia, calcined hartſhorn, cornu cervi calcinatum, bone-aſhes.

VI. Sorbentia affecting the liver, ſtomach, and other viſcera. Ruſt of iron, filings of iron, ſalt of ſteel, ſal martis, blue vitriol, white vitriol, calomel, emetic tartar, ſugar of lead, white arſenic.

VII. Sorbentia affecting venereal ulcers. Mercury diſſolved or corroded by the following acids:

1. Diſſolved in vitriolic acid, called turpeth mineral, or hydrargyrus vitriolatus.

2. Diſſolved in nitrous acid, called hydrargyrus nitratus ruber.

3. Diſſolved in muriatic acid, mercurius corroſivus ſublimatus, or hydrargyrus muriatus.

4. Corroded by muriatic acid. Calomel.

5. Precipitated from muriatic acid, mercurius precipitatus albus, calx hydrargyri alba.

6. Corroded by carbonic acid? The black powder on crude mercury.

7. Calcined, or united with oxygen.

8. United with animal fat, mercurial ointment.

9. United with ſulphur. Cinnabar.

10. Partially united with ſulphur. Æthiops mineral.

11. Divided by calcareous earth. Hydrargyrus cum cretâ.

12. Divided by vegetable mucilage, by ſugar, by balſams.

VIII. Sorbentia affecting the whole ſyſtem. Evacuations by veneſection and catharſis, and then by the exhibition of opium.

IX. Sorbentia externally applied.

1. Solutions of mercury, lead, zinc, copper, iron, arſenic; or metallic calces applied in dry powder, as ceruſſa, lapis calaminaris.

2. Bitter vegetables in decoctions and in dry powders, applied externally, as Peruvian bark, oak bark, leaves of wormwood, of tanſey, camomile flowers or leaves.

3. Electric ſparks, or ſhocks.

X. Bandage ſpread with emplaſtrum e minio, or with carpenter's glue mixed with one twentieth part of honey.

XI. Portland's powder its continued uſe pernicious, and of hops in beer.

ART.

## Art. V.

## INVERTENTIA.

I. Those things, which invert the natural order of the ſucceſſive irritative motions, are termed invertentia.

1. Emetics invert the motions of the ſtomach, duodenum, and œſophagus.

2. Violent cathartics invert the motions of the lacteals, and inteſtinal lymphatics.

3. Violent errhines invert the naſal lymphatics, and thoſe of the frontal and maxillary ſinuſes. And medicines producing nauſea, invert the motions of the lymphatics about the fauces.

4. Medicines producing much pale urine, as a certain quantity of alcohol, invert the motions of the urinary abſorbents; if the doſe of alcohol is greater, it inverts the ſtomach, producing the drunken ſickneſs.

5. Medicines producing cold ſweats, palpitation of the heart, globus hyſtericus; as violent evacuations, ſome poiſons, fear, anxiety, act by inverting the natural order of the vaſcular motions.

## II. Observations on the Invertentia.

I. 1. The action of vomiting seems originally to have been occasioned by disagreeable sensation from the distention or acrimony of the aliment; in the same manner as when any disgustful material is taken into the mouth, as a bitter drug, and is rejected by the retrograde motions of the tongue and lips; as explained in Class IV. 1. 1. 2. and mentioned in Sect. XXXV. 1. 3. Or the disagreeable sensation may thus excite the power of volition, which may also contribute to the retrograde actions of the stomach and œsophagus, as when cows bring up the contents of their first stomach to re-masticate it. To either of these is to be attributed the action of mild emetics, which soon cease to operate, and leave the stomach stronger, or more irritable, after their operation; owing to the accumulation of the sensorial power of irritation during its torpid or inverted action. Such appears to be the operation of ipecacuanha, or of antimonium tartarizatum, in small doses.

2. But there is reason to believe, that the stronger emetics, as digitalis, first stimulate the absorbent vessels of the stomach into greater action; and that the inverted motions of these absorbents next occur, pouring the lymph, lately taken up, or obtained from other lymphatic branches, into the stomach: the quantity of which in some diseases, as in the cholera morbus, is inconceivable. This inverted motion, first of the absorbents of the stomach, and afterwards of the stomach itself, seems to originate from the exhaustion or debility, which succeeds the unnatural degree of action, into which they had been previously stimulated. An unusual defect of stimulus, as of food without spice or wine in the stomachs of those, who have been much

accuſtomed to ſpice or wine, will induce ſickneſs or vomiting; in this caſe the defective energy of the ſtomach is owing to defect of accuſtomed ſtimulus; while the action of vomiting from digitalis is owing to a deficiency of ſenſorial power, which is previouſly exhauſted by the exceſs of its ſtimulus. See Sect. XXXV. 1. 3. and Claſs IV. 1. 1. 2.

For firſt, no increaſe of heat ariſes from this action of vomiting; which always occurs, when the ſecerning ſyſtem is ſtimulated into action. Secondly, the motions of the abſorbent veſſels are as liable to inverſion as the ſtomach itſelf; which laſt, with the œſophagus, may be conſidered as the abſorbent mouth and belly of that great gland, the inteſtinal canal. Thirdly, the claſs of ſorbentia, as bitters and metallic ſalts, given in large doſes, become invertentia, and vomit, or purge. And laſtly, the ſickneſs and vomiting induced by large potations of wine, or opium, does not occur till next day in ſome people, in none till ſome time after their ingurgitation. And tincture of digitalis in the doſe of 30 or 60 drops, though applied in ſolution, is a conſiderable time before it produces its effect; though vomiting is inſtantaneouſly induced by a nauſeous idea, or a nauſeous taſte in the mouth. At the ſame time there ſeem to be ſome materials, which can immediately ſtimulate the ſtomach into ſuch powerful action, as to be immediately ſucceeded by paralyſis of it, and conſequent continued fever, or immediate death; and this without exciting ſenſation, that is, without our perceiving it. Of theſe are the contagious matter of ſome fevers ſwallowed with the ſaliva, and probably a few grains of arſenic taken in ſolution. See Suppl. I. 8. 8. Art. IV. 2. 6. 9.

3. Some branches of the lymphatic ſyſtem become inverted by their ſympathy with other branches, which are only ſtimulated into too violent abſorption. Thus when the ſtomach and duodenum are much ſtimulated by alcohol, by nitre, or by worms, in ſome perſons the

the urinary lymphatics have their motion inverted, and pour that material into the bladder, which is abſorbed from the inteſtines. Hence the drunken diabetes is produced; and hence chyle is ſeen in the urine in worm caſes.

When on the contrary ſome branches of the abſorbent ſyſtems have their motions inverted in conſequence of the previous exhauſtion of their ſenſorial power by any violent ſtimulus, other branches of it have their abſorbent power greatly increaſed. Hence continued vomiting, or violent cathartics, produce great abſorption from the cellular membrane in caſes of dropſy; and the fluids thus abſorbed are poured into the ſtomach and inteſtines by the inverted motions of the lacteals and lymphatics. See Sect. XXIX. 4. and 5.

4. The quantity of the doſe of an emetic is not of ſo great conſequence as of other medicines, as the greateſt part of it is rejected with the firſt effort. All emetics are ſaid to act with greater certainty when given in a morning, if an opiate had been given the night before. For the ſenſorial power of irritation of the ſtomach had thus been in ſome meaſure previouſly exhauſted by the ſtimulus of the opium, which thus facilitates the action of the emetic; and which, when the doſe of opium has been large, is frequently followed on the next day by ſpontaneous ſickneſs and vomitings, as after violent intoxication.

Ipecacuanha is the moſt certain in its effect from five grains to thirty; white vitriol is the moſt expeditious in its effect, from twenty grains to thirty diſſolved in warm water; but emetic tartar, antimonium tartarizatum, from one grain to four to ſane people, and from thence to twenty to inſane patients, will anſwer moſt of the uſeful purpoſes of emetics; but nothing equals the digitalis purpurea for the purpoſe of abſorbing water from the cellular membrane in the anaſarca pulmonum, or hydrops pectoris. See Art. II. 3. 7.

II. Violent

II. Violent cathartics. 1. Where violent cathartics are required, as in dropfies, the fquill in dried powder made into fmall pills of a grain, or a grain and a half, one to be given every hour till they operate brifkly, is very efficacious; or half a grain of emetic tartar diffolved in an ounce of peppermint-water, and given every hour, till it operates. Scammony, and other ftrong purges, are liable to produce hypercatharfis, if they are not nicely prepared, and accurately weighed, and are thence dangerous in common practice. Gamboge is uncertain in its effects, it has otherwife the good property of being taftelefs; and on that account fome preparation of it might be ufeful for children, by which its dofe could be afcertained, and its effects rendered more uniform.

2. In inflammations of the bowels with conftipation calomel, given in the dofe from ten to twenty grains after due venefection, is moft efficacious; and if made into very fmall pills is not liable to be rejected by vomiting, which generally attends thofe cafes. When this fails, a grain of aloes every hour will find its way, if the bowel is not deftroyed; and fometimes, I believe, if it be, when the mortification is not extenfive. If the vomiting continues after the pain ceafes, and efpecially if the bowels become tumid with air, which founds on being ftruck with the finger, thefe patients feldom recover. Opiates given along with the cathartics I believe to be frequently injurious in inflammation of the bowels, though they may thus be given with advantage in the faturnine colic; the pain and conftipation in which difeafe are owing to torpor or inactivity, and not to too great action.

III. Violent errhines and fialagogues. 1. Turpeth mineral in the quantity of one grain mixed with ten grains of fugar anfwers every purpofe to be expected from errhines. Their operation is by inverting

 the

the motions of the lymphatics of the membrane, which lines the noſtrils, and the caverns of the forehead and cheeks; and may thence poſſibly be of ſervice in the hydrocephalus internus.

Some other violent errhines, as the powder of white hellebore, or Cayan pepper, diluted with ſome leſs acrid powder, are ſaid to cure ſome cold or nervous head-achs; which may be effected by inflaming the noſtrils, and thus introducing the ſenſorial power of ſenſation, as well as increaſing that of irritation; and thus to produce violent action of the membranes of the noſtrils, and of the frontal and maxillary ſinuſes, which may by aſſociation excite into action the torpid membranes, which occaſion the head-ach.

2. A copious ſalivation without any increaſe of heat often attends hyſteric diſeaſes, and fevers with debility, owing to an inverſion of the lymphatics of the mouth, ſee Claſs I. 1. 2. 6. The ſame occurs in the nauſea, which precedes vomiting; and is alſo excitable by diſagreeable taſtes, as by ſquills, or by nauſeous ſmells, or by nauſeous ideas. Theſe are very ſimilar to the occaſional diſcharge of a thin fluid from the noſtrils of ſome people, which recurs at certain periods, and differs from defective abſorption.

IV. Violent diuretics. 1. If nitre be given from a dram to half an ounce in a morning at repeated draughts, the patient becomes ſickiſh, and much pale water is thrown into the bladder by the inverted action of the urinary lymphatics. Hence the abſorption in ulcers is increaſed and the cure forwarded, as obſerved by Dr. Rowley.

2. Cantharides taken inwardly ſo ſtimulate the neck of the bladder as to increaſe the diſcharge of mucus, which appears in the urine; but I once ſaw a large doſe taken by miſtake, not leſs than half an ounce or an ounce of the tincture, by which I ſuppoſe the urinary

lymphatics

lymphatics were thrown into violent inverted motions, for the patient drank repeated draughts of ſubtepid water to the quantity of a gallon or two in a few hours; and during the greateſt part of that time he was not I believe two entire minutes together without making water. A little blood was ſeen in his water the next day, and a ſoreneſs continued a day longer without any other inconvenience.

3. The decoction of foxglove ſhould alſo be mentioned here, as great effuſions of urine frequently follow its exhibition. See Art. IV. 2. 3. 7. And an infuſion or tincture of tobacco as recommended by Dr. Fowler of York.

4. Alcohol, and opium, if taken ſo as to induce ſlight intoxication, and the body be kept cool, and much diluting liquids taken along with them, have ſimilar effect in producing for a time a greater flow of urine, as moſt intemperate drinkers muſt occaſionally have obſerved. This circumſtance ſeems to have introduced the uſe of gin, and other vinous ſpirits as a diuretic, unfortunately in the gravel, amongſt ignorant people; which diſeaſe is generally produced by fermented or ſpirituous liquors, and always increaſed by them.

5. Fear and anxiety are well known to produce a great frequency of making water. A perſon, who believed he had made a bad purchaſe concerning an eſtate, told me, that he made five or ſix pints of water during a ſleepleſs night, which ſucceeded his bargain; and it is uſual, where young men are waiting in an anti-room to be examined for college preferment, to ſee the chamber-pot often wanted.

V. Cold ſweats about the head, neck, and arms, frequently attend thoſe, whoſe lungs are oppreſſed, as in ſome dropſies and aſthma. A cold ſweat is alſo frequently the harbinger of death. Theſe are from the inverted motions of the cutaneous lymphatic branches of thoſe parts.

### III. CATALOGUE OF INVERTENTIA.

I. Emetics, ipecacuanha, emetic tartar, antimonium tartariſatum, ſquill, ſcilla maritima, carduus benedictus, cnicus acarna, chamœmile, anthemis nobilis, white vitriol, vitriolum zinci, foxglove, digitalis purpurea, clyſters of tobacco.

II. Violent cathartics, emetic tartar, ſquill, buckthorn, rhamnus catharticus, ſcammonium, convolvulus ſcammonia, gamboge, elaterium, colocynth, cucumis colocynthis, veratrum.

III. Violent errhines and ſialagogues, Turpeth mineral, hydragyrus vitriolatus, aſarum europæum, euphorbium, capſicum, veratrum, nauſeous ſmells, nauſeous ideas.

IV. Violent diuretics, nitre, ſquill, ſeneka, cantharides, alcohol, foxglove, tobacco, anxiety.

V. Cold ſudorifics, poiſons, fear, approaching death.

## Art. VI.

## REVERTENTIA.

I. Those things, which reſtore the natural order of the inverted irritative motions, are termed Revertentia.

1. As muſk, caſtor, aſafœtida, valerian, eſſential oils.

2. Externally the vapour of burnt feathers, of volatile ſalts, or oils, bliſters, ſinapiſms.

Theſe reclaim the inverted motions without increaſing the heat of the body above its natural ſtate, if given in their proper doſes, as in the globus hyſtericus, and palpitation of the heart.

The incitantia revert theſe morbid motions more certainly, as opium and alcohol: and reſtore the natural heat more; but if they induce any degree of intoxication, they are ſucceeded by debility, when their ſtimulus ceaſes.

### II. Observations on the Revertentia.

I. The hyſteric diſeaſe is attended with inverted motions feebly exerted of the œſophagus, inteſtinal canal and lymphatics of the bladder. Hence the borborigmi, or rumbling of the bowels, owing to their fluid contents deſcending as the air beneath aſcends. The globus hyſtericus conſiſts in the retrograde motion of the œſophagus, and the great flow of urine from that of the lymphatics ſpread on the neck of the bladder; and a copious ſalivation ſometimes happens to



theſe

theſe patients from the inverſion of the lymphatics of the mouth ; and palpitation of the heart owing to weak or incipient inverſion of its motions ; and ſyncope, when this occurs in its greateſt degree.

Theſe hyſteric affections are not neceſſarily attended with pain ; though it ſometimes happens, that pains, which originate from quieſcence, afflict theſe patients, as the hemicrania, which has erroneouſly been termed the clavus hyſtericus ; but which is owing ſolely to the inaction of the membranes of that part, like the pains attending the cold fits of intermittents, and which frequently returns like them at very regular periods of time.

Many of the above ſymptoms are relieved by muſk, caſtor, the fœtid gums, valerian, oleum animale, oil of amber, which act in the uſual doſe without heating the body. The pains, which ſometimes attend theſe conſtitutions, are relieved by the ſecernentia, as eſſential oils in common tooth-ach, and balſam of Peru in the flatulent colic. But the incitantia, as opium, or vinous ſpirit, reclaim theſe morbid inverted motions with more certainty, than the fœtids ; and remove the pains, which attend theſe conſtitutions, with more certainty than the ſecernentia ; but if given in large doſes, a debility and return of the hyſteric ſymptoms occurs, when the effect of the opium or alcohol ceaſes. Opiates and fœtids joined ſeem beſt to anſwer the purpoſe of alleviating the preſent ſymptoms ; and the ſorbentia, by ſtimulating the lymphatics and lacteals into continued action, prevent a relapſe of their inverſion, as Peruvian bark, and ruſt of iron. See Claſs I. 3. 1. 10.

II. Vomiting conſiſts in the inverted order of the motions of the ſtomach, and œſophagus ; and is alſo attended with the inverted motions of a part of the duodenum, when bile is ejected ; and of the lymphatics of the ſtomach and fauces, when nauſea attends, and when much lymph is evacuated. Permanent vomiting is for a time relieved by the incitantia, as opium or alcohol ; but is liable to return, when their

their action ceases. A blister on the back, or on the stomach, is more efficacious for restraining vomiting by their stimulating into action the external skin, and by sympathy affecting the membranes of the stomach. In some fevers attended with incessant vomiting Sydenham advised the patient to put his head under the bed-clothes, till a sweat appeared on the skin, as explained in Class IV. 1. 1. 3.

In chronical vomiting I have observed crude mercury of good effect in the dose of half an ounce twice a day. The vomitings, or vain efforts to vomit, which sometimes attend hysteric or epileptic patients, are frequently instantly relieved for a time by applying flour of mustard-seed and water to the small of the leg; and removing it, as soon as the pain becomes considerable. If sinapisms lie on too long, especially in paralytic cases, they are liable to produce troublesome ulcers. A plaster or cataplasm, with opium and camphor on the region of the stomach, will sometimes revert its retrograde motions.

III. Violent catharsis, as in diarrhœa or dysentery, is attended with inverted motions of the lymphatics of the intestines, and is generally owing to some stimulating material. This is counteracted by plenty of mucilaginous liquids, as solutions of gum arabic, or small chicken broth, to wash away or dilute the stimulating material, which causes the disease. And then by the use of the intestinal sorbentia, Art. IV. 2. 5. as rhubarb, decoction of logwood, calcined hartshorn, Armenian bole; and lastly, by the incitantia, as opium.

IV. The diabœtes consists in the inverted motions of the urinary lymphatics, which is generally I suppose owing to the too great action of some other branch of the absorbent system. The urinary branch should be stimulated by cantharides, turpentine, resin (which when taken in larger doses may possibly excite it into inverted action), by the sorbentia and opium. The intestinal lymphatics should

be

be rendered leſs active by torpentia, as calcareous earth, earth of alum; and thoſe of the ſkin by oil externally applied over the whole body; and by the warm-bath, which ſhould be of 96 or 98 degrees of heat, and the patient ſhould ſit in it every day for half an hour.

V. Inverted motions of the inteſtinal canal with all the lymphatics, which open into it, conſtitute the ileus, or iliac paſſion; in which diſeaſe it ſometimes happens, that clyſters are returned by the mouth. After veneſection from ten grains to twenty of calomel made into very ſmall pills; if this is rejected, a grain of aloe every hour; a bliſter; crude mercury; warm-bath; if a clyſter of iced water?

Many other inverted motions of different parts of the ſyſtem are deſcribed in Claſs I. 3. and which are to be treated in a manner ſimilar to thoſe above deſcribed. It muſt be noted, that the medicines mentioned under number one in the catalogue of revertentia are the true articles belonging to this claſs of medicines. Thoſe enumerated in the other four diviſions are chiefly ſuch things as tend to remove the ſtimulating cauſes, which have induced the inverſion of the motions of the part, as acrimonious contents, or inflammation, of the bowels in diarrhœa, diabetes, or in ileus. But it is probable after theſe remote cauſes are deſtroyed, that the fetid gums, muſk, caſtor, and balſams, might be given with advantage in all theſe caſes.

III. Cata-

### III. Catalogue of Revertentia.

I. Inverted motions, which attend the hyſteric diſeaſe, are reclaimed, 1. By muſk, caſtor. 2. By aſafœtida, galbanum, ſagapænum, ammoniacum, valerian. 3. Eſſential oils of cinnamon, nutmeg, cloves, infuſion of penny-royal, mentha, pulegium, peppermint, mentha piperita, ether, camphor. 4. Spirit of hartſhorn, oleum animale, ſpunge burnt to charcoal, black-ſnuffs of candles, which conſiſt principally of animal charcoal, wood-ſoot, oil of amber. 5. The incitantia, as opium, alcohol, vinegar. 6. Externally the ſmoke of burnt feathers, oil of amber, volatile ſalt applied to the noſtrils, bliſters, ſinapiſms.

II. Inverted motions of the ſtomach are reclaimed by opium, alcohol, bliſters, crude mercury, ſinapiſms, camphor and opium externally, clyſters with aſafœtida.

III. Inverted motions of the inteſtinal lymphatics are reclaimed by mucilaginous diluents, and by inteſtinal ſorbentia, as rhubarb, logwood, calcined hartſhorn, Armenian bole; and laſtly by incitantia, as opium.

IV. Inverted motions of the urinary lymphatics are reclaimed by cantharides, turpentine, roſin, the ſorbentia, and opium, with calcareous earth, and earth of alum, by oil externally, warm-bath.

V. Inverted motions of the inteſtinal canal are reclaimed by calomel, aloe, crude mercury, bliſters, warm-bath, clyſters with aſafœtida, clyſters of iced water? or of ſpring water further cooled by ſalt diſſolved in water contained in an exterior veſſel? Where there exiſts an introſuſception of the bowel in children, could the patient be held up for a time by the feet with his head downwards, or be laid with his body on an inclined plane with his head downwards, and crude mercury be injected as a clyſter to the quantity of two or three pounds?

## Art. VII.

## TORPENTIA.

I. Those things, which diminiſh the exertion of the irritative motions, are termed torpentia.

1. As mucus, mucilage, water, bland oils, and whatever poſſeſſes leſs ſtimulus than our uſual food. Diminution of heat, light, ſound, oxygen, and of all other ſtimuli; veneſection, nauſea, and anxiety.

2. Thoſe things which chemically deſtroy acrimony, as calcareous earth, ſoap, tin, alcalies, in cardialgia; or which prevent chemical acrimony, as acid of vitriol in cardialgia, which prevents the fermentation of the aliment in the ſtomach, and its conſequent acidity. Secondly, which deſtroy worms, as calomel, iron filings or ruſt of iron, in the round worms; or amalgama of quickſilver and tin, or tin in very large doſes, in the tape-worms. Will ether in clyſters deſtroy aſcarides? Thirdly, by chemically deſtroying extraneous bodies, as cauſtic alcali, lime, mild alcali in the ſtone. Fourthly, thoſe things which lubricate the veſſels, along which extraneous bodies ſlide, as oil in the ſtone in the urethra, and to expedite the expectoration of hardened mucus; or which leſſen the friction of the contents in the inteſtinal canal in dyſentery or aphtha, as calcined hartſhorn, clay, Armenian bole, chalk, bone-aſhes. Fifthly, ſuch things as ſoften or extend the cuticle over tumors, or phlegmons, as warm water, poultices, fomentations, or by confining the perſpirable

matter on the part by cabbage-leaves, oil, fat, bee's-wax, plaſters, oiled ſilk, externally applied.

Theſe decreaſe the natural heat and remove pains occaſioned by exceſs of irritative motions.

## II. Observations on the Torpentia.

I. As the torpentia conſiſt of ſuch materials as are leſs ſtimulating than our uſual diet, it is evident, that where this claſs of medicines is uſed, ſome regard muſt be had to the uſual manner of living of the patient both in reſpect to quantity and quality. Hence wounds in thoſe, who have been accuſtomed to the uſe of much wine, are very liable to mortify, unleſs the uſual potation of wine be allowed the patient. And in theſe habits I have ſeen a delirium in a fever cured almoſt immediately by wine; which was occaſioned by the too mild regimen directed by the attendants. On the contrary in great inflammation, the ſubduction of food, and of ſpirituous drink, contributes much to the cure of the diſeaſe. As by theſe means both the ſtimulus from diſtention of the veſſels, as well as that from the acrimony of the fluids, is decreaſed; but in both theſe reſpects the previous habits of diet of the patients muſt be attended to. Thus if tea be made ſtronger, than the patient has uſually drank it, it belongs to the article ſorbentia; if weaker, it belongs to the torpentia.

II. 2. Water in a quantity greater than uſual diminiſhes the action of the ſyſtem not only by diluting our fluids, and thence leſſening their ſtimulus, but by lubricating the ſolids; for not only the parts

parts of our ſolids have their ſliding over each other facilitated by the interpoſition of aqueous particles; but the particles of mucaginous or ſaccharine ſolutions ſlide eaſier over each other by being mixed with a greater portion of water, and thence ſtimulate the veſſels leſs.

At the ſame time it muſt be obſerved, that the particles of water themſelves, and of animal gluten diſſolved in water, as the glue uſed by carpenters, ſlide eaſier over each other by an additional quantity of the fluid matter of heat.

Theſe two fluids of heat and of water may be eſteemed the univerſal ſolvents or lubricants in reſpect to animal bodies, and thus facilitate the circulation, and the ſecretion of the various glands. At the ſame time it is poſſible, that theſe two fluids may occaſionally aſſume an aerial form, as in the cavity of the cheſt, and by compreſſing the lungs may cauſe one kind of aſthma, which is relieved by breathing colder air. An increaſed quantity of heat by adding ſtimulus to every part of the ſyſtem belongs to the article Incitantia.

III. 3. 1. The application of cold to the ſkin, which is only another expreſſion for the diminution of the degree of heat we are accuſtomed to, benumbs the cutaneous abſorbents into inaction; and by ſympathy the urinary and inteſtinal abſorbents become alſo quieſcent. The ſecerning veſſels continuing their action ſomewhat longer, from the warmth of the blood. Hence the uſual ſecretions are poured into the bladder and inteſtines, and no abſorption is retaken from them. Hence ſprinkling the ſkin with cold water increaſes the quantity of urine, which is pale; and of ſtool, which is fluid; theſe have erroneouſly been aſcribed to increaſed ſecretion, or to obſtructed perſpiration.

The thin diſcharge from the noſtrils of ſome people in cold weather is owing to the torpid ſtate of the abſorbent veſſels of the membrana ſneideriana, which as above are benumbed ſooner than thoſe, which perform the ſecretion of the mucus.

 The

The quick anhelation, and palpitation of the heart, of thoſe, who are immerſed in cold water, depends on the quieſcence of the external abſorbent veſſels and capillaries. Hence the cutaneous circulation is diminiſhed, and by aſſociation an almoſt univerſal torpor of the ſyſtem is induced; thence the heart becomes incapable to puſh forwards its blood through all the inactive capillaries and glands; and as the terminating veſſels of the pulmonary artery ſuffer a ſimilar inaction by aſſociation, the blood is with difficulty puſhed through the lungs.

Some have imagined, that a ſpaſmodic conſtriction of the ſmaller veſſels took place, and have thus accounted for their reſiſtance to the force of the heart. But there ſeems no neceſſity to introduce this imaginary ſpaſm; ſince thoſe, who are converſant in injecting bodies, find it neceſſary firſt to put them into warm water to take away the ſtiffneſs of the cold dead veſſels; which become inflexible like the other muſcles of dead animals, and prevent the injected fluid from paſſing.

All the ſame ſymptoms occur in the cold fits of intermittents; in theſe the coldneſs and paleneſs of the ſkin with thirſt evince the diminution of cutaneous abſorption; and the dryneſs of ulcers, and ſmall ſecretion of urine, evince the torpor of the ſecerning ſyſtem; and the anhelation, and coldneſs of the breath, ſhew the terminations of the pulmonary artery to be likewiſe affected with torpor.

After theſe veſſels of the whole ſurface of the body both abſorbent and ſecretory have been for a time torpid by the application of cold water; and all the internal ſecerning and abſorbent ones have been made torpid from their aſſociation with the external; as ſoon as their uſual ſtimulus of warmth is renewed, they are thrown into more than their uſual energy of action; as the hands become hot and painful on approaching the fire after having been immerſed ſome time in ſnow. Hence the face becomes of a red colour in a cold day on turning from the wind, and the inſenſible perſpiration increaſed by repeatedly

peatedly going into frosty air, but not continuing in it too long at a time.

2. When by the too great warmth of a room or of clothes the secretion of perspirable matter is much increased, the strength of the patient is much exhausted by this unnecessary exertion of the capillary system, and thence of the whole secerning and arterial system by association. The diminution of external heat immediately induces a torpor or quiescence of these unnecessary exertions, and the patient instantly feels himself strengthened, and exhilarated; the animal power, which was thus wasted in vain, being now applied to more useful purposes. Thus when the limbs on one side are disabled by a stroke of the palsy, those of the other side are perpetually in motion. And hence all people bear riding and other exercises best in cold weather.

Patients in fevers, where the skin is hot, are immediately strengthened by cold air; which is therefore of great use in fevers attended with debility and heat; but may perhaps be of temporary disservice, if too hastily applied in some situations of fevers attended with internal topical inflammation, as in peripneumony or pleurisy, where the arterial strength is too great already, and the increased action of the external capillaries being destroyed by the cold, the action of the internal inflamed part may be suddenly increased, unless venesection and other evacuations are applied at the same time. Yet in most cases the application of cold is nevertheless salutary, as by decreasing the heat of the particles of blood in the cutaneous vessels, the stimulus of them, and the distention of the vessels becomes considerably lessened. In external inflammations, as the small-pox, and perhaps the gout and rheumatism, the application of cold air must be of great service by decreasing the action of the inflamed skin, though the contrary is too frequently the practice in those diseases. It must be observed, that for all these purposes the application of it should be continued a long

8 time,

time, otherwiſe an increaſed exertion follows the temporary torpor, before the diſeaſe is deſtroyed.

3. After immerſion in cold water or in cold air the whole ſyſtem becomes more exciteable by the natural degree of ſtimulus, as appears from the ſubſequent glow on the ſkin of people otherwiſe pale; and even by a degree of ſtimulus leſs than natural, as appears by their becoming warm in a ſhort time during their continuance in a bath, of about 80 degrees of heat, as in Buxton bath. See Sect. XII. 2. 1. XXXII. 3. 3.

This increaſed exertion happens to the abſorbent veſſels more particularly, as they are firſt and moſt affected by theſe temporary diminutions of heat; and hence like the medicines, which promote abſorption, the cold-bath contributes to ſtrengthen the conſtitution, that is to increaſe its irritability; for the diſeaſes attended with weakneſs, as nervous fevers and hyſteric diſeaſes, are ſhewn in Section XXXII. 2. 1. to proceed from a want of irritability, not from an exceſs of it. Hence the digeſtion is greater in froſty weather, and the quantity of perſpiration. For theſe purpoſes the application of cold muſt not be continued too long. For in riding a journey in cold weather, when the feet are long kept too cold, the digeſtion is impaired, and cardialgia produced.

4. If the diminution of external heat be too great, produced too haſtily, or continued too long, the torpor of the ſyſtem either becomes ſo great, that the animal ceaſes to live; or ſo great an energy of motion or orgaſm of the veſſels ſucceeds, as to produce fever or inflammation. This moſt frequently happens after the body has been temporarily heated by exerciſe, warm rooms, anger, or intemperance. Hence colds are produced in the external air by reſting after exerciſe, or by drinking cold water. See Claſs I. 2. 2. 1.

Frequent cold immerſions harden or invigorate the conſtitution,

 which

which they effect by habituating the body to bear a diminution of heat on its surface without being thrown into such extensive torpor or quiescence by the consent of the vessels of the skin with the pulmonary and glandular system; as those experience, who frequently use the cold-bath. At first they have great anhelation and palpitation of heart at their ingress into cold water; but by the habit of a few weeks they are able to bear this diminution of heat with little or no inconvenience; for the power of volition has some influence over the muscles subservient to respiration, and by its counter efforts gradually prevents the quick breathing, and diminishes the associations of the pulmonary vessels with the cutaneous ones. And thus though the same quantity of heat is subducted from the skin, yet the torpor of the pulmonary vessels and internal glands does not follow. Hence during cold immersion less sensorial power is accumulated, and in consequence, less exertion of it succeeds on emerging from the bath. Whence such people are esteemed hardy, and bear the common variations of atmospheric temperature without inconvenience. See Sect. XXXII. 3. 2.

IV. Venesection has a just title to be classed amongst the torpentia in cases of fever with arterial strength, known by the fulness and hardness of the pulse. In these cases the heat becomes less by its use, and all exuberant secretions, as of bile or sweat, are diminished, and room is made in the blood-vessels for the absorption of mild fluids; and hence the absorption also of new vessels, or extravasated fluids, the produce of inflammation, is promoted. Hence venesection is properly classed amongst the sorbentia, as like other evacuations it promotes general absorption, restrains hæmorrhages, and cures those pains, which originate from the too great action of the secerning vessels, or from the torpor of the absorbents. I have more than once been witness to the sudden removal of nervous head-achs by venesection,

tion, though the patient was already exhausted, pale, and feeble; and to its great use in convulsions and madness, whether the patient was strong or weak; which diseases are the consequence of nervous pains; and to its stopping long debilitating hæmorrhages from the uterus, when other means had been in vain essayed. In inflammatory pains, and inflammatory hæmorrhages, every one justly applies to it, as the certain and only cure.

V. When the circulation is carried on too violently, as in inflammatory fevers, those medicines, which invert the motions of some parts of the system, retard the motions of some other parts, which are associated with them. Hence small doses of emetic tartar, and ipecacuanha, and large doses of nitre, by producing nausea debilitate and lessen the energy of the circulation, and are thence useful in inflammatory diseases. It must be added, that if nitre be swallowed in powder, or soon after it is dissolved, it contributes to lessen the circulation by the cold it generates, like ice-water, or the external application of cold air.

VI. The respiration of air mixed with a greater proportion of azote than is found in the common atmosphere, or of air mixed with hydrogen, or with carbonic acid gas, so that the quantity of oxygen might be less than usual, would probably act in cases of inflammation with great advantage. In consumptions this might be most conveniently and effectually applied, if a phthisical patient could reside day and night in a porter or ale brewery, where great quantities of those liquors were perpetually fermenting in vats or open barrels; or in some great manufactory of wines from raisins or from sugar.

Externally the application of carbonic acid gas to cancers and other ulcers instead of atmospheric air may prevent their enlargement, by preventing

preventing the union of oxygen with matter, and thus producing a new contagious animal acid.

### III. Catalogue of Torpentia.

1. Veneſection. Arteriotomy.

2. Cold water, cold air, reſpiration of air with leſs oxygen.

3. Vegetable mucilages.
   - *a.* Seeds.—Barley, oats, rice, young peas, flax, cucumber, melon, &c.
   - *b.* Gums.—Arabic, Tragacanth, Senegal, of cherry-trees.
   - *c.* Roots.—Turnip, potatoe, althea, orchis, ſnow-drop.
   - *d.* Herbs.—Spinach, brocoli, mercury.

4. Vegetable acids, lemon, orange, currants, gooſeberries, apples, grape, &c. &c.

5. Animal mucus, hartſhorn jelly, veal broth, chicken water, oil? fat? cream?

6. Mineral acids, of vitriol, nitre, ſea-ſalt.

7. Silence, darkneſs.

8. Invertentia in ſmall doſes, nitre, emetic tartar, ipecacuanha given ſo as to induce nauſea.

9. Antacids.—Soap, tin, alcalies, earths.

10. Medicines preventative of fermentation, acid of vitriol.

11. Anthelmintics.—Indian pink, tin, iron, cowhage, amalgama, ſmoak of tobacco.

12. Lithonthriptics, lixiv. ſaponarium, aqua calcis, fixable air.

13. Externally, warm bath, and poultices, oil, fat, wax, plaſters, oiled ſilk, carbonic acid gas on cancers, and other ulcers.

---

## ADDENDA.

*Page* 625, *line* 1, *after* ‘ *number*’ *pleaſe to add*, ‘ except when the patient has naturally a pulſe ſlower than uſual in his healthy ſtate.’

*Page* 197, *after line* 8, *pleaſe to add*, ‘ Where the difficulty of breathing is very urgent in the croup, bronchotomy is recommended by Mr. Field.’ Memoir of a Medical Society, London, 1773, Vol. IV.

ADDITION.

# ADDITION.

## INABILITY TO EMPTY THE BLADDER.

To be introduced at the end of Clafs III. 2. 1. 6. on Paralyfis Veficæ Urinariæ.

An inability to empty the bladder frequently occurs to elderly men, and is often fatal. This fometimes arifes from their having too long been reftrained from making water from accidental confinement in public fociety, or otherwife; whence the bladder has become fo far diftended as to become paralytic; and not only this, but the neck of the bladder has become contracted fo as to refift the introduction of the catheter. In this deplorable cafe it has frequently happened, that the forcible efforts to introduce the catheter have perforated the urethra; and the inftrument has been fuppofed to pafs into the bladder when it has only paffed into the cellular membrane along the fide of it; of which I believe I have feen two or three inftances; and afterwards the part has become fo much inflamed as to render the introduction of the catheter into the bladder impracticable.

In this fituation the patients are in imminent danger, and fome have advifed a trocar to be introduced into the bladder from the rectum; which I believe is generally followed by an incurable ulcer. One patient, whom I saw in this situation, began to make a spoonful of water after six or seven days, and gradually in a few days emptied his bladder to about half its size, and recovered; but I believe he never afterwards was able completely to evacuate it.

In this situation I lately advised about two pounds of crude quicksilver to be poured down a glass tube, which was part of a barometer tube, drawn less at one end, and about two feet long, into the urethra, as the patient lay on his back; which I had previously performed upon a horse; this easily passed, as was supposed, into the bladder; on standing erect it did not return, but on kneeling down, and lying horizontally on his hands, the mercury readily returned; and on this account it was believed to have passed into the bladder, as it so easily returned, when the neck of the bladder was lower than the fundus of it. But nevertheless as no urine followed the mercury, though the bladder was violently distended, I was led to believe, that the urethra had been perforated by the previous efforts to introduce a catheter and bougee; and that the mercury had passed on the outside of the bladder into the cellular membrane.

As the urethra is so liable to be perforated by the forcible efforts to introduce the catheter, when the bladder is violently distended in this deplorable disease, I should strongly recommend the injection of a pound or two of crude mercury into the urethra to open by its weight the neck of the bladder previous to any violent or very frequent essays with a catheter whether of metal or of elastic resin.

LINES,

# LINES,

TO BE PLACED AT THE END OF

# ZOONOMIA.

*BY A FRIEND.*

---

*JAMQUE OPUS EXEGI.*

---

The work is done!—nor Folly's active rage,
Nor Envy's self, shall blot the golden page;
Time shall admire, his mellowing touch employ,
And mend the immortal tablet, not destroy.

# INDEX

OF THE

# ARTICLES.

A.



Arſenic,

## I.



## L.



## M.



## N.

## O.



## P.



## Q.



## R.



## S.



Salt,

## T.



## U.

## V.



## W.



## Z.



THE END.

Printed in the United States